W0042406

Gmelin Handbook of Inorganic and Organometallic Chemistry

8th Edition

Gmelin Handbook of Inorganic and Organometallic Chemistry

8th Edition

Gmelin Handbuch der Anorganischen Chemie

Achte, völlig neu bearbeitete Auflage

PREPARED AND ISSUED BY Gmelin-Institut für Anorganische Chemie der Max-Planck-Gesellschaft zur Förderung der Wissenschaften
Director: Ekkehard Fluck

FOUNDED BY Leopold Gmelin

8TH EDITION 8th Edition begun under the auspices of the Deutsche Chemische Gesellschaft by R. J. Meyer

CONTINUED BY E. H. E. Pietsch and A. Kotowski, and by Margot Becke-Goehring

Springer-Verlag Berlin Heidelberg GmbH 1993

Gmelin-Institut für Anorganische Chemie
der Max-Planck-Gesellschaft zur Förderung der Wissenschaften

ADVISORY BOARD

Min.-Rat Dr. H. Bechte (Bundesministerium für Forschung und Technologie, Bonn), Prof. Dr. K. Dehnicke (Philipps-Universität, Marburg), Prof. Dr. H. Grünewald (Bayer AG, Leverkusen), Prof. Dr. H. Harnisch, Chairman (Hoechst AG, Frankfurt/Main-Höchst), Prof. Dr. H. Nöth (Ludwig-Maximilians-Universität, München), Prof. Dr. H. Offermanns (Degussa AG, Frankfurt/Main), Prof. Dr. A. Simon (Max-Planck-Institut für Festkörperforschung, Stuttgart), Prof. Dr. Dr. h.c. mult. G. Wilke (Max-Planck-Institut für Kohlenforschung, Mülheim/Ruhr), Prof. Dr. H. F. Zacher (Präsident der Max-Planck-Gesellschaft, München)

DIRECTOR

Prof. Dr. Dr. h.c. Ekkehard Fluck

CORRESPONDENT MEMBERS OF THE SCIENTIFIC STAFF	Dr. U. Krüerke, Dr. R. H. Miller, Dr. A. R. Pebler, Dr. K. Rumpf
EMERITUS MEMBER OF THE INSTITUTE	Prof. Dr. Dr. E.h. Margot Becke
CORRESPONDENT MEMBERS OF THE INSTITUTE	Prof. Dr. Dr. h.c. Hans Bock, Prof. Dr. Dr. Alois Haas, Sc. D. (Cantab.)

GMELIN HANDBOOK

Dr. J. von Jouanne

Dr. L. Berg, Dr. H. Bergmann, Dr. J. Faust, J. Füssel, Dr. H. Katscher, Dr. R. Keim, Dipl.-Phys. D. Koschel, Dr. A. Kubny, Dr. P. Merlet, Dr. M. Mirbach, Prof. Dr. W. Petz, Dr. H. Schäfer, Dr. F. A. Schröder, Dr. A. Slawisch, Dr. W. Töpper

Dr. R. Albrecht, Dr. G. Bär, D. Barthel, Dr. N. Baumann, Dr. K. Behrends, Dr. W. Behrendt, D. Benzaid, Dr. R. Bohrer, K. D. Bonn, Dipl.-Chem. U. Boßlet, Dr. U. Busch, A.-K. Castro, Dipl.-Ing. V. A. Chavizon, E. Cloos, A. Dittmar, Dipl.-Geol. R. Ditz, R. Dowideit, Dr. H.-J. Fachmann, B. Fischer, Dr. D. Fischer, Dipl.-Ing. N. Gagel, Dr. K. Greiner, Dipl.-Bibl. W. Grieser, Dr. R. Haubold, Dipl.-Min. H. Hein, H.-P. Hente, H. W. Herold, U. Hettwer, G. Hoell, Dr. G. Hönes, Dr. W. Hoffmann, G. Horndasch, Dr. W. Huisl, Dr. M. Irmler, B. Jaeger, Dr. R. Jotter, Dipl.-Chem. P. Kämpf, Dr. B. Kalbskopf, H.-G. Karrenberg, Dipl.-Phys. H. Keller-Rudek, Dipl.-Chem. C. Koeppel, Dr. M. Körfer, R. Kolb, Dr. M. Kotowski, E. Kranz, E. Krawczyk, Dipl.-Chem. I. Kreuzbichler, Dr. V. Kruppa, Dr. W. Kurtz, M. Langer, Dr. B. Ledüc, H. Mathis, E. Meinhard, M. Meßer, C. Metz, K. Meyer, E. Mlitzke, Dipl.-Chem. B. Mohsin, Dr. U. Neu-Becker, K. Nöring, Dipl.-Min. U. Nohl, Dr. U. Ohms-Bredemann, Dr. H. Pscheidl, Dipl.-Phys. H.-J. Richter-Ditten, E. Rudolph, G. Rudolph, Dr. B. Sarbas, Dr. R. Schemm, Dr. D. Schiöberg, V. Schlicht, Dipl.-Chem. D. Schneider, E. Schneider, A. Schwärzel, Dr. B. Schwager, R. Simeone, Dr. F. Stein, Dr. C. Strametz, Dr. G. Swoboda, Dr. D. Tille, A. Tuttas, Dipl.-Phys. J. Wagner, R. Wagner, M. Walter, Dr. E. Warkentin, Dr. C. Weber, Dr. A. Wietelmann, Dr. M. Winter, Dr. B. Wöbke, K. Wolff

GMELIN ONLINE

Dr. R. Deplanque

Dr. P. Kuhn, Dr. G. Olbrich

Dr. R. Baier, Dr. B. Becker, Dipl.-Chem. E. Best, Dr. H.-U. Böhmer, Dipl.-Phys. R. Bost, Dr. A. Brandl, Dr. R. Braun, Dr. T. Buck, Dipl.-Chem. R. Durban, R. Hanz, Dipl.-Phys. C. Heinrich-Sterzel, Dr. S. Kalwellis-Mohn, Dr. A. Kirchhoff, Dipl.-Chem. H. Köttelwesch, Dr. M. Kunz, Dipl.-Chem. R. Maass, Dr. K. Müller, Dr. A. Nebel, Dipl.-Chem. R. Nohl, Dr. M. Nohlen, H. Reinhardt, Dr. B. Rempfer, Dipl.-Ing. H. Vanecek

Volumes published on "Nitrogen" (Syst. No. 4)

Main Volume 1
History. Occurrence. The Element — 1934

Main Volume 2
Compounds of Nitrogen with Hydrogen — 1935

Main Volume 3
Compounds of Nitrogen with Oxygen — 1936

Main Volume 4
Compounds with Oxygen — 1936

Supplement Volume B 1
Compounds with Noble Gases and Hydrogen — 1993 (present volume)

Supplement Volume B 2
Compounds with Noble Gases and Hydrogen — 1993

Gmelin Handbook of Inorganic and Organometallic Chemistry

8th Edition

N
Nitrogen

Supplement Volume B 1

Compounds with Noble Gases and Hydrogen

With 11 illustrations

AUTHORS Walter Hack (Göttingen)
Reinhard Haubold, Claudia Heinrich-Sterzel,
Hannelore Keller-Rudek, Ulrike Ohms-Bredemann, Dag Schiöberg,
Carol Strametz

EDITORS Dieter Koschel, Peter Merlet, Ulrike Ohms-Bredemann,
Joachim Wagner

CHIEF EDITOR Peter Merlet

System Number 4

Springer-Verlag Berlin Heidelberg GmbH 1993

LITERATURE CLOSING DATE: MID 1992
IN MANY CASES MORE RECENT DATA HAVE BEEN CONSIDERED

Library of Congress Catalog Card Number: Agr 25-1383

ISBN 978-3-662-06335-4 ISBN 978-3-662-06333-0 (eBook)
DOI 10.1007/978-3-662-06333-0

This work is subject to copyright. All rights are reserved, whether the whole or part of the material is concerned, specifically those of translation, reprinting, reuse of illustrations, broadcasting, reproduction by photocopying machine or similar means, and storage in data banks. Under § 54 of the German Copyright Law where copies are made for other than private use, a fee is payable to "Verwertungsgesellschaft Wort", Munich.

© by Springer-Verlag Berlin Heidelberg 1993
Originally published by Springer-Verlag, Berlin · Heidelberg · New York · London · Paris · Tokyo · Hong Kong · Barcelona in 1993
Softcover reprint of the hardcover 8th edition 1993

The use of registered names, trademarks, etc., in this publication does not imply, even in the absence of a specific statement, that such names are exempt from the relevant protective laws and regulations and therefore free for general use.

Preface

"Nitrogen" Suppl. Vol. B1 describes the compounds of nitrogen with noble gases and, in the major part, binary compounds composed of one nitrogen atom and hydrogen. Nitrogen-hydrogen compounds with two and more nitrogen atoms are covered in "Nitrogen" Suppl. Vol. B2.

There is some information on various nitrogen-noble gas species, to a large extent because of the interest in their bonding behavior. Experimental data have been obtained chiefly for some singly charged cations, particularly those formed by argon like ArN^+ and ArN_2^+. The existence of others has only been established by mass spectrometry.

The binary compounds of nitrogen and hydrogen comprise NH, NH_2, NH_4, NH_5, the corresponding ions, and some adducts. NH_3 and NH_4^+ are not treated. The predominant part of the volume covers the description of the molecules NH and NH_2. Both species are present in photolytic processes in the atmosphere. They play an essential role in combustion systems regardless of whether the nitrogen stems from the nitrogen-containing fuel or from the air. Thus, much work has been devoted to the understanding of the nitrogen chemistry in combustion and in the atmosphere. The production and detection methods as well as the reactions have been comprehensively described. In addition detailed information is given on the spectral behavior, the knowledge of which is important for detecting the molecules and for studying their kinetics.

Frankfurt am Main
November 1993

Peter Merlet

Table of Contents

Compounds of Nitrogen

1 Compounds of Nitrogen with Noble Gases

Species composed of nitrogen and noble gases are described in the following chapter. They are variably called compound, van der Waals complex, exciplex, or cluster and are uniformely written like a true chemical compound regardless of the kind of interaction between nitrogen and noble gas. Neutral species are treated first; the more important ions, particularly singly charged cations, follow.

There is not much information on neutral nitrogen-noble gas compounds except some indications that a few weakly bound van der Waals complexes and exciplexes do exist.

Nitrogen-noble gas monocations have been detected mass-spectrometrically for the noble gases He, Ne, Ar, Kr, and Xe; some of them have also been studied spectroscopically. The major portion of this chapter deals with cations formed by argon. They comprise simple molecular cations such as ArN^+, ArN_2^+ as well as cluster ions of the type $Ar_mN_{2n}^+$ with m up to 15 and n up to 23.

Quantum chemical calculations predict that most dications are metastable with respect to charge-separation reactions. But some exhibit significant dissociation barriers which should make them even detectable. In fact, the doubly charged XN^{2+} cations with X = Ne, Ar, Kr, and Xe could recently be identified in a charge-stripping mass-spectrometric experiment.

Only a few theoretical studies on trications have been made. They predict unbound or metastable ionic states.

HeN^-, the only nitrogen-noble gas anion described, is predicted to be unbound.

1.1 Neutral Nitrogen-Noble Gas Compounds

HeN

CAS Registry Number: *[71159-49-4]*

Ab initio MO SCF calculations showed that the ground state $HeN(^2\Pi)$ is unbound [1]. CNDO/2 calculations of the binding energy and equilibrium distance were reported in [2].

NeN

CAS Registry Number: –

So far, there seems to be no experimental evidence for the existence of NeN. Neon crystals doped with nitrogen only showed emission lines of nitrogen in the luminescence spectrum in contrast to nitrogen-doped argon crystals where emission from an ArN* exciplex was observed [3].

ArN

CAS Registry Number: *[68078-25-1]*

The vacuum ultraviolet emission spectrum of solid argon doped with nitrogen showed a broad band with a maximum at 163 nm which is attributed to a transition from an ArN* exciplex to the unbound ground state ($^4\Sigma \rightarrow {}^4\Sigma$). This band has its maximum intensity at a concentration of 0.25 mol% N_2. ArN* is thought to be formed by a neutral argon atom and an excited nitrogen atom via $Ar(^1S) + N(^4P) \rightarrow ArN(^4\Sigma)$. A minimum distance of 1.93 Å and a depth of the energy potential curve of 94.6 kJ/mol were estimated [3].

ArN_2

CAS Registry Numbers: *[53169-18-9], [52692-00-9]*

ArN_2 van der Waals molecules were formed by expanding typically a mixture of 300 Torr N_2, 150 Torr Ar, and 1100 Torr He through a 10 μm supersonic nozzle and detected by monitoring the decomposition process of the corresponding cation [4, 5]; cf. p. 5. Ar_mN_{2n} clusters with m, n ≥ 1 are also accessible by this method [6]. ArN_2 was also identified in mixtures of argon and nitrogen at high pressures and temperatures near the normal boiling point of argon. A discrete infrared absorption of ArN_2 which was recorded with a long path cell maintained at 87 K broadened and diminished with increasing temperature. The spectrum of ArN_2 showed similar features as the N_2 spectrum. The NN stretching frequency was shifted by 0.2 cm^{-1} to lower wavenumbers. The prominent fine structure in this region was assigned to the hindered internal rotation of the N_2 moiety in the complex with a rotational barrier of 238 J/mol. A T-shaped equilibrium geometry with the N-N bond perpendicular to the N_2-Ar van der Waals bond axis was concluded from the fine structure. The bond length N_2-Ar was estimated to be 3.9 Å [7].

An ArN_2 species in Ar-N_2 mixtures was mass spectrometrically observed earlier [8].

XeN

CAS Registry Number: *[119188-64-6]*

A XeN* exciplex was formed by microwave discharge through a mixture of $N(^2P)$ and Xe atoms. Emission bands observed at 1040 and 350 nm were attributed to a bound-bound transition and to a transition into a repulsive state according to $XeN(^2P) \rightarrow XeN(^2D) + h\nu(IR)$ and $XeN(^2P) \rightarrow XeN(^4S) + h\nu(UV)$. The ratio of the band intensities was found to be ~20:1 [9].

The emission band system at 475 to 492 nm observed in a discharge through a Xe-N_2 mixture was assigned to XeN transitions, because this band system only appeared in discharges of nitrogen-xenon mixtures [10 to 12].

References:

[1] Liebman, J. F.; Allen, L. C. (J. Am. Chem. Soc. **92** [1970] 3539/43).
[2] Macias, A.; Marin, C. (An. Quim. **73** [1977] 472/7).
[3] Poltoratskii, Yu. B.; Fugol', I. Ya. (Fiz. Nizk. Temp. [Kiev] **4** [1978] 783/9; Sov. J. Low Temp. Phys. [Engl. Transl.] **4** [1978] 373/5).
[4] Stephan, K.; Märk, T. D. (Chem. Phys. Lett. **87** [1982] 226/8).
[5] Stephan, K.; Märk, T. D. (Int. J. Mass Spectrom. Ion Phys. **47** [1983] 195/8).
[6] Ding, A.; Futrell, J. H.; Cassidy, R. A.; Cordis, L.; Hesslich, J. (Surf. Sci. **156** [1985] 282/91).

[7] Henderson, G.; Ewing, G. E. (Mol. Phys. **27** [1974] 903/15).
[8] Leckenby, R. E.; Robbins, E. J. (Proc. R. Soc. London A **291** [1966] 389/412, 398).
[9] Vilesov, A. F.; Pravilov, A. M.; Smirnova, L. G. (Opt. Spektrosk. **65** [1988] 896/8; Opt. Spectrosc. [Engl. Transl.] **65** [1988] 529/30).
[10] Herman, L.; Herman, R. (Nature **191** [1961] 346/7).
[11] Herman, L.; Herman, R. (Nature **193** [1962] 156/7).
[12] Herman, L.; Herman, R. (J. Phys. Radium [8] **24** [1963] 73/5).

1.2 Singly Charged Cations

HeN^+

CAS Registry Number: *[11092-10-7]*

Cluster ions of HeN^+, probably $(HeN)^+(N_2)_n$ and/or $(HeN_3)^+(N_2)$, were observed as satellites in the secondary ion mass spectrum of solid nitrogen first bombarded with low-energy helium atoms and later probed with heavier ions, for example argon ions [1]. Theoretical studies were performed to examine the structure and stability of HeN^+. The most relevant studies so far used the fourth-order Møller-Plesset perturbation theory (MP4) [2 to 4]. The following table lists values for the internuclear distances r_e, the excitation energy ΔE, the dissociation energy D, and the vibrational frequency ν which were calculated for the ground-state $HeN^+(X\ ^3\Sigma^-)$ ($\rightarrow He(^1S)+N^+(^3P)$) and the excited-state $HeN^+(^3\Pi)$ ($\rightarrow He(^1S)+N^+(^3D)$) [4]:

state	r_e in Å	ΔE in kJ/mol	D_0 in kJ/mol	D_e in kJ/mol	ν in cm^{-1}
$HeN^+(X\ ^3\Sigma^-)$	1.749	–	15.5	17.2	250
$HeN^+(^3\Pi)$	1.007	829.4	–	288.3	–

HeN^+ in its ground state represents a van der Waals complex, whereas in the excited state HeN^+ is covalently bound [4]. An earlier computation at an ab initio SCF level likewise yielded a weak bond in $HeN^+(X\ ^3\Sigma^-)$ of about 7 kJ/mol [5]. Other ab initio MO SCF calculations yielded a weakly bound ground state, one highly and two very weakly bound excited states [6]. CNDO/2 calculations of the binding energy and equilibrium distance were reported in [7].

HeN_2^+

CAS Registry Number: *[117268-50-5]*

HeN_2^+ ions were created in a pulsed supersonic expansion jet (N_2/He ratio 1:200) crossed close to the nozzle by an electron beam [8].

The photodissociation spectrum of HeN_2^+ recorded in the 391-nm region exhibits two major vibrational bands: the stronger band is almost coincident with the $B\ ^2\Sigma_u^+ \leftarrow X\ ^2\Sigma_g^+$ origin transition of N_2^+, while the weaker one, displaced 195 cm^{-1} to higher energy from the more intense band, is a hot band of HeN_2^+ which is associated with the 1-1 transition of N_2^+. The similarity of this spectrum with that of N_2^+ was interpreted by the ion being essentially a free internal rotor in both X and B electronic states. These results [8] confirmed the predicted absence of an appreciable barrier to rotation [9].

The two-dimensional interaction potential of the $N_2^+(X\ ^2\Sigma_g^+)$-He(X ^{1}S) system was obtained from highly correlated ab initio MCSCF-CI calculations. The well depth, determined to be about 140 cm^{-1} (0.017 eV), confirms a possible bound HeN_2^+ ion. The well depth is nearly independent of the angle formed between the axis of the He atom to the center of mass of N_2^+ and the N-N bond axis. Rovibrational calculations were performed and yielded band origins, rotational constants, and stretching frequencies. Dipole moments of the ion were calculated relative to the center of mass [9].

Studies of the vibrational excitation of N_2^+ ions in collisions with He gave a rough estimate of 0.026 eV for the ion-molecule interaction energy in HeN_2^+ [10].

NeN$^+$

CAS Registry Number: *[11092-18-5]*

NeN$^+$ was obtained from an equimolar mixture of neon and nitrogen by electron ionization and was observed as the precursor of NeN^{2+} in a charge-stripping mass spectrometer experiment [11].

Complete active space (CAS) SCF plus second-order configuration interaction (SOCI) calculations [12] and MP4 calculations [13, 14] were performed. The following table lists values for the internuclear distance r_e, dissociation energy D, ionization energy E_i, and spectroscopic constants for NeN$^+$($^3\Sigma^-$) ($\rightarrow$ Ne(^{1}S) + N$^+$ (^{3}P)):

r_e in Å	D_e	D_0 in kJ/mol	E_i in eV	ν	ω_e	$\omega_e x_e$ all in cm^{-1}	B_e	Ref.
1.746	45.3	—	26.6	—	390.5	10.944	0.657	[12]
1.767	38.4	36.0	26.3	393	—	—	—	[14]

Earlier ab initio MO SCF calculations were performed for NeN$^+$ in five electronic states, the weakly bound ground state, one highly, one weakly bound, and two unbound excited states [15]; see also [6]. CNDO/2 calculations of the binding energy and equilibrium distance were reported in [7].

ArN$^+$

CAS Registry Number: *[71159-51-8]*

The formation of ArN$^+$, observed mass spectrometrically in mixtures of Ar with N_2 [16 to 18], probably occurs via $(N_2^+)^* + Ar \rightarrow ArN^+ + N$ [16, 17]. ArN$^+$ ions form in collisions of nitrogen with photoionized argon in the center-of-mass energy range of 8.2 to 41.2 eV. ArN$^+$ is probably formed via a nearly collinear arrangement Ar$^+\cdots$N-N [19]. In secondary ion mass spectra of solid N_2 with a 4-keV argon ion beam, weak peaks were assigned to ArN$^+$ and ArN_2^+, and a very weak peak to $Ar(N_2)_2^+$ [1].

ArN$^+$ was also formed by collision-induced dissociation of ArN_2^+ and $ArNH_3^+$ [18].

ArN$^+$ was predicted to be covalently bound in its ground state [14]. Molecular constants were calculated for ground-state ArN$^+$(X $^3\Sigma^-$) ($\rightarrow$Ar + N$^+$) using the higher-order Møller-Plesset perturbation theory [14, 20] or the CASSCF [20] and CASSCF/SOCI procedures [21]. The following table lists values for the internuclear distance r_e, dissociation energy D, and spectroscopic constants:

r_e in Å	D_0 in kJ/mol	ω_e	$\omega_e x_e$	B_e all in cm^{-1}	α_e	Ref.
1.869	182	532	2.6	0.47	0.0043	[21]
1.905	121	515	5.0	0.45	0.0044	[20]

CNDO/2 calculations of binding energy and equilibrium distance were reported in [7].

An appearance potential of 22.2 eV was determined [17]. The ionization energy, $E_i = 20.7 \pm 2.5$ eV, was obtained from charge-stripping experiments [11].

Potential energy curves were calculated with ab initio methods (CASSCF/SOCI) for the bound triplet state A $^3\Pi$ and the two repulsive triplet states B $^3\Pi$ and C $^3\Sigma^-$ [21].

The photoabsorption spectrum was studied in the visible wavelength region by photofragment kinetic energy spectroscopy. Photofragments observed after dissociating ArN^+ with photons of 1.8 to 2.5 eV by crossing or merging the ion beam with a laser beam were Ar^+/N and Ar/N^+. The N^+ signal was several orders of magnitude smaller than the Ar^+ signal [21]. The preference for the channel $ArN^+ \rightarrow Ar^+ + N$ was also revealed by a study of the collision-induced dissociation reactions of ArN^+ which yielded a kinetic energy release of 0.91 eV for the channel $ArN^+ \rightarrow Ar^+ + N$ and of 0.185 eV for the channel $ArN^+ \rightarrow Ar + N^+$ [18]. From the calculated potential energy curves for the triplet states X $^3\Sigma^-$, A $^3\Pi$, B $^3\Pi$, and C $^3\Sigma^-$, it was concluded that N^+ fragments could only result from the A $^3\Pi \leftarrow$ X $^3\Sigma^-$ transition of ArN^+, whereas the C $^3\Sigma^- \leftarrow$ X $^3\Sigma^-$ transition was tentatively assigned to the Ar^+ formation. The dissociation energy, $D_0 = 2.16 \pm 0.10$ eV, was determined for the process $ArN^+ \rightarrow Ar + N^+$ [21]. The value compares favorably with $D_e = 2.3$ eV [15] and $D_0 =$ 2.09 eV, obtained from the appearance potential of ArN^+ (see above) [21]. The dissociation energies calculated at the MP4 level agree quite well [14, 20].

The abundance of ArN^+ ions, which can interfere during the analytic determination of metal ions by inductively coupled plasma mass spectrometry, can be reduced by using xenon as a target gas via the exothermic reaction $ArN^+ + Xe \rightarrow XeN^+ + Ar$ [22].

ArN_2^+

CAS Registry Numbers: *[12322-21-3], [140447-93-4]*

ArN_2^+ ions are formed in Ar–N_2 mixtures by electron impact in the ion source of a mass spectrometer, probably via the reactions $Ar^+ + N_2 \rightarrow Ar + N_2^+$, $N_2^+ + Ar \rightleftharpoons (ArN_2^+)^*$, $(ArN_2^+)^* + M \rightarrow ArN_2^+ + M$ [23], or $Ar^* + N_2 \rightarrow ArN_2^+ + e^-$ [16, 17]. Another route is the production of neutral ArN_2 species together with other heterogeneous clusters by supersonic expansion of nitrogen in argon and subsequent ionization by electrons or photons [24, 25]. The highest yield was achieved at 25% N_2 [25]. For the occurrence of ArN_2^+ in a secondary ion mass spectrum, see p. 4.

A linear structure for $(Ar\text{-}NN)^+$ was predicted by applying an impulsive model for the photodissociation studies on ArN_2^+ [23] and on the entropy change in the reaction $Ar + (N_2)_2^+ \rightleftharpoons ArN_2^+ + N_2$ [26]. Ab initio calculations were performed on linear ArN_2^+ with $C_{\infty v}$ [27, 28] and on T-shaped ArN_2^+ with C_{2v} symmetry [28]. A collinear $(N\text{-}Ar\text{-}N)^+$ species was predicted to be highly unstable [28]. Calculations at the MP2/6-311G* level yielded the distances r(NN) = 1.079 Å and r(ArN) = 2.293 Å for the linear species [27]. Calculations at the HF/6-31G level gave for the linear species r(NN) = 1.084 Å and r(ArN) = 2.224 Å and for the T-shaped species r(NN) = 1.093 Å and a distance of Ar to the N_2 center of mass of 3.407 Å. The linear $C_{\infty v}$ form was found to be more stable than the T-shaped one: 165.7 kJ/mol at the Hartree-Fock level of theory and 78.7 kJ/mol at the MP4 level. Harmonic vibrational frequencies calculated for the $C_{\infty v}$ form were 212, 208, and 2414 cm^{-1} [28].

Appearance potentials of 14.6 [16] and 15.1 eV [17] were found by mass spectrometry.

Dissociation energies of 119.7 kJ/mol for Ar^+/N_2 formation and of 103.3 kJ/mol for Ar/N_2^+ formation were derived from photodissociation experiments [23]. Other experimental values for the formation of Ar/N_2^+ were 102.5 kJ/mol [29], $\geq 109 \pm 21$ kJ/mol [16], and ≥ 63 kJ/mol

[17]. Dissociation energies were calculated at the MP4 level to be 111.7 kJ/mol for Ar^+/N_2 and 82.8 kJ/mol for Ar/N_2^+ [28].

The equilibrium reaction $Ar+(N_2)_2^+ \rightleftharpoons ArN_2^+ + N_2$ was studied mass spectrometrically between 300 and 500 K [23], between 313 to 738 K [29], and at 300 K as a function of the center-of-mass interaction energy [26]. The following thermochemical data were derived:

T in K	ΔH° in kJ/mol	ΔS° in $J \cdot mol^{-1} \cdot K^{-1}$	Ref.
428	-5.4 ± 1.7	6.3 ± 2	[23]
298	-2.30 ± 0.12	14.9 ± 0.3	[29]
300	$+8.7 \pm 1.0$	19 ± 2	[26]

Values were calculated for 298 K at the MP4/6-31G level: $\Delta H° = -8.8$ kcal/mol, $\Delta S° = 10.1$ $cal \cdot mol^{-1} \cdot K^{-1}$ [28]. The equilibrium is shifted to the right based on the results of [23, 28, 29], but contrary to the results of [26]. The rate constant for the forward reaction was determined to be about 3.5×10^{-10} $cm^3 \cdot molecule^{-1} \cdot s^{-1}$ at 300 K [26]. Under equilibrium conditions in a gas flow system, a forward rate constant of $k = 1.4 \times 10^{-12}$ $cm^3 \cdot molecule^{-1} \cdot s^{-1}$ and a reverse rate constant of $k \leq 3 \times 10^{-13}$ $cm^3 \cdot molecule^{-1} \cdot s^{-1}$ were determined at 300 K [30]. A high-pressure mass spectrometric study revealed a rate constant of 1×10^{-10} $cm^3 \cdot molecule^{-1} \cdot s^{-1}$ for the forward reaction [16].

A reaction rate of $(1.5 \pm 0.5) \times 10^{-9}$ $cm^3 \cdot molecule^{-1} \cdot s^{-1}$ was determined for the reaction $Ar^* + N_2 \rightarrow ArN_2^+ + e^-$ in an earlier mass spectrometric study [31]. $\Delta_r H° = -112.1 \pm 1.3$ kJ/mol and $\Delta_r S° = -81.6 \pm 2.1$ $J \cdot mol^{-1} \cdot K^{-1}$ at 298 K were derived for the reaction $N_2^+ + Ar \rightarrow ArN_2^+$ [27] using the corresponding data of the reactions $(N_2)_2^+ + Ar \rightarrow ArN_2^+ + N_2$ [23] and $N_2^+ + 2\,N_2 \rightarrow (N_2)_2^+ + N_2$ [32].

The collision-induced and metastable dissociation of ArN_2^+ leads to N_2^+/Ar and Ar^+/N_2 fragment channels [18, 24, 33]. In one case, some ArN^+ was also produced [18]. Metastable fragmentation of ArN_2^+, prepared by electron impact ionization of neutral ArN_2, was found to proceed with a rate of $\sim 65\ s^{-1}$ for Ar^+/N_2 and of $\sim 100\ s^{-1}$ for N_2^+/Ar production (at 70 eV electron energy). The dissociation rates of the two processes did not vary by more than 10% within the electron energy range 20 to 180 eV [24]. Only collision-induced dissociation rather than metastable fragmentation was found in a later experiment, where ArN_2^+ was formed by ion-molecule reactions in a high-pressure ion source [34].

Photodissociation experiments were performed by crossing an ArN_2^+ beam with a laser beam. Absolute photodissociation cross sections σ were measured in two studies [23, 35], the results of which cover the range 660 to 350 nm: σ steadily increases from $\sim 0.04 \times 10^{-18}$ cm^2 at 660 nm to $\sim 8 \times 10^{-18}$ cm^2 at 350 nm (see figure 5 in [23]). Two product channels $ArN_2^+ + h\nu \rightarrow Ar^+ + N_2$ and $ArN_2^+ + h\nu \rightarrow Ar + N_2^+$ were identified. The ArN_2^+ ions seem to be excited to a repulsive upper state and then dissociate to the ground-state products $Ar^+ + N_2$ and $Ar + N_2^+$ in a ratio of 3:1 [23].

For the reaction $ArN_2^+ + Ar \rightarrow Ar_2^+ + N_2$ mass spectrometric measurements in a drift tube yielded $\Delta_r H° = 9.04 \pm 0.13$ kJ/mol and $\Delta_r S° = 7.82 \pm 0.33$ $J \cdot mol^{-1} \cdot K^{-1}$ at 298 K [29]. The reaction $ArN_2^+ + Ar \rightleftharpoons Ar_2^+ + N_2$ studied in a gas flow system was found to have an equilibrium constant of 0.045 at 300 K with a forward rate constant $k \gg 1 \times 10^{-12}$ $cm^3 \cdot molecule^{-1} \cdot s^{-1}$ and a reverse rate constant $k \gg 2 \times 10^{-11}$ $cm^3 \cdot molecule^{-1} \cdot s^{-1}$ [30].

ArN_2^+ ions can form $Ar_2N_2^+$ ions by clustering with argon atoms; see below.

ArN_{2n}^+ ($2 \leq n \leq 23$)

CAS Registry Number: $Ar(N_2)_2^+$ *[84692-39-7]*

$Ar(N_2)_2^+$ clusters are formed together with other heterogeneous clusters by electron impact ionization of the equivalent neutral clusters prepared by supersonic expansion of argon-nitrogen mixtures [33]. For the observation of $Ar(N_2)_2^+$ in a secondary ion mass spectrum, see p. 4.

ArN_{2n}^+ ($2 \leq n \leq 23$) clusters were produced by the impact of an Ar beam (8 keV) on solid N_2-Ar mixtures (10:1) held at 4 or 12 K. It was suggested that the clusters are a mixture of two isomers: $N_4^+(ArN_{2n-6})$ and $ArN_2^+(N_{2n-4})$. The larger ones are more likely to contain N_4^+ as central ion than the smaller ones which may contain $Ar(N_2)_2^+$ as central ion [36].

The decomposition of $Ar(N_2)_2^+$, studied with a reversed double-focusing mass spectrometer, proceeds via $Ar(N_2)_2^+ \rightarrow ArN_2^+ + N_2$ and $Ar(N_2)_2^+ \rightarrow Ar + (N_2)_2^+$ by both collision-induced and metastable dissociation [33].

Laser-induced, collision-induced, and metastable decomposition reactions of the larger cluster ions were examined with a triple-quadrupole mass spectrometer. Various phenomena observed during the decomposition of the clusters suggested four kinds of energy storage: thermal, vibrational [$N_2(v=1)$], electronic, and chemical. In smaller clusters ($n \leq 9$) containing argon in the inner shell, two moles of N_2 are lost via the reaction $Ar + N_4^+ \rightarrow N_2 + ArN_2^+$, which is indicative for the chemical metastability of these clusters [36].

$Ar_2N_2^+$, $Ar_3N_2^+$

CAS Registry Numbers: $Ar_2N_2^+$ *[136048-01-6], [140447-94-5]*; $Ar_3N_2^+$ *[140622-66-8]*

$Ar_2N_2^+$ clusters were formed together with other heterogeneous argon-nitrogen clusters by supersonic expansion of an argon-nitrogen gas mixture and subsequent photoionization [25] or ionization by electron impact [25, 33, 37]. The formation of $Ar_2N_2^+$, $Ar_3N_2^+$, and other $Ar_mN_2^+$ ($m>3$) clusters was observed in gaseous mixtures of several Torr of argon with 10^{-6} Torr of nitrogen, ionized by a pulsed 2-keV electron beam. The formation of $Ar_2N_2^+$ was only observed at ion source temperatures below 150 K. The rate of the reaction $ArN_2^+ + 2\ Ar \rightarrow Ar_2N_2^+ + Ar$ was found to be too slow to establish an equilibrium between $Ar_2N_2^+$ and ArN_2^+. But with the free energy estimated to be -8.8 kJ/mol at 101.0 K and assuming $\Delta S = -71\ J \cdot mol^{-1} \cdot K^{-1}$, one obtains $\Delta H \leq -15.9$ kJ/mol. The enthalpy change for the reaction $Ar_2N_2^+ + 2\ Ar \rightarrow Ar_3N_2^+ + Ar$ was determined to be -7.3 kJ/mol assuming that the entropy change is $-57\ J \cdot mol^{-1} \cdot K^{-1}$. For the reaction $Ar_3N_2^+ + 2\ Ar \rightarrow Ar_4N_2^+ + Ar$ values of $\Delta H° = -7.0$ kJ/mol and $\Delta S° = -75\ J \cdot mol^{-1} \cdot K^{-1}$ were determined [27].

A linear structure was derived for $Ar_2N_2^+$ with ab initio calculations at the ROHF level. The arrangement Ar-Ar-N-N with the distances r(ArAr) = 3.066 Å, r(ArN) = 2.351 Å, r(NN) = 1.070 Å was found to be more stable than the arrangement Ar-N-N′-Ar′ with the distances r(ArN) = 2.291 Å, r(NN′) = 1.070 Å, r(N′Ar′) = 4.212 Å [27]. A linear arrangement Ar-Ar-N-N with a delocalization of the positive charge was also concluded from the results of an $Ar_2N_2^+$ photodissociation study [37]. Two T-shaped structures were proposed for $Ar_3N_2^+$: one has a long linear Ar-Ar′-N-N chain (r(ArAr′) = 3.069 Å, r(Ar′N) = 2.352 Å, r(NN) = 1.070 Å) and a short, nearly rectangular Ar′-Ar″ bond (r(Ar′Ar″) = 3.966 Å) and the other, which is slightly less stable, has a long, linear Ar-N-N′-Ar chain (r(ArN) = 2.304 Å, r(NN′) = 1.085 Å, r(N′Ar) = 4.226 Å) and a short, also nearly rectangular Ar″-N′ bond (r(Ar″N′) = 4.214 Å) [27].

The threshold of the photoionization curve for $Ar_2N_2^+$ lies at 14.5 eV [25].

The absolute cross sections for photodissociation of $Ar_2N_2^+$ were measured to be between 470 and 550 nm. A maximum of $(2.1 \pm 0.3) \times 10^{-16}$ cm^2 was found at ~500 nm. The spectral features suggested that direct dissociation takes place and that the photodissociation spectrum is practically identical with the photoabsorption spectrum of $Ar_2N_2^+$. The visible absorption band closely resembles the $^2\Sigma_g^+ \leftarrow {}^2\Sigma_u^+$ absorption band of Ar_3^+. The peak was blue-shifted by ~20 nm. The product species were analyzed by photofragment, time-of-flight mass spectrometry and revealed at least three dissociation channels: (1) $Ar_2N_2^+ + h\nu \rightarrow N_2^+ + 2\,Ar$ (or Ar_2); (2) $Ar_2N_2^+ + h\nu \rightarrow Ar^+ + Ar + N_2$ (or ArN_2); (3) $Ar_2N_2^+ + h\nu \rightarrow Ar_2^+ + N_2$ [37]. The decomposition of $Ar_2N_2^+$ clusters, when studied with a reversed double focusing mass spectrometer, was found to proceed via the two channels $Ar_2N_2^+ \rightarrow ArN_2^+ + Ar$ and $Ar_2N_2^+ \rightarrow Ar^+ + ArN_2$ by both collision-induced and metastable dissociation [33].

$Ar_mN_{2n}^+$ ($2 \leq m \leq 15$, $1 \leq n \leq 23$)

Clusters of the composition $Ar_mN_2^+$ with m up to 15 were produced by electron-impact ionization of Ar_mN_2 clusters which were formed by the cluster-gas exchange reaction $Ar_{m+1} + N_2 \rightarrow Ar_mN_2 + Ar$ under single collision conditions [38]. The formation of the clusters $Ar_mN_2^+$ with m up to 9 was observed in gaseous mixtures of some Torr of argon with 10^{-6} Torr of nitrogen which were subsequently ionized by a pulsed 2-keV electron beam [27]. Clusters of the composition $Ar_mN_{2n}^+$ with $3 \leq m \leq 14$ and $1 \leq n \leq 4$ were formed after electron impact on the equivalent neutral clusters produced by supersonic expansion of 25% N_2 in Ar [25]. The intensity drop in the mass spectrum for the cluster ions $Ar_mN_2^+$ with $m > 12$, which agreed with the observations reported in [25], was interpreted by the formation of a stable shell of 12 argon atoms around the central N_2^+ ion [38].

Clusters of the composition $Ar_2N_{2n-4}^+$ with $n \geq 4$ were produced by the impact of an Ar beam (8 keV) on solid N_2-Ar mixtures (10:1) held at 4 or 12 K and were analyzed mass spectrometrically [36].

KrN^+, KrN_2^+

CAS Registry Numbers: KrN^+ *[60280-83-3]*; $NKrN^+$ *[60280-82-2]*; $KrNN^+$ *[61312-92-3]*

KrN^+ and KrN_2^+ ions were observed mass spectrometrically in gaseous mixtures of krypton with nitrogen [16, 39]; they are probably formed by the reactions $Kr + (N_2^+)^* \rightarrow KrN^+ + N$ and $Kr^* + N_2 \rightarrow KrN_2^+ + e^-$ [39]. KrN^+ was detected in mixtures of 98.4% N_2 and 1.6% Kr, while KrN_2^+ was observed in nitrogen with ~0.05% Kr after electron impact ionization in a drift-tube mass spectrometer [35]. KrN^+ ions were detected in the flowing afterglow of krypton plasmas with 0.5 to 2% nitrogen via $Kr_2^+ + N \rightarrow KrN^+ + Kr$, while KrN_2^+ ions were found in the flowing afterglow of nitrogen plasmas containing about 0.05% krypton [40].

Total photodissociation cross sections of KrN^+ and KrN_2^+ were measured between 565 and 670 nm with a drift-tube mass spectrometer and a tunable dye laser. The photodissociation proceeds via $KrN^+ + h\nu \rightarrow Kr^+ + N$ and $KrN_2^+ + h\nu \rightarrow Kr^+ + N_2$. The absolute cross sections steadily increased for KrN^+ and decreased for KrN_2^+ towards smaller wavelengths. The maximum absolute cross sections of about 1.0×10^{-18} cm^2 for KrN^+ and 0.48×10^{-18} cm^2 for KrN_2^+ were obtained at 575 and 655 nm, respectively [35].

An ionization energy of $E_i = 21.6 \pm 1.8$ eV was obtained for KrN^+ by charge-stripping mass spectrometry, where the monocation KrN^+ was used as precursor for KrN^{2+} [11]. Appearance potentials of 22.2 eV for KrN^+ and 13.2 eV for KrN_2^+ were determined by mass spectrometry [39].

KrN^+ reacts with atomic nitrogen via $KrN^+ + N \rightarrow Kr^+ + N_2$ in a nitrogen-krypton plasma leaving the N_2 molecule in an unspecified energy state [40].

XeN^+, XeN_2^+

CAS Registry Number: XeN^+ *[12504-95-9]*

XeN^+ ions were observed mass spectrometrically in mixtures of xenon with nitrogen [16, 35] as a product of the reaction $(N_2^+)^* + Xe \rightarrow XeN^+ + N$ [16]. The ionization energy $E_i = 21.1 \pm 2.2$ eV for XeN^+ was obtained by charge-stripping mass spectrometry, where the monocation XeN^+ was used as precursor for XeN^{2+} [11]. On the basis of ab initio MO SCF calculations on HeN^+ and NeN^+, it was assumed that XeN^+ salts could be prepared [6].

An upper limit of 10^{-20} cm^2 was determined for the photodissociation cross section of XeN^+ at 600 nm [35].

XeN_2^+ was observed mass spectrometrically in Xe–N_2 mixtures [16].

References:

[1] Jonkman, H. T.; Michl, J. (J. Am. Chem. Soc. **103** [1981] 733/7).
[2] Koch, W.; Frenking, G.; Gauss, J.; Cremer, D.; Collins, J. R. (J. Am. Chem. Soc. **109** [1987] 5917/34).
[3] Koch, W.; Frenking, G. (J. Chem. Soc. Chem. Commun. **1986** 1095/6).
[4] Frenking, G.; Koch, W.; Cremer, D.; Gauss, J.; Liebman, J. F. (J. Phys. Chem. **93** [1989] 3397/410).
[5] Cooper, D. L.; Wilson, S. (Mol. Phys. **44** [1981] 161/72).
[6] Liebman, J. F.; Allen, L. C. (J. Am. Chem. Soc. **92** [1970] 3539/43).
[7] Macias, A.; Marin, C. (An. Quim. **73** [1977] 472/7).
[8] Bieske, E. J.; Soliva, A.; Welker, M. A.; Maier, J. P. (J. Chem. Phys. **93** [1990] 4477/8).
[9] Miller, S.; Tennyson, J.; Follmeg, B.; Rosmus, P.; Werner, H. J. (J. Chem. Phys. **89** [1988] 2178/84).
[10] Kriegel, M.; Richter, R.; Lindinger, W.; Barbier, L.; Ferguson, E. E. (J. Chem. Phys. **88** [1988] 213/9).

[11] Jonathan, P.; Boyd, R. K.; Brenton, A. G.; Beynon, J. H. (Chem. Phys. **110** [1986] 239/46).
[12] Koch, W.; Liu, B.; Frenking, G. (J. Chem. Phys. **92** [1990] 2464/8).
[13] Frenking, G.; Koch, W. (Int. J. Mass Spectrom. Ion Processes **82** [1988] 335/8).
[14] Frenking, G.; Koch, W.; Cremer, D.; Gauss, J.; Liebman, J. F. (J. Phys. Chem. **93** [1989] 3410/8).
[15] Liebman, J. F.; Allen, L. C. (Int. J. Mass Spectrom. Ion Phys. **7** [1971] 27/31).
[16] Munson, M. S. B.; Field, F. H.; Franklin, J. L. (J. Chem. Phys. **37** [1962] 1790/9).
[17] Kaul, W.; Fuchs, R. (Z. Naturforsch. **15a** [1960] 326/35).
[18] Jonathan, P.; Brenton, A. G.; Beynon, J. H.; Boyd, R. K. (Int. J. Mass Spectrom. Ion Processes **71** [1986] 257/82).
[19] Flesch, G. D.; Ng, C. Y. (J. Chem. Phys. **92** [1990] 2876/82).
[20] Wong, M. W.; Radom, L. (J. Phys. Chem. **93** [1989] 6303/8).

[21] Broström, L.; Larsson, M.; Mannervik, S.; Sonnek, D. (J. Chem. Phys. **94** [1991] 2734/40).
[22] Rowan, J. T.; Houk, R. S. (Appl. Spectrosc. **43** [1989] 976/80).
[23] Kim, H.-S.; Bowers, M. T. (J. Chem. Phys. **93** [1990] 1158/64).
[24] Stephan, K.; Märk, T. D. (Chem. Phys. Lett. **87** [1982] 226/8).
[25] Ding, A.; Futrell, J. H.; Cassidy, R. A.; Cordis, L.; Hesslich, J. (Surf. Sci. **156** [1985] 282/91).
[26] Tichy, M.; Twiddy, N. D.; Wareing, D. P.; Adams, N. G.; Smith, D. (Int. J. Mass Spectrom. Ion Processes **81** [1987] 235/46).

[27] Hiraoka, K.; Mori, T.; Yamabe, S. (Chem. Phys. Lett. **189** [1992] 7/12).
[28] Frecer, V.; Jain, D. C.; Sapse, A.-M. (J. Phys. Chem. **95** [1991] 9263/6).
[29] Teng, H. H.; Conway, D. C. (J. Chem. Phys. **59** [1973] 2316/23).
[30] Vinogradov, P. S.; Dmitriev, O. V.; Veretennikov, I. N. (Khim. Fiz. **10** [1991] 501/10).
[31] Holcombe, N. T.; Lampe, F. W. (J. Chem. Phys. **57** [1972] 449/54).
[32] Hiraoka, K.; Nakajima, G. (J. Chem. Phys. **88** [1988] 7709).
[33] Stephan, K.; Märk, T. D. (Int. J. Mass Spectrom. Ion Phys. **47** [1983] 195/8).
[34] Illies, A. J.; Bowers, M. T. (Org. Mass Spectrom. **18** [1983] 553/60).
[35] Miller, T. M.; Ling, J. H.; Saxon, R. P.; Moseley, J. T. (Phys. Rev. [3] A **13** [1976] 2171/7).
[36] Magnera, T. F.; David, D. E.; Michl, J. (J. Chem. Soc. Faraday Trans. **86** [1990] 2427/40).
[37] Nagata, T.; Kondow, T. (Z. Phys. D At. Mol. Clusters **20** [1991] 153/5).
[38] Ozaki, Y.; Fukuyama, T. (Int. J. Mass Spectrom. Ion Processes **88** [1989] 227/39).
[39] Kaul, W.; Taubert, R. (Z. Naturforsch. **17a** [1962] 88/9).
[40] Tracy, C. J.; Oskam, H. J. (Phys. Rev. [3] A **14** [1976] 1371/4).

1.3 Doubly Charged Cations

HeN^{2+}

CAS Registry Number: *[80896-02-2]*

Ab initio calculations using the fourth-order Møller-Plesset perturbation theory (MP4(SDTQ)) were performed on HeN^{2+} in its ground (X $^2\Pi$) and excited states ($^4\Sigma^-$) which both were found to be covalently bound. HeN^{2+} is metastable and dissociates via the exothermic charge-separation reactions $HeN^{2+}(X\ ^2\Pi) \rightarrow He^+(^2S) + N^+(^3P)$ and $HeN^{2+}(^4\Sigma^-) \rightarrow He^+(^2S) + N^+(^3D)$. There is a barrier to dissociation of 201.7 kJ/mol for ground-state HeN^{2+}. Other calculated data are [1]:

state	r_e in Å	ΔE	D_0	D_e	ν in cm^{-1}
		all in kJ/mol			
$HeN^{2+}(X\ ^2\Pi)$	1.321	–	−291.6	−285.8	1011
$HeN^{2+}(^4\Sigma^-)$	1.060	325.0	−622.6	−611.7	1809

Earlier ab initio MO SCF calculations on $HeN^{2+}(X\ ^2\Pi)$ gave a well depth of about 170 kJ/mol for the potential curve based on dissociation to $He(^1S) + N^{2+}(^2P)$ [2].

HeN_2^{2+}

CAS Registry Number: *[80896-04-4]*

Ab initio SCF calculations of $(HeNN)^{2+}$ gave a dissociation energy of 359 kJ/mol relative to He and N_2^{2+}. The molecule is nonlinear with a bond angle NNHe of 92.3° and bond distances $r_e(NN) = 1.44$ and $r_e(NHe) = 1.50$ Å [2].

He_2N^{2+}

CAS Registry Number: *[106007-04-9]*

Ab initio calculations at the MP4 and MP2 levels were performed on $(HeNHe)^{2+}$. He_2N^{2+} is metastable with respect to dissociation via $He_2N^{2+}(^2B_1) \rightarrow He^+(^2S) + HeN^+(^3\Sigma^-)$ which is exothermic by 158.6 kJ/mol [3, 4]. Other dissociation channels and their reaction energies are: $He_2N^{2+}(^2B_1) \rightarrow N^{2+}(^2P) + 2\ He(^1S)$ with $\Delta E_e = 360.7$ kJ/mol, $He_2N^{2+}(^2B_1) \rightarrow N^+(^3P) +$

$He^+(^2S) + He(^1S)$ with $\Delta E_e = -127.2$ kJ/mol [5]. A bond distance r(HeN) = 1.326 Å and an angle HeNHe of 87.9° were determined for the ion in C_{2v} symmetry. Vibrational frequencies of 950 (B_2), 937 (A_1), and 623 cm^{-1} (A_1) were derived [3].

$He_2N_2^{2+}$

CAS Registry Number: *[109909-11-7]*

Ab initio calculations at the MP2 level were performed on $(HeNNHe)^{2+}$. The following molecular parameters were calculated for the ion in the ^1A_g state with C_{2h} symmetry, i.e. an ion with a planar trans configuration: r(HeN) = 1.359, r(NN) = 1.228 Å, bond angle HeNN = 90.9°. Harmonic vibrational frequencies in cm^{-1} were derived: 1608 (A_g), 707 (A_g), 507 (B_u), 501 (A_g), 479 (A_u), and 304 (B_u) [3].

NeN^{2+}

CAS Registry Number: *[106676-71-5]*

NeN^{2+} was observed upon the charge-stripping of its precursor NeN^+ [8]. NeN^{2+} is probably formed only from ground-state NeN^+. The formation from excited states of NeN^+ may be impossible due to an unfavorable Franck-Condon factor between the excited states of NeN^+ and the ground state of NeN^{2+} [6]. Ab initio calculations at the CASSCF-CI level were performed on ground-state $NeN^{2+}(X\,^2\Pi)$ and its low-lying excited states. The ground-state ion is found to be metastable towards charge-separating dissociation: $NeN^{2+}(X\,^2\Pi) \rightarrow Ne^+(^2P) + N^+(^3P)$ with the reaction energy $\Delta E = -441$ kJ/mol. There is a barrier of dissociation of 91 kJ/mol. The equilibrium distance, $r_e = 1.59$ Å, and the spectroscopic constants, $\omega_e = 716.2$, $\omega_e x_e = 9.025$, and $B_e = 0.877$ cm^{-1}, were calculated for NeN^{2+} [7]. The vibrational frequency $\nu = 936$ cm^{-1} was calculated at the MP2 level [6].

The potential energy curve for ground-state NeN^{2+} has its quasi-bound maximum 6.1 eV above the dissociation limit. The translational energy release associated with the $NeN^{2+} \rightarrow Ne^+ + N^+$ process was determined by ion kinetic energy spectroscopy to be 6.5 ± 0.5 eV [8] and calculated to be 5.51 eV [7].

NeN_2^{2+}

CAS Registry Number: *[128358-94-1]*

Ab initio calculations at the HF and at the MP2 level were carried out on linear $NeNN^{2+}$ as part of a study on triatomic, asymmetrical A=B=C systems with 22 electrons. The bond lengths r(NeN) = 1.341 and r(NN) = 1.092 Å were calculated at the HF level (1.561 and 1.263 Å at the MP2 level). Calculated vibrational frequencies at the HF level (in cm^{-1}) are: $\nu_1(\Sigma_g) = 909$, $\nu_2(\Pi) = 116$, $\nu_3(\Sigma_u) = 2710$ [9].

Ne_2N^{2+}

CAS Registry Number: *[127085-64-7]*

A bond distance r(NeN) = 1.602 Å and a bond angle NeNNe of 98° were calculated with ab initio methods at the MP2 level for $(NeNNe)^{2+}$ in its $^2B_1$ state (C_{2v} symmetry). An exothermic dissociation energy of −234 kJ/mol was calculated at the MP4(SDTQ) level for the charge-separation process $Ne_2N^{2+}(^2B_1) \rightarrow N^+(^3P) + Ne^+(^2P) + Ne(^1S)$. An endothermic energy of 558 kJ/mol was calculated for the process $Ne_2N^{2+}(^2B_1) \rightarrow N^{2+}(^2P) + 2\,Ne(^1S)$ [5].

ArN^{2+}

CAS Registry Number: *[121705-52-0]*

ArN^{2+} was formed by charge-stripping of the precursor ArN^+ [8]. Ab initio calculations were performed on ArN^{2+} in its ground state at the complete active space (CAS) SCF, MP3 and MP4 levels of theory. ArN^{2+} was found to be metastable with an exothermic dissociation energy of −449 kJ/mol for the process $ArN^{2+} \rightarrow Ar^+ + N^+$. The barrier to dissociation was determined to be 24 kJ/mol. The equilibrium distance, $r_e(ArN) = 1.657$ Å, was calculated at the MP3 level. The spectroscopic constants (in cm^{-1}), $\omega_e = 562$, $\omega_e x_e = 14.3$, $B_e = 0.54$, and $\alpha_e = 0.0104$, were calculated at the CASSCF level [10]. An approximate potential curve with a maximum at 5.4 eV (521 kJ/mol) above the dissociation limit was calculated earlier for ArN^{2+} [8].

The transition structure of the process $ArN^{2+} \rightarrow Ar^+ + N^+$ was calculated at the HF level to have a bond length of $r_{TS} = 2.190$ Å, dissociation barrier of $D_e = 28$ kJ/mol ($D_0 = 24$ kJ/mol), and a kinetic energy release of $T = 5.7$ eV [10].

Ar_2N^{2+}

CAS Registry Number: *[127085-65-8]*

A bond distance ArN of 1.742 Å and a bond angle ArNAr of 108.4° were calculated with ab initio methods at the MP2 level for $(ArNAr)^{2+}$ in its $^2B_1$ state (C_{2v} symmetry). An exothermic dissociation energy of −207 kJ/mol was calculated at the MP4(SDTQ) level for the charge-separation process $Ar_2N^{2+}(^2B_1) \rightarrow N^+(^3P) + Ar^+(^2P) + Ar(^1S)$. An endothermic energy of 1156 kJ/mol was derived for the process $Ar_2N^{2+}(^2B_1) \rightarrow N^{2+}(^2P) + 2\,Ar(^1S)$ [5].

KrN^{2+}

CAS Registry Number: *[106676-70-4]*

The dication KrN^{2+} was formed by a charge-stripping process from KrN^+. The appearance potential for the unimolecular charge-separation process $KrN^{2+} \rightarrow Kr^+ + N^+$ was measured to be 40 ± 1.5 eV. The translational energy release, $T_e = 6.0 \pm 0.2$ eV, connected with this dissociation process was determined by ion kinetic energy spectroscopy [8].

XeN^{2+}

CAS Registry Number: *[106676-69-1]*

The dication XeN^{2+} was formed by charge-stripping of its precursor XeN^+. The appearance potential for the unimolecular charge-separation process $XeN^{2+} \rightarrow Xe^+ + N^+$ was measured to be 32 ± 2 eV. The translational energy release, $T_e = 6.1 \pm 0.3$ eV, connected with this dissociation process was determined by ion kinetic energy spectroscopy [8].

References:

[1] Frenking, G.; Koch, W.; Cremer, D.; Gauss, J.; Liebman, J. F. (J. Phys. Chem. **93** [1989] 3397/410).
[2] Cooper, D. L.; Wilson, S. (Mol. Phys. **44** [1981] 161/72).
[3] Koch, W.; Frenking, G.; Gauss, J.; Cremer, D.; Collins, J. R. (J. Am. Chem. Soc. **109** [1987] 5917/34).
[4] Koch, W.; Frenking, G. (J. Chem. Soc. Chem. Commun. **1986** 1095/6).
[5] Frenking, G.; Koch, W.; Reichel, F.; Cremer, D. (J. Am. Chem. Soc. **112** [1990] 4240/56).
[6] Frenking, G.; Koch, W. (Int. J. Mass Spectrom. Ion Processes **82** [1988] 335/8).

[7] Koch, W.; Liu, B.; Frenking, G. (J. Chem. Phys. **92** [1990] 2464/8).
[8] Jonathan, P.; Boyd, R. K.; Brenton, A. G.; Beynon, J. H. (Chem. Phys. **110** [1986] 239/46).
[9] Pyykkö, P.; Zhao, Y.-F. (J. Phys. Chem. **94** [1990] 7753/9).
[10] Wong, M. W.; Radom, L. (J. Phys. Chem. **93** [1989] 6303/8).

1.4 Triply Charged Cations

NeN^{3+}

CAS Registry Number: *[124395-76-2]*

Calculations at the HF level and with the second-order Møller-Plesset perturbation theory (MP2) yielded a potential curve minimum for the species: $r_e = 1.415$ Å, $\nu = 774$ cm^{-1} at the HF level, $r_e = 1.499$ Å, $\nu = 455$ cm^{-1} at the MP2 level [1]; the use of higher orders of the Møller-Plesset perturbation theory yielded a purely repulsive ground-state potential curve for NeN^{3+} [1, 2].

ArN^{3+}

CAS Registry Number: *[121705-53-1]*

Ab initio calculations were performed on ArN^{3+} in its ground state at the complete active space (CAS) SCF, MP3, and MP4 levels of theory [3, 4]. ArN^{3+} was found to be metastable with an exothermic fragmentation energy of -927 kJ/mol for the process $ArN^{3+} \rightarrow Ar^{2+} + N^+$. The barrier to dissociation was determined to be 16 kJ/mol. The equilibrium distance, $r_e(ArN) = 1.504$ Å, was calculated at the MP3 level [3] and is close to a bond length of 1.533 Å obtained with a version of the coupled cluster theory including double substitution where the effect of single and triple substitutions is incorporated into fourth-order perturbation theory (ST4CCD). ST4CCD was assumed to provide the best estimate for a multiply charged system [4]. The spectroscopic constants (in cm^{-1}), $\omega_e = 565$, $\omega_e x_e = 19.7$, $B_e = 0.59$, and $\alpha_e = 0.0160$, were calculated at the CASSCF level [3, 4].

The transition structure of the process $ArN^{3+} \rightarrow Ar^{2+} + N^+$ was calculated at the HF level to have a bond length of $r_{TS} = 2.069$ Å, a dissociation barrier of $D_e = 22$ kJ/mol ($D_0 = 16$ kJ/mol), and a kinetic energy release of $T = 10.9$ eV [3].

References:

[1] Pyykkö, P. (Mol. Phys. **67** [1989] 871/8).
[2] Wong, M. W.; Nobes, R. H.; Bouma, W. J.; Radom, L. (J. Chem. Phys. **91** [1989] 2971/9).
[3] Wong, M. W.; Radom, L. (J. Phys. Chem. **93** [1989] 6303/8).
[4] Wong, M. W.; Radom, L. (J. Phys. Chem. **94** [1990] 638/44).

1.5 Nitrogen-Helium Anion, HeN^-

Ab initio MO SCF calculations confirmed that HeN^- in the states $^3\Sigma(\pi^2)$ and $^1\Delta(\pi^2)$ is unbound as expected.

Reference:

Liebman, J. F.; Allen, L. C. (J. Am. Chem. Soc. **92** [1970] 3539/43).

2 Compounds of Nitrogen with Hydrogen

The following part deals with binary compounds composed of one nitrogen atom and hydrogen (mononitrogen compounds). They are arranged in the order of increasing number of hydrogen atoms. Thus, this volume deals with the species NH, NH_2, NH_4, NH_5, and corresponding ions and some adducts. NH_3 and NH_4^+ are beyond the scope of this volume.

Binary compounds composed of two or more nitrogen atoms and hydrogen are covered in the volume "Nitrogen" Suppl. Vol. B 2.

2.1 Mononitrogen Compounds

2.1.1 The Imidogen Radical, NH

Other names: Nitrogen monohydride, nitrene, azene, azanediyl, azanylidene, imino

CAS Registry Numbers: NH *[13774-92-0]*, ND *[15123-00-9]*, ^{15}NH *[34089-09-3]*, ^{13}NH *[57437-96-4]*

2.1.1.1 Production and Detection of NH Radicals

Walter Hack
Max-Planck-Institut für Strömungsforschung
Göttingen

Introduction

The imidogen radical is found in a wide variety of environments and has been observed in various of astrophysical sources. It is the key for understanding the fate of nitrogen in combustion systems regardless of whether the nitrogen originates from nitrogen-containing fuel or from the air. NH was first observed in a nonlaboratory source in the spectrum of the sun [1]. Since then NH was detected again in the spectrum of the sun and in comets [2 to 4]. NH is important in understanding the photochemistry of ammoniacal planetary atmospheres and interstellar chemistry [5]. It is also observable in the Earth's atmosphere [6]. The imidogen radical was one of the first radicals which was detected in stellar atmospheres [3, 7 to 14].

NH is one of the few species with a triplet electronic ground state ($^3\Sigma^-$), isoelectronic with $O(^3P)$ and $CH_2(\tilde{X}\ ^3B_1)$. The first excited metastable singlet state ($a\ ^1\Delta$), corresponding to $O(^1D)$ and $CH_2(\tilde{a}\ ^1A_1)$, is often regarded as a differentiated species, because the reactivity of a given species, in general, depends on its electronic structure. Thus the kinetics of NH(X) and NH(a) can be expected to be significantly different and each has to be studied independently.

In practice it turns out to be more complicated to find a clean source for NH(X) than for NH(a) (see below). Nevertheless, the reactions of NH(X) and NH(a) can be studied directly in independent experiments. Detection methods are quite often state-specific, thus there is no need to emphasize that NH(X), NH(a), and NH in other electronic states can be detected characteristically (see p. 25).

NH can be produced in the gas phase as well as in a liquid or in the solid state (matrices). In a collision-free situation, the excited electronic state has a maximum lifetime, since collisions depopulate the excited states. Thus the situation in the gas phase is significantly different from that in the condensed phase. Due to the efficient quenching in condensed media, the radical sources in the different situations are described in different sections.

References on pp. 17/9.

2.1.1.1.1 NH Radical Sources in the Gas Phase

In principal it is possible to generate NH(X, A,... a, b, c,...) in two consecutive steps: first, to produce the species NH, and, second, to bring it into the desired electronic state by excitation or deactivation. In many experiments, however, a single step mechanism is used to produce NH directly in the desired quantum state. In thermal systems at temperatures below about 1000 K, the electronic ground state is the dominant state. At very high temperatures or in nonthermal systems (photolytic, radiolytic systems), higher electronic states can be populated. Thus the NH sources are divided into those for NH(X) and those for NH(a, b,.... A,...), i.e., NH in higher electronic states.

2.1.1.1.1.1 NH(X), NH in the Electronic Ground State

Early experiments to produce NH, either by thermal decomposition of various precursors (NH_3 [15], NH_2Cl [16, 17], N_2H_4 [18 to 20], HNCO [21], HN_3 [22 to 25]), or by photolysis of these compounds (NH_3 [26], N_2H_4 [27], HNCO [28 to 32]), were not state-specific. In particular, in discharges of different compounds or mixtures of compounds (N_2-H_2 [33, 34], HN_3 [35]), NH was not produced and detected in a specific state, since the existence of NH was either concluded only from the final products or NH was detected indirectly, with a long time delay between the production and detection, so that quenching could have occurred. In several experiments NH was detected even after trapping processes. On the other hand, if emission of NH out of a specific state is observed, it is not at all obvious that most of the NH radicals are in this state.

The controlled formation of NH radicals in their electronic ground state is not an easy task. The thermal decomposition of HN_3 was recommended as an NH radical source in the early 30's [22]. This, however, is true only at sufficiently high temperatures — well above 800 K. The thermal decomposition of HN_3 at lower temperatures ($530<T/K<750$), however, was not recommended as an NH(X) source due to secondary reactions of NH with HN_3, which proceed rapidly compared to the formation reaction at these low temperatures [36].

In pyrolysis experiments, NH_3 [22, 36 to 44] and HN_3 [21, 45 to 49], which produces only NH(a $^1\Delta$) in the photolysis (see p. 19), can be applied. The pyrolysis of HN_3 and HNCO has two low-lying product channels, (i) the spin-allowed NH(a) + N_2(X) and CO(X), respectively, and (ii) the spin-forbidden pathway NH(X) + N_2(X) (CO(X)). In shock tube studies, mainly NH(X) is formed.

In the thermal decomposition of N_2H_4 in shock waves, NH(X) is formed in secondary reactions [20]. In shock-heated NH_3-noble gas mixtures at high temperatures ($T>3000$ K) [37 to 39] and in a high-temperature plasma ($T=3200$ K), emission from NH(A) was observed [40]. At lower temperatures in shock waves NH(X) was observed [41]. HN_3 and HNCO can be pyrolyzed at significantly lower temperatures ($T>1200$ K); under these conditions mainly NH(X) is formed [44, 45].

The VUV photolysis of ammonia [26, 50 to 68] leads to the formation of NH(X $^3\Sigma^-$), but simultaneously NH in other electronic states, in particular a $^1\Delta$ and b $^1\Sigma$, is formed, since the low-lying singlet states of NH_3 do not correlate with NH(X) and $H_2(^1\Sigma_g^+)$ [69, 70] but with NH(X) + $H_2(^3\Sigma_u^+$, i.e. 2 H). Thus short wavelengths ($\lambda<155$ nm) have to be used to obtain NH(X), and the simultaneously formed electronically excited NH* radicals have to be quenched to the electronic ground state. Similar problems arise, if other precursor molecules such as N_2H_4 [27, 54, 71 to 76] or CH_3NH_2 and $C_2H_5NH_2$ [77] are inserted.

At very short wavelengths ($\lambda = 121.6$ nm) HN_3 can be photolyzed via $HN_3 + h\nu \rightarrow NH(X\ ^3\Sigma) + N_2(B\ ^3\Pi_g)$ [78]. The reactions of the electronically excited nitrogen molecules with HN_3, however, will preclude this photodecomposition process, or the equivalent photoreaction of HNCO, being used as an NH(X) source.

The two-photon laser photolysis of NH_3, N_2H_4, and CH_3NH_2, with NH_2 as an intermediate state, does not lead to the formation of NH in the electronic ground state but to electronically excited NH radicals [79 to 82]. The multiphoton dissociation of NH_3 at $\lambda = 193$ nm, however, produces mainly highly excited electronic ground-state NH [83].

When NH_3 and ND_3 were photolyzed in the infrared region (CO_2 laser) electronically excited fragments were also observed [84, 85]. In the infrared multiphoton dissociation (IRMPD) of ND_3, the imidogen radical is formed in both the spin-forbidden ($X\ ^3\Sigma^-$) and the spin-allowed ($a\ ^1\Delta$) states [86, 87]. In a CO_2 laser pyrolysis experiment, in which a mixture of SF_6, CF_4, and NH_3 was irradiated, ground-state NH radicals were produced [88].

In the vibrational overtone excitation in $HN_3(v_1 = 5, 6)$, NH in the electronic ground state can be produced in a spin-forbidden dissociation [89 to 91].

In the chemical reaction $H + N_3 \rightarrow NH + N_2$ [92], and in the analogous reaction $H + NCO \rightarrow NH + CO$ [93], NH is formed in the $a\ ^1\Delta$ state as well as in the $^3\Sigma^-$ electronic ground state [92, 93].

NH(X) is formed in the reaction of electronically excited N atoms with H_2: $N(^2D) + H_2 \rightarrow NH(X\ ^3\Sigma^-) + H$ [94, 95]. The presence of NH radicals in an Ar-H_2-N_2 plasma is probably due to that reaction [96]. Ground-state N atoms were assumed to produce the NH(X) radical in the reaction with $NH_2(\tilde{X})$ radicals [97], but it turned out later that the reaction pathway leading to $N_2 + 2\ H$ is by far the dominant one [98].

Translational hot H atoms, formed in the HI photolysis, react with N_2O to form NH(X) [99]. NH(X) was also observed when the $N_2O \cdot HBr$ van der Waals complex was photolyzed [100, 101].

NH(X) is produced in thermal systems in the chemical reactions

$$NH_2 + \{H, OH\} \rightleftharpoons NH + \{H_2, H_2O\}$$

(see p. 126), and it can be obtained in the laboratory-based chemical reaction

$$F(^2P) + NH_2(\tilde{X}\ ^2A_1) \rightarrow NH(X\ ^3\Sigma^-) + HF(^1\Sigma^+)$$

in which NH(X) as well as NH(a) are formed [102, 103]. The latter reaction above proceeds, as concluded from the rotational state distribution in HF [104, 105], on a triplet surface as an abstraction reaction; most of the excess energy $\Delta_RH = -179.7$ kJ/mol is found as the vibrational energy of $HF(v \leq 4)$. If the reaction would proceed on a singlet surface with an $NH_2F(\tilde{X})^{\ddagger}$ intermediate, the formation of $NH(a\ ^1\Delta)$, which is exothermic by $\Delta_RH = -29$ kJ/mol, would be expected.

Finally, the NH(X) can be formed in a two-step mechanism. In UV photolysis of HN_3 or HNCO, $NH(a\ ^1\Delta)$ is the primary product (see p. 19); thus the photolysis can be used as an NH(X) source only in two steps: formation of NH(a) followed by quenching to NH(X) by Xe or N_2.

References:

[1] Fowler, A.; Gregory, C. C. L. (Philos. Trans. R. Soc. London A **218** [1919] 251/72).
[2] Swings, P.; Elvey, C. T.; Babcock, H. W. (Astrophys. J. **94** [1941] 320/43).
[3] Jennings, D. E. (J. Quant. Spectrosc. Radiat. Transfer **40** [1988] 221/38).
[4] Litvak, M. M.; Rodriguez-Kuiper, E. N. (Astrophys. J. **253** [1982] 622/33).
[5] Huntress, W. T. (Chem. Soc. Rev. **6** [1977] 295/323).
[6] Brewer, A. W.; Davies, P. A.; Kerr, J. B. (Nature [London] **240** [1972] 35/6).
[7] Shaw, R. W. (Astrophys. J. **83** [1936] 225/37).
[8] Roach, F. E. (Astrophys. J. **89** [1939] 99/115).
[9] Schade, A. (Bull. Astron. Inst. Neth. **17** [1964] 311/57).
[10] Schmitt, J. L. (Publ. Astron. Soc. Pac. **81** [1969] 657/64).

[11] Lambert, D. L.; Beer, F. (Astrophys. J. **177** [1972] 541/5).
[12] Yorka, S. B. (Astron. J. **88** [1983] 1816/24).
[13] Lambert, D. L.; Brown, J. A.; Hinkle, K. H.; Johnson, H. R. (Astrophys. J. **284** [1984] 223/37).
[14] Ridgway, S. T.; Carbon, D. F.; Hall, D. N. B. (Astrophys. J. Suppl. **54** [1984] 177/210).
[15] Guenebaut, H.; Latour, M. (J. Chim. Phys. Phys. Chim. Biol. **59** [1962] 970/9).
[16] Jander, J.; Fischer, J. (Angew. Chem. **71** [1959] 626/7).
[17] Jander, J.; Fischer, J. (Z. Anorg. Allg. Chem. **313** [1961] 37/47).
[18] Eberstein, I. J.; Glassman, I. (Symp. Int. Combust. Proc. **10** [1965] 365/74).
[19] Tsivenko, V. I.; Myasnikov, I. A. (Zh. Fiz. Khim. **47** [1973] 871/3; Russ. J. Phys. Chem. [Engl. Transl.] **47** [1973] 493/5).
[20] Diesen, R. W. (J. Chem. Phys. **39** [1963] 2121/8).

[21] Back, R. A.; Childs, J. (Can. J. Chem. **46** [1968] 1023/4).
[22] Meyer, R.; Schumacher, H. J. (Z. Phys. Chem. A **170** [1934] 33/40).
[23] Rice, F. O.; Freamo, M. (J. Am. Chem. Soc. **73** [1951] 5529/30).
[24] Rice, F. O.; Grelechi, C. (J. Am. Chem. Soc. **79** [1957] 1880/1).
[25] Rice, F. O.; Luchenbach, T. A. (J. Am. Chem. Soc. **82** [1960] 2681/2).
[26] Groth, W. E. (Z. Phys. Chem. B **37** [1937] 315/22).
[27] Ramsay, D. A. (J. Phys. Chem. **57** [1953] 415/7).
[28] Mui, J. Y. P.; Back, R. A. (Can. J. Chem. **41** [1963] 826/33).
[29] Dixon, R. N. (Can. J. Phys. **38** [1960] 10/6).
[30] Dixon, R. N. (Philos. Trans. R. Soc. London A **252** [1960] 165/92).

[31] Brash, J. L.; Back, R. A. (Can. J. Chem. **43** [1965] 1778/83).
[32] Bradley, J. N.; Gilbert, J. R.; Svejda, P. (Trans. Faraday Soc. **64** [1968] 911/8).
[33] Takahashi, S. (Mem. Def. Acad. Math. Phys. Chem. Eng. [Yokosuka, Jpn.] **9** [1969] 351/68).
[34] Guenebaut, H.; Pannetier, G.; Goudmand, P. (C. R. Hebd. Seances Acad. Sci. **251** [1960] 1480/2).
[35] Rice, F. O.; Freamo, M. (J. Am. Chem. Soc. **75** [1953] 548/9).
[36] Richardson, W. C.; Setser, D. W. (Can. J. Chem. **47** [1969] 2725/7).
[37] Cann, M. W. P.; Kash, S. W. (J. Chem. Phys. **41** [1964] 3055/60).
[38] Avery, H. E.; Bradley, J. N.; Tuffnell, R. (Trans. Faraday Soc. **60** [1964] 335/64).
[39] Genich, A. P.; Zhirnov, A. A.; Manelis, G. B. (Kinet. Katal. **16** [1975] 841/5; Kinet. Catal. [Engl. Transl.] **16** [1975] 729/32).
[40] Levitskii, A. A.; Ovsyannikov, A. A.; Polak, L. S. (Commun. 3rd Symp. Int. Chim. Plasmas, Limoges, Fr., 1977, 9 pp.; C.A. **90** [1979] No. 44465).

[41] Dove, J. E.; Nip, W. S. (Can. J. Chem. **57** [1979] 689/701).
[42] Davidson, D. F.; Kohse-Höinghaus, K.; Chang, A. Y.; Hanson, R. K. (Int. J. Chem. Kinet. **22** [1990] 513/35).

[43] Zaslonko, I. S.; Kogarko, S. M.; Mozzhukhin, E. V. (Kinet. Katal. **13** [1972] 829/35; Kinet. Catal. [Engl. Transl.] **13** [1972] 745/50).
[44] Kajimoto, O.; Yamamoto, Y.; Fueno, T. (J. Phys. Chem. **83** [1979] 429/35).
[45] Kajimoto, O.; Kondo, O.; Okada, K.; Fujikane, J.; Fueno, T. (Bull. Chem. Soc. Jpn. **58** [1985] 3469/74).
[46] Mertens, J. D.; Chang, A. Y.; Hanson, R. K.; Bowman, C. T. (Int. J. Chem. Kinet. **21** [1989] 1049/67).
[47] Wu, C. H.; Wang, H.-T.; Lin, M. C.; Fifer, R. A. (J. Phys. Chem. **94** [1990] 3344/7).
[48] He, Y.; Liu, X.; Lin, M.; Melius, C. F. (Int. J. Chem. Kinet. **23** [1991] 1129/49).
[49] Yokoyama, K.; Sakane, Y.; Fueno, T. (Bull. Chem. Soc. Jpn. **64** [1991] 1738/42).
[50] Bayes, K. D.; Becker, K. H.; Welge, K. H. (Z. Naturforsch. **17a** [1962] 676/80).

[51] Stuhl, F.; Welge, K. H. (Z. Naturforsch. **17a** [1962] 676/80).
[52] Stuhl, F.; Welge, K. H. (Z. Naturforsch. **18a** [1963] 900/6).
[53] Becker, K. H.; Welge, K. H. (Z. Naturforsch. **18a** [1963] 600/3).
[54] Becker, K. H.; Welge, K. H. (Z. Naturforsch. **19a** [1964] 1006/15).
[55] Dagdigian, P. J. (J. Chem. Phys. **90** [1989] 6110/5).
[56] Welge, K. H. (J. Chem. Phys. **45** [1966] 4373/4).
[57] Bayes, K. D.; Becker, K. H.; Welge, K. H. (AD-664932 [1967]; C.A. **69** [1968] No. 48222).
[58] Okabe, H.; Lenzi, M. (J. Chem. Phys. **47** [1967] 5241/6).
[59] Schurath, U.; Schindler, R. N. (Ber. Bunsen-Ges. Phys. Chem. **72** [1968] 1027/9).
[60] Mantei, K. A.; Bair, E. J. (J. Chem. Phys. **49** [1968] 3248/56).

[61] Hansen, I.; Höinghaus, K.; Zetzsch, C.; Stuhl, F. (Chem. Phys. Lett. **42** [1976] 370/2).
[62] Zetzsch, C. (Habilitationsschr. Univ. Bochum 1977, 242 pp.).
[63] Hansen, I.; Höinghaus, K.; Zetzsch, C.; Stuhl, F. (NBS Spec. Publ. [U.S.] No. 526 [1978] 334/6).
[64] Zetzsch, C. (J. Photochem. **9** [1978] 151/3).
[65] Donnelly, V. M.; Baronavski, A. P.; McDonald, J. R. (Chem. Phys. **43** [1979] 271/81).
[66] Nguyen Xuan, C.; Di Stefano, G.; Lenzi, M.; Margani, A. (J. Chem. Phys. **74** [1981] 6219/23).
[67] Zetzsch, C.; Stuhl, F. (Ber. Bunsen-Ges. Phys. Chem. **85** [1981] 564/8).
[68] Slanger, T. G.; Black, G. (J. Chem. Phys. **77** [1982] 2432/7).
[69] Runau, R.; Peyerimhoff, S. D.; Buenker, J. (J. Mol. Spectrosc. **68** [1977] 253/68).
[70] Manz, U.; Reinsch, E. A.; Rosmus, P.; Werner, H.-J.; Seil, S. O. (J. Chem. Soc. Faraday Trans. **87** [1991] 1809/14).

[71] Vinogradov, I. P.; Vilesov, F. I. (Zh. Fiz. Khim. **51** [1977] 2017/21; Russ. J. Phys. Chem. [Engl. Transl.] **51** [1977] 1178/80).
[72] Hawkins, W. G.; Houston, P. L. (J. Phys. Chem. **86** [1982] 704/9).
[73] Nishi, N.; Shinohara, H.; Hanazaki, I. (Reza Kenkyu **10** [1982] 394/9).
[74] Vinogradov, I. P.; Firsor, V. V. (Opt. Spectrosk. **53** [1982] 46/9; Opt. Spectrosc. [Engl. Transl.] **53** [1982] 26/8).
[75] Harrison, J. A.; Whyte, A. R.; Phillips, L. F. (Chem. Phys. Lett. **129** [1986] 346/52).
[76] Lindberg, P.; Raybone, D.; Salthaouse, J. A.; Watkinson, T. M.; Whitehead, J. C. (Mol. Phys. **62** [1987] 1297/306).
[77] Kawasaki, M.; Tanaka, I. (J. Phys. Chem. **78** [1974] 1784/9).
[78] Hikida, T.; Maruyama, Y.; Saito, Y.; Mori, Y. (Chem. Phys. **121** [1988] 63/71).
[79] Ni, T.; Yu, S.; Ma, X.; Kong, F. (Chem. Phys. Lett. **126** [1986] 413/8).
[80] Kenner, R. D.; Rohrer, F.; Browarzik, R. K.; Kaes, A.; Stuhl, F. (Chem. Phys. **118** [1987] 141/52).

[81] Kenner, R. D.; Browarzik, R. K.; Stuhl, F. (Chem. Phys. **121** [1988] 457/71).

[82] Van Dijk, C. A.; Sandholm, S. T.; Davis, D. D.; Bradshaw, J. D. (J. Phys. Chem. **93** [1989] 6363/7).
[83] Chappell, E. L.; Jeffries, J. B.; Crosley, D. R. (J. Chem. Phys. **97** [1992] 2400/5).
[84] Kelly, P. J. (UCRL-51893 [1975] 304 pp.; C.A. **85** [1976] No. 184594).
[85] Masanet, J.; Deson, J.; Lalo, C.; Lempereur, F.; Tardieu de Maleissye, J. (J. Photochem. **36** [1987] 1/10).
[86] Alexander, M. H.; Werner, H. J.; Dagdigian, P. J. (J. Chem. Phys. **89** [1988] 1388/400).
[87] Stephenson, J. C.; Casassa, C.; King, D. S. (J. Chem. Phys. **89** [1988] 1378/87).
[88] Garland, N. L.; Jeffries, J. B.; Crosley, D. R.; Smith, G. P.; Copeland, R. A. (J. Chem. Phys. **84** [1986] 4970/5).
[89] Foy, B. R.; Casassa, M. P.; Stephenson, J. C.; King, D. S. (J. Chem. Phys. **89** [1988] 608/9).
[90] Foy, B. R.; Casassa, M. P.; Stephenson, J. C.; King, D. S. (J. Chem. Phys. **90** [1989] 7037/45).
[91] Foy, B. R.; Casassa, M. P.; Stephenson, J. C.; King, D. S. (J. Chem. Phys. **92** [1990] 2782/9).
[92] Chen, J.; Quiñones, E.; Dagdigian, P. J. (J. Chem. Phys. **93** [1990] 4033/42).
[93] Quiñones, E.; Chen, J.; Dagdigian, P. J. (Chem. Phys. Lett. **174** [1990] 65/70).
[94] Cheah, C. T.; Clyne, M. A. A. (J. Chem. Soc. Faraday Trans. II **76** [1980] 1543/60).
[95] Dodd, J. A.; Lipson, S. J.; Flanagan, D. J.; Blumberg, W. A. M.; Person, J. C.; Green, B. D. (J. Chem. Phys. **94** [1991] 4301/10).
[96] Gorbatenko, V. P.; Evgenov, I. V.; Pegov, V. S. (Fiz. Khim. Obrab. Mater. No. 5 [1990] 58/62).
[97] Whyte, A. R.; Phillips, L. F. (Chem. Phys. Lett. **102** [1983] 451/4).
[98] Whyte, A. R.; Phillips, L. F. (J. Phys. Chem. **88** [1984] 5670/3).
[99] Shin, S. K.; Chen, Y.; Oh, D.; Wittig, C. (Philos. Trans. R. Soc. London A **332** [1990] 361/74).
[100] Hoffmann, G.; Oh, D.; Wittig, C. (J. Chem. Soc. Faraday Trans. II **85** [1989] 1141/53).
[101] Hoffmann, G.; Oh, D.; Iams, H.; Wittig, C. (Chem. Phys. Lett. **155** [1989] 356/62).
[102] Foner, S. N.; Hudson, R. L. (J. Chem. Phys. **74** [1981] 5017/21).
[103] Dransfeld, P.; Hack, W.; Kurzke, H.; Temps, F.; Wagner, H. Gg. (Symp. Int. Combust. Proc. **20** [1985] 655/63).
[104] Donaldson, D. J.; Sloan, J. J.; Goddard, J. D. (J. Chem. Phys. **82** [1985] 4524/36).
[105] Goddard, J. D.; Donaldson, D. J.; Sloan, J. J. (Chem. Phys. **114** [1987] 321/9).

2.1.1.1.1.2 NH(a, b, c,... A,...), NH in Electronically Excited States of the Singlet and Triplet Manifold

There are four known singlet valence states of NH: a $^1\Delta$, b $^1\Sigma^-$, c $^1\Pi$, and d $^1\Sigma^-$. As far as chemical kinetics is concerned, the **a $^1\Delta$** state is the most interesting one, whereas for the other three states only quenching rates have been determined. The singlet ladder of the NH starts with the a $^1\Delta$ state (see p. 32). This state, which has a long radiative lifetime ($\tau_0 = 1.7$ s) since the transition to the $^3\Sigma^-$ electronic ground state is twofold forbidden, is of great kinetic interest. The imidogen radical in the lowest metastable singlet state (a $^1\Delta$) is mainly produced in the UV photodissociation of HN_3 [1 to 25] and HNCO [6, 24, 26 to 38]. NH_3 laser photolysis ($\lambda = 193$ nm) also produces NH(a) in a one-photon process but with a low quantum yield ($\Phi = 0.016$) [14]. The near-UV photolysis of HN_3 ($185 \leq \lambda/nm \leq 290$) is a clean NH(a $^1\Delta$) source [4] due to the spin-allowed process

$$HN_3(\tilde{X}\ ^1A') + h\nu \rightarrow NH(a\ ^1\Delta) + N_2(X\ ^1\Sigma_g^+)$$

The N_2 fragment is energetically constrained to the X $^1\Sigma_g^+$ state; H atom formation due to $HN_3 + h\nu \rightarrow H + N_3$ is of minor importance [1]. However, it should be mentioned that in this NH(a) source other reactive species (H atoms) are reported to be formed in a one-photon process in the laser photolysis of HN_3 with quantum yields of $\Phi = 0.15$ ($\lambda = 193$ nm) and $\Phi = 0.24$ ($\lambda = 248$ nm) [39]. At shorter wavelengths (Kr, Xe resonance lines) the formation of NH(c) was observed [40]. The spin-forbidden formation of NH(A $^3\Pi$) was due to secondary reactions of N_2^* with HN_3 (see p. 22) [40 to 42]. A detailed analysis of the resulting primary photodissociation products showed for the electronic states: NH(a) ≥ 0.998; NH(X) $< 8 \times 10^{-4}$; NH(b) $< 1 \times 10^{-3}$; and NH(A) $\leq 5 \times 10^{-5}$ [6].

The initial vibrational population of NH(a) was also studied experimentally. The NH(a) is formed only in its vibrational ground state if HN_3 is photolyzed at $\lambda = 308$ nm. At $\lambda = 226$ nm, however, NH(a) is, in contradiction to earlier observations [5, 6], not only formed in its vibrational ground state ($\lambda = 266$ nm) as stated in the literature [5, 6], but also in the higher vibrational states ($0 \leq v \leq 4$) for the photolysis wavelengths ($\lambda = 193$, 248, and 266 nm) [14, 17, 18, 21 to 24]. Recently, the vibrational distribution in NH(a) was studied, not only for two photolysis wavelengths, but continuously for the complete spectral range between 220 and 290 nm [25]. The vibrational distribution varies strongly with the photolysis wavelength, with an onset for the production of both the v = 1 and v = 2 states of NH(a) between 308 and 248 nm, far above the thermodynamic threshold. For the formation of v = 1 relative to v = 0, a pronounced maximum is observed at approximately 240 nm [25].

Also the rotational distribution of NH(a) was observed [8, 13, 16]. The photodissociation dynamics

$$HN_3(\tilde{X}\ ^1A') + h\nu \rightarrow HN_3(\tilde{A}\ ^1A'') \rightarrow NH(a\ ^1\Delta) + N_2(X\ ^1\Sigma_g^+)$$

has been studied in great detail experimentally [16, 43] and theoretically [44]. Not only were the above mentioned scalar properties E_{vib} and E_{rot} of the ejected NH(a) fragment measured, but also E_{trans}, and moreover the vectorial properties (angular distribution, rotational alignment, and correlation between translational and rotational motion) were analyzed [43].

The NH(a) formation in the system

$$HNCO(\tilde{X}\ ^1A') + h\nu \rightarrow NH(a\ ^1\Delta) + CO(X\ ^1\Sigma^+)$$

is analogous to the NH(a) formation from the isoelectronic HN_3 [45]. NH(a) and CO are the major fragments ($\leq 90\%$ [35]) in the photodissociation at $\lambda = 193$ nm, in which NH(a) can be characterized with a Boltzmann temperature of $T_{rot} \approx 1100$ K [33]. From the initial translational energy of NH(a) measured by photofragment translational spectroscopy, it was concluded that NH(a) is formed without a barrier [38], an observation which is in agreement with the negligible temperature dependence of the quenching rate constant k(NH(a) + CO) (see p. 120). The spin conservation, as discussed above for HN_3, is responsible for the formation of NH(a $^1\Delta$). In the Hg(^{3}P) photosensitized decomposition of HNCO, the major products are H + NCO [46].

NH(a) can also be obtained in the ammonia photolysis $NH_3 + h\nu(\lambda \leq 224\ nm) \rightarrow NH(a) + H_2$ [47 to 50], in which the NH(a) yield increases with decreasing wavelength between 147 and 105 nm. But it is not the dominant dissociation pathway.

NH(a) can be produced in the chemical reactions $O(^1D) + NH_3 \rightarrow NH(a) + H_2O$ [51] and $O(^1D) + HCN \rightarrow NH(a) + CO(X)$ [52]. The reaction $F + NH_2 \rightarrow NH + HF$ in part proceeds on the singlet surface of the NH_2F molecule and thus NH(a) can be formed [53].

The higher singlet states are produced either directly from photolysis or from excitation of NH(a). In this respect the NH(c $^1\Pi$) is of special importance, since this state is used in the NH(a) LIF detection method (see below and p. 27).

The next state in the singlet ladder is the **b $^1\Sigma^+$** state. NH in this state can be obtained in the NH_3 VUV photolysis [54 to 57] and ND(b) from ND_3 VUV photolysis, respectively [56]. The vacuum ultraviolet radiation is produced in Ar discharge separated from the photolysis volume by an LiF window. The formation of NH(b $^1\Sigma^+$) was also observed in the ArF laser photolysis of HN_3 [58]. The NH(b $^1\Sigma^+$) formation was observed in a discharge flow reactor in the reaction of H atoms with N_3 radicals [59].

The **c $^1\Pi$** state is mainly obtained by exciting the allowed transition (c $^1\Pi \leftarrow$ a $^1\Delta$), which is done in all NH(a) laser-induced fluorescence studies. In the photolysis of CH_3NH_2 and $C_2H_5NH_2$ at $\lambda = 123.6$ nm, the production of NH(c) can be concluded from the observed NH(c) emission [60]. Electron bombardment of Ar-NH_3, N_2H_4, and CH_3NH_2 mixtures, respectively, at energies above the threshold (≥ 7.8 eV) gave rise to NH(c) emission [61]. In the energy transfer reaction of Ar*, Kr*, and Xe*(^3P_j) with HNCO and HN_3, the emission from NH(c) is observed [62]. Electron impact dissociation of HNCO, HN_3, and NH_3 leads to the emission of NH(c $^1\Pi$) [63 to 65]; the precursor molecules can also be $C_2H_5NH_2$, CH_3NH_2, or N_2H_4 [63]. At short wavelengths ($\lambda = 121.6$ nm) the direct formation of NH(c) in the NH_3, HN_3, and HNCO photolysis was observed [66], i.e., with Lyman α radiation or in the case of HNCO with synchrotron radiation ($107 \leq \lambda/\text{nm} \leq 180$). The quantum yield for the NH(c $^1\Pi$) formation varies smoothly between 0.9% for $\lambda = 107$ nm and 0.02% for $\lambda = 135$ nm [67]. Small amounts (at most 2%) of NH(c) formation are also reported for longer photolysis wavelengths ($325 \leq \lambda/\text{nm} \leq 335$) for the isocyanic and hydrazoic acid precursors [41, 45]. In the laser photolysis of HN_3 at $\lambda = 193$ nm, NH(c) is formed with a quantum yield of $\Phi = 6.3 \times 10^{-4}$ [15].

The lowest triplet state above the electronic ground state, **A $^3\Pi$**, is used for the LIF detection of NH(X), and is thus of interest in that system (see p. 25). Since the A-X transition is allowed, NH(A $^3\Pi$) can be formed by exciting the electronic ground state with light at $\lambda = 336$ nm [68]. Besides the two-step mechanism (formation of NH(X) followed by excitation), it can be obtained directly by NH_3 photolysis. There are, however, also ways to produce NH in the A $^3\Pi$ state directly from stable precursor molecules.

In the VUV photolysis of isocyanic acid with synchrotron radiation ($107 \leq \lambda/\text{nm} \leq 180$), NH(A $^3\Pi$) is observed as a direct photofragment [67]. The highest quantum yield ($\Phi = 0.3\%$) is obtained at 140 nm [67]. Also in the VUV photolysis (Kr lines, ($\lambda = 123.6$, 116.5 nm) [45] and Lyman α line ($\lambda = 121.6$ nm) [66]) of HNCO, emission from NH(A $^3\Pi$) was observed. In the laser photolysis ($\lambda = 193$ nm) of CH_3NH_2 [69, 70], HN_3, and N_2H_4 [70] as well as NH_3 and ND_3 [71], strong emission from NH(A) and ND(A), respectively, was observed. The vibrational and rotational distribution in NH(A $^3\Pi$, v, J), formed in the ArF laser photolysis of N_2H_4, was studied to determine the photolysis mechanism as a two-photon process [72, 73]. The photodissociation of NH_3 to produce NH(A) was examined in the photolysis wavelength range 60 to 134 nm. The quantum yield for producing NH(A) was found to be less than 0.2% at $\lambda \geq 97.6$ nm. At wavelengths below 70 nm the photoionization quantum yield reaches unity [74]. In the VUV ($\lambda \geq 105$ nm) photolysis of N_2H_4, a weak formation of NH(A) appeared at wavelengths shorter than 118 nm. From this onset, the primary process $N_2H_4 + h\nu \rightarrow NH(A) + H + NH_2(\tilde{X})$ was concluded [75]. In the excimer laser photolysis ($\lambda =$ 193 nm) of N_2H_4 and N_2D_4, emission of NH(A) and ND(A), respectively, formed by two-photon processes, was observed [73].

In the multiphoton dissociation of DN_3 in the infrared spectral region (IRMPD) with CO_2 lasers, ND(A) emission is observed [76]. The ND(A) is, however, formed in a secondary reaction of ND(a) with vibrationally excited DN_3.

NH can be formed directly in the A $^3\Pi$ state by energy transfer of Ar*, Kr*, Xe*(^3P_j) to HNCO or HN_3 [62]. In the energy transfer process of Kr* with ammonia, only the channel $Kr(^3P_2) + NH_3 \rightarrow NH(A\ ^3\Pi) + H_2 + Kr(^1S_0)$ is observed, in contrast to $Ar^* + NH_3$ in which also the emission of NH(c $^1\Pi$) was observed [77]. Also in the system $He(^3S_1) + NH_3$, both NH(A$^3\Pi$) and NH(c $^1\Pi$) were obtained as primary products of the energy transfer process [78]. Energy transfer in the reaction $N_2(A\ ^3\Sigma_u^+) + HN_3 \rightarrow NH(A\ ^3\Pi) + 2\ N_2$ is responsible for NH(A) formation [40, 79, 80]. A reaction with N_2^* can give rise to NH(A) formation in the reaction $HN_3 + N_2^* \rightarrow NH(A) + 2\ N_2$ in the VUV photolysis of HN_3, since the N_2^* (N_2(B)) is formed as a primary photoproduct of HN_3 [41].

The electron impact dissociation of NH_3, HN_3, HNCO, N_2H_4, $C_2H_5NH_2$, and CH_3NH_2 leads to NH also in the A $^3\Pi$ state [63 to 65, 81]. HNCS dissociates via $HNCS \rightarrow NH(A\ ^3\Pi) + CS(a\ ^3\Sigma^+)$ when colliding with electrons having an energy larger than 9.2 eV [82].

NH(A $^3\Pi$) can be formed as a primary product in various chemical reactions [83 to 89]. The fast reaction of electronically excited O atoms, $O(^1D) + HNCO \rightarrow NH(A) + CO_2$, leads to the direct formation of NH(A) with a rate constant of $k(300\ K) = 2.8 \times 10^{13}\ cm^3 \cdot mol^{-1} \cdot s^{-1}$ [83]. The reaction of N^+ ions with H_2, CH_4, C_2H_4, C_2H_6, and C_3H_8 was used to produce NH(A $^3\Pi$, v) in a molecular beam [84 to 86]. The reaction of the CH(X) radical with NO, with a room-temperature rate constant of $k = 1.2 \times 10^{14}\ cm^3 \cdot mol^{-1} \cdot s^{-1}$ [90] was shown to produce NH in the A $^3\Pi$ state [87, 88], and the reaction CD + NO produces ND(A) [89]. The CH(X) radical reaction is one of the main sources of NH in flames (see p. 126). NO, excited at $\lambda = 193$ nm, reacts with H atoms and produces NH(A) radicals [91]. In the reaction of electronically excited N atoms, N(^{2}D, ^{2}P), with HBr or HI, NH(A $^3\Pi$) is formed as detected by its emission [92]. At high temperatures, $T \cong 3500$ K, the HNCO pyrolysis in an incident shock wave was accompanied by NH(A) UV emission [93].

Whether a radical source is suitable, depends on the purpose for which it is used. For kinetic measurements in particular, if information about the reaction products are wanted, a clean and state-specific source is needed; whereas for photophysical or spectroscopic investigations in general any system containing the species in the state under consideration is sufficient.

References:

[1] Konar, R. S.; Matsumoto, S.; Darwent, B. B. (Trans. Faraday Soc. **67** [1971] 1698/706).
[2] Stepanov, P. I.; Zamanskii, V. M.; Moskvitina, E. N.; Kuzyakov, Y. Y. (Vestn. Mosk. Univ. Ser. 2 Khim. **14** [1973] 306/9).
[3] Paur, R. J.; Bair, E. J. (J. Photochem. **1** [1973] 255/65).
[4] Paur, R. J.; Bair, E. J. (Int. J. Chem. Kinet. **8** [1976] 139/52).
[5] McDonald, J. R.; Miller, R. G.; Baronavski, A. P. (Chem. Phys. Lett. **51** [1977] 57/60).
[6] Baronavski, A. P.; Miller, R. G.; McDonald, J. R. (Chem. Phys. **30** [1978] 119/31).
[7] Piper, L. G.; Krech, R. H.; Taylor, R. L. (J. Chem. Phys. **73** [1980] 791/800).
[8] DeKoven, B. M.; Baronavski, A. P. (Chem. Phys. Lett. **86** [1982] 392/6).
[9] Kodama, S. (Bull. Chem. Soc. Jpn. **56** [1983] 2348/54).
[10] Kodama, S. (Bull. Chem. Soc. Jpn. **56** [1983] 2355/62).
[11] Kodama, S. (Bull. Chem. Soc. Jpn. **56** [1983] 2363/70).
[12] Rohrer, F.; Stuhl, F. (Chem. Phys. Lett. **111** [1984] 234/7).
[13] Hall, J. L.; Adams, H.; Kasper, J. V. V.; Curl, R. F.; Tittel, F. K. (J. Opt. Soc. Am. B Opt. Phys. **2** [1985] 781/5).
[14] Kenner, R. D.; Rohrer, F.; Stuhl, F. (J. Chem. Phys. **86** [1987] 2036/43).
[15] Rohrer, F.; Stuhl, F. (J. Chem. Phys. **88** [1988] 4788/99).
[16] Gericke, K.-H.; Theinl, R.; Comes, F. J. (Chem. Phys. Lett. **164** [1989] 605/11).

[17] Hack, W.; Mill, T. (J. Mol. Spectrosc. **144** [1990] 358/65).
[18] Mill, T. (Ber. Max-Planck-Inst. Strömungsforsch. **1990** No. 14, 104 pp.).
[19] Gericke, K.-H.; Theinl, R.; Comes, F. J. (J. Chem. Phys. **92** [1990] 6548/55).
[20] Chu, J. J.; Marcus, P.; Dagdigian, P. J. (J. Chem. Phys. **93** [1990] 257/67).

[21] Nelson, H. H.; McDonald, J. R. (J. Chem. Phys. **93** [1990] 8777/83).
[22] Nelson, H. H.; McDonald, J. R.; Alexander, M. H. (J. Phys. Chem. **94** [1990] 3291/4).
[23] Hack, W.; Mill, T. (J. Phys. Chem. **97** [1993] 5599/606).
[24] Bohn, B.; Stuhl, F. (J. Phys. Chem. **97** [1993] 4891/8).
[25] Hawley, M.; Baronavski, A. P.; Nelson, H. H. (J. Chem. Phys. [1993] in press).
[26] Holland, R.; Style, D. W. G.; Dixon, R. N.; Ramsay, D. A. (Nature **182** [1958] 336/7).
[27] Dixon, R. N. (Can. J. Phys. **37** [1959] 1171/86).
[28] Dixon, R. N. (Can. J. Phys. **38** [1960] 10/6).
[29] Dixon, R. N. (Philos. Trans. R. Soc. London A **252** [1960] 165/92).
[30] Back, R. A. (J. Chem. Phys. **40** [1964] 3493/6).

[31] Brash, J. L.; Back, R. A. (Can. J. Chem. **43** [1965] 1778/83).
[32] Back, R. A.; Childs, J. (Can. J. Chem. **46** [1968] 1023/4).
[33] Drozdoski, W. S.; Baronavski, A. P.; McDonald, J. R. (Chem. Phys. Lett. **64** [1979] 421/5).
[34] Fujimoto, G. T.; Umstead, M. E.; Lin, M. C. (Chem. Phys. **65** [1982] 197/203).
[35] Spiglanin, T. A.; Perry, R. A.; Chandler, D. W. (J. Phys. Chem. **90** [1986] 6184/9).
[36] Spiglanin, T. A.; Chandler, D. W. (J. Chem. Phys. **87** [1987] 1577/81).
[37] Spiglanin, T. A.; Perry, R. A.; Chandler, D. W. (J. Chem. Phys. **87** [1987] 1568/76).
[38] Spiglanin, T. A.; Chandler, D. W. (Chem. Phys. Lett. **141** [1987] 428/32).
[39] Gericke, K.-H.; Lock, M.; Comes, F. J. (Chem. Phys. Lett. **186** [1991] 427/30).
[40] Welge, K. H. (J. Chem. Phys. **45** [1966] 4373/4).

[41] Okabe, H. (J. Chem. Phys. **49** [1968] 2726/33).
[42] Maruyama, Y.; Hikida, T.; Mori, Y. (Chem. Phys. Lett. **116** [1985] 371/5).
[43] Gericke, K.-H.; Lock, M.; Fasold, R.; Comes, F. J. (J. Chem. Phys. **96** [1992] 422/32).
[44] Alexander, M. H.; Werner, H. J.; Dagdigian, P. J. (J. Chem. Phys. **89** [1988] 1388/400).
[45] Okabe, H. (J. Chem. Phys. **53** [1970] 3507/15).
[46] Friswell, N. J.; Back, R. A. (Can. J. Chem. **46** [1968] 527/30).
[47] Schurath, U.; Tiedemann, P.; Schindler, R. N. (J. Phys. Chem. **72** [1969] 456/9).
[48] Groth, W. E.; Schurath, U.; Schindler, R. N. (J. Phys. Chem. **72** [1968] 3914/20).
[49] McNesby, J. R.; Tanaka, I.; Okabe, H. (J. Chem. Phys. **36** [1962] 605/7).
[50] Slanger, T. G.; Black, G. (J. Chem. Phys. **77** [1982] 2432/7).

[51] Sanders, N. D.; Butler, J. E.; McDonald, J. R. (J. Chem. Phys. **73** [1980] 5381/3).
[52] Carpenter, B. K.; Goldstein, N.; Karn, A.; Wiesenfeld, J. R. (J. Chem. Phys. **81** [1984] 1785/93).
[53] Dransfeld, P.; Hack, W.; Kurzke, H.; Temps, F.; Wagner, H. Gg. (Symp. Int. Combust. Proc. **20** [1985] 655/63).
[54] Zetzsch, C. (Habilitationsschr. Univ. Bochum 1977, 242 pp.).
[55] Gelernt, B.; Filseth, S. V.; Carrington, T. (Chem. Phys. Lett. **36** [1975] 238/41).
[56] Gelernt, B.; Filseth, S. V.; Carrington, T. (J. Chem. Phys. **65** [1976] 4940/4).
[57] Van Dijk, C. A.; Sandholm, S. T.; Davies, D. D.; Bradshaw, J. D. (J. Phys. Chem. **93** [1989] 6363/7).
[58] Blumenstein, U.; Rohrer, F.; Stuhl, F. (Chem. Phys. Lett. **107** [1984] 347/50).
[59] Kajimoto, O.; Kawajiri, T.; Fueno, T. (Chem. Phys. Lett. **76** [1980] 315/8).
[60] Kawasaki, M.; Tanaka, I. (J. Phys. Chem. **78** [1974] 1784/9).

[61] Fujita, I. (Z. Phys. Chem. [Munich] **149** [1986] 17/25).
[62] Stedman, D. H. (J. Chem. Phys. **52** [1970] 3966/70).
[63] Fukui, K.; Fujita, I.; Kuwata, K. (J. Phys. Chem. **81** [1977] 1252/7).
[64] Tokue, I.; Ito, Y. (Chem. Phys. **79** [1983] 383/9).
[65] Tokue, I.; Ito, Y. (Chem. Phys. **89** [1984] 51/7).
[66] Hikida, T.; Maruyama, Y.; Saito, Y.; Mori, Y. (Chem. Phys. **121** [1988] 63/71).
[67] Uno, K.; Hikida, T.; Hiraya, A.; Shobatake, K. (Chem. Phys. Lett. **166** [1990] 475/9).
[68] Garland, N. L.; Jeffries, J. B.; Crosley, D. R.; Smith, G. P.; Copeland, R. A. (J. Chem. Phys. **84** [1986] 4970/5).
[69] Nishi, N.; Shinohara, H.; Hanazaki, I. (Reza Kenkyu **10** [1982] 394/9).
[70] Haak, H. K.; Stuhl, F. (J. Phys. Chem. **88** [1984] 3627/33).

[71] Haak, H. K.; Stuhl, F. (J. Phys. Chem. **88** [1984] 2201/4).
[72] Hawkins, W. G.; Houston, P. L. (J. Phys. Chem. **86** [1982] 704/9).
[73] Lindberg, P.; Raybone, D.; Salthouse, J. A.; Watkinson, T. M.; Whitehead, J. C. (Mol. Phys. **62** [1987] 1297/306).
[74] Wu, C. Y. R. (J. Chem. Phys. **86** [1987] 5584/6).
[75] Biehl, H.; Stuhl, F. (J. Photochem. Photobiol. A **59** [1991] 135/42).
[76] Simpson, T. B.; Mazur, E.; Lehmann, K. K.; Burak, I.; Bloembergen, N. (J. Chem. Phys. **79** [1983] 3373/81).
[77] Sekiya, H.; Nishiyama, N.; Tsuji, M.; Nishimura, Y. (J. Chem. Phys. **88** [1988] 5249/51).
[78] Someda, K.; Kondow, T.; Kuchitsu, K. (J. Phys. Chem. **92** [1988] 368/74).
[79] Stedman, D. H; Setser, D. W. (Chem. Phys. Lett. **2** [1968] 542/4).
[80] Dessaux, D.; Picavet-Bernard, G.; Goudmand, P. (C. R. Seances Acad. Sci. C **276** [1973] 635/8).

[81] Fukui, K.; Fujita, I.; Kuwata, K. (Bull. Chem. Soc. Jpn. **45** [1972] 2278/80).
[82] Kajimoto, O.; Kondo, O.; Okabe, K.; Fujikane, J.; Fueno, T. (Bull. Chem. Soc. Jpn. **58** [1985] 3469/74).
[83] Ongstad, A. P.; Liu, X.; Coombe, R. D. (J. Phys. Chem. **92** [1988] 5578/80).
[84] Kusunoki, I.; Ottinger, Ch.; Simonis, J. (Chem. Phys. Lett. **41** [1976] 601/5).
[85] Kusunoki, I.; Ottinger, Ch. (J. Chem. Phys. **70** [1979] 710/21).
[86] Kusunoki, I.; Ottinger, Ch. (J. Chem. Phys. **70** [1979] 699/709).
[87] Lichtin, D. A.; Berman, M. R.; Lin, M. C. (Chem. Phys. Lett. **108** [1984] 18/24).
[88] Nishiyama, N.; Sekiya, H.; Yamaguchi, S.; Tsuji, M.; Nishimura, Y. (J. Phys. Chem. **90** [1986] 1491/3).
[89] Nishiyama, N.; Sekiya, H.; Tsuji, M.; Nishimura, Y. (Chem. Phys. **112** [1987] 265/70).
[90] Wagal, S. S.; Carrington, T.; Filsath, S. V.; Sadowski, C. M. (Chem. Phys. **69** [1982] 61/70).

[91] Phillips, L. F. (J. Photochem. **21** [1983] 365/7).
[92] Tabayashi, K.; Ohshima, S.; Shhobatake, K. (J. Chem. Phys. **80** [1984] 5335/7).
[93] Yokuyama, K.; Sakane, Y.; Fueno, T. (Bull. Chem. Soc. Jpn. **64** [1991] 1738/42).

2.1.1.1.2 NH Production in Condensed Media

The condensed phase can be an aqueous solution [1, 2], or liquid saturated hydrocarbons (C_2H_6 [3, 4], C_5H_{12} [5], cyclopropane [6], cyclohexane [7]), or liquid olefins [8] and acetylenes [9], as well as alcohols [10] and liquid ammonia [11 to 14] at various temperatures. Most often, however, a low-temperature inert gas matrix is the condensed phase in which isolated NH radicals are produced [15 to 30]. Also NH adsorbed on zeolite [31] has to be mentioned in this context.

The NH radicals can be produced in the gas phase in a discharge (NH_3/Ar) and then condensed at a cold surface ($T \cong 20$ K) [19, 30], or they can be produced directly in the liquid or solid phase. In the latter cases, UV photolysis of precursor molecules such as NH_2Cl [15, 18], HN_3 [2 to 10, 17, 20, 23 to 26, 29, 32 to 34], HNCO [23, 24, 26], NH_4ClO_4 [35] or the VUV photolysis of ammonia [27, 28] was used to produce NH radicals. The irradiation of N_3^- ions in aqueous solutions at $\lambda = 254$ nm gives rise to NH(a) radicals via the chemical reaction of electronically excited N_3^- ions with water [1]. The γ radiation of Cu^+ ions in liquid NH_3 was used to produce NH(X) radicals [36]. Also the direct γ irradiation of liquid NH_3 [12] and the radiolysis of liquid NH_3 and ND_3 (electrons of about 8 MeV) produces NH and ND, respectively [11, 13, 14].

Due to the fast quenching processes, the NH radicals appear in the stable ($^3\Sigma^-$) electronic ground state or in the metastable ($^1\Delta$) electronic state, which in the case of a large excess of reactant (e.g., in C_2H_6(l)) reacts before it is quenched to the ground state (see p. 127).

NH radicals in Ne, Ar, Kr, and Xe matrices have been studied in great detail as far as the lifetimes are concerned [23] but also with regard to trapping. From the high-resolution a $^1\Delta \rightarrow$ X $^3\Sigma$ emission spectrum, two NH trapping sites in the matrix with different symmetry were identified [28].

References on pp. 27/30.

2.1.1.1.3 Detection Methods

NH was first detected by its A $^3\Pi$-X $^3\Sigma^-$ transition at the wavelength of 336 nm exactly one hundred years ago in 1893 [37]. Since then this optical transition was observed in absorption as well as in emission in a large number of experimental studies for NH [38 to 58] and for the isotopic species ND [59 to 62]. Also theoretical investigations have dealt with this transition [58, 62, 63]. Several detection methods are based on the A $^3\Pi$-X $^3\Sigma^-$ transition. With the A$\rightarrow$X emission the presence of electronically excited NH in the system is indicated. The A$\leftarrow$X absorption can be applied if NH(X) has to be detected. A significantly higher sensitivity is obtained if radiation-induced fluorescence is applied, in which the A$\leftarrow$X excitation is done by resonance radiation [45] or laser radiation, i.e., laser-excited fluorescence (LEF), more often called laser-induced fluorescence (LIF) [46, 48, 50 to 52, 61 to 63]. To determine absolute concentrations of NH radicals, the absolute absorption intensities, i.e., the transition probabilities, have to be known; they were determined in [64] and in [65] for the A-X transition. If the LIF method is used, the quenching rate constants and quencher concentrations in environments like flames [55, 56] have to be known. Even for the determination of relative NH radical concentrations, e.g., along a flame axis, the assumption of constant quenching rates is necessary [48] if the quenching rates are not known. Semiquantitative concentration determinations in flames, accurate to within a factor of three, are claimed to be possible with the LIF method [49]. In the saturated LIF method, in which the lifetime of the upper state is no longer determined by spontaneous emission and quenching, but by stimulated emission, an LIF signal independent of quencher concentrations can be obtained. This method was used to determine NH concentrations in flames [50]. Another elegant method to avoid quenching is to excite a predissociating electronic state, the lifetime of which is shorter, due to predissociation, than the time between collisions, and thus the fluorescence quantum yield is low, but independent of quenching up to pressures well above atmospheric pressures. This method has successfully been applied to other radicals like OH and NO but not yet to NH. There are, however, rovibronic states in the NH(A) suitable for this method.

A coherent technique, which has been applied as a tool to study NH radicals in flames, is the degenerate four wave mixing technique (DFWM) [66, 67]. In this method [68], the interaction of three input laser beams of identical frequency, ω, with a nonlinear medium produces a fourth coherent signal beam (the DFWM signal) also having the frequency ω and propagating counter and collinear with one of the incoming beams, the probe beam. The advantage of this technique is that a highly collimated signal beam is obtained [67].

To detect NH in the electronic ground state, the infrared absorption was applied in particular in solid matrices [21, 69] but also in the gas phase using a color center laser [70] and different frequency laser systems [71]. On the other hand, also the IR emission of NH(X $^3\Sigma^-$, v) was investigated experimentally and theoretically [72]. IR emission of the vibrational states of the X $^3\Sigma^-$ electronic state were also observed with Fourier transform emission spectroscopy [73] with low [74] and high resolution [75].

In the far infrared, the Zeemann splitting gives rise to transitions on which a very sensitive detection method is based, the so-called laser magnetic resonance (LMR) technique. The LMR absorption has been used to detect NH(X, v=0) at wavelengths near 0.3 mm [76] and NH(X, v=0, 1) [77], and also ^{15}NH(X, v=0) and ND(v=0, 1) [77]. The far-infrared laser magnetic resonance (FIR-LMR) technique is not limited to triplet states; also the NH(a $^1\Delta$) state was detected [78] as well as NH(a, v=1) [79]. (For further NH(a) detection methods see below). Transitions, initiated in the microwave region between two Zeeman components of the same state which are degenerate at zero field, are applied to detect NH(a $^1\Delta$) in the gas phase with the electron spin resonance (ESR) technique [80]. The high sensitivity in both methods is due to an intracavity device and magnetic field modulation.

The pure rotational spectrum in the far infrared (zero field) was achieved with a tunable IR laser [81]; this might, combined with multipass absorption devices, be an interesting method to detect NH(X) radicals.

A universal detection method is the mass spectrometric detection technique using electron impact ionization, which was used to detect NH in flow reactors [82 to 86] and even in shock tube experiments [87]. This method is in general not very sensitive and not state-specific, although for NH it was possible to discriminate between NH(X) and NH(a) formed in the reaction $F+NH_2$ [86]. The sensitivity, and simultaneously the specificity, can be significantly improved by using photoionization. The energy to ionize a molecule, radical, or atom results from a VUV photon [88] or from intense laser radiation from several photons with or without resonant intermediate states. The resonance-enhanced multiphoton ionization (REMPI) method combined with mass spectrometry was used to detect NH (ND) but until now only in the a $^1\Delta$ state [89 to 95]. NH was actually the first radical which was detected by REMPI. With 3 photons in the wavelength range of 394 to 397 nm, the NH is excited to a high-lying Rydberg state, out of which it is ionized with another photon (3+1 REMPI) [92]. New high-lying Rydberg states in NH and ND have been found with 2+1 REMPI [93 to 96].

NH radicals in higher electronic states can be detected by emission as mentioned above for the A $^3\Pi$ state at 336 nm [54, 58, 62, 97]; this A $^3\Pi$-X $^3\Sigma$ emission was also observed with Fourier transform spectroscopy [47, 53]. Those electronically excited NH states, which have allowed transitions to lower electronic states, can easily detected by emission, but also a spin-forbidden, dipole-allowed radiative transition b $^1\Sigma^+ \rightarrow$ X $^3\Sigma^-$ was observed [98], and even the highly forbidden intercombination transition a $^1\Delta \rightarrow$ X $^3\Sigma^-$ was observed directly in emission [99]. These emissions, however, were of importance more for the spectroscopy of NH to determine the excitation energy of NH(a, b) than for detecting NH(a) or NH(b) (see below).

NH in its metastable (a $^1\Delta$) state (the radiative lifetime was determined by ab initio methods in [100]) is mainly detected by LIF using the c $^1\Pi$–a $^1\Delta$ transition at 324 nm [101 to 103]. This method is in common use for kinetic experiments [104 to 112, 118, 119]. The c→a emission has been intensively studied since the 1930's to detect NH(c) radicals [113 to 115]. The transition c→a was observed in high resolution with a hollow cathode discharge lamp [116]. Also the ND(c–a) system was analyzed, and Franck-Condon factors were calculated [117]. Improved transition probabilities, which are important for quantitative spectroscopy, were determined for the c–a transition [120, 121]. Also for the c–a transition, the collision-free LIF via a predissociating state can be used by choosing an appropriate state in the c $^1\Pi$ ladder.

Fourier transform emission spectra for the transition c–a for NH(c, v = 0, 1) were recorded [122]. The photophysics of the c $^1\Pi$ state, which is essential for the LIF detection method, was studied experimentally in [122, 123] and theoretically in [63].

There are four electronically excited singlet valence states in NH: a $^1\Delta$, b $^1\Sigma^-$, c $^1\Pi$, and d $^1\Sigma^+$ and three Rydberg states: f $^1\Pi$, g $^1\Delta$, and h $^1\Sigma^+$; the e $^1\Pi$ Rydberg state is probably repulsive [93]. All transitions, at least for the valence states, are known; the Rydberg states play a role in the MPI spectra. Only a few transitions provide valuable optical probes and thus will be discussed here only briefly.

The excitation of NH(c $^1\Pi$) to the d $^1\Sigma^+$ state at 253.0 nm and the emission of the d $^1\Sigma$ state [124 to 127] (see p. 82) is mainly of spectroscopic interest as is also the d $^1\Sigma^+$–b $^1\Sigma^+$ VUV transition at 162.0 nm [128] (see p. 82).

The b $^1\Sigma^+$ state can be detected by the b $^1\Sigma \rightarrow$ X $^3\Sigma$ transition [98] (see above); high-resolution observation of this system has been made [129] (see p. 69). The radiative lifetime of NH(b $^1\Sigma$) has been determined using ab initio methods [100]. The c–b system at 450.2 nm has been studied [130] (see p. 71) and was used to detect NH(b) and ND(b) via the c→a ultraviolet fluorescence after the c←b excitation [131]. The theoretical investigations for the spin-forbidden, dipole-allowed radiative transition and the spin-allowed c–b transition resulted in a radiative lifetime of τ_0(b $^1\Sigma^+$) = 97 ms [63].

Finally, an indirect NH(a) detection method should be mentioned that was used in the NH(a) – HN_3 system. The reaction of NH(a) with HN_3 produces $NH_2(\tilde{A})$: NH(a) + $HN_3 \rightarrow NH_2(\tilde{A})$ + $N_3(\tilde{X}\ ^2\Pi)$. The chemiluminescence of $NH_2(\tilde{A}\ ^2A_1)$ produced in this reaction is used to detect NH(a) [7].

References:

[1] Burak, I.; Treinin, A. (J. Am. Chem. Soc. **87** [1965] 4031/6).
[2] Kawai, J.; Tsunashima, S.; Sato, S. (Chem. Lett. No. 6 [1983] 823/6).
[3] Tsunashima, S.; Hamada, J.; Hotta, M.; Sato, S. (Bull. Chem. Soc. Jpn. **53** [1980] 2443/7).
[4] Kitamura, T.; Tsunashima, S.; Sato, S. (Kokagaku Toronkai Koen Yoshishu **1979** 230/1; C.A. **93** [1980] No. 149351).
[5] Kawai, J.; Tsunashima, S.; Sato, S. (Nippon Kagaku Kaishi No. 1 [1984] 32/6).
[6] Hamada, J.; Tsunashima, S.; Sato, S. (Bull. Chem. Soc. Jpn. **55** [1982] 1739/42).
[7] McDonald, J. R.; Miller, R. G.; Baronavski, A. P. (Chem. Phys. **30** [1978] 133/45).
[8] Hamada, J.; Tsunashima, S.; Sato, S. (Bull. Chem. Soc. Jpn. **56** [1983] 662/6).
[9] Kawai, J.; Tsunashima, S.; Sato, S. (Chem. Phys. Lett. **110** [1984] 655/8).
[10] Tsunashima, S.; Kitamura, T.; Sato, S. (Bull. Chem. Soc. Jpn. **54** [1981] 2869/71).

[11] Khaikin, G. I.; Zhigunov, V. A. (Khim. Vys. Energ. **9** [1975] 211/3).
[12] Delcourt, M. O.; Belloni, J.; Saito, E. (J. Chem. Phys. **80** [1976] 1101/5).

[13] Belloni, J.; Billiau, F.; Cordier, P.; Delaire, J. A.; Delcourt, M. O. (J. Phys. Chem. **82** [1978] 532/6).
[14] Belloni, J.; Cordier, P.; Delaire, J. A.; Delcourt, M. O. (J. Phys. Chem. **82** [1978] 537/9).
[15] Jander, J.; Fischer, J. (Angew. Chem. **71** [1959] 626/7).
[16] Keyser, L. F.; Robinson, G. W. (J. Am. Chem. Soc. **82** [1960] 5245/6).
[17] Papazian, H. A. (J. Chem. Phys. **32** [1960] 456/60).
[18] Jander, J.; Fischer, J. (Z. Anorg. Allg. Chem. **313** [1961] 37/47).
[19] Glasel, J. A. (Proc. Natl. Acad. Sci. U.S. A **47** [1961] 174/80).
[20] Jacox, M. E.; Milligan, D. E. (J. Am. Chem. Soc. **85** [1963] 278/82).

[21] Milligan, D. E.; Jacox, M. E. (J. Chem. Phys. **41** [1964] 2838/41).
[22] Milligan, D. E.; Jacox, M. E. (J. Chem. Phys. **47** [1967] 5157/68).
[23] Esser, H.; Langen, J.; Schurath, U. (Ber. Bunsen-Ges. Phys. Chem. **87** [1983] 636/43).
[24] Esser, H.; Langen, J.; Schurath, U. (Bull. Soc. Chim. Belg. **92** [1983] 672).
[25] Collins, S. T.; Pimentel, G. C. (J. Phys. Chem. **88** [1984] 4258/64).
[26] Ramsthaler-Sommer, A.; Eberhardt, K. E.; Schurath, U. (J. Chem. Phys. **85** [1986] 3760/9).
[27] Ramsthaler-Sommer, A.; Becker, A. C.; van Riesenbeck, N.; Lodemann, K.-P.; Schurath, U. (Chem. Phys. **140** [1990] 331/8).
[28] Blindauer, C.; van Riesenbeck, N.; Seranski, K.; Winter, M.; Becker, A. C.; Schurath, U. (Chem. Phys. **150** [1991] 93/108).
[29] Yokoyama, K.; Kitaibe, H.; Fueno, T. (Bull. Chem. Soc. Jpn. **64** [1991] 1731/7).
[30] Hizhnyakov, V.; Seranski, K.; Schurath, U. (Chem. Phys. **162** [1992] 249/56).

[31] Uyama, H.; Uchibura, T.; Niijma, H.; Matsumoto, O. (Chem. Lett. **1987** No. 4, pp. 555/8).
[32] Tsunashima, S.; Hotta, M.; Hamada, J.; Sato, S. (Proc. Yamada 3rd Conf. Free Radicals, Sandra, Jpn., 1979, pp. 269/70).
[33] Tsunashima, S.; Hotta, M.; Sato, S. (Chem. Phys. Lett. **64** [1979] 435/9).
[34] Kawai, J.; Tsunashima, S.; Sato, S. (Bull. Chem. Soc. Jpn. **55** [1982] 3312/6).
[35] Chen, K.; Wang, G.; Kuo, C.; Shyy, I.; Chang, Y. (Chem. Phys. Lett. **167** [1990] 351/5).
[36] Delcourt, M. O.; Belloni, J. (Radiochem. Radioanal. Lett. **13** [1973] 329/38).
[37] Eder, J. M. (Dtsch. Wien Akad. **60** [1893] 1).
[38] Fowler, A.; Gregory, C. C. L. (Philos. Trans. R. Soc. London A **218** [1919] 251/72).
[39] Funke, G. (Z. Phys. **96** [1935] 787/98).
[40] Shaw, R. W. (Astrophys. J. **83** [1936] 225/37).

[41] Funke, G. (Z. Phys. **101** [1936] 104/12).
[42] Swings, P.; Elvey, C. T.; Babcock, H. W. (Astrophys. J. **94** [1941] 320/43).
[43] Dixon, R. N. (Can. J. Phys. **37** [1959] 1171/86).
[44] Schmitt, J. L. (Publ. Astron. Soc. Pac. **81** [1969] 657/64).
[45] Hansen, I.; Höinghaus, K.; Zetzsch, C.; Stuhl, F. (Chem. Phys. Lett. **42** [1976] 370/2).
[46] Anderson, W. R.; Crosley, D. R. (Chem. Phys. Lett. **62** [1979] 275/8).
[47] Sakai, H.; Hansen, P.; Esplia, M.; Johansson, R.; Peltola, M.; Strong, J. (Appl. Opt. **21** [1982] 228/34).
[48] Vanderhoff, J. A.; Anderson, W. R.; Kotlar, A. J.; Beyer, R. A. (Symp. Int. Combust. Proc. **20** [1984] 1299/306).
[49] Crosley, D. R. (Combust. Flame **78** [1989] 153/67).
[50] Salmon, J. T.; Lucht, R. P.; Sweeny, D. W.; Laurendeau, N. M. (Symp. Int. Combust. Proc. **20** [1984/85] 1187/93).

[51] Fairchild, P. W.; Smith, G. P.; Crosley, D. R.; Jeffries, J. B. (Chem. Phys. Lett. **107** [1984] 181/6).

[52] Ubachs, W.; Ter Meulen, J. J.; Dymanus, A. (Can. J. Phys. **62** [1984] 1374/91).
[53] Brazier, C. R.; Ram, R. S.; Bernath, P. F. (J. Mol. Spectrosc. **120** [1986] 381/402).
[54] Roose, T. R.; Hanson, R. K.; Kruger, C. H. (Symp. Int. Combust. **18** [1981] 853/62).
[55] Anderson, R. R.; Decker, L. J.; Kotlar, A. J. (Combust. Flame **48** [1982] 179/90).
[56] Chou, M. S.; Dean, A. M.; Stern, D. (J. Chem. Phys. **76** [1982] 5334/40).
[57] Litvak, M. M.; Rodriguez-Kuiper, E. N. (Astrophys. J. **253** [1982] 622/33).
[58] Gustafsson, O.; Kindvall, G.; Larsson, M.; Olsson, B. J.; Sigray, P. (Chem. Phys. Lett. **138** [1987] 185/94).
[59] Shimauchi, M. (Sci. Light [Tokyo] **15** [1966] 161/5).
[60] Bollmark, P.; Kopp, I.; Rydh, B. (J. Mol. Spectrosc. **34** [1970] 487/99).

[61] Patel-Misra, D.; Sauder, D. G.; Dagdigian, P. J. (Chem. Phys. Lett. **174** [1990] 113/8).
[62] Patel-Misra, D.; Parlant, G.; Sauder, D. G.; Yarkony, D. R. (J. Chem. Phys. **94** [1991] 1913/22).
[63] Yarkony, D. R. (J. Chem. Phys. **91** [1989] 4745/57).
[64] Harrington, J. A.; Modica, A. P.; Libby, D. R. (J. Quant. Spectrosc. Radiat. Transfer **6** [1966] 799/805).
[65] Lents, J. M., Jr. (J. Quant. Spectrosc. Radiat. Transfer **13** [1973] 297/310).
[66] Dreier, T.; Rakestraw, D. J. (Appl. Phys. B **50** [1990] 479/85).
[67] Rakestraw, D. J.; Thorne, L. R.; Dreier, T. (Symp. Int. Combust. Proc. **23** [1990] 1901/7).
[68] Reintjes, F. J. (Nonlinear Optical Parametric Processes in Liquids and Gases, Academic, New York 1984).
[69] Rosengren, K.; Pimentel, G. C. (J. Chem. Phys. **43** [1965] 507/16).
[70] Hall, J. L.; Adams, H.; Kasper, J. V. V.; Curl, R. F.; Tittel, F. K. (J. Opt. Soc. Am. B Opt. Phys. **2** [1985] 781/5).

[71] Benrath, P. F.; Amano, T. (J. Mol. Spectrosc. **95** [1982] 359/64).
[72] Chackerian, C., Jr.; Guelachvili, G.; Lopez-Piñeiro, A.; Tipping, R. H. (J. Chem. Phys. **90** [1989] 641/9).
[73] Elhanine, M.; Farrenq, R.; Guelachvili, G. (Proc. SPIE-Int. Soc. Opt. Eng. **1575** [1992] 314/5; C.A. **116** [1992] No. 244082).
[74] Green, B. D.; Caledonia, G. E. (J. Chem. Phys. **77** [1982] 3821/3).
[75] Boudjaadar, D.; Brion, J.; Chollet, P.; Guelachvili, G.; Vervloet, M. (J. Mol. Spectrosc. **119** [1986] 352/66).
[76] Radford, H. E.; Litvak, M. M. (Chem. Phys. Lett. **34** [1975] 561/4).
[77] Wayne, F. D.; Radford, H. E. (Mol. Phys. **32** [1976] 1407/22).
[78] Leopold, K. R.; Evenson, K. M.; Brown, J. M. (J. Chem. Phys. **85** [1986] 324/30).
[79] Vasconcellos, E. C. C.; Davidson, S. A.; Brown, J. M.; Leopold, K. R.; Evenson, K. M. (J. Mol. Spectrosc. **122** [1987] 242/5).
[80] Bradburn, G. R.; Lilenfeld, H. V. (J. Phys. Chem. **95** [1991] 555/8).

[81] Van den Heuvel, F. C.; Meerts, W. L.; Dymanus, A. (Chem. Phys. Lett. **92** [1982] 215/8).
[82] Reed, R. I.; Snedden, W. (J. Chem. Soc. **1959** 4132/3).
[83] Foner, S. N.; Hudson, R. L. (J. Chem. Phys. **45** [1966] 40/8).
[84] Melton, C. E. (J. Chem. Phys. **45** [1966] 4414/24).
[85] Bradley, J. N.; Gilbert, J. R.; Park, A. J. (Adv. Mass Spectrom. **4** [1968] 669/75).
[86] Foner, S. N.; Hudson, R. L. (J. Chem. Phys. **74** [1981] 5017/21).
[87] Dove, J. E.; Nip, W. S. (Can. J. Chem. **57** [1979] 689/701).
[88] Engelking, P. C.; Lineberger, W. C. (J. Chem. Phys. **65** [1976] 4323/4).
[89] Nieman, G. C.; Colson, S. D. (J. Chem. Phys. **68** [1978] 5656/7).
[90] Glownia, J. H.; Riley, S. J.; Colson, S. D.; Nieman, G. C. (J. Chem. Phys. **73** [1980] 4296/309).

[91] Chu, J. J.; Marcus, P.; Dagdigian, P. J. (J. Chem. Phys. **93** [1990] 257/67).
[92] Johnson, R. D., III.; Hudgens, J. W. (J. Chem. Phys. **92** [1990] 6420/5).
[93] Clement, S. G.; Ashfold, M. N. R.; Western, C. M. (J. Chem. Soc. Faraday Trans. **88** [1992] 3121/8).
[94] Clement, S. G.; Ashfold, M. N. R.; Western, C. M.; De Beer, E.; De Lange, C. A.; Westwood, N. P. C. (J. Chem. Phys. **96** [1992] 4963/73).
[95] Wang, K.; Stephens, J. A.; McKoy, V.; De Beer, E.; De Lange, C. A; Westwood, N. P. C. (J. Chem. Phys. **97** [1992] 211/21).
[96] De Beer, E.; Born, M.; De Lange, C. A.; Westwood, N. P. C. (Chem. Phys. Lett. **186** [1991] 40/6).
[97] Vanderhoff, J. A. (Combust. Flame **84** [1991] 73/92).
[98] Masanet, J.; Gilles, A.; Vermeil, C. (J. Photochem. **3** [1974, 1975] 417/29).
[99] Rohrer, F.; Stuhl, F. (Chem. Phys. Lett. **111** [1984] 234/7).
[100] Marian, C. M.; Klotz, R. (Chem. Phys. **95** [1985] 213/23).

[101] Ubachs, W.; Meyer, G.; Ter Meulen, J. J.; Dymanus, A. (J. Mol. Spectrosc. **115** [1986] 88/104).
[102] McDonald, J. R.; Miller, R. G.; Baronavski, A. P. (Chem. Phys. Lett. **51** [1977] 57/60).
[103] Piper, L. G.; Krech, R. H.; Taylor, R. L. (J. Chem. Phys. **73** [1980] 791/800).
[104] Hack, W.; Wilms, A. (Ber. Max-Planck-Inst. Strömungsforsch. **1987** No. 20).
[105] Hack, W.; Wilms, A. (J. Phys. Chem. **93** [1989] 3540/6).
[106] Hack, W.; Wilms, A. (Z. Phys. Chem. **161** [1989] 107/21).
[107] Bower, R. D.; Jacoby, M. T.; Blauer, J. A. (J. Chem. Phys. **86** [1987] 1954/6).
[108] Wilms, A. (Diss. Univ. Göttingen **1987**).
[109] Nelson, H. H.; McDonald, J. R. (J. Chem. Phys. **93** [1990] 8777/83).
[110] Nelson, H. H.; McDonald, J. R.; Alexander, M. H. (J. Phys. Chem. **94** [1990] 3291/4).

[111] Adams, J. S.; Pasternack, L. (J. Phys. Chem. **95** [1991] 2975/82).
[112] Rathmann, K. (Diss. Univ. Göttingen [1992]).
[113] Pearse, R. W. (Proc. R. Soc. London A **143** [1933] 112/23).
[114] Dieke, G. H.; Blue, R. W. (Phys. Rev. [2] **45** [1934] 395/400).
[115] Nakamura, G.; Shidei, T. (Jpn. J. Phys. **10** [1935] 5/10).
[116] Ramsay, D. A.; Starre, P. (J. Mol. Spectrosc. **93** [1982] 445/6).
[117] Cheung, W. Y.; Gelernt, B.; Carrington, T. (Chem. Phys. Lett. **66** [1979] 287/90).
[118] Hack, W.; Rathmann, K. (J. Phys. Chem. **94** [1990] 3636/9).
[119] Hack, W.; Rathmann, K. (J. Phys. Chem. **94** [1990] 4155/61).
[120] Hack, W.; Mill, T. (J. Phys. Chem. **97** [1993] 5599/606).

[121] Mill, T. (Ber. Max-Planck-Inst. Strömungsforsch. **1990** No. 14, 104 pp.).
[122] Ram, R. S.; Bernath, P. F. (J. Opt. Soc. Am. B Opt. Phys. **3** [1986] 1170/4).
[123] Parlant, G.; Dagdigian, P. J.; Yarkony, D. R. (J. Chem. Phys. **94** [1991] 2364/7).
[124] Lunt, W.; Pearse, R. W. B.; Smith, E. C. W. (Proc. R. Soc. London A **155** [1936] 173/82).
[125] Whittaker, F. L. (Proc. Phys. Soc. [London] **90** [1967] 535/41).
[126] Narasimham, N. A.; Krishnamurty, G. (Proc. Indian Acad. Sci. A **64** [1966] 97/110).
[127] Krishnamurty, G.; Narasimham, N. A. (J. Mol. Spectrosc. **29** [1969] 410/4).
[128] Graham, W. R. M.; Lew, H. (Can. J. Phys. **56** [1978] 85/99).
[129] Cossart, D. (J. Chim. Phys. Phys. Chim. Biol. **76** [1979] 1045/50).
[130] Whittaker, F. L. (J. Phys. B: At. Mol. Phys. **1** [1968] 977/82).

[131] Gelernt, B.; Smith, A. L. (Chem. Phys. Lett. **60** [1979] 261/4).

2.1.1.2 Molecular Properties

Since the beginning of systematic quantum-chemical studies on diatomic hydrides in the late fifties up to mid of 1992, more than 340 calculations of varying scope and accuracy have been published, including the NH molecule and its positive and negative ions. Some 300 of them are ab initio calculations, beginning with simple LCAO MO approaches and continuing with high-quality Hartree-Fock (SCF MO), configuration interaction (CI), perturbation theory (MP, MBPT), and, more recently, density functional procedures. The calculations have been performed with the intent to test computational methods and to compute properties of NH per se. About 65% of the ab initio studies published up to 1991 have been quoted in the various series of

Bibliographies on Quantum-Chemical Calculations:

Krauss, M.; Compendium of ab initio Calculations of Molecular Energies and Properties; NBS-TN-438 [1967] 1/139.

Richards, W. G.; Walker, T. E. H.; Hinkley, R. K.; A Bibliography of ab initio Molecular Wave Functions, Clarendon Press, Oxford 1971.

Richards, W. G.; Walker, T. E. H.; Farnell, L.; Scott, P. R.; Bibliography of ab initio Molecular Wave Functions. Supplement for 1970-1974, Clarendon Press, Oxford 1974.

Richards, W. G.; Scott, P. R.; Colburn, E. A.; Marchington, A. F.; Bibliography of ab initio Molecular Wave Functions. Supplement for 1974-1977, Clarendon Press, Oxford 1978.

Richards, W. G.; Scott, P. R.; Sackwild, V.; Robins, S. A.; A Bibliography of Ab Initio Molecular Wave Functions. Supplement for 1978-1980, Clarendon Press, Oxford 1981.

Ohno, K.; Morokuma, K.; Quantum Chemistry Literature Data Base – Bibliography of Ab Initio Calculations for 1978-1980, Elsevier, Amsterdam 1982.

Annual Supplements appeared in the following volumes of the Journal of Molecular Structure: **91** [1982], **106** [1983], **119** [1984], **134** [1985], **148** [1986], **154** [1987], **182** [1988], **203** [1989], **211** [1990], **252** [1991], **278** [1992].

In the following sections on molecular properties and spectra of NH, only a few relevant calculations will be occasionally quoted in order to supplement the experimental results or in cases where experimental results either are not available or are disagreeing or uncertain. Otherwise, the reader is referred to the bibliographies.

2.1.1.2.1 Electron Configuration. Electronic States

The eight electrons of the NH molecule combine to the lowest electron configuration, $(1\sigma)^2\ (2\sigma)^2\ (3\sigma)^2\ (1\pi)^2$, where 1σ is the N1s core orbital and 2σ, 3σ, and 1π are essentially the N2s, N2pσ-H1s bonding, and N2pπ orbitals, respectively. This configuration gives rise to the ground state X $^3\Sigma^-$, which correlates with the ground-state atoms $N(^4S)+H(^2S)$, and to two excited metastable states, a $^1\Delta$ and b $^1\Sigma^+$, which correlate with $N(^2D)+H(^2S)$ and $N(^2P)+H(^2S)$. Excitation of one or two 3σ electrons into the 1π orbital results in the excited states $(1\sigma)^2\ (2\sigma)^2\ (3\sigma)^1\ (1\pi)^3$ A $^3\Pi_i$, c $^1\Pi$ and $(1\sigma)^2\ (2\sigma)^2\ (3\sigma)^0\ (1\pi)^4$ d $^1\Sigma^+$ which correlate with $N(^2D)+H(^2S)$ and $N(2\ ^2P)+H(^2S)$, respectively; see e.g. [1].

These six **valence states** are known from experiment. The ground state X $^3\Sigma^-$ has been observed by far-IR absorption, far-IR laser magnetic resonance (LMR), and by IR absorption and emission spectroscopy. Moreover, it has been identified with the lower state of the triplet system A $^3\Pi_i \leftrightarrow$ X $^3\Sigma^-$, observed in absorption and emission in the UV around 336 nm and with the lower states of the forbidden a $^1\Delta \leftrightarrow X\ ^3\Sigma^-$ and b $^1\Sigma^+ \leftrightarrow X\ ^3\Sigma^-$ transitions in the visible region at 795 and 471 nm. The metastable states a $^1\Delta$ and b $^1\Sigma^+$ and the remaining singlet states c $^1\Pi$ and d $^1\Sigma^+$ have been identified with the lower and upper states of the UV and visible singlet systems c $^1\Pi \leftrightarrow$ a $^1\Delta$ at 324 nm, c $^1\Pi \rightarrow$ b $^1\Sigma^+$ and d $^1\Sigma^+ \rightarrow$ b $^1\Sigma^+$ at 450 and 162 nm, and d $^1\Sigma^+ \rightarrow$ c $^1\Pi$ at 253 nm. The d $^1\Sigma^+$ state has also been identified as the intermediate state responsible for the two-photon resonance enhancement in the multiphoton ionization spectrum of NH(a $^1\Delta$) at excitation wavelengths around 285 nm. Term values T_e or T_0, derived from the UV-visible absorption, emission, and resonance-enhanced multiphoton ionization (REMPI) spectra, are given in the following table (for details concerning the method of derivation and additional references, see the remarks below the table):

state	NH T_0(in cm^{-1})	T_e(in cm^{-1})	ND T_0(in cm^{-1})	T_e(in cm^{-1})	remark
a $^1\Delta$	12655.2 (12641.7)	–	12620.5	–	a)
b $^1\Sigma^+$	21238.5(20)	–	21224.7(20)	–	b)
A $^3\Pi_i$	29761.1829(1)	29790.7	29790.712(5)	29810.6	c)
c $^1\Pi$	43359.3 (43345.8)	44420.6 (44407.1)	43469.9	43811.8	d)
d $^1\Sigma^+$	82858.0 (82844.5)	83160.5 (83147.0)	82918.1	83140.2	e)

a) Derived using the relation $T_0(a\ ^1\Delta) = \nu_{00}(b\ ^1\Sigma^+ \rightarrow X\ ^3\Sigma^-) + \nu_{00}(c\ ^1\Pi \rightarrow b\ ^1\Sigma^+) - \nu_{00}(c\ ^1\Pi \rightarrow a\ ^1\Delta)$ and recent results for the band origins of the b→X [2, 3], c→b [4] (or for NH the more recent result for $\nu_{00}(d \rightarrow b) - \nu_{00}(d \rightarrow c)$ [5], yielding the T_0 value given in parentheses), and c→a [6, 7] spectra. The earlier data for $\nu_{00}(b \rightarrow X)$ and $\nu_{00}(c \rightarrow a)$ (and $\nu_{00}(b \rightarrow c)$ from [4]) gave $T_0 = 1.561$ and 1.565 eV (12590 and 12623 cm^{-1}) for NH and ND, respectively [8, 9], which Huber and Herzberg [10] converted to the uncertain equilibrium values $T_e = 12566$ and 12596 cm^{-1}. – $T_0 \approx 12500\ cm^{-1}$ was obtained [10] from the difference of the threshold energies for the production of NH(c $^1\Pi$)+CO(X $^1\Sigma^+$) and NH(X $^3\Sigma^-$)+ CO(a $^3\Pi$) by HNCO photolysis [11], and $T_0 = 1.579 \pm 0.017\ eV = 12735 \pm 137\ cm^{-1}$ from the photoelectron spectrum of the NH^- ion [12]. The a $^1\Delta \rightarrow X\ ^3\Sigma^-$ transition of NH was directly observed at 794.5 nm (Q branch), thus $T_0 \approx 12570\ cm^{-1}$ [13].

b) Band origins of the b→X spectra [2]; similar values, 21238.26(7) and 21224.87(4) cm^{-1}, were obtained by [3]. The earlier results, $T_0 = 21231(11)\ cm^{-1}$ for NH and 21244(6) cm^{-1} for ND [9], were used to derive the equilibrium values $T_e = 21202$ and 21298 cm^{-1} [10].

c) Band origins of the A $^3\Pi_i \rightarrow X\ ^3\Sigma^-$ spectra: for NH measured by [14], for ND derived by [15] from earlier data [16, 17] using the same Hamiltonian as [14]. These were converted to equilibrium values T_e using the vibrational constants for the A and X states given in [14] and [15]. – Earlier studies resulted in $T_0 \approx 29777\ cm^{-1}$ for NH [16, 18 to 20] and $T_0 \approx 29799\ cm^{-1}$ for ND [16, 21, 22] which Huber and Herzberg [10] converted to $T_e = 29807.4$ and 29820 cm^{-1}, respectively.

d) Derived using the relation $T_0(c\ ^1\Pi) = \nu_{00}(c\ ^1\Pi \leftrightarrow a\ ^1\Delta) + \nu_{00}(a\ ^1\Delta \rightarrow X\ ^3\Sigma^-)$ and $\nu_{00}(c \leftrightarrow a) = 30704.1\ cm^{-1}$ for NH from high-resolution c $^1\Pi \leftrightarrow$ a $^1\Delta$ spectra [6, 23, 24], $\nu_{00}(c \rightarrow a) = 30849.06\ cm^{-1}$ for ND from the emission spectrum [7], and the a $^1\Delta - X\ ^3\Sigma^-$ energy separations given in the first row of the table. The equilibrium values were obtained using the $\Delta G''_{1/2}$ and $\Delta G'_{1/2}$ values for NH(a $^1\Delta$) and NH(c $^1\Pi$) given by [23] and using the vibrational constants ω_e and $\omega_e x_e$ for ND(a $^1\Delta$) and ND(c $^1\Pi$) given by [25] and [5]. – Huber and Herzberg [10] used c→b [4, 5, 26] and c↔a [7, 27, 28] data to derive $T_e = 43744\ cm^{-1}$ (uncertain value) for NH and $T_e = 43786\ cm^{-1}$ for ND.

e) Derived using the relation $T_0(d\,^1\Sigma^+) = \nu_{00}(d\,^1\Sigma^+ \leftarrow a\,^1\Delta) + \nu_{00}(a\,^1\Delta \rightarrow X\,^3\Sigma^-)$ and $\nu_{00}(d \leftarrow a) = 70202.8(5)$ and 70297.6(5) cm^{-1} from the REMPI spectra of NH(a $^1\Delta$) and ND(a $^1\Delta$), respectively [29] and the a $^1\Delta - X\,^3\Sigma^-$ energy separations as given in the first row of the table. The equilibrium values were obtained using reliable vibrational constants ω_e and $\omega_e x_e$ for the X $^3\Sigma^-$ [14, 15] and d $^1\Sigma^+$ [5] states of NH and ND. – Huber and Herzberg [10] used d→c [5, 30 to 33] and d→b [5, 34] data to derive $T_e = 83160$ and 83168 cm^{-1} for NH and ND.

The ground state X $^3\Sigma^-$ and the experimentally observed and a number of unobserved excited valence states have been the subject of numerous quantum-chemical calculations with the scope of either actually deriving molecular properties of NH or testing computational methods. A configuration interaction (CI) study of $^3\Sigma^-$, $^3\Pi$, and $^5\Sigma^-$ states of NH predicted, besides the known X $^3\Sigma^-$ and A $^3\Pi_i$ states and five states of mainly Rydberg character (see below), a repulsive 1 $^5\Sigma^-$ state dissociating into N(^{4}S) + H(^{2}S) (with Rydberg character at a small internuclear distance r and a valence character increasing with r), a repulsive 2 $^3\Sigma^-$ state dissociating into N(^{2}D) + H(^{2}S), and a repulsive (or extremely slightly bound with a shallow minimum outside the Franck-Condon region of the ground state) 2 $^3\Pi$ state dissociating into N(^{2}P) + H(^{2}S); their vertical excitation energies from the X $^3\Sigma^-$ minimum at $r_e = 1.97\ a_0$ are $\Delta E = 8.88$, 9.59, and 8.86 eV; the CI coefficients of the dominant configurations for the triplet states at various internuclear distances from 1.25 to 20.0 a_0 are listed in the original paper [35]. Multiconfiguration (MC) SCF wave functions were used to describe the b $^1\Sigma^+$, d $^1\Sigma^+$, A $^3\Pi_i$, and 2 $^3\Pi$ states; the latter has been predicted to be slightly bound [36 to 38]. Calculations based on a quasi-degenerate many-body perturbation theory (QDMBPT) resulted in the known bound valence states (see also [39, 40]) and repulsive states $^5\Sigma^-$ (N(^{4}S) + H(^{2}S)), $^3\Sigma^-$, $^3\Delta$, $^1\Sigma^-$ (N(^{2}D) + H(^{2}S)), and $^3\Pi$, $^1\Pi$, $^3\Sigma^+$ (N(^{2}P) + H(^{2}S)) with vertical excitation energies from the X $^3\Sigma^-$minimum between ~10 and 14 eV (read from a figure) [41]. An earlier CI study resulted in the same (six observed and seven unobserved) states, but found 2 $^3\Sigma^-$ (N(^{2}D) + H(^{2}S)) and 2 $^3\Pi$ (N(^{2}P) + H(^{2}S)) to be bound with energies of 5.5 and 6.6 eV, respectively (T_0 values) [42]. Furthermore, term values and/or excitation energies for the a, b, c, d, and A states were calculated using various CI methods, i.e., valence CI [43], SCF MO near Hartree-Fock limit plus limited CI [44, 45], POL-CI (polarization-CI) [46], CEPA (coupled electron pair approximation) [47], FORS-IACC (full optimized reaction space including the intra-atomic correlation correction) [48], MRD CI (multireference double excitation CI) [49], MC SCF (multiconfiguration SCF) [50], and, more recently, by an alternative method based on algebraic instead of differential solutions of the Schrödinger equation [51, 52].

Rydberg states arise by exciting a 1π or 3σ electron into orbitals of the nsσ, npσ, npπ, ndσ, ndπ, ndδ series which converge to the ionic ground and excited states $(1\sigma)^2\,(2\sigma)^2\,(3\sigma)^2\,(1\pi)^1$ X $^2\Pi_r$ and $(1\sigma)^2\,(2\sigma)^2\,(3\sigma)^1\,(1\pi)^2$ a $^4\Sigma^-$, A $^2\Sigma^-$, B $^2\Delta$, C $^2\Sigma^+$. States expected for n = 3 pertaining to the

{NH^+ X $^2\Pi_r$}	and	{NH^+ a $^4\Sigma^-$, A $^2\Sigma^-$, B $^2\Delta$, C $^2\Sigma^+$}

cores are:

3sσ, 3pσ, 3dσ $^{1,3}\Pi$	3sσ, 3pσ, 3dσ $^{1,3}\Sigma^+$, $^1\Sigma^-$, $^3\Sigma^-$(2), $^{1,3}\Delta$, $^5\Sigma^-$
3pπ, 3dπ $^{1,3}\Sigma^-$, $^{1,3}\Sigma^+$, $^{1,3}\Delta$	3pπ, 3dπ $^1\Pi$(3), $^3\Pi_r$(2), $^3\Pi_i$(2), $^{1,3}\Phi$, $^5\Pi_r$
3dδ $^{1,3}\Phi$, $^{1,3}\Pi$	3dδ $^{1,3}\Sigma^+$, $^{1,3}\Sigma^-$, $^1\Delta$(2), $^3\Delta$(3), $^{1,3}\Gamma$, $^5\Delta$.

Some of these first members with n = 3 and {NH^+ X, a, B} cores were detected as intermediate states, responsible for the two-photon resonance enhancement in the multiphoton ionization (REMPI) spectra of NH and ND ({NH^+ X} 3sσ $^{1,3}\Pi$ states are expected or predicted to be repulsive). REMPI of NH, ND(a $^1\Delta$) gives rise to singlet intermediate Rydberg states; five states, f $^1\Pi$, g $^1\Delta$, h $^1\Sigma^+$ [53 to 55], i $^1\Pi$, and j $^1\Delta$ [56], could be characterized to a certain extent. The f, g, and h states have been identified with the $\{NH^+ X\}$ 3pσ $^1\Pi$, 3pπ $^1\Delta$, and 3pπ $^1\Sigma^+$ states [55], where for the f state an orbital evolution with increasing internuclear distance from predominant 3p to 3s character is predicted theoretically [35]. The i $^1\Pi$ state was found to be of mixed $\{NH^+ X\}$ 3pσ/3dσ character, and the j $^1\Delta$ state to arise from core mixing of $\{NH^+ B\}$ 3sσ with $\{NH^+ X\}$ 3dπ; additionally, perturbations of j-state vibrational levels appearing in the REMPI spectra were attributed to additional Rydberg states with $^1\Sigma$(3dπ), $^1\Pi$(3dσ or 3dδ), and $^1\Delta$ (or $^1\Pi$ or $^1\Phi$) symmetry [56]. REMPI of NH, ND(X $^3\Sigma^-$) gives rise to triplet Rydberg states; among these have been detected and partially characterized the states B $^3\Pi$, C $^3\Sigma^-$ [57, 58], D $^3\Pi$, E $^3\Sigma^-$, and F $^3\Sigma^-$ [58]. The B, D, and E states were found to be "regular" Rydberg states, the B state being the triplet counterpart of the f state ($\{NH^+ X\}$ 3pσ), the D state being dominated by the $\{NH^+ a\}$ 3pπ configuration, and the E state arising probably from the $\{NH^+ X\}$ 3pπ configuration. The C state ($\{NH^+ X\}$ 3pπ?), which is observed only for ND as a perturber of the B state, and the F state ($\{NH^+ a\}$ 3pσ?) were found to possess significant valence character and were associated [58] with two states obtained from an ab initio CI calculation and labeled 3 $^3\Sigma^-$ and 4 $^3\Sigma^-$ (see below) [35]. The term values T_0 obtained from the band origins ν_{00} of the f $^1\Pi$, g $^1\Delta$, h $^1\Sigma^+$, i $^1\Pi$, j $^1\Delta \leftarrow$ a $^1\Delta$ [55, 56] and B $^3\Pi$, C $^3\Sigma^-$, D $^3\Pi$, E $^3\Sigma^-$, F $^3\Sigma^- \leftarrow$ X $^3\Sigma^-$ [58] transitions are as follows, where T_0(a $^1\Delta$) = 12655.2 (12641.7) cm^{-1} for NH and T_0(a $^1\Delta$) = 12620.5 cm^{-1} for ND (cf. p. 32) have been adopted:

state	T_0 in cm^{-1} for NH	T_0 in cm^{-1} for ND
f $^1\Pi$	86433.0 (86419.5)	86454.6
g $^1\Delta$	88197.2 (88183.7)	88223.4
h $^1\Sigma^+$	89525.5 (89512.0)	89531.7
i $^1\Pi$	92187.3 (92173.8)	92177.7
j $^1\Delta$	95649.2 (95635.7) [a]	95710.2 [b]
B $^3\Pi$	85538.6	85588.1
C $^3\Sigma^-$	–	85734.3
D $^3\Pi$	89277.1	89394.3
E $^3\Sigma^-$	–	89829.3
F $^3\Sigma^-$	90276.8	90311.8

[a] A $^1\Delta$ perturber of the v′=0 level of j $^1\Delta$ has T_0=95916.2 (95902.7) cm^{-1}. – [b] $^1\Pi$, $^1\Sigma$ perturbers of the v′=0 level of j $^1\Delta$ have T_0=95642.5, 95741.5 cm^{-1}, $^1\Delta$ perturber of the v′=1 level has T_0=97732.1 cm^{-1}.

A theoretical (ab initio CI) study of $^3\Sigma^-$, $^3\Pi$, and $^5\Sigma^-$ states of NH found, besides the known X $^3\Sigma^-$ and A $^3\Pi_i$ valence states and three states of mainly valence character (see above), five states of mainly Rydberg character; their vertical excitation energies ΔE (in eV) from the X $^3\Sigma^-$ minimum at r_e=1.97 a_0 and dissociation limits are as follows (the CI coefficients of the dominant configurations contributing to the triplet states at various internuclear distances from 1.25 to 20.0 a_0 are listed in the original paper) [35]:

3 $^3\Sigma^-$	10.70	N(^{4}S)+H(^{2}S)	2 $^5\Sigma^-$	10.70	N(^{4}S)+H(^{2}S)
4 $^3\Sigma^-$	11.23	N(^{4}S)+H(^{2}P)	3 $^5\Sigma^-$	12.10	N(^{4}S)+H(^{2}P)
3 $^3\Pi$	10.78	N(^{4}S)+H(^{2}P)			

Averaged **electron densities** around the N and H nuclei in the X $^3\Sigma^-$, a $^1\Delta$, A $^3\Pi_i$, c $^1\Pi$, v=0 states of NH were derived from the measured (high-resolution LIF of A $^3\Pi_i \leftarrow$ X $^3\Sigma^-$ and c $^1\Pi \leftarrow$ a $^1\Delta$ transitions hyperfine and quadrupole coupling constants (cf. pp. 40/2);

the values confirm the picture of an N2pσ-H1sσ-bonding orbital and two (X $^3\Sigma^-$, a $^1\Delta$) or three (A $^3\Pi_i$, c $^1\Pi$) 2pπ orbitals located mainly on the N nucleus [6, 59].

References:

[1] Herzberg, G. (Molecular Spectra and Molecular Structure, Vol. 1, Spectra of Diatomic Molecules, Van Nostrand, Princeton, N. J., 1950, pp. 336/7, 341, 368/71).
[2] Zetzsch, C.; Stuhl, F. (Ber. Bunsenges. Phys. Chem. **80** [1976] 1348/54).
[3] Cossart, D. (J. Chim. Phys. Phys.-Chim. Biol. **76** [1979] 1045/50).
[4] Whittaker, F. L. (J. Phys. B **1** [1968] 977/82).
[5] Graham, W. R. M.; Lew, H. (Can. J. Phys. **56** [1978] 85/99).
[6] Ubachs, W.; Meyer, G.; Ter Meulen, J. J.; Dymanus, A. (J. Mol. Spectrosc. **115** [1986] 88/104).
[7] Hanson, H.; Kopp, I.; Kronekvist, M.; Aslund, N. (Arkiv Fys. **30** [1965] 1/8).
[8] Gilles, A.; Masanet, J.; Vermeil, C. (Chem. Phys. Lett. **25** [1974] 346/7).
[9] Masanet, J.; Gilles, A.; Vermeil, C. (J. Photochem. **3** [1974/75] 417/29).
[10] Huber, K. P.; Herzberg, G. (Molecular Spectra and Molecular Structure, Vol. 4, Constants of Diatomic Molecules, Van Nostrand Reinhold, New York 1979, pp. 456/61).

[11] Okabe, H. (J. Chem. Phys. **53** [1970] 3507/15).
[12] Engelking, P. C.; Lineberger, W. C. (J. Chem. Phys. **65** [1976] 4323/4).
[13] Rohrer, F.; Stuhl, F. (Chem. Phys. Lett. **111** [1984] 234/7).
[14] Brazier, C. R.; Ram, R. S.; Bernath, P. F. (J. Mol. Spectrosc. **120** [1986] 381/402).
[15] Patel-Misra, D.; Sauder, D. G.; Dagdigian, P. J. (Chem. Phys. Lett. **174** [1990] 113/8).
[16] Bollmark, P.; Kopp, I.; Rydh, B. (J. Mol. Spectrosc. **34** [1970] 487/99).
[17] Kopp, I.; Kronekvist, M.; Aslund, N. (Arkiv Fys. **30** [1965] 9/17).
[18] Murai, T.; Shimauchi, M. (Sci. Light [Tokyo] **15** [1966] 48/67).
[19] Malicet, J.; Brion, J.; Guenebaut, H. (J. Chim. Phys. Phys. Chim. Biol. **67** [1970] 25/30).
[20] Dixon, R. N. (Can. J. Phys. **37** [1959] 1171/86).

[21] Krishnamurty, G.; Narasimham, N. A. (Proc. Indian Acad. Sci. A **67** [1968] 50/60).
[22] Krishnamurty, G.; Narasimham, N. A. (Proc. 1st Int. Conf. Spectrosc., Bombay 1967, Vol. 1, pp. 142/4).
[23] Ram, R. S.; Bernath, P. F. (J. Opt. Soc. Am. B Opt. Phys. **3** [1986] 1170/4).
[24] Ramsay, D. A.; Sarre, P. J. (J. Mol. Spectrosc. **93** [1982] 445/6).
[25] Cheung, W. Y.; Gelernt, B.; Carrington, T. (Chem. Phys. Lett. **66** [1979] 287/90).
[26] Lunt, W.; Pearse, R. W. B.; Smith, E. C. W. (Proc. R. Soc. [London] A **151** [1935] 602/9).
[27] Pearse, R. W. B. (Proc. R. Soc. [London] A **143** [1934] 112/23).
[28] Shimauchi, M. (Sci. Light [Tokyo] **13** [1965] 53/63).
[29] Ashfold, M. N. R.; Clement, S. G.; Howe, J. D.; Western, C. M. (J. Chem. Soc. Faraday Trans. **87** [1991] 2515/23).
[30] Lunt, W.; Pearse, R. W. B.; Smith, E. C. W. (Proc. Roy. Soc. [London] A **155** [1936] 173/82).

[31] Whittaker, F. L. (Proc. Phys. Soc. [London] **90** [1967] 535/41).
[32] Krishnamurty, G.; Narasimham, N. A. (J. Mol. Spectrosc. **29** [1969] 410/4).
[33] Narasimham, N. A.; Krishnamurty, G. (Proc. Indian Acad. Sci. A **64** [1966] 97/110).
[34] Whittaker, F. L. (Can. J. Phys. **47** [1969] 1291/3).
[35] Goldfield, E. M.; Kirby, K. P. (J. Chem. Phys. **87** [1987] 3986/94).
[36] Banerjee, A.; Grein, F. (J. Chem. Phys. **66** [1977] 1054/62).
[37] Banerjee, A.; Grein, F. (Chem. Phys. **35** [1978] 119/27).
[38] Banerjee, A. K. (Diss. Univ. New Brunswick, Can., 1977 from Diss. Abstr. Int. B **38** [1977] 1677).

[39] Sun, H.; Sheppard, M. G.; Freed, K. F. (J. Chem. Phys. **74** [1981] 6842/8).
[40] Sun, H.; Freed, K. F. (J. Chem. Phys. **76** [1982] 5051/9).

[41] Park, J. K.; Sun, H. (Bull. Korean Chem. Soc. **11** [1990] 34/41).
[42] Kouba, J.; Öhrn, Y. (J. Chem. Phys. **52** [1970] 5387/94).
[43] O'Neil, S. V.; Schaefer, H. F., III (J. Chem. Phys. **55** [1971] 394/401).
[44] Liu, H. P. D.; Verhaegen, G. (Int. J. Quantum Chem. Symp. No. 5 [1971] 103/18).
[45] Liu, H. P. D.; Legentil, J.; Verhaegen, G. (Sel. Top. Mol. Phys. Proc. Int. Symp., Ludwigsburg, FRG, 1970 [1972], pp. 19/33).
[46] Hay, P. J.; Dunning, T. H., Jr. (J. Chem. Phys. **64** [1976] 5077/87).
[47] Staemmler, V.; Jaquet, R. (Theor. Chim. Acta **59** [1981] 501/15).
[48] Schmidt, M. W.; Lam, B. M. T.; Elbert, S. T.; Ruedenberg, K. (Theor. Chim. Acta **68** [1985] 69/86).
[49] Marian, C. M.; Klotz, R. (Chem. Phys. **95** [1985] 213/23).
[50] Yarkony, D. R. (J. Chem. Phys. **91** [1989] 4745/57).

[51] Frank, A.; Lemus, R.; Iachello, F. (J. Chem. Phys. **91** [1989] 29/41).
[52] Lemus, R.; Frank, A. (Phys. Rev. Lett. **66** [1991] 2863/6).
[53] Johnson, R. D., III; Hudgens, J. W. (J. Chem. Phys. **92** [1990] 6420/5).
[54] De Beer, E.; Born, M.; De Lange, C. A.; Westwood, N. P. C. (Chem. Phys. Lett. **186** [1991] 40/6).
[55] Clement, S. G.; Ashfold, M. N. R.; Western, C. M. (J. Chem. Soc. Faraday Trans. **88** [1992] 3121/8).
[56] Clement, S. G.; Ashfold, M. N. R.; Western, C. M.; De Beer, E.; De Lange, C. A.; Westwood, N. P. C. (J. Chem. Phys. **96** [1992] 4963/73).
[57] Clement, S. G.; Ashfold, M. N. R.; Western, C. M.; Johnson, R. D., III; Hudgens, J. W. (J. Chem. Phys. **96** [1992] 5538/40).
[58] Clement, S. G.; Ashfold, M. N. R.; Western, C. M.; Johnson, R. D., III; Hudgens, J. W. (J. Chem. Phys. **97** [1992] 7064/72).
[59] Ubachs, W.; Ter Meulen, J. J.; Dymanus, A. (Can. J. Phys. **62** [1984] 1374/91).

2.1.1.2.2 Ionization Potentials

Removal of the 1π electron from NH($X\ ^3\Sigma^-$) leads to the ionic ground state $X\ ^2\Pi_r$, removal of the 3σ electron to the excited ionic states $a\ ^4\Sigma^-$, $A\ ^2\Sigma^-$, $B\ ^2\Delta$, and $C\ ^2\Sigma^+$. Only a few experimental data for the first, third, and fourth ionization potentials E_i of gaseous NH are available. Resonance-enhanced multiphoton ionization (REMPI) of NH coupled with photoelectron spectroscopy (PES) yielded the most accurate results so far [1] and confirmed the values for the first E_i obtained by electron-impact mass spectrometry (EIMS) [2] and by He I PES of NH [3]. Values for the second and third E_i to be observed in the He I PES of NH were predicted [3] from the optical emission spectra of NH^+ [4]. Adiabatic and vertical E_i's (in eV) are compared in the following table:

ionic state	E_i(ad) [1]	E_i(ad) [2]	E_i(vert) [3]	E_i(ad) [3]
$X\ ^2\Pi_r$	13.476 ± 0.002	13.47 ± 0.05	13.49 ± 0.01	–
$a\ ^4\Sigma^-$	–	–	–	~13.53
$A\ ^2\Sigma^-$	16.16 ± 0.04	–	–	~16.16
$B\ ^2\Delta$	16.34 ± 0.04	–	–	–
method	REMPI-PES	EIMS	He I PES	predicted
remark	a)	b)	c)	–

a) Taken from a (2+1)-REMPI-PES study of NH(a $^1\Delta$) at 246 to 286 nm with the Rydberg states f $^1\Pi$, g $^1\Delta$, and h $^1\Sigma$ as intermediate states [1].

b) Ionization curves of NH(X $^3\Sigma^-$) and NH(a $^1\Delta$) (20:1, from sequential F-atom reactions with NH_3) at electron energies of 11.5 to 14.3 eV [2] combined with the spectroscopic a $^1\Delta-$X $^3\Sigma^-$ separation [5, 6]. Earlier electron impact data are $E_i=13.1\pm0.2$ eV (ionization curves of NH(X $^3\Sigma^-$) and NH(a $^1\Delta$) following pulsed electrical discharges in NH_3; a $^1\Delta-$X $^3\Sigma^-$ separation not yet exactly known) [7] and $E_i=13.10\pm0.05$ eV (appearance potentials of NH^+ formed by NH_3 dissociation combined with N−H and H−H bond energies) [8]. The quoted uncertainty of 0.05 eV of the latter result was questioned by [2], considering the quoted uncertainties of 0.1 eV in the appearance potentials.

c) In the He I PES of NH_2 produced by the reaction $F+NH_3\rightarrow NH_2+HF$, a weak peak at 13.49 eV was tentatively assigned to the first E_i of NH, generated by the secondary reaction $F+NH_2\rightarrow NH+HF$ [3].

In good agreement with the directly observed values for the first ionization potential, $E_i=13.52$ eV was derived from the reaction enthalpy of $N^++H_2\rightarrow NH^++H$ (cf. p. 130) and by using a value of 3.45 eV for the dissociation energy of the NH molecule (estimate from experimental $D_0^0\leq3.47$ eV [9] and theoretical $D_0^0=3.34$ eV [10]); this in turn proofs the accuracy of the D_0^0(NH) value (cf. p. 56) [11].

A binding energy of 397.8 eV for the N1s electron of NH was obtained from X-ray PES recorded upon adsorption and decomposition of NH_3 on a tungsten (110) surface [12].

Theoretical values for the first E_i result from ab initio SCF MO [13], CI [13 to 16], and perturbation [17 to 19] calculations; the CI and perturbation calculations yield values between 13.0 and 13.5 eV.

References:

[1] De Beer, E.; Born, M.; De Lange, C. A.; Westwood, N. P. C. (Chem. Phys. Lett. **186** [1991] 40/6).

[2] Foner, S. N.; Hudson, R. L. (J. Chem. Phys. **74** [1981] 5017/21).

[3] Dunlavey, S. J.; Dyke, J. M.; Jonathan, N.; Morris, A. (Mol. Phys. **39** [1980] 1121/35).

[4] Colin, R.; Douglas, A. E. (Can. J. Phys. **46** [1968] 61/73).

[5] Masanet, J.; Gilles, A.; Vermeil, C. (J. Photochem. **3** [1974/75] 417/29).

[6] Gilles, A.; Masanet, J.; Vermeil, C. (Chem. Phys. Lett. **25** [1974] 346/7).

[7] Foner, S. N.; Hudson, R. L. (J. Chem. Phys. **45** [1966] 40/8).

[8] Reed, R. I.; Snedden, W. (J. Chem. Soc. **1959** 4132/3).

[9] Huber, K. P.; Herzberg, G. (Molecular Spectra and Molecular Structure, Vol. 4, Constants of Diatomic Molecules, Van Nostrand Reinhold, New York 1979, pp. 456/61).

[10] Liu, H. P. D.; Verhaegen, G. (J. Chem. Phys. **53** [1970] 735/45).

[11] Adams, N. G.; Smith, D. (Chem. Phys. Lett. **117** [1985] 67/70).

[12] Grunze, M.; Brundle, C. R.; Tománek, D. (Surf. Sci. **119** [1982] 133/49).

[13] Pope, S. A.; Hillier, I. H.; Guest, M. F. (Faraday Symp. Chem. Soc. No. 19 [1984] 109/23).

[14] Pope, S. A.; Hillier, I. H.; Guest, M. F.; Kendric, J. (Chem. Phys. Lett. **95** [1983] 247/9).

[15] Power, D.; Brint, P.; Spalding, T. R. (J. Mol. Struct. **110** [1984] 155/66 [THEOCHEM **19**]).

[16] Rosmus, P.; Meyer, W. (J. Chem. Phys. **66** [1977] 13/9).

[17] Pople, J. A.; Curtiss, L. A. (J. Phys. Chem. **91** [1987] 155/62).

[18] Pople, J. A.; Curtiss, L. A. (J. Phys. Chem. **91** [1987] 3637/9).

[19] Sun, H.; Sheppard, M. G.; Freed, K. F. (J. Chem. Phys. **74** [1981] 6842/8).

2.1.1.2.3 Electron Affinity

Autodetachment spectra of $^{14}NH^-$ and $^{15}NH^-$ ion beams upon IR laser excitation of the fundamental vibration-rotation transitions around 3000 cm^{-1} and photodetachment of $^{14}NH^-$ beams by Ar^+ ion laser radiation at 488 nm (cf. pp. 157/8) gave the following electron affinities EA (in eV):

$^{14}NH(X\ ^3\Sigma^-)$	$^{14}NH(a\ ^1\Delta)$	$^{15}NH(X\ ^3\Sigma^-)$	remark	Ref.
0.374362 ± 0.000005	—	0.3761 ± 0.0025	a)	[1 to 3]
0.370 ± 0.004	—	—	b)	[4]
0.381 ± 0.014	1.960 ± 0.010	—	c)	[5]
0.38 ± 0.03	—	—	d)	[6]

a) Autodetachment spectra of $^{14}NH^-$ [1, 3] and $^{15}NH^-$ [2, 3] ion beams produced from HN_3 in a hot-cathode discharge source. The spectroscopic results $E_i(^{14}NH^-) = EA(^{14}NH) = 3019.4077 \pm 0.0413\ cm^{-1}$ and $E_i(^{15}NH^-) = EA(^{15}NH) = 3033.84 \pm 20.78\ cm^{-1}$ have been converted to eV units using the factor 1 eV = 8065.479 cm^{-1}.

b) First autodetachment study of NH^- (from an HN_3 discharge source) and of a negative ion beam in general, giving EA $\leq$ 0.374 eV as an upper limit; the quoted value was estimated [4] upon taking the lower limit, EA $\geq$ 0.367 eV, from the photodetachment result of [5] (see c)).

c) Photodetachment of NH^- from an HN_3 discharge source; two peaks assigned to the transitions into the ground and first excited state of NH, both accessible with Ar^+ ion laser radiation [5].

d) First tentative assignment of a weak photodetachment peak due to mass 15 in the negative ion beam, extracted from an NH_3 discharge source [6].

The electron affinity cannot be reliably predicted theoretically from the total energies of NH and NH^- (SCF, CI, PNO-CI, CEPA, MP2, MP3, MP4) [7 to 10] or the energy of the first virtual MO (4σ) of NH (MBPT) [11], since the change in electron correlation energy caused by the attachment of an electron is of the same order of magnitude as the electron affinity itself (see also [12]). Calculations on NH and NH^- using the composite methods G1 and its improved version G2, which treat the electron correlation by the Møller-Plesset perturbation theory (MP4) and quadratic configuration interaction (QCISD(T)), gave EA = 0.28 eV [13] and 0.30 eV [14], respectively, and an MP4 calculation using the isogyric comparison with H_2 yielded EA = 0.30 eV [15].

References:

[1] Al-Za'al, M.; Miller, H. C.; Farley, J. W. (Phys. Rev. A Gen. Phys. **35** [1987] 1099/112).
[2] Miller, H. C.; Farley, J. W. (J. Chem. Phys. **86** [1987] 1167/71).
[3] Miller, H. C. (Diss. Univ. Oregon 1988, pp. 1/243, 48/132, 82; Diss. Abstr. Int. B **49** [1989] 5366/7).
[4] Neumark, D. M.; Lykke, K. R.; Andersen, T.; Lineberger, W. C. (J. Chem. Phys. **83** [1985] 4364/73).
[5] Engelking, P. C.; Lineberger, W. C. (J. Chem. Phys. **65** [1976] 4323/4).
[6] Celotta, R. J.; Bennett, R. A.; Hall, J. L. (J. Chem. Phys. **60** [1974] 1740/5).
[7] Cade, P. E. (Proc. Phys. Soc. [London] **91** [1967] 842/54).
[8] Rosmus, P.; Meyer, W. (J. Chem. Phys. **69** [1978] 2745/51).
[9] Frenking, G.; Koch, W. (J. Chem. Phys. **84** [1986] 3224/9).
[10] Novoa, J. J.; Mota, F.; Arnau, F. (J. Phys. Chem. **95** [1991] 3096/105).

[11] Sun, H.; Sheppard, M. G.; Freed, K. F. (J. Chem. Phys. **74** [1981] 6842/8).
[12] Zittel, P. F.; Lineberger, W. C. (J. Chem. Phys. **65** [1976] 1932/6).
[13] Pople, J. A.; Head-Gordon, M.; Fox, D. J.; Raghavachari, K.; Curtiss, L. A. (J. Chem. Phys. **90** [1989] 5622/9).
[14] Curtiss, L. A.; Raghavachari, K.; Trucks, G. W.; Pople, J. A. (J. Chem. Phys. **94** [1991] 7221/30).
[15] Pople, J. A.; Schleyer, P. von R.; Kaneti, J.; Spitznagel, G. W. (Chem. Phys. Lett. **145** [1988] 359/64).

2.1.1.2.4 Dipole Moment. Dipole Moment Function. Multipole Moments

The only experimental data for the dipole moment μ were obtained from Stark effect measurements on various vibrational-rotational lines in the UV emission spectra, A $^3\Pi_i \rightarrow$ X $^3\Sigma^-$ and c $^1\Pi \rightarrow$ a $^1\Delta$; the excited NH molecules were generated by electric discharges in NH_3. The following results were obtained for the states involved (ab initio calculations indicate the polarization $N^- H^+$ if $\mu > 0$):

state	$\lvert\mu\rvert$ in D	Ref.	state	$\lvert\mu\rvert$ in D	Ref.
X $^3\Sigma^-$	1.389 ± 0.075	[1]	a $^1\Delta$	1.49 ± 0.06	[2]
A $^3\Pi_i$	1.31 ± 0.03	[2]	c $^1\Pi$	1.70 ± 0.07	[2]

Numerous ($\sim$30) quantum-chemical calculations of the dipole moment μ, its derivative with respect to the internuclear distance $d\mu/dr$, and the dipole moment function $\mu(r)$ for NH in the electronic ground state and in various excited states are available; see the bibliography of quantum-chemical calculations on p. 31. Early results for μ (before 1970) are listed in [1]. High-quality, ab initio Hartree-Fock calculations give values of $\sim$1.61 to 1.65 D for μ(NH, X $^3\Sigma^-$) (see for example [3 to 7]), whereas the inclusion of electron correlation slightly reduces the values to 1.47 D; see for example [4, 7 to 12]. Dipole moments for excited valence states were calculated by applying CI methods: A $^3\Pi_i$ [9, 11, 12], a $^1\Delta$ [9, 12, 13], b $^1\Sigma^+$ [9, 10, 12, 13], c $^1\Pi$ [9, 12], d $^1\Sigma^+$ [12], 1 $^5\Sigma^-$ [12]. The dipole moment function of NH(X $^3\Sigma^-$) is tabulated in [4, 8], and dipole moment functions for NH(X $^3\Sigma^-$, A $^3\Pi_i$), and three unobserved states 2 $^3\Sigma^-$, 2 $^3\Pi_i$, and 1 $^5\Sigma^-$ are depicted in [11].

For the higher multipole moments of NH(X $^3\Sigma^-$), only theoretical data from ab initio Hartree-Fock calculations [5 to 7, 14] and an ACPF (averaged coupled-pair functional) calculation [7] are available. Values for the quadrupole, octopole, and hexadecapole moments Θ, Ω, and Φ calculated relative to the center of mass at the equilibrium internuclear distance r_e are as follows ($e = 1.60217733 \times 10^{-19}$ C, $a_0 = 5.29177249 \times 10^{-11}$ m):

method	Θ in $e \cdot a_0^2$	Ω in $e \cdot a_0^3$	Φ in $e \cdot a_0^4$	Ref.
SCF MO	0.448151	1.8886	4.24778	[5]
numerical Hartree-Fock	0.499092	1.891634	4.700893	[6]
SCF MO	0.4974	1.9000	–	[7]
ACPF*)	0.4384	1.7111	–	[7]

*) Θ and Ω also for NH(a $^1\Delta$) and NH(b $^1\Sigma^+$).

References:

[1] Scarl, E. A.; Dalby, F. W. (Can. J. Phys. **52** [1974] 1429/37).
[2] Irwin, T. A. R.; Dalby, F. W. (Can. J. Phys. **43** [1965] 1766/75).

[3] Cade, P. E.; Huo, W. M. (J. Chem. Phys. **45** [1966] 1063/5).
[4] Meyer, W.; Rosmus, P. (J. Chem. Phys. **63** [1975] 2356/75).
[5] Maroulis, G.; Sana, M.; Leroy, G. (J. Mol. Struct. **122** [1985] 269/80 [THEOCHEM **23**]).
[6] Laaksonen, L.; Müller-Plathe, F.; Diercksen, G. H. F. (J. Chem. Phys. **89** [1988] 4903/8).
[7] Jansen, G.; Hess, B. A. (Chem. Phys. Lett. **192** [1992] 21/8).
[8] Das, G.; Wahl, A. C.; Stevens, W. J. (J. Chem. Phys. **61** [1974] 433/4).
[9] Hay, P. J.; Dunning, T. H., Jr. (J. Chem. Phys. **64** [1976] 5077/87).
[10] Marian, C. M.; Klotz, R. (Chem. Phys. **95** [1985] 213/23).

[11] Goldfield, E. M.; Kirby, K. P. (J. Chem. Phys. **87** [1987] 3986/94).
[12] Yarkony, D. R. (J. Chem. Phys. **91** [1989] 4745/57).
[13] Schmidt, M. W.; Lam, M. T. B; Elbert, S. T.; Ruedenberg, K. (Theor. Chim. Acta **68** [1985] 69/86).
[14] Guseinov, I. I.; Mursalov, T. M.; Sadykhov, F. S.; Aliev, V. T. (Zh. Strukt. Khim. **29** No. 4 [1988] 162; J. Struct. Chem. [USSR] **29** [1988] 626).

2.1.1.2.5 Polarizability

A recent quantum-chemical ab initio ACPF (averaged coupled-pair functional) calculation gave the parallel and perpendicular components of the polarizability of NH(X $^3\Sigma^-$), $\alpha_{\|} = 11.770\ a_0^3$ (= 1.744 Å^3) and $\alpha_\perp = 9.016\ a_0^3$ (= 1.336 Å^3); the average molecular polarizability thus is $\alpha = (\alpha_{\|} + 2\alpha_\perp)/3 = 9.934\ a_0^3$ (= 1.472 Å^3). Slightly higher values were obtained for the two lowest excited states, a $^1\Delta$ and b $^1\Sigma^+$, of NH [1]. Using the time-dependent coupled Hartree-Fock method, the frequency-dependent polarizability $\alpha(\omega)$ was calculated; for the isotropic static polarizability, a value of 9.3655 a_0^3 was obtained [2].

References:

[1] Jansen, G.; Hess, B. A. (Chem. Phys. Lett. **192** [1992] 21/8).
[2] Hettema, H.; Wormer, P. E. S. (J. Chem. Phys. **93** [1990] 3389/96).

2.1.1.2.6 Spectroscopic Constants

2.1.1.2.6.1 Hyperfine and Quadrupole Coupling Constants. Zeeman Parameters

Hyperfine structure (hfs) has been resolved in the far-IR laser magnetic resonance (LMR) spectra of NH (ND) in its ground state X $^3\Sigma^-$ [1, 2] and lowest excited state a $^1\Delta$ [3 to 5], in the high-resolution far-IR absorption spectrum of NH(X $^3\Sigma^-$) [6], and in the high-resolution laser-induced fluorescence (LIF) spectra of the A $^3\Pi_i \leftarrow$ X $^3\Sigma^-$ [7] and c $^1\Pi \leftarrow$ a $^1\Delta$ [8] transitions. Analysis of the spectra by using Hamiltonians, which include in addition to the terms for rotation, centrifugal distortion, and fine structure also those for hfs, quadrupole, and Zeeman interactions, enabled the corresponding parameters to be derived. The hfs parameters b and c introduced by Frosch and Foley [9] characterize the interactions between the electron spin (S = 1 for X $^3\Sigma^-$ and A $^3\Pi_i$) and the nuclear spins of H (I = 1/2), D (I = 1), ^{14}N (I = 1), and ^{15}N (I = 1/2), where $b + c/3 = \alpha$ and $c/3 = \beta$ are the Fermi contact term (or isotropic hf coupling constant) and the dipole-dipole term, respectively; the parameters a and d depict the hf interactions between the electron orbital angular momentum ($\Lambda = 2$ for a $^1\Delta$, $\Lambda = 1$ for A $^3\Pi_i$ and c $^1\Pi$) and the nuclear spins. The quadrupole interaction of the ^{14}N nucleus is described by the coupling constants eQq_1 and eQq_2, and the Zeeman interactions between the electron spin, the orbital angular momentum, and the rotational angular momentum and the magnetic field by the corresponding g factors g_S, g_L, and g_R. For the theoretical background, see [10].

Results for ^{14}NH, ^{15}NH, and ^{14}ND in the vibronic ground state X $^3\Sigma^-$, v=0 are as follows (in MHz):

constant	^{14}NH	^{14}NH	^{14}NH	^{15}NH	^{14}ND
b_N	41.7±0.5	41.86(33)	42.8(16)	60.1(56)	45.1(30)
c_N	−66.3±0.6	−67.94(61)	−68.4(30)	93.9(66)	−74.7(60)
$b_N+c_N/3$	19.6±0.4	19.22(18)	20.0(6)	−28.8(34)	20.2(10)
$b_{H(D)}$	−96.5±1.4	−96.80(57)	−100.2(26)	−95.3(54)	−15.5(30)
$c_{H(D)}$	90.6±1.9	91.70(160)	88.8(48)	86.4(66)	15.3(60)
$b_{H(D)}+c_{H(D)}/3$	−66.3±1.2	−66.23(32)	−70.6(10)	−66.5(32)	−10.2(10)
$eQq_1(^{14}N)$	−5.0±1.2	−	−	−	−
remark	a)	b)	c)	c)	c)
Ref.	[7]	[6]	[2]	[2]	[2]

a) More than 300 hf lines analyzed in the LIF spectrum of the A $^3\Pi_i \leftarrow$ X $^3\Sigma^-$, v=0←0 transition at 336 nm [7].

b) Analysis of five and nine hf lines of the N=1←0, J=2←1 and N=1←0, J=1←1 transitions, respectively, of NH(X $^3\Sigma^-$, v=0) observed in the high-resolution far-IR absorption spectrum around 1 THz (300 μm) [6].

c) Submillimeter-LMR using six laser lines at 163 to 315 μm; analysis of fifteen resonances for ^{14}NH, v=0, N=1←0, eight resonances for ^{14}ND, v=0, N=2←1, 3←2, and three resonances for ^{15}NH, v=0, N=1←0. The quadrupole interactions were set to zero, and the g factors fixed at $g_S=2.002100$, $g_R=-0.000100$, and $\Delta g'=\Delta g-g_L^e=-0.001690$ (Δg is a diamagnetic correction; g_L^e is an orbital g factor due to coupling-in of excited electronic states) [2]. Previously incorrect results for ^{14}NH, v=0, N=1←0 [1] thus were improved. Hfs constants for $^{14}NH(X\ ^3\Sigma^-$, v=1) (N=1←0, eight resonances) and $^{14}ND(X\ ^3\Sigma^-$, v=1) (N=2←1, 3←2, two resonances) were also derived [2].

Ground-state hfs parameters have been calculated by applying various quantum-chemical ab initio methods. Those using highly correlated wave functions are generally in satisfactory agreement with the experiment, giving the following isotropic hf coupling constants $\alpha=b+c/3$ (in MHz):

α_N	19.1	18.0	18.8	16.6	18.9	17.6
α_H	−65.8	−63.5	−63.3	−64.5	−66.4	−53.6
method*)	QCISD(T)	UCCD(ST)	MC SCF	MRD CI	MBPT	SD CI
Ref.	[11]	[12]	[13]	[14, 15]	[16]	[17]

*) QCISD(T)=quadratic CI including single, double, and triple excitations, UCCD(ST)= coupled-cluster doubles method based on the unrestricted Hartree-Fock method and corrected for single and triple replacements, MC SCF=multiconfiguration SCF, MRD CI=multi-reference singles and doubles CI, MBPT=many-body perturbation theory, SD CI=singles and doubles CI.

The experimental results for the excited state A $^3\Pi_i$, v=0 (see remark a) in the above table for the X $^3\Sigma^-$ state) [7] are (in MHz):

a_N = 89.6±2.3	a_H = 74.1±1.5	$eQq_1(^{14}N)$ = 7.1±1.5
b_N = 153.6±0.4	b_H = 270.8±0.6	$eQq_2(^{14}N)$ = 21.9±2.4
c_N = 15.2±2.6	c_H = 90.5±4.5	
$b_N+c_N/3$ = 158.7±0.9	$b_H+c_H/3$ = 301.0±1.6	
d_N = 66.4±0.4	d_H = 26.0±0.6	

The results for the excited states a $^1\Delta$, v=0 and c $^1\Pi$, v=0 are as follows (in MHz):

constant	NH(a $^1\Delta$)	ND(a $^1\Delta$)	NH(a $^1\Delta$)	NH(c $^1\Pi$)
a_N	109.65(85)	109.63(22)	110.0±0.5	106.3±1.8
$a_{H(D)}$	70.9(14)	11.03(23)	69.6±1.2	58.2±4.3
$eQq_1(^{14}N)$	[−4.0]	−4.0(15)	−8.0±2.6	11.4±2.3
$eQq_2(^{14}N)$	−	−	−	−32.0±3.2
g_R	−0.00158(6)	−0.0086(10)	−	−
g_L	[1.00103]	1.000506(17)	−	−
remark	a)	a)	b)	b)
Ref.	[3, 4]	[3, 4]	[8]	[8]

a) Far-IR LMR using four laser lines at 102 to 190 μm; analysis of 18 resonances for NH(a $^1\Delta$, v=0, J=3←2) and 66 and 24 resonances, respectively, for ND(a $^1\Delta$, v=0, J=3←2 and 5←4); the values for NH(a $^1\Delta$) in brackets were obtained from the fit of the ND(a $^1\Delta$) data [3, 4]. LMR of NH(a $^1\Delta$, v=1, J=3←2) at 105.5 μm gave the constants a_N, a_H, and g_R for the v=1 level of NH(a $^1\Delta$) [5].

b) Analysis of 68 hf lines in the LIF spectrum of the c $^1\Pi$←a $^1\Delta$, v=0←0 transition at 324 nm [7].

References:

[1] Radford, H. E.; Litvak, M. M. (Chem. Phys. Lett. **34** [1975] 561/4).
[2] Wayne, F. D.; Radford, H. E. (Mol. Phys. **32** [1976] 1407/22).
[3] Leopold, K. R.; Evenson, K. M.; Brown, J. M. (J. Chem. Phys. **85** [1986] 324/30).
[4] Leopold, K. R.; Zink, L. R.; Evenson, K. M.; Jennings, D. A.; Brown, J. M. (NBS Spec. Publ. [U. S.] **716** [1986] 452/8).
[5] Vasconcellos, E. C. C.; Davidson, S. A.; Brown, J. M.; Leopold, K. R.; Evenson, K. M. (J. Mol. Spectrosc. **122** [1987] 242/5).
[6] Van den Heuvel, F. C.; Meerts, W. L.; Dymanus, A. (Chem. Phys. Lett. **92** [1982] 215/8).
[7] Ubachs, W.; Ter Meulen, J. J.; Dymanus, A. (Can. J. Phys. **62** [1984] 1374/91).
[8] Ubachs, W.; Meyer, G.; Ter Meulen, J. J.; Dymanus, A. (J. Mol. Spectrosc. **115** [1986] 88/104).
[9] Frosch, R. A.; Foley, H. M. (Phys. Rev. [2] **88** [1952] 1337/49).
[10] Hirota, E. (High-Resolution Spectroscopy of Transient Molecules, in: Springer Ser. Chem. Phys. **40** [1985] pp. 33/42, 62/71).

[11] Carmichael, I. (J.Phys. Chem. **95** [1991] 108/11).
[12] Carmichael, I. (J. Phys. Chem. **94** [1990] 5734/40).
[13] Chipman, D. M. (J. Chem. Phys. **91** [1989] 5455/65).
[14] Engels, B.; Peyerimhoff, S. D. (J. Phys. B At. Mol. Opt. Phys. **21** [1988] 3459/71).
[15] Engels, B.; Peyerimhoff, S. D. (Mol. Phys. **67** [1989] 583/600).
[16] Kristiansen, P.; Veseth, L. (J. Chem. Phys. **84** [1986] 6336/44).
[17] Bender, C. F.; Davidson, E. R. (Phys. Rev. [2] **183** [1969] 23/30).

2.1.1.2.6.2 Fine-Structure Constants

The fine structure in purely rotational, rotational-vibrational, and electronic spectra of NH (ND) arises from the interaction of the unpaired electron spin with the orbital angular momentum (spin-orbit coupling constant A for the $^3\Pi$ states), from the interaction of the unpaired electron spins with each other and with the rotational angular momentum (spin-

spin coupling and spin-rotation coupling constants λ and γ for the $^3\Sigma^-$ and $^3\Pi$ states), and from Λ-type doubling (constants p, q, and o for the $^3\Pi$, $^1\Pi$, and $^1\Delta$ states).

For the electronic **ground state** X $^3\Sigma^-$ of NH and ND, the spin-rotation and spin-spin coupling constants were derived from the far-IR absorption [1] and LMR [2] spectra, from IR absorption [3] and emission [4] spectra in the region of the fundamental band, from IR emission spectra in the region of the $\Delta v=1$ ($v''=0$ to 4) sequence [5], and from the A $^3\Pi_i \leftrightarrow$ X $^3\Sigma^-$ absorption and emission transitions with v′ and v″ up to 2 [6 to 15]. Results of highest precision for γ and λ and their centrifugal distortion terms γ_D and λ_D for NH(X $^3\Sigma^-$, v=0) and ND(X $^3\Sigma^-$, v=0) are as follows (in cm^{-1}):

constant	NH	^{14}NH	^{14}ND	NH	NH	ND
$\gamma\cdot10^2$	−5.4785(6)	−5.466(8)	−2.94(2)	−5.221(83)	−5.4844(22)	−2.776(94)
$\gamma_D\cdot10^5$	−	−	−	−1.68(11)	1.5098(75)	−
λ	0.920006(14)	0.9197(2)	0.9184(4)	0.91984(33)	0.920063(148)	0.9221(57)
$\lambda_D\cdot10^5$	−	−	−	7.1(13)	−0.909(142)	−
remark	a)	b)	b)	c)	d)	e)
Ref.	[1]	[2]	[2]	[5]	[6]	[7]

a) Analysis of the rotational hyperfine spectrum ($N=1\leftarrow0$, $J=2\leftarrow1$ and $1\leftarrow1$ transitions) of NH(X $^3\Sigma^-$, v=0) in the far IR around 1 THz; original values: $\gamma=-1642.4(17)$ MHz and $\lambda=27581.1(43)$ MHz [1].

b) Submillimeter-LMR spectra of ^{14}NH(X $^3\Sigma^-$, v=0, 1), ^{14}ND(X $^3\Sigma^-$, v=0, 1), and ^{15}NH(X $^3\Sigma^-$, v=0) (cf. p. 41, remark c) of the table which lists the hf constants); results for ^{14}NH(X $^3\Sigma^-$, v=1) are also given [2].

c) Fourier transform IR emission spectrum of NH(X $^3\Sigma^-$) in the region of the $\Delta v=1$ sequence, $v=1\rightarrow0$ to $5\rightarrow4$; results are given for the v=0 to 5 levels, and polynomial expansions in $(v+1/2)^n$ (n=0 to 4) terms have been derived leading to the equilibrium values $\gamma_e=-5.6724(14)\cdot10^{-2}$, $\gamma_{D,e}=-1.822(56)\cdot10^{-5}$, $\lambda_e=0.91911(12)$, and $\lambda_{D,e}=8.5(8)\cdot10^{-5}$ [5]. The centrifugal distortion constants quoted here are denoted 2 γ_D and 2 λ_D in the original paper; their definitions are obviously $\gamma_{eff}=\gamma-2\,\gamma_D\cdot J(J+1)$ and $\lambda_{eff}=\lambda-2\,\lambda_D\cdot J(J+1)$; compare values in column d).

d) Fourier transform emission spectrum of the A $^3\Pi_i\rightarrow$X $^3\Sigma^-$, $\Delta v=0$, ±1 ($v'\leq2$, $v''\leq2$) transitions of NH; definition of the centrifugal distortion constants is $\lambda_{eff}=\lambda+\lambda_D\cdot J(J+1)$ and $\gamma_{eff}=\gamma+\gamma_D\cdot J(J+1)+\gamma_H\cdot J^2(J+1)^2$; $\gamma_H=-1.366(89)\cdot10^{-9}$. Results for the v=1 and 2 levels of NH(X $^3\Sigma^-$) are also given [6].

e) Fitting the data for the A $^3\Pi_i\rightarrow$X $^3\Sigma^-$, $v=0\rightarrow0$, $1\rightarrow1$ [9, 10], and $2\rightarrow2$ [16] emission bands of ND to the same Hamiltonian used by [6]; results for the v=1 and 2 levels are also given [7].

Ab initio SCF MO and CI calculations of the spin-spin coupling constant λ are available in [2, 17 to 19] and [20, 21], respectively.

Fine-structure constants for the **excited valence states** were derived from the A $^3\Pi_i\leftrightarrow$ X $^3\Sigma^-$ [6 to 16], c $^1\Pi\leftrightarrow$a $^1\Delta$ [22 to 26], c $^1\Pi\rightarrow$b $^1\Sigma^+$ [27], and d $^1\Sigma^+\rightarrow$c $^1\Pi$ [28, 29] spectra. The following results for the A $^3\Pi_i$, v=0 levels of NH and ND were very precisely determined (in cm^{-1}):

constant	NH	ND	constant	NH	ND
A	−34.61976(15)	−34.5987(84)	$p\cdot10^2$	5.5222(25)	2.770(100)
$A_D\cdot10^5$	[−8.1]	–	$p_D\cdot10^5$	−1.7795(151)	–
$\gamma\cdot10^2$	2.9830(24)	−1.66(11)	$q\cdot10^2$	−3.15870(40)	−0.8540(83)
$\gamma_D\cdot10^6$	−5.406(52)	–	$q_D\cdot10^5$	1.3822(29)	–
λ	−0.19968(22)	−0.2107(79)	o	1.28447(20)	1.3022(69)
$\lambda_D\cdot10^5$	−1.630(156)	–	$o_D\cdot10^4$	−1.3440(136)	–
remark	a)	b)		a)	b)
Ref.	[6]	[7]		[6]	[7]

a) Fourier transform emission spectrum of the A $^3\Pi_i \rightarrow X\ ^3\Sigma^-$, $\Delta v=0, \pm1$ ($v' \le 2$, $v'' \le 2$) transitions of NH; results are given also for the v=1 and 2 levels. The higher centrifugal distortion terms of the Λ-doubling constants p_H, q_H, q_L, and o_H were also derived; the A_D value is calculated. Assuming a vibrational dependence of the spin-orbit coupling constant of the form $A_v=A_e-\alpha_A(v+1/2)$, the values $A_e=-34.6038(23)$ cm^{-1} and $\alpha_A=-0.0314(26)$ cm^{-1} were derived [6].

b) Fitting the data for the A $^3\Pi_i \rightarrow X\ ^3\Sigma^-$, $v=0\rightarrow0$, $1\rightarrow1$ [9, 10], and $2\rightarrow2$ [16] emission bands of ND to the same Hamiltonian used by [6]; results for the v=1 and 2 levels are also given. Combining with own data from LIF measurements in the $1\leftarrow0$, $2\leftarrow1$, and $3\leftarrow2$ bands gave (cf. remark a)) the values $A_e=-34.5362(45)$ cm^{-1} and $\alpha_A=-0.125(15)$ cm^{-1} [7].

An ab initio MC SCF calculation of the spin-orbit coupling constant [30] and an SCF calculation of the Λ-doubling constant [18] are available.

The high-resolution Fourier transform emission spectrum of the c $^1\Pi \rightarrow a\ ^1\Delta$, $v=1\rightarrow0$, $0\rightarrow0$, $0\rightarrow1$ transitions [23] and the high-resolution LIF spectrum of the c $^1\Pi \leftarrow a\ ^1\Delta$, $v=0\leftarrow0$ transition [22] gave the following Λ-type doubling constants (in cm^{-1}):

NH(a $^1\Delta$, v=0) $q\cdot10^7$	NH(c $^1\Pi$, v=0) $q\cdot10^3$	$q_D\cdot10^6$	$q_H\cdot10^8$	remark	Ref.
−1.62(23)	16.711(38)	−3.92(38)	0.216(88)	a)	[23]
−1.61(10)	16.74(1)	−4.54(13)	0.404(43)	b)	[22]

a) Values for the v=1 levels of both states are also available.

b) Values converted to cm^{-1} units from q=502.2(3) MHz, $q_D=-0.136(4)$ MHz, $q_H=0.121(13)$ kHz for c $^1\Pi$ and q=−4.82(31) kHz for a $^1\Delta$.

In another analysis of the c $^1\Pi \rightarrow a\ ^1\Delta$, $v=0\rightarrow0$ band of NH and in the analysis of the $0\rightarrow0$ and $0\rightarrow1$ bands of ND, Λ-type doubling of the a $^1\Delta$ state was neglected. For the c $^1\Pi$, v=0 levels of NH and ND, $q=16.60(18)\cdot10^{-3}$, $q_D=-2.8(9)\cdot10^{-6}$ cm^{-1} [24], and $q=4.47(2)\cdot10^{-3}$ cm^{-1} [25] were derived. The analyses of the d $^1\Sigma^+ \rightarrow c\ ^1\Pi$ and c $^1\Pi \rightarrow$ b $^1\Sigma^+$ emission spectra gave for the c $^1\Pi$, v=0 levels of NH and ND q(in 10^{-3} cm^{-1})=16.5 and 4.36 [28], 15.4(1) and 4.42(3) [29], 15.55(7) and 4.56(4) [27]; results for the v=1 level of NH and the v=1 and 2 levels of ND [28, 29] are also available.

For the triplet **Rydberg states** detected by resonance-enhanced multiphoton ionization (REMPI) of NH, ND(X $^3\Sigma^-$) the following spin-orbit, spin-rotation, spin-spin coupling, and Λ-doubling constants were derived for the v=0 levels (in cm^{-1}) [31] (preliminary results for the B $^3\Pi$ state in [32]):

constant	B $^3\Pi$ NH	ND	D $^3\Pi$ NH	ND	C $^3\Sigma^-$ ND	E $^3\Sigma^-$ ND	F $^3\Sigma^-$ NH	ND
A	39.9(5)	39.6(2)	1.2(4)	4.2(4)	–	–	–	–
γ	0.52(9)	$0^{a)}$	0.11(3)	0	0	0	0	0
λ	$0^{b)}$	0.5(2)	0	−0.5(3)	0	−0.4(2)	0	0
p	0	0.17(4)	0	−0.21(5)	–	–	–	–
q	0	$0.045(9)^{c)}$	0.89(1)	0.278(9)	–	–	–	–

a) 0.09(2) for v = 1. – b) 1.4(2) for v = 1. – c) For v = 1.

For the singlet Rydberg state j $^1\Pi$, v = 0, REMPI of ND(a $^1\Delta$) gave |q| = 0.0368(5) cm^{-1} [33].

References:

[1] Van den Heuvel, F. C.; Meerts, W. L.; Dymanus, A. (Chem. Phys. Lett. **92** [1982] 215/8).
[2] Wayne, F. D.; Radford, H. E. (Mol. Phys. **32** [1976] 1407/22).
[3] Bernath, P. F.; Amano, T. (J. Mol. Spectrosc. **95** [1982] 359/64).
[4] Sakai, H.; Hansen, P.; Esplin, M.; Johansson, R.; Peltola, M.; Strong, J. (Appl. Opt. **21** [1982] 228/34).
[5] Boudjaadar, D.; Brion, J.; Chollet, P.; Guelachvili, G.; Vervloet, M. (J. Mol. Spectrosc. **119** [1986] 352/66).
[6] Brazier, C. R.; Ram, R. S.; Bernath, P. F. (J. Mol. Spectrosc. **120** [1986] 381/402).
[7] Patel-Misra, D.; Sauder, D. G.; Dagdigian, P. J. (Chem. Phys. Lett. **174** [1990] 113/8).
[8] Malicet, J.; Brion, J.; Guenebaut, H. (J. Chim. Phys. Phys. Chim. Biol. **67** [1970] 25/30).
[9] Bollmark, P.; Kopp, I.; Rydh, B. (J. Mol. Spectrosc. **34** [1970] 487/99).
[10] Kopp, I.; Kronekvist, M.; Aslund, N. (Arkiv Fys. **30** [1965] 9/17).

[11] Krishnamurty, G.; Narasimham, N. A. (Proc. Indian Acad. Sci. A **67** [1968] 50/60).
[12] Krishnamurty, G.; Narasimham, N. A. (Proc. 1st Int. Conf. Spectrosc., Bombay 1967, Vol. 1, pp. 142/4).
[13] Shimauchi, M. (Sci. Light [Tokyo] **15** [1966] 161/5).
[14] Dixon, R. N. (Can. J. Phys. **37** [1959] 1171/86).
[15] Veseth, L. (J. Phys. B **5** [1972] 229/41).
[16] Shimauchi, M. (Sci. Light [Tokyo] **16** [1967] 185/90).
[17] Wayne, F. D.; Colbourn, E. A. (Mol. Phys. **34** [1977] 1141/55).
[18] Palmiere, P.; Sink, M. L. (J. Chem. Phys. **65** [1976] 3641/6).
[19] Klobukowski, M. (Chem. Phys. Lett. **183** [1991] 417/22).
[20] Jensen, J. O.; Yarkony, D. R. (Chem. Phys. Lett. **141** [1987] 391/6).

[21] Yarkony, D. R. (J. Chem. Phys. **91** [1989] 4745/57).
[22] Ubachs, W.; Meyer, G.; Ter Meulen, J. J.; Dymanus, A. (J. Mol. Spectrosc. **115** [1986] 88/104).
[23] Ram, R. S.; Bernath, P. F. (J. Opt. Soc. Am. B Opt. Phys. **3** [1986] 1170/4).
[24] Ramsay, D. A.; Sarre, P. J. (J. Mol. Spectrosc. **93** [1982] 445/6).
[25] Hanson, H.; Kopp, I.; Kronekvist, M.; Aslund, N. (Arkiv Fys. **30** [1965] 1/8).
[26] Shimauchi, M. (Sci. Light [Tokyo] **13** [1965] 53/63).
[27] Whittaker, F. L. (J. Phys. B **1** [1968] 977/82).
[28] Graham, W. R. M.; Lew, H. (Can. J. Phys. **56** [1978] 85/99).
[29] Whittaker, F. L. (Proc. Phys. Soc. [London] **90** [1967] 535/41).
[30] Koseki, S.; Schmidt, M. W.; Gordon, M. S. (J. Phys. Chem. **96** [1992] 10768/72).

[31] Clement, S. G.; Ashfold, M. N. R.; Western, C. M.; Johnson, R. D., III; Hudgens, J. W. (J. Chem. Phys. **97** [1992] 7064/72).
[32] Clement, S. G.; Ashfold, M. N. R.; Western, C. M.; Johnson, R. D., III; Hudgens, J. W. (J. Chem. Phys. **96** [1992] 5538/40).
[33] Clement, S. G.; Ashfold, M. N. R.; Western, C. M.; De Beer, E.; De Lange, C. A.; Westwood, N. P. C. (J. Chem. Phys. **96** [1992] 4963/73).

2.1.1.2.6.3 Rotational and Vibrational Constants. Internuclear Distance

Electronic Ground State and Excited Valence States. Purely rotational, rotational-vibrational, and numerous vibronic spectra have been analyzed to derive the rotational and centrifugal distortion constants B_e or B_v and D_e, H_e, L_e or D_v, H_v, L_v, the corresponding rotation-vibration interaction constants α_e, γ_e, β_e, η_e, and the vibrational constants ω_e, $\omega_e x_e$, $\omega_e y_e$, $\omega_e z_e$ of NH and ND in their various valence states. The equilibrium internuclear distances r_e were obtained by converting the rotational constant B_e. The high-quality results are compiled in the following two tables: Table 1 gives the constants for the ground state X $^3\Sigma^-$ and the lowest excited state a $^1\Delta$, obtained from far-IR and IR spectra, and Table 2 the constants obtained from the A $^3\Pi_i \leftrightarrow$X $^3\Sigma^-$, c $^1\Pi \leftrightarrow$a $^1\Delta$, c $^1\Pi \rightarrow$b $^1\Sigma^+$, d $^1\Sigma^+ \rightarrow$ b $^1\Sigma^+$, and d $^1\Sigma^+ \rightarrow$c $^1\Pi$ emission and absorption systems in the near-IR, visible, and UV. Details concerning the various analyses as well as additional references can be found in the remarks below both tables.

Geometry optimization for determining r_e and evaluating the rotational and vibrational constants are included in numerous quantum-chemical ab initio calculations on NH. These are essentially the same studies that are quoted in the next section on potential energy functions (see table on p. 55). Further references may be found in the bibliographies on quantum-chemical calculations given on p. 31.

Table 1
NH, ND in the States X $^3\Sigma^-$ and a $^1\Delta$. Rotational and Vibrational Constants from Purely Rotational and Rotational-Vibrational Spectra in the Far IR and IR.
Rotational and vibrational constants in cm^{-1}, equilibrium internuclear distance r_e in Å.

constant	NH(X $^3\Sigma^-$)				ND(X $^3\Sigma^-$)
$B_e\{B_0\}$	{16.343282(22)}	{16.3433(2)}	16.667745(82)	16.66683(16)	{8.7815(5)}
α_e	–	–	0.65006(20)	0.64699(32)	–
$\gamma_e \cdot 10^3$	–	–	2.020(146)	–	–
$D_e\{D_0\} \cdot 10^4$	[{17.14(3)}]	[{16.85}]	17.1611(48)	17.15(11)	[{4.833}]
$\beta_e \cdot 10^4$	–	–	0.2713	–	–
$H_e\{H_0\} \cdot 10^7$	–	–	1.295(11)	–	–
$\omega_e\{\Delta G_{1/2}\}$	–	–	3282.583(33)	{3125.5724(20)}	–
$\omega_e x_e$	–	–	78.915(27)	–	–
$\omega_e y_e$	–	–	0.395(9)	–	–
$\omega_e z_e$	–	–	−0.047(1)	–	–
r_e	–	1.03722(2)	1.03755	1.03756(6)	–
remark	a)	b)	c)	d)	b)
Ref.	[1]	[2]	[3]	[4]	[2]

constant	NH(a $^1\Delta$)		ND(a $^1\Delta$)
$\{B_0\}$	{16.4461503(32)}	{16.43273(29)}	{8.83111819(10)}
$\{D_0\} \cdot 10^4$	[{16.812(36)}]	{16.812(36)}	{4.79792924(30)}
$\{H_0\} \cdot 10^7$	–	{1.45(10)}	–
$\{L_0\} \cdot 10^{11}$	–	{1.3(16)}	–
$\{\Delta G_{1/2}\}$	–	{3182.7768(37)}	–
remark	e)	f)	e)
Ref.	[5, 6]	[7]	[5, 6]

a) Rotational hyperfine absorption spectrum of the $N=1\leftarrow0$, $J=2\leftarrow1$ and $1\leftarrow1$ transitions around 1 THz. The analysis has been performed with the centrifugal distortion constant fixed at the optical value reevaluated by [8] from the data of [9]. Original result: $B_0=489959.26(68)$ MHz [1].

b) Submillimeter-LMR spectra of $^{14}NH(X\ ^3\Sigma^-, v=0, 1)$ and $ND(X\ ^3\Sigma^-, v=0)$ using laser wavelengths between 163 and 315 μm. The centrifugal distortion constants were constrained to optical values; that for NH is from [9], while the origin of D_0 for ND is not clear ($D_0\cdot10^4=4.855$ and 4.883 were published by [10] and [8], respectively, cited by [2]). Results for $^{14}NH(X\ ^3\Sigma^-, v=1)$ are also available [2].

c) High-resolution Fourier transform IR emission spectrum of the $\Delta v=1$ sequence from $v=1\rightarrow0$ to $5\rightarrow4$. B_v, D_v, and H_v for $v=0$ to 5 are listed in [2]; they have been fitted to polynomial expansions up to $(v+1/2)^4$ to give the equilibrium values quoted above. The higher-order vibrational interaction constants of B, D, and H are also available. A misprint in [3, table VII] gives α_e too low by a factor of ten. – A Fourier transform IR emission spectrum of the $1\rightarrow0$ to $4\rightarrow3$ transitions (presumably) was recorded at lower resolution and analyzed by [11, 12].

d) High-resolution IR absorption spectrum of the $v=1\leftarrow0$ transition. Values above are given with three times the standard error [4].

e) Far-IR LMR of NH, $ND(a\ ^1\Delta, v=0)$. Original results: $B_0=493043.182(95)$ MHz (D_0 constrained to the IR-spectroscopic value of [7]) for NH and $B_0=264750.263(30)$, $D_0=14.38383(91)$ MHz for ND [5, 6]. Extension of the experiment to $NH(a\ ^1\Delta, v=1)$ gave the rotational constants B_1 and D_1 [13].

f) High-resolution IR absorption spectrum in the $v=1\leftarrow0$ band; B_1, D_1, H_1, and L_1 values are also available [7].

Table 2
NH, ND in the States X $^3\Sigma^-$, A $^3\Pi_i$, a $^1\Delta$, b $^1\Sigma^+$, c $^1\Pi$, d $^1\Sigma^+$. Rotational and Vibrational Constants from Rovibronic Spectra in the Near IR, Visible, and UV.
Rotational and vibrational constants in cm^{-1}; internuclear distance r_e in Å.

constant	NH(X $^3\Sigma^-$)	ND(X $^3\Sigma^-$)	NH(A $^3\Pi_i$)	ND(A $^3\Pi_i$)	NH(a $^1\Delta$)	NH(a $^1\Delta$)	NH(a $^1\Delta$)	ND(a $^1\Delta$)
$B_e\{B_0\}$	16.666970(7)	8.90820(25)	16.681963(8)	8.90318(20)	{16.43197(20)}	{16.432551(76)}	16.7442(10)	8.9564(882)
α_e	0.647567(32)	0.25366(40)	0.712880(35)	0.27962(30)	–	–	0.6256(25)	0.2454(516)
$\gamma_e \cdot 10^3$	0.365(20)	0.34(35)	−16.160(24)	−4.21(30)	–	–	3.65(13)	–
$D_e\{D_0\} \cdot 10^4$	17.1431(31)	–	17.822(28)	–	{16.657(26)}	{16.7309(57)}	16.9	–
$\beta_e \cdot 10^4$	−0.2309(36)	–	0.146(31)	–	–	–	−0.32	–
$H_e\{H_0\} \cdot 10^8$	12.7(3)	–	11.6(8)	–	{8.87(99)}	{11.54(22)}	–	–
$\eta_e \cdot 10^8$	−0.7(3)	–	−1.6(8)	–	–	–	–	–
$L_e\{L_0\} \cdot 10^{10}$	–	–	–	–	{1.141(89)}	{0.060(34)}	–	–
$\omega_e\{\Delta G_{1/2}\}$	3281(20)	2399.27(32)	3232(2)	2364.9(11)	–	{3182.7879(16)}	3332.37(63)	2435.15(1794)
$\omega_e x_e$	78(1)	40.87(16)	98(1)	51.82(14)	–	–	76.03(55)	39.38(592)
$\omega_e y_e$	–	–	–	–	–	–	0.84(17)	–
$\omega_e z_e$	–	–	–	–	–	–	0.055(19)	–
r_e	1.03722	–	1.03675	–	–	–	1.03483(3)	–
remark	a)	b)	a)	b)	c)	d)	e)	f)
Ref.	[14]	[15]	[14]	[15]	[16]	[17]	[18, 19, 46]	[20]

Table 2 (continued)

constant	NH(b $^1\Sigma^+$)	NH(b $^1\Sigma^+$)	ND(b $^1\Sigma^+$)	NH(c $^1\Pi$)	NH(c $^1\Pi$)	NH(c $^1\Pi$)	ND(c $^1\Pi$)	NH(d $^1\Sigma^+$)	ND(d $^1\Sigma^+$)
$B_e\{B_0\}$	16.705	16.7326(9)	8.9472(6)	{14.14203(22)}	{14.15158(10)}	14.537	7.833	14.390	7.693
α_e	0.591	0.6049(3)	0.2383(8)	–	–	0.593	0.379	0.621	0.257
$\gamma_e \cdot 10^3$	–	–	–	–	–	−0.347	−0.047	–	0.0033
$D_e\{D_0\} \cdot 10^4$	16.0	16.54(7)	4.64(2)	{22.220(27)}	{22.462(11)}	19.7	6.29	16.0	4.8
$\beta_e \cdot 10^4$	–	−0.29(2)	−0.084(26)	–	–	–	–	–	–
$\{H_0\} \cdot 10^8$	–	–	–	{−20.61(94)}	{−5.89(47)}	–	–	–	–
$\{L_0\} \cdot 10^{10}$	–	–	–	{0.999(74)}	{−2.814(73)}	–	–	–	–
$\omega_e\{\Delta G_{1/2}\}$	3352.43	3354.7	2451.3	–	{2121.4557(69)}	2551.2	1756.5	2672.6	1953.7
$\omega_e x_e$	74.2	74.4	39.7	–	–	214.3	50.95	71.2	38.2
$\omega_e y_e$	0.70	–	–	–	–	–	−10.37	–	–
$\omega_e z_e$	−0.035	–	–	–	–	–	–	–	–
r_e	1.0360	–	–	–	–	1.1106	1.1055	1.1163	1.1156
remark	g)	h)	h)	c)	d)	g)	g)	g)	g)
Ref.	[21]	[22, 23]	[16]	[17]	[21]	[21]	[21]	[21]	[21]

a) High-resolution Fourier transform emission spectrum of the A $^3\Pi_i \rightarrow X\ ^3\Sigma^-$, v=0→0, 1→1, 2→2, 1→0, 2→1, 0→1, and 1→2 bands [14]. – For earlier analyses of the A↔X system of NH, see [9, 24, 25].

b) High-resolution LIF spectrum of the A $^3\Pi_i \leftarrow X\ ^3\Sigma^-$, v=1←0, 2←1, 3←2 transitions. Combination of the results with earlier emission data for the 0→0, 1→1 [8, 10], and 2→2 [26] bands of ND enabled the equilibrium constants to be derived [15]. – Further rotational analyses of the A $^3\Pi_i \rightarrow X\ ^3\Sigma^-$ spectrum are in [27 to 29].

c) High-resolution LIF spectrum of the c $^1\Pi \leftarrow$ a $^1\Delta$, v=0←0 band [16]. B_v and D_v values for NH(a $^1\Delta$, v=2 and 3) were derived from LIF in the 0←2, 1←2, and 1←3 bands [30].

d) High-resolution Fourier transform emission spectrum of the c $^1\Pi \rightarrow$ a $^1\Delta$, 1→0, 0→0, 0→1 bands [17]; IR-spectroscopic data for the a $^1\Delta$ state from [7] have been included in the analysis. The 0→0 [31], 0→1 [32], 0→0, 1, 2 and 1→0, 1, 2 [33] bands were analyzed earlier.

e) High-resolution LIF spectrum of the c $^1\Pi \leftarrow$ a $^1\Delta$, v=0←2, 3 [18] and v=1←0, 1, 2, 3, 4 [19, 46] bands. Data of [17], [7], and [16] have been included in deriving the equilibrium constants.

f) Emission spectrum of the c $^1\Pi \rightarrow$ a $^1\Delta$, v=0→2 and 1→2 bands of ND [20]; combination with data for the 0→0, 1→0, and 0→1 bands of [34] and B_v and D_v (v=0 and 2) values of [21]. B_2 and D_2 values, supposed to supersede those of [20], were derived from the (2+1) REMPI spectrum of ND(a $^1\Delta$) [35]. – Analyses of the 0→0 [31] and of the 0→0, 1, 2, the 1→0, 1, 2, 3, and the 2→0, 1 bands [33] have also been carried out.

g) Analyses of UV emission spectra, d $^1\Sigma^+ \rightarrow$ c $^1\Pi$, v=0→0, 0→1, 1→0, 1→1, 2→1, 2→2 for NH and (additionally 2→3, 3→3) ND, d $^1\Sigma^+ \rightarrow$ b $^1\Sigma^+$, v=0→1, 2; 1→2, 3; 2→3, 4, 5; 3→4, 5, 6; 5→6, 7, 8; 6→8, 9 for NH [21]. Bands of the d→c system with v′, v″≤2 were analyzed earlier [36, 37]. The (2+1) REMPI spectra of NH, ND(a $^1\Delta$) (cf. p. 52) give for the d $^1\Sigma^+$ state the rotational constants (in cm^{-1}): B_0=14.0971(65) and 7.5727(43), $D_0 \cdot 10^4$=16.80(28) and 4.82(10) (also B_1 and D_1) [38], B_0=14.089(4) and 7.5662(7), $D_0 \cdot 10^4$=16.4(2) and 4.71(1) (also B_1, B_2 and D_1, D_2; literature data for the a $^1\Delta$ state used) [39].

h) Analyses of the d $^1\Sigma^+ \rightarrow$ b $^1\Sigma^+$, v=0→1, 0→2, 1→3 bands of NH, the d $^1\Sigma^+ \rightarrow$ b $^1\Sigma^+$, v=0→1, 1→2, 0→2, 1→3, 2→4 bands of ND [23] and the c $^1\Pi \rightarrow$ b $^1\Sigma^+$, v=0→0, 0→1 bands of NH and ND [22]. Rotational constants for the b $^1\Sigma^+$, v=0 and 1 levels of NH and ND were also derived from the b $^1\Sigma^+ \rightarrow$ X $^3\Sigma^-$ emission [40 to 42].

Rydberg States. Vibrationally and rotationally resolved, (2+1)-resonance-enhanced multiphoton ionization (REMPI) spectra and REMPI-PES of NH, ND(a $^1\Delta$) and NH, ND(X $^3\Sigma^-$) radicals in the UV (cf. pp. 83/7) revealed the existence of some singlet and triplet Rydberg states. Using literature data for the spectroscopic constants of the a $^1\Delta$ and X $^3\Sigma^-$ states, the rotational constants B and centrifugal distortion constants D were derived by [35, 43, 44]. In the analysis of [38], B and D values for both the upper and the lower states of the two-photon transitions were derived. B_0 and D_0 values for the ground-vibrational level are as follows (in cm^{-1}):

state	NH B_0	NH $D_0 \cdot 10^4$	ND B_0	ND $D_0 \cdot 10^4$	Ref.
f $^1\Pi$	14.238(2)	16.98(9)	7.730(1)	5.09(2)	[43] a)
	14.2281(65)	16.29(33)	7.7268(45)	4.94(13)	[38]
g $^1\Delta$	14.950(5)	23.2(2)	8.064(3)	6.31(9)	[43] a)
	14.9936(85)	23.20(47)	7.7669(158)	11.39(79)	[38] b)
h $^1\Sigma^+$	15.360(9)	25.0(4)	8.200(2)	5.53(5)	[43] a)
	15.2584(210)	20.77(212)	8.1524(118)	3.49(78)	[38]
i $^1\Pi$	13.566(10)	12.6(5)	7.361(1)	4.03(4)	[35]
j $^1\Delta$	12.932(2)	11.9(7)	7.051(4)	5.96(7)	[35] c)
B $^3\Pi$	13.75(8)	250(30)	7.63(3)	65(7)	[44] d)
C $^3\Sigma^-$	–	–	6.56(3)	40(6)	[44]
D $^3\Pi$	13.86(2)	21(4)	7.63(1)	12(2)	[44]
E $^3\Sigma^-$	–	–	7.40(4)	0	[44]
F $^3\Sigma^-$	9.17(2)	−84(3)	4.682(7)	−13.0(5)	[44]

a) Gives also values for f $^1\Pi$, v=1 of ND, g $^1\Delta$, v=1, 2 of NH, v=1, 2, 3 of ND, and h $^1\Sigma^+$, v=2 of ND. – b) Gives also values for g $^1\Delta$, v=1 of NH and ND. – c) Gives also values for j $^1\Delta$, v=1, 2 of NH and ND and for three states, $^1\Sigma$, $^1\Pi$, and $^1\Delta$, identified as perturbers of the j $^1\Delta$ state. – d) Gives also values for B $^3\Pi$, v=1 of NH and ND.

For the states labeled 3 $^3\Sigma^-$, 4 $^3\Sigma^-$, 3 $^3\Pi$, 2 $^5\Sigma^-$, and 3 $^5\Sigma^-$ and recognized by an ab initio CI calculation to have mainly Rydberg character, the spectroscopic constants B_e, α_e, ω_e, $\omega_e x_e$, and the internuclear distances r_e were derived [45].

References:

[1] Van den Heuvel, F. C.; Meerts, W. L.; Dymanus, A. (Chem. Phys. Lett. **92** [1982] 215/8).
[2] Wayne, F. D.; Radford, H. E. (Mol. Phys. **32** [1976] 1407/22).
[3] Boudjaadar, D.; Brion, J.; Chollet, P.; Guelachvili, G.; Vervloet, M. (J. Mol. Spectrosc. **119** [1986] 352/66).
[4] Bernath, P. F.; Amano, T. (J. Mol. Spectrosc. **95** [1982] 359/64).
[5] Leopold, K. R.; Evenson, K. M.; Brown, J. M. (J. Chem. Phys. **85** [1986] 324/30).
[6] Leopold, K. R.; Zink, L. R.; Evenson, K. M.; Jennings, D. A.; Brown, J. M. (NBS Spec. Publ. [U. S.] 716 [1986] 452/8).
[7] Hall, J. L.; Adams, H.; Kasper, J. V. V.; Curl, R. F.; Tittel, F. K. (J. Opt. Soc. Am. B Opt. Phys. **2** [1985] 781/5).
[8] Bollmark, P.; Kopp, I.; Rydh, B. (J. Mol. Spectrosc. **34** [1970] 487/99).
[9] Dixon, R. N. (Can. J. Phys. **37** [1959] 1171/86).
[10] Kopp, I.; Kronekvist, M.; Aslund, N. (Arkiv Fys. **30** [1965] 9/17).
[11] Sakai, H.; Hansen, P.; Esplin, M.; Johansson, R.; Peltola, M.; Strong, J. (Appl. Opt. **21** [1982] 228/34).

[12] Hansen, P.; Sakai, H.; Esplin, M. (Proc. Soc. Photo.-Opt. Instrum. Eng. **191** [1979] 15/20).
[13] Vasconcellos, E. C. C.; Davidson, S. A.; Brown, J. M.; Leopold, K. R.; Evenson, K. M. (J. Mol. Spectrosc. **122** [1987] 242/5).
[14] Brazier, C. R.; Ram, R. S.; Bernath, P. F. (J. Mol. Spectrosc. **120** [1986] 381/402).
[15] Patel-Misra, D.; Sauder, D. G.; Dagdigian, P. J. (Chem. Phys. Lett. **174** [1990] 113/8).
[16] Ubachs, W.; Meyer, G.; Ter Meulen, J. J.; Dymanus, A. (J. Mol. Spectrosc. **115** [1986] 88/104).
[17] Ram, R. S.; Bernath, P. F. (J. Opt. Soc. Am. B Opt. Phys. **3** [1986] 1170/4).
[18] Hack, W.; Mill, T. (J. Mol. Spectrosc. **144** [1990] 358/65).
[19] Mill, T. (Ber. Max-Planck-Inst. Strömungsforsch. **14** [1990] 1/104, 21/30).
[20] Cheung, W. Y.; Gelernt, B.; Carrington, T. (Chem. Phys. Lett. **66** [1979] 287/90).

[21] Graham, W. R. M.; Lew, H. (Can. J. Phys. **56** [1978] 85/99).
[22] Whittaker, F. L. (J. Phys. B **1** [1968] 977/82).
[23] Whittaker, F. L. (Can. J. Phys. **47** [1969] 1291/3).
[24] Malicet, J.; Brion, J.; Guenebaut, H. (J. Chim. Phys. Phys. Chim. Biol. **67** [1970] 25/30).
[25] Murai, T.; Shimauchi, M. (Sci. Light [Tokyo] **15** [1966] 48/67).
[26] Shimauchi, M. (Sci. Light [Tokyo] **16** [1967] 185/90).
[27] Shimauchi, M. (Sci. Light [Tokyo] **15** [1966] 161/5).
[28] Krishnamurty, G.; Narasimham, N. A. (Proc. Indian Acad. Sci. A **67** [1968] 50/60).
[29] Krishnamurty, G.; Narasimham, N. A. (Proc. 1st Int. Conf. Spectrosc., Bombay 1967, Vol. 1, pp. 142/4).
[30] Nelson, H. H.; McDonald, J. R. (J. Chem. Phys. **93** [1990] 8777/83).

[31] Shimauchi, M. (Sci. Light [Tokyo] **13** [1965] 53/63).
[32] Ramsay, D. A.; Sarre, P. J. (J. Mol. Spectrosc. **93** [1982] 445/6).
[33] Zetzsch, C. (J. Photochem. **9** [1978] 151/3).
[34] Hanson, H.; Kopp, I.; Kronekvist, M.; Aslund, N. (Arkiv Fys. **30** [1965] 1/8).
[35] Clement, S. G.; Ashfold, M. N. R.; Western, C. M.; De Beer, E.; De Lange, C. A.; Westwood, N. P. C. (J. Chem. Phys. **96** [1992] 4963/73).
[36] Whittaker, F. L. (Proc. Phys. Soc. [London] **90** [1967] 535/41).
[37] Narasimham, N. A.; Krishnamurty, G. (Proc. Indian Acad. Sci. A **64** [1966] 97/110).
[38] Johnson, R. D., III; Hudgens, J. W. (J. Chem. Phys. **92** [1990] 6420/5).
[39] Ashfold, M. N. R.; Clement, S. G.; Howe, J. D.; Western, C. M. (J. Chem. Soc. Faraday Trans. **87** [1991] 2515/23).
[40] Cossart, D. (J. Chim. Phys. Phys.-Chim. Biol. **76** [1979] 1045/50).

[41] Zetzsch, C.; Stuhl, F. (Ber. Bunsenges. Phys. Chem. **80** [1976] 1348/54).
[42] Zetzsch, C.; Stuhl, F. (Chem. Phys. Lett. **33** [1975] 375/7).
[43] Clement, S. G.; Ashfold, M. N. R.; Western, C. M. (J. Chem. Soc. Faraday Trans. **88** [1992] 3121/8).
[44] Clement, S. G.; Ashfold, M. N. R.; Western, C. M.; Johnson, R. D., III; Hudgens, J. W. (J. Chem. Phys. **97** [1992] 7064/72).
[45] Goldfield, E. M.; Kirby, K. P. (J. Chem. Phys. **87** [1987] 3986/94).
[46] Hack, W.; Mill, T. (J. Phys. Chem. **97** [1993] 5599/606).

2.1.1.2.7 Potential Energy Functions

Morse potential functions for the six valence states X $^3\Sigma^-$, a $^1\Delta$, b $^1\Sigma^+$, A $^3\Pi_i$, c $^1\Pi$, d $^1\Sigma^+$ of NH were calculated [1] from the spectroscopic constants derived prior to 1973 [2, 3]; crossings of the A $^3\Pi_i$ state at $r_c = 1.36$ Å and of the c $^1\Pi$ state at $r_c \approx r_e = 1.11$ Å with a possibly weakly bonding $^5\Sigma^-$ state, which arises from the ground-state asymptote $N(^4S) + H(^2S)$, have been derived from the measured lifetimes and the observed predissociation of A- and c-state rovibrational levels [4]; crossing of the d $^1\Sigma^+$ state with a repulsive $^1\Pi$ state was found to cause predissociation of d-state levels observed in the d $^1\Sigma^+ \rightarrow$ c $^1\Pi$ emission spectrum [5] (see however p. 82). The potential curves are shown in **Fig. 1**; also indicated are the observed vibrational levels v=0 and 1 for the bound states and the highest observed rotational levels for the c $^1\Pi$ (v=0 and 1) state which lie above the dissociation limit [1].

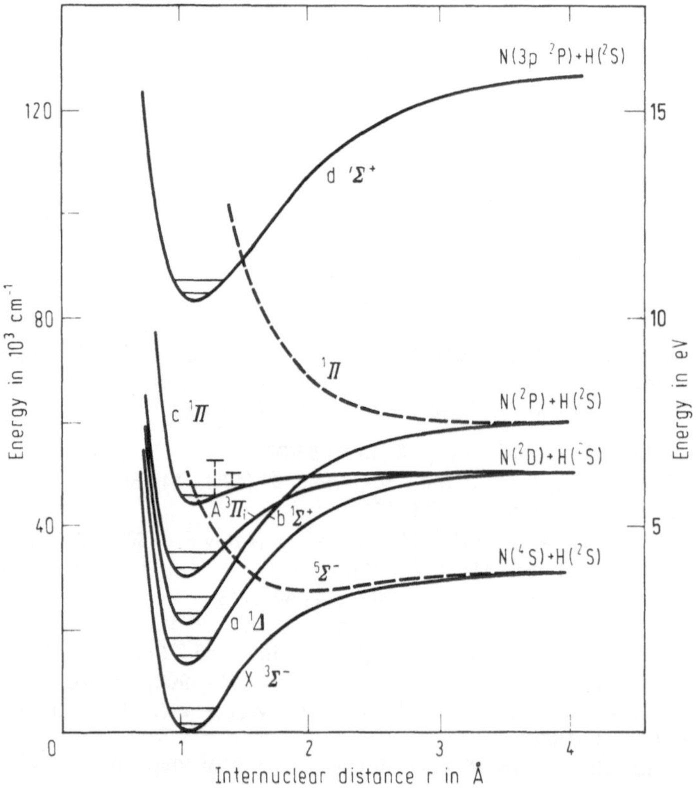

Fig. 1. Potential energy curves for NH. The vibrational levels v=0 and 1 are indicated for all states, the highest observed rotational levels for the c $^1\Pi$ state [1].

The classical turning points, r_{min} and r_{max}, of the vibrational levels have been calculated for the ground state X $^3\Sigma^-$ (v=0 to 6; molecular constants from [6]) assuming a Rydberg-Klein-Rees-Vanderslice potential [7], for the X $^3\Sigma^-$ and A $^3\Pi_i$ states (v' and v''=0 to 8; molecular constants from [8]) assuming Morse and modified Rydberg-Klein-Rees (RKR) potentials [9], and for the b $^1\Sigma^+$ and d $^1\Sigma^+$ states (v' and v''=0 to 3; molecular constants from [10]) assuming modified RKR and simplified potentials [11] and [12], respectively.

The harmonic and anharmonic force constants f_i, defined by the potential energy function $V(r) = 1/2! \cdot f_2(r-r_e)^2 + 1/3! \cdot f_3(r-r_e)^3 + \cdots\cdots$ were calculated from the ground-state spectroscopic constants taken from [13]: $f_2 = 5.41$ aJ/Å^2, $f_3 = -35.0$ aJ/Å^3 [14].

A few empirical potential functions have been tested by comparing calculated and experimental molecular constants, such as α_e, $\omega_e x_e$, and the binding energy D_e [15, 16].

Numerous quantum-chemical ab initio studies include the calculation of ground- and excited-state potential functions; tabulated energy values or graphical representations E(r) are given:

states	figure for r (in a_0)	table for r (in a_0)	method [a]	Ref.
$X^3\Sigma^-$, $a^1\Delta$, $b^1\Sigma^+$, $A^3\Pi_i$, $c^1\Pi$, $^5\Sigma^-$	1.3 to 5.0 and ∞	1.3 to 5.0 and ∞	QDMBPT	[17]
$2^3\Sigma^-$, $2^1\Pi$, $2^3\Pi$, $^3\Delta$, $^1\Sigma^-$, $^3\Sigma^+$	1.3 to 5.0 and ∞	–	QDMBPT	[17]
$X^3\Sigma^-$, $a^1\Delta$, $b^1\Sigma^+$, $A^3\Pi_i$, $c^1\Pi$, $d^1\Sigma^+$, $^5\Sigma^-$	–	1.4 to 2.7	CI	[18]
$X^3\Sigma^-$	b)	–	MC SCF	[19]
$X^3\Sigma^-$, $A^3\Pi_i$	1.0 to 6.0	–	CAS SCF	[20]
$X^3\Sigma^-$, $A^3\Pi_i$, $2^3\Sigma^-$, $2^3\Pi$, $1^5\Sigma^-$	1.0 to 6.0	1.25 to 6.0 and 20	CI	[21]
$3^3\Sigma^-$, $4^3\Sigma^-$, $3^3\Pi$, $2^5\Sigma^-$, $3^5\Sigma^-$	1.0 to 6.0	–	CI	[21]
$X^3\Sigma^-$	1.4 to 5.5	–	MRD CI	[22]
$X^3\Sigma^-$	0 to 5 (UHF)	1.5 to 7.5	MC SCF	[23]
$X^3\Sigma^-$	1.6 to 10	1.6 to 7 and 10, 20	UMP2	[24]
$A^3\Pi_i$, $2^3\Pi$, $b^1\Sigma^+$, $d^1\Sigma^+$	1 to 15	1 to 15	MC SCF	[25 to 27]
$X^3\Sigma^-$, $a^1\Delta$, $b^1\Sigma^+$, $A^3\Pi_i$, $c^1\Pi$	1.5 to 6	–	POL-CI	[28]
$X^3\Sigma^-$	–	1.0 to 3.5 and 10	CI	[29]
$X^3\Sigma^-$, $a^1\Delta$, $b^1\Sigma^+$	1.3 to 3.8	–	CI	[30]
$X^3\Sigma^-$, $a^1\Delta$, $b^1\Sigma^+$, $A^3\Pi_i$, $c^1\Pi$, $d^1\Sigma^+$, $^5\Sigma^-$, $2^3\Sigma^-$, $2^1\Pi$, $2^3\Pi$, $^3\Delta$, $^1\Sigma^-$, $^3\Sigma^+$	1.5 to 6 and ∞	1.5 to 6 and ∞	CI	[31]

[a] QDMBPT = quasi-degenerate many-body perturbation theory, (POL-) CI = (polarization-) configuration interaction, MC SCF = multiconfiguration SCF, CAS SCF = complete active space SCF, MRD CI = multireference double-excitation CI, UHF = unrestricted Hartree-Fock, UMP2 = unrestricted Møller-Plesset perturbation method of second order. — [b] An analytical, combined polynomial-exponential form for E(x) with $x = (r - r_e)/r$ and a figure for $x = -0.98$ to $+0.04$.

References:

[1] Zetzsch, C. (Ber. Bunsenges. Phys. Chem. **82** [1978] 639/43).
[2] Rosen, B. (International Tables of Selected Constants, Vol. 17, Spectroscopic Data Relative to Diatomic Molecules, Pergamon, Oxford 1970, p. 274).
[3] Barrow, R. F. (Diatomic Molecules, Bibliography of Spectroscopic Data I-1973, Pergamon, Oxford 1974 from [1]).
[4] Smith, W. H.; Brzozowski, J.; Erman, P. (J. Chem. Phys. **64** [1976] 4628/33).
[5] Krishnamurty, G.; Narasimham, N. A. (J. Mol. Spectrosc. **29** [1969] 410/4).
[6] Brazier, C. R.; Ram, R. S.; Bernath, P. F. (J. Mol. Spectrosc. **120** [1986] 381/402).
[7] Reddy, R. R.; Viswanath, R. (Astrophys. Space Sci. **155** [1989] 39/43).
[8] Pannetier, G.; Guenebaut, H. (Acta Chim. Acad. Sci. Hung. **18** [1959] 347/64).
[9] Singh, M.; Chaturvedi, J. P. (J. Quant. Spectrosc. Radiat. Transfer **37** [1987] 81/4).
[10] Whittaker, F. L. (Can. J. Phys. **47** [1969] 1291/3).

[11] Rao, T. V. R.; Lakshman, S. V. J. (Indian J. Pure Appl. Phys. **11** [1973] 539/40).
[12] Dongre, M. B. (Indian J. Phys. **50** [1976] 409/15).
[13] Dixon, R. N. (Can. J. Phys. **37** [1959] 1171/86).
[14] Mills, I. M. (Theor. Chem. [London] **1** [1974] 110/59).
[15] Ramani, K.; Ghodgaonkar, A. M. (Indian J. Chem. A **21** [1982] 803/4).
[16] Pandey, R. P.; Pandey, J. D. (Indian J. Chem. A **20** [1981] 592/3).

[17] Park, J. K.; Sun, H. (Bull. Korean Chem. Soc. **11** [1990] 34/41).
[18] Yarkony, D. R. (J. Chem. Phys. **91** [1989] 4745/57).
[19] Biskupic, S.; Klein, R. (J. Mol. Struct. **170** [1988] 27/31 [THEOCHEM **47**]).
[20] Gustafsson, O.; Kindvall, G.; Larsson, M.; Olsson, B. J.; Sigray, P. (Chem. Phys. Lett. **138** [1987] 185/94).

[21] Goldfield, E. M.; Kirby, K. P. (J. Chem. Phys. **87** [1987] 3986/94).
[22] Wright, J. S.; Williams, R. J. (J. Chem. Phys. **79** [1983] 2893/902).
[23] Neisius, D.; Verhaegen, G. (Chem. Phys. Lett. **89** [1982] 228/33).
[24] Klimo, V.; Tino, J. (Mol. Phys. **41** [1980] 483/90).
[25] Banerjee, A.; Grein, F. (Chem. Phys. **35** [1978] 119/27).
[26] Banerjee, A.; Grein, F. (J. Chem. Phys. **66** [1977] 1054/62).
[27] Banerjee, A. K. (Diss. Univ. New Brunswick, Can., 1977 from Diss. Abstr. Int. B **38** [1977] 1677).
[28] Hay, P. J.; Dunning, T. H., Jr. (J. Chem. Phys. **64** [1976] 5077/87).
[29] Das, G.; Wahl, A. C.; Stevens, W. J. (J. Chem. Phys. **61** [1974] 433/4).
[30] O'Neil, S. V.; Schaefer, H. F., III (J. Chem. Phys. **55** [1971] 394/401).

[31] Kouba, J.; Öhrn, Y. (J. Chem. Phys. **52** [1970] 5387/94).

2.1.1.2.8 Dissociation Energy. Heat of Formation. Thermodynamic Functions

The **heat of formation**, $\Delta_fH_0^\circ$, and the **dissociation energy,** D_0°, of NH in its ground state X $^3\Sigma^-$ have long been subjects of considerable controversy. These quantities, which are related to each other via the thermochemical cycle $\Delta_fH_0^\circ(N)+\Delta_fH_0^\circ(H)=\Delta_fH_0^\circ(NH)+D_0^\circ(NH)$ and the well-established heats of formation of N and H (e.g. $\Delta_fH_0^\circ(N)+\Delta_fH_0^\circ(H)=470.82+216.035=686.855$ kJ/mol = 7.119 eV [1, pp. 1211 and 1530]), have been measured by various spectroscopic and thermochemical techniques. Published values are between 396 and 320 kJ/mol for $\Delta_fH_0^\circ$ and between 3.0 and 3.8 eV (291 and 367 kJ/mol) for D_0°. A few more or less critical reviews, however, which tried to resolve some of the controversy, confined $\Delta_fH_0^\circ$ to the range 351 to 369 kJ/mol and D_0° to the range 3.29 to 3.48 eV (317 to 336 kJ/mol); the most recent and convincing recommendation is $\Delta_fH_0^\circ=357\pm1$ kJ/mol, $D_0^\circ=3.42\pm0.01$ eV (see the following table; details are given in the subsequent remarks):

$\Delta_fH_0^\circ$ in kJ/mol	D_0° in eV (kJ/mol)	year	authors	remark
339 ± 10	~3.6 (348)	1971	JANAF Tables [2]	a)
377 ± 21 or 17	3.21 (310)	1974, 1982, 1985	JANAF Tables [1, 3, 4]	b)
352 ± 10	3.47 (335)	1979	Piper [5]	c)
≥352	≤3.47 (335)	1979	Huber and Herzberg [6]	d)
351.0	3.48 (336)	1982	NBS Tables [7]	e)
352 to 369	3.29 to 3.47 (317 to 335)	1985	Hofzumahaus and Stuhl [8]	f)
357 ± 1	3.42 (330)	1989	Anderson [9]	g)

a) The earliest studies, i.e., a Birge-Sponer extrapolation of ground-state vibrational levels [10] combined with thermochemical data [11, 12] and the dissociative ionization of HN_3 [13] and NH_3 [14] by electron impact, placed D_0° to values around 3.6 eV. Two electron impact results and two semiempirical calculations yielded the same value which led to

$\Delta_fH_0^o = \Delta_fH_{298.15}^o = 339 \pm 10$ kJ/mol being recommended in the 1971 JANAF Tables [2]. More recently, a spectroscopic value of $D_0^o \approx 3.60$ eV was derived from the Fourier transform IR emission spectrum of the $v+1 \rightarrow v$ bands of NH($X\ ^3\Sigma^-$) with v=0 to 4; the band centers were fitted to a polynomial expansion of G(v) up to fourth order of (v+1/2), and D_0^o was calculated by assuming that it corresponds to the v value (≈ 17) for which dG/dv=0 [15].

b) A lower dissociation energy, $D_0^o = 3.21$ eV, is consistent with the results of intensity measurements on A $^3\Pi_i \leftarrow X\ ^3\Sigma^-$ absorption lines in shock-wave-excited NH_3-Kr and N_2-H_2-Kr mixtures [16], A-state lifetime measurements [17, 18], A $^3\Pi_i$ (v'=1, 0)$\rightarrow$X $^3\Sigma^-$ and c $^1\Pi$ (v'=0)$\rightarrow$a $^1\Delta$ emission spectra excited (up to rotationally predissociating levels) by the reactions of Ar, Kr, Xe metastables with HN_3, NH_3, HNCO [19], and measurements of the NH, NH_3, and OH concentrations in NH_3-O_2-N_2 flames [20]. This later revised value (see next paragraph) was subsequently adopted by the 1974, 1982, and 1985 JANAF Tables which recommend $\Delta_fH_0^o = \Delta_fH_{298.15}^o = 377.2 \pm 21$ kJ/mol [3], 377 ± 17 kJ/mol [4] for a standard-state pressure of 1 atm and $\Delta_fH_0^o = 376.51 \pm 16.7$ kJ/mol, $\Delta_fH_{298.15}^o = 376.56 \pm 16.7$ kJ/mol for a standard-state pressure of 0.1 MPa [1]. Strange to say the authors used the lower $\Delta_fH_0^o$ value for NH from their 1971 tables to derive $\Delta_fH_0^o = \Delta_fH_{298.15}^o = 375.3 \pm 21$ kJ/mol for the ND molecule [1, 4]. – An upper limit of 377 kJ/mol for $\Delta_fH_0^o$ was found to agree with the threshold energy for NH formation by VUV photolysis of HNCS [21].

c) In a convincing critical review of all results prior to 1979, Piper [5] revealed and corrected errors and faulty assumptions in the analyses of [16, 19, 20] and reanalyzed a number of experiments on dissociative excitation and ionization of NH-containing molecules, such as formation of NH(A $^3\Pi_i$) or NH(c $^1\Pi$) by electron impact on HN_3, NH_3, HNCO, $(CH_2)_2NH$, CH_3NH_2, N_2H_4 and observing the A $^3\Pi_i \rightarrow$ X $^3\Sigma^-$ or c $^1\Pi \rightarrow$ a $^1\Delta$ emission [22], photoionization of NH_3 resulting in NH(c $^1\Pi \rightarrow$ a $^1\Delta$) fluorescence [23, 24], formation of NH(X $^3\Sigma^-$)+N_2^+ by electron impact on HN_3 [13], formation of $NH^+ + H_2$ or $NH^+ + 2\,H$ by electron impact on NH_3 [14], and formation of NH(X $^3\Sigma^-$)+N_2(B $^3\Pi_g$, v'=11) by collisions of Kr($^3P_2$) metastables with HN_3 [19]. The most reliable of these experiments gave an upper limit for $\Delta_fH_0^o$ of 3.63 ± 0.10 eV. By including the results of spectroscopic analyses of the NH(c $^1\Pi$) predissociation [25, 26], a value of $\Delta_fH_0^o = 3.65 \pm 0.10$ eV $= 352 \pm 10$ kJ/mol (thus $D_0^o = 3.47$ eV) was recommended [5].

d) Huber and Herzberg [6] adopted only the spectroscopic results, $D_0^o \leq 3.47$ eV for NH and $D_0^o \leq 3.54$ eV for ND (thus $\Delta_fH_0^o \geq 352$ and ≥ 345 kJ/mol), derived from the limiting curves for predissociation by rotation of the NH(c $^1\Pi$, v=0) and ND(c $^1\Pi$, v=0 to 3) levels [25].

e) The values recommended in the NBS Tables [7], $\Delta_fH_0^o = 351.0$ kJ/mol, $\Delta_fH_{298.15}^o =$ 351.5 kJ/mol, and $\Delta_fG_{298.15}^o = 345.6$ kJ/mol at a standard-state pressure of 0.1 MPa, are given without any reference.

f) A rather uncritical review of Hofzumahaus and Stuhl [8] gives a list of $\Delta_fH_0^o$ and D_0^o data published between 1958 and 1984; the list includes, besides the more recent results from VUV photolysis of NH_3 [27] and from NH_3-O_2 flame equilibria [28, 29], those ([13, 14, 16, 19, 20, 24 to 26]) already considered by Piper [5], but without taking notice of his reanalyses and corrections (especially those for the low D_0^o and high $\Delta_fH_0^o$ values ([16, 19, 20]) quoted in the JANAF Tables up to the 3rd edition); based on their own experimental result, $\Delta_fH_0^o \leq 367 \pm 2$ kJ/mol derived from the rotational population in NH(A $^3\Pi_i$, v'=0) generated by ArF laser photolysis of NH_3, and on the previous recommendations [4 to 6], Hofzumahaus and Stuhl [8] suggest $\Delta_fH_0^o$ to be in the range 352 to 369 kJ/mol and D_0^o in the range 317 to 335 kJ/mol = 3.29 to 3.47 eV.

g) In the most recent (critical) review, Anderson [9] summarizes and discusses spectroscopic and thermochemical results for $\Delta_fH_0^o$ published since 1978 (the earlier results are discussed elsewhere [30]) and corrects results from flame experiments. Accordingly, the proper value for $\Delta_fH_0^\circ$ is bracketed between 347 and 369 kJ/mol: A firm lower limit, $\Delta_fH_0^\circ \geq 352 \pm 5$ kJ/mol, is established by the predissociation limit of the c $^1\Pi$ state ($D_0^\circ \leq 3.47$ eV), derived from d $^1\Sigma^+ \rightarrow$ c $^1\Pi$ emission spectra [25]; a similar value, but incorrectly concluded to be an upper limit, $\Delta_fH_0^\circ < 354$ kJ/mol ($D_0^\circ > 3.45$ eV), was derived from earlier emission data by [26]. Upper limits of $\Delta_fH_0^\circ$ are obtained from the energetics of dissociation reactions which produce NH, whereby it is necessary to know exactly the distribution of internal energies in the dissociation products. Such a sensible thermodynamic analysis is provided by a study of the ArF laser photolysis (193 nm) of NH_3, i.e., of the reaction $NH_3 + 2h\nu \rightarrow NH(A\ ^3\Pi_i) + 2\ H$; the A-state rotational distribution gave $\Delta_fH_0^\circ \leq 367 \pm 2$ kJ/mol [8]. Other VUV photolysis studies on NH_3 provided less accurate upper limits: $\Delta_fH_0^\circ \leq 377$ to 382 kJ/mol was derived from the maximum rotational line (predissociation) observed in the c $^1\Pi \rightarrow$ a $^1\Delta$ emission spectrum [31], whereas the threshold energies for NH(c $^1\Pi$) and NH(b $^1\Sigma^+$) formation yielded $\Delta_fH_0^\circ \leq 396 \pm 8$ kJ/mol [32], $\leq 368 \pm 6$ kJ/mol [33], and $\leq 323 \pm 7$ kJ/mol [27] (the latter value is considered much too low, presumably due to experimental insufficiencies). A reanalysis of the experiments on NH_3-O_2 flame equilibria by [20], [34], and [28, 29] gave $\Delta_fH_0^\circ = 362 \pm 16$, 361 ± 6, and 370 ± 5 kJ/mol, respectively, which were found to be inconclusive. Two values, obtained by the quite reliable spectroscopic and thermochemical studies referred to above, fall into the range 347 to 369 kJ/mol. They were derived by combining the heat of formation of the NH^+ ion and the ionization potential of NH, $\Delta_fH_0^\circ(NH^+) = \Delta_fH_0^\circ(NH) + E_i(NH)$, where $\Delta_fH_0^\circ(NH^+)$ was obtained in two independent measurements; they show excellent agreement in spite of their small error limits and were considered the best data currently available: $\Delta_fH_0^\circ = 356 \pm 2$ kJ/mol ($D_0^\circ = 3.43 \pm 0.02$ eV) was obtained from the threshold energy of NH^+ formation upon photoionization of NH_2 [35], $\Delta_fH_0^\circ = 359 \pm 3$ kJ/mol ($D_0^\circ = 3.40 \pm 0.03$ eV) from the reaction enthalpy of $N^+ + H_2 \rightarrow NH^+ + H$ [36]; in both cases, the PES value $E_i(NH) = 13.49$ eV [37] has been used. The weighted average, $\Delta_fH_0^\circ = 357 \pm 1$ kJ/mol ($D_0^\circ = 3.42$ eV), was therefore recommended by Anderson [9]. — The reaction kinetics of $N^+ + H_2 \rightarrow NH^+ + H$, studied by another technique, confirmed the result of [36] yielding $D_0^\circ = 3.42 \pm 0.01$ eV [38].

Fitting a Rydberg-Klein-Rees-Vanderslice potential function for the ground state of NH (cf. p. 54) to the empirical Lippincott potential function (see [39]) resulted in $D_0 = 3.45 \pm 0.14$ eV [40].

A great number of quantum-chemical ab initio calculations include the evaluation of D_e or D_0 for the ground state, the excited valence and Rydberg states, and in a few cases the evaluation of $\Delta_fH_0^\circ$. Most of these studies have been quoted in the preceding section on potential energy functions (see table on p. 55). D_e values for the ground state X $^3\Sigma^-$, obtained from highly correlated wave functions, are in the range 3.2 to 3.7 eV, e.g., 3.54 eV (QDMBPT) [41], 3.35 eV (MP4) [42], 3.35 and 3.38 eV (combined MP4 and CI) [43, 44], 3.35 eV (CI) [45], 3.344 eV (CAS SCF) [46], 3.43 eV (MRD CI) [47, 48], 3.40 eV (CEPA) [49], and are in satisfactory agreement with the most reliable experimental results (3.32 to 3.52 eV, see above). The recently recommended values for the heat of formation are in fair agreement with the theoretical value $\Delta_fH_0^\circ = 364$ kJ/mol obtained by MP4 calculations [42, 50]. More references may be found in a recent compilation of theoretical and experimental D_e values [41] and in the bibliographies on quantum-chemical calculations (see p. 31).

The **heat capacity** C_p°, **thermodynamic functions** S°, $-(G^\circ - H_{298}^\circ)/T$, $H^\circ - H_{298}^\circ$, and the equilibrium constant K_f for the formation of NH as an ideal gas from the elements have been calculated for a standard-state pressure of 1 atm and tabulated for 298.15 K and between 0 and 6000 K at 100 K intervals in the 2nd edition of the JANAF Tables [2 to 4]; results of [2] are based on $D_0^\circ(NH) = 3.6$ eV, and the ground-state configuration $^3\Sigma^-$

and spectroscopic constants were used to establish the partition function; the subsequent JANAF tabulations are based on $D^o_0(NH) = 3.21$ eV (cf. remarks b), c) on p. 57), and electronically excited states are included in the partition function, i.e., a $^1\Delta$, b $^1\Sigma^+$, A $^3\Pi_i$ in [3], a $^1\Delta$, b $^1\Sigma^+$, A $^3\Pi_i$, c $^1\Pi$, d $^1\Sigma^+$ in [4]. The results of [4] were converted to a standard-state pressure of 0.1 MPa and tabulated in the 3rd edition of the JANAF Tables [1]. Analogous tabulations for the ND molecule [1, 4] are based on a $\Delta_fH^o_{298.15}$ value and molecular constants adjusted from those of the NH molecule [2]. Selected values for NH from [1] are as follows:

T in K	C^o_p	S^o	$-(G^o - H^o_{298})/T$	$H^o - H^o_{298}$ in kJ/mol	log K_f
	in $J \cdot mol^{-1} \cdot K^{-1}$				
0	0.0	0.0	∞	−8.617	∞
100	29.126	149.428	207.157	−5.773	−195.613
200	29.133	169.618	183.918	−2.860	−97.295
298.15	29.147	181.253	181.253	0.0	−64.921
400	29.175	189.821	182.397	2.970	−48.123
600	29.461	201.690	186.980	8.826	−31.732
800	30.219	210.259	191.775	14.787	−23.538
1000	31.258	217.111	196.179	20.933	−18.622
1500	33.730	230.275	205.470	37.207	−12.068
2000	35.473	240.233	212.965	54.535	−8.791
3000	37.821	255.088	224.671	91.251	−5.511
4000	39.725	266.234	233.724	130.042	−3.867
5000	41.480	275.289	241.158	170.652	−2.878
6000	43.185	283.003	247.505	212.986	−2.214

The NBS Tables [7] recommend $C^o_{p,298.15} = 29.192$ $J \cdot mol^{-1} \cdot K^{-1}$, $S^o_{298.15} = 181.23$ $J \cdot mol^{-1} \cdot K^{-1}$, and $H^o_{298.15} - H^o_0 = 8.602$ kJ/mol at a standard-state pressure of 0.1 MPa (without giving the literature sources).

Values for C^o_p, S^o, $-(G^o - H^o_0)$, $H^o - H^o_0$, and log K (based on $\Delta_fH^o_0 = 338.852$ kJ/mol) are tabulated in [51] between 100 and 6000 K at 100 K intervals, between 6200 and 10000 K at 200 K intervals, and between 10500 and 20000 K at 500 K intervals.

Using the 1982 JANAF [4] thermodynamic and Huber-Herzberg [6] ground-state spectroscopic (including D^o_0) data, the dissociation function of NH between 1000 and 6000 K has been derived [52]. With the 1971 JANAF data [2], the coefficients of a power series approximation for the Gibbs free-energy function, $-(G^o - H^o_{298})/T = \varphi_0 + \varphi \ln x + \varphi_{-2} x^{-2} + \varphi_{-1} x^{-1} + \varphi_1 x + \varphi_2 x^2 + \varphi_3 x^3$ where $x = 10^{-4}$ T (T ≤ 6000 K), were calculated [53].

Earlier tabulations of thermodynamic functions are in [54 to 56].

Partition functions [57 to 60] and equilibrium constants [58] for temperatures up to 10000 K were derived mostly for their astrophysical interest.

References:

[1] Chase, M. W., Jr.; Davies, C. A.; Downey, J. R., Jr.; Frurip, D. J.; McDonald, R. A.; Syverud, A. N. (JANAF Thermochemical Tables, 3rd Ed., J. Phys. Chem. Ref. Data **14** Suppl. 1 [1985] 1/1856, 999, 1236).

[2] Stull, D. R.; Prophet, H. (NSRDS-NBS-37 [1971]).

[3] Chase, M. W.; Curnutt, J. L.; Hu, A. T.; Prophet, H.; Syverud, A. N.; Walker, L. C. (J. Phys. Chem. Ref. Data **3** [1974] 311/480, 438).
[4] Chase, M. W., Jr.; Curnutt, J. L.; Downey, J. R., Jr.; McDonald, R. A.; Syverud, A. N.; Valenzuela, E. A. (J. Phys. Chem. Ref. Data **11** [1982] 695/940, 784, 844).
[5] Piper, L. G. (J. Chem. Phys. **70** [1979] 3417/9).
[6] Huber, K. P.; Herzberg, G. (Molecular Spectra and Molecular Structure, Vol. 4, Constants of Diatomic Molecules, Van Nostrand Reinhold, New York 1979, pp. 456/61).
[7] Wagman, D. D.; Evans, W. H.; Parker, V. B.; et al. (J. Phys. Chem. Ref. Data **11** Suppl. No. 2 [1982] 2-65).
[8] Hofzumahaus, A.; Stuhl, F. (J. Chem. Phys. **82** [1985] 5519/26).
[9] Anderson, W. R. (J. Phys. Chem. **93** [1989] 530/6).
[10] Pannetier, G.; Gaydon, A. G. (J. Chim. Phys. **48** [1951] 221/4).

[11] Altshuller, A. P. (J. Chem. Phys. **22** [1954] 1947/8).
[12] Clyne, M. A. A.; Thrush, B. A. (Proc. Chem. Soc. [London] **1962** 227).
[13] Franklin, J. L.; Dibeler, V. H.; Reese, R. M.; Krauss, M. (J. Am. Chem. Soc. **80** [1958] 298/302).
[14] Reed, R. I.; Snedden, W. (J. Chem. Soc. **1959** 4132/3).
[15] Boudjaadar, D.; Brion, J.; Chollet, P.; Guelachvili, G.; Vervloet, M. (J. Mol. Spectrosc. **119** [1986] 352/66).
[16] Seal, K. E.; Gaydon, A. G. (Proc. Phys. Soc. [London] **89** [1966] 459/66).
[17] Bennett, R. G.; Dalby, F. W. (J. Chem. Phys. **32** [1960] 1716/9).
[18] Fink, E.; Welge, K. H. (Z. Naturforsch. **19a** [1964] 1193/201).
[19] Stedman, D. H. (J. Chem. Phys. **52** [1970] 3966/70).
[20] Kaskan, W. E.; Nadler, M. P. (J. Chem. Phys. **56** [1972] 2220/4).

[21] Lenzi, M.; Mele, A.; Paci, M. (Gazz. Chim. Ital. **103** [1973] 977/87).
[22] Fukui, K.; Fujita, I.; Kuwata, K. (J. Phys. Chem. **81** [1977] 1252/7).
[23] Okabe, H.; Lenzi, M. (J. Chem. Phys. **47** [1967] 5241/6).
[24] Okabe, H. (Photochemistry of Small Molecules, Wiley-Interscience, New York 1978, p. 376).
[25] Graham, W. R. M.; Lew, H. (Can. J. Phys. **56** [1978] 85/99).
[26] Zetzsch, C. (Ber. Bunsenges. Phys. Chem. **82** [1978] 639/43).
[27] Quinton, A. M.; Simons, J. P. (J. Chem. Soc. Faraday Trans. II **78** [1982] 1261/9).
[28] Dean, A. M.; Chou, M.-S.; Stern, D. (Int. J. Chem. Kinet. **16** [1984] 633/53).
[29] Dean, A. M.; Chou, M.-S.; Stern, D. (ACS Symp. Ser. **249** [1984] 71/86).
[30] Anderson, W. R. (BRL-TR-2921 [1988] from [9]).

[31] Washida, N.; Inoue, G.; Suzuki, M.; Kajimoto, O. (Chem. Phys. Lett. **114** [1985] 274/8).
[32] Nguyen Xuan, C.; Di Stefano, G.; Lenzi, M.; Margani, A. (J. Chem. Phys. **74** [1981] 6219/23).
[33] Suto, M.; Lee, L. C. (J. Chem. Phys. **78** [1983] 4515/22).
[34] Fisher, C. J. (Combust. Flame **30** [1977] 143/9).
[35] Gibson, S. T.; Greene, J. P.; Berkowitz, J. (J. Chem. Phys. **83** [1985] 4319/28).
[36] Ervin, K. M.; Armentrout, P. B. (J. Chem. Phys. **86** [1987] 2659/73).
[37] Dunlavey, S. J.; Dyke, J. M.; Jonathan, N.; Morris, A. (Mol. Phys. **39** [1980] 1121/35).
[38] Marquette, J. B.; Rebrion, C.; Rowe, B. R. (J. Chem. Phys. **89** [1988] 2041/7).
[39] Steele, D.; Lippincott, E. R.; Vanderslice, J. T. (Rev. Mod. Phys. **34** [1962] 239/51).
[40] Reddy, R. R.; Viswanath, R. (Astrophys. Space Sci. **155** [1989] 39/43).

[41] Park, J. K.; Sun, H. (Bull. Korean Chem. Soc. **11** [1990] 34/41).
[42] Pople, J. A.; Luke, B. T.; Frisch, M. J.; Binkley, J. S. (J. Chem. Phys. **89** [1985] 2198/203).

[43] Pople, J. A.; Head-Gordon, M.; Fox, D. J.; Raghavachari, K.; Curtiss, L. A. (J. Chem. Phys. **90** [1989] 5622/9).
[44] Curtiss, L. A.; Raghavachari, K.; Trucks, G. W.; Pople, J. A. (J. Chem. Phys. **94** [1991] 7221/30).
[45] Goldfield, E. M.; Kirby, K. P. (J. Chem. Phys. **87** [1987] 3986/94).
[46] Bauschlicher, C. W., Jr.; Langhoff, S. R. (Chem. Phys. Lett. **135** [1987] 67/72).
[47] Wright, J. S.; Williams, R. J. (J. Chem. Phys. **78** [1983] 5264/6).
[48] Wright, J. S.; Williams, R. J. (J. Chem. Phys. **79** [1983] 2893/902).
[49] Meyer, W.; Rosmus, P. (J. Chem. Phys. **63** [1975] 2356/75).
[50] Melius, C. F.; Ho, P. (J. Phys. Chem. **95** [1991] 1410/9).

[51] Glushko, V. P.; Gurvich, L. V.; Bergman, G. A.; Veits, I. V.; Medvedev, V. A.; Khachkuruzov, G. A.; Yungman, V. S. (Thermodynamic Properties of Individual Substances, Vol. 1, Book 2, Nauka, Moscow 1978, pp. 223/4).
[52] Sharp, C. M. (NATO ASI Ser. C **157** [1985] 661/72).
[53] Rozhdestvenskii, I. B.; Gutov, V. N.; Zhigul'skaya, N. A. (Sb. Tr. Energ. Inst. im. G. M. Krzhizhanovskogo No. 7 [1973] 88/121).
[54] Bouvier, A. (Compt. Rend. **258** [1964] 5210/2).
[55] Yungman, V. S.; Gurvich, L. V.; Rtishcheva, N. P. (Tr. Gos. Inst. Prikl. Khim. No. **49** [1962] 20/37).
[56] Mader, C. L. (AECU-4508 [1959] 1/206; N.S.A. **14** [1960] No. 4592).
[57] Rossi, S. C. F.; Maciel, W. J.; Benevides-Soares, P. (Astron. Astrophys. **148** [1985] 93/6).
[58] Sauval, A. J.; Tatum, J. B. (Astrophys. J. Suppl. Ser. **56** [1984] 193/209).
[59] Tatum, J. B. (Publ. Dom. Astrophys. Observ. Victoria B. C. **13** [1966] 17 pp.).
[60] Shipley, K. L. (J. Appl. Phys. **40** [1969] 3037/43).

2.1.1.3 Spectra

2.1.1.3.1 Electron Paramagnetic Resonance (EPR) Spectrum. Stern-Gerlach (SG) Deflection Spectrum

Most recently, the EPR spectrum of gaseous, metastable NH(a $^1\Delta$, J=2) has been recorded in an $F+NH_3$ flame at 9.0935 GHz and magnetic field strengths of 0.9 to 1.0 T; the 24 lines (two outer quartets and two inner octets) expected for a Hamiltonian that includes rotation, hyperfine, and Zeeman terms were well resolved [1].

The EPR spectrum, recorded upon HN_3 decomposition by rf discharges and exhibiting a doublet at $g\approx2.008$ with the hyperfine splitting A=2.2 mT, was assigned to transitions between the Zeeman levels of NH(X $^3\Sigma^-$, J=0) [2]. Similar spectra, $g\approx2.010$, A=2.38 mT, recorded upon photolytic decomposition of HN_3 in noble-gas matrices at 4.2 K, were tentatively attributed to matrix-isolated NH(X $^3\Sigma^-$) species [3, 4]; upon DN_3 photolysis, the expected triplet EPR spectrum of ND(X $^3\Sigma^-$) was observed [3].

The SG magnetic deflection spectra of NH(X $^3\Sigma^-$) at a rotational temperature of 3 K (spin-rotational levels N=0, J=1 and N=1, J=2) and in magnetic fields up to 2 T were calculated (using LMR spectroscopic constants from [5]; see next paragraph); experiments were carried out only on the $^3\Sigma^-$ molecule O_2 [6].

References:

[1] Bradburn, G. R.; Lilenfeld, H. V. (J. Phys. Chem. **95** [1991] 555/8).
[2] Ferraro, W. C. (Diss. Univ. of British Columbia 1964 from [3]).

[3] Fischer, P. H. H.; Charles, S. W.; McDowell, C. A. (J. Chem. Phys. **46** [1967] 2162/6).
[4] Coope, J. A. R.; Farmer, J. B.; Gardner, C. L.; McDowell, C. A. (J. Chem. Phys. **42** [1965] 2628/9).
[5] Wayne, F. D.; Radford, H. E. (Mol. Phys. **32** [1976] 1407/22).
[6] Herrick, D. R.; Robin, M. B.; Gedanken, A. (J. Mol. Spectrosc. **133** [1989] 61/81).

2.1.1.3.2 Far-Infrared Spectrum

Absorption Spectrum at 0.3 mm. The rotational hyperfine (hf) spectrum of NH(X $^3\Sigma^-$, v=0) (obtained from a glow discharge in NH_3) was recorded in the region of the N=1←0, J=2←1 and N=1←0, J=1←1 transitions around 0.3 mm (≈1 THz) with a tunable laser-sideband spectrometer. The hyperfine-free rotational frequencies were ν_0(J=2←1) = 974474.85(80) MHz and ν_0(J=1←1) = 999974.03(80) MHz. Five lines with $\Delta\nu$ (= $\nu_{obs} - \nu_0$) = +3.47(80) to −38.28(60) MHz were assigned to the partially overlapping hf components of the J=2←1 transition (ΔF=1 and 0, F″=5/2, 3/2, 1/2), nine lines with $\Delta\nu$ = +79.94(120) to −66.11(60) MHz were assigned to the partially overlapping hf components of the J=1←1 transition (ΔF=1, 0, and −1, F″=5/2, 3/2, 1/2; $J+I_N=F_1$, $F_1+I_H=F$ with $I_N=1$, $I_H=1/2$). The hfs and fine-structure constants for NH(X $^3\Sigma^-$, v=0) were derived in good agreement with data from far-IR LMR (see below) and laser-induced fluorescence (LIF) in the UV (cf. p. 74) [1].

Laser Magnetic Resonance (LMR) at 0.32 to 0.10 mm. NH and ND radicals for LMR studies were produced by reacting F atoms (from MW discharges in CF_4 or F_2) with NH_3 or ND_3. A great number of rotational transitions between the magnetic sublevels M_J of the rotational levels J, Zeeman-tuned by magnetic fields of 0.1 to 2 T into resonance with a number of fixed far-IR laser lines, were detected for the lowest vibronic states of the radicals. Resonances were found for the N=1←0, 2←1, and 3←2 transitions in the ground state X $^3\Sigma^-$, v=0, 1 (J=N, N±1; Hund's case (b)) and for the J=3←2 and 5←4 transitions in the a $^1\Delta$, v=0, 1 state (Hund's case (a)). Hyperfine interaction occurred between the nuclear spins of ^{14}N (I=1), ^{15}N (I=1/2), ^{1}H (I=1/2), ^{2}H (I=1) and the electron spin (S=1 for X $^3\Sigma^-$) or the electron orbital angular momentum (Λ=2 for a $^1\Delta$). The splitting of most of the $M'_J \leftarrow M''_J$ resonances into doublets of triplets for ^{14}NH, into triplets of triplets for ^{14}ND, and into doublets of doublets in the case of ^{15}NH was resolved. The following spectra are available (λ_L = laser line, H = magnetic field strength, σ, π = number of hyperfine lines for laser light polarized perpendicular ($\Delta M_J = \pm1$) or parallel ($\Delta M_J = 0$) to the magnetic field):

species	N′←N″	J′←J″	λ_L in μm	H in T	σ	π	Ref.
^{14}NH(X $^3\Sigma^-$, v=0)	1←0	1←1, 2←1	302.3	0.39 to 1.11	3	1	[2, 3]*)
	1←0	0←1, 2←1	314.8	0.22 to 1.09	3	1	
	1←0	1←1, 2←1	301.3	0.20 to 0.89	3	1	[2]
	1←0	2←1	311.1	0.34 to 0.59	2	1	
^{14}NH(X $^3\Sigma^-$, v=1)	1←0	1←1, 2←1	314.8	0.45 to 1.39	3	1	[2]
	1←0	1←1	311.1	0.87 to 1.71	2	1	
	2←1	1←1	163.0	0.26	−	1	
^{14}ND(X $^3\Sigma^-$, v=0)	2←1	1←1	302.3	0.69 to 0.83	2	1	[2]
	2←1	1←1, 3←1	301.3	0.61 to 1.31	3	1	
	3←2	4←2, 4←3, 2←3	192.9	0.26 to 1.36	4	3	
^{14}ND(X $^3\Sigma^-$, v=1)	2←1	1←1	302.3	0.16	−	1	[2]
	3←2	3←3	192.9	0.25	−	1	

Table (continued)

species	N′←N″	J′←J″	λ_L in μm	H in T	σ	π	Ref.
$^{15}NH(X\ ^3\Sigma^-, v=0)$	1←0	1←1	302.3	0.16 to 0.34	2	1	[2]
$^{14}NH(a\ ^1\Delta, v=0)$	–	3←2	102.2	1.02 to 1.55	12	6	[4, 5]
$^{14}ND(a\ ^1\Delta, v=0)$	–	3←2	189.8	0.54 to 1.70	36	18	[4, 5]
	–	3←2	187.8	0.65 to 1.98	30	18	
	–	5←4	113.7	0.77 to 1.57	24	18	
$^{14}ND(a\ ^1\Delta, v=1)$	–	3←2	105.5	0.023 to 0.125	29	16	[6]

*) First detection of the submillimeter rotational spectrum of NH [3]; results corrected in [2].

Stimulated Emission at 26.4 to 20.0 μm. Purely rotational laser oscillations were observed for NH in the X $^3\Sigma^-$, a $^1\Delta$, b $^1\Sigma^+$, and A $^3\Pi_i$ states upon flash photolysis of HN_3 in a large excess of noble gases (Ne, Ar, Kr; ratio 1:100). The wavenumbers ν and assignments are as follows [7]:

state	transition	ν in cm^{-1}	state	transition	ν in cm^{-1}
X $^3\Sigma^-$, v=0	R(12), N=12	410.3	b $^1\Sigma^+$, v=0	R(14)	470.9
	R(11), N=12	410.6	b $^1\Sigma^+$, v=1	R(14)	452.3
				R(15)	479.4
a $^1\Delta$, v=0	R(11)	382.8			
	R(12)	412.6	A $^3\Pi_i$, v=0	R(11), N=12	408.1
	R(13)	441.7		R(12), N=12	408.4
	R(14)	470.3		R(13), N=12	408.8
	R(15)	498.6		R(10), N=11	378.7
a $^1\Delta$, v=1	R(13)	423.3		R(11), N=11	379.0

The major primary photolysis product of HN_3 is NH in the a $^1\Delta$, v=0 state, but smaller concentrations reach various vibrational levels in the c $^1\Pi$ state. These are assumed to be responsible for the population of vibrational levels in the X, a, b, and A states and also partially for the population of the a $^1\Delta$, v=0 level, by radiative transitions and collisional deactivation. Computer simulations have been carried out for the various relaxation mechanisms that result in population inversion and thus purely rotational lasing [7].

Einstein Coefficients. Using the spectroscopic constants for the X $^3\Sigma^-$, v=0 and A $^3\Pi_i$, v=0 states derived from the A←X absorption spectrum [8] and assuming Hund's case (b) to hold for these states, the Einstein A coefficients were calculated for 23 purely rotational transitions between the X $^3\Sigma^-$, v=0, N=0 to 5 levels and for 79 purely rotational transitions between the A $^3\Pi_i$, v=0, N=1 to 5 (and Λ-doublets) levels [9].

References:

[1] Van den Heuvel, F. C.; Meerts, W. L.; Dymanus, A. (Chem. Phys. Lett. **92** [1982] 215/8).
[2] Wayne, F. D.; Radford, H. E. (Mol. Phys. **32** [1976] 1407/22).
[3] Radford, H. E.; Litvak, M. M. (Chem. Phys. Lett. **34** [1975] 561/4).
[4] Leopold, K. R.; Evenson, K. M.; Brown, J. M. (J. Chem. Phys. **85** [1986] 324/30).
[5] Leopold, K. R.; Zink, L. R.; Evenson, K. M.; Jennings, D. A.; Brown, J. M. (NBS Spec. Publ. [U. S.] 716 [1986] 452/8).

[6] Vasconcellos, E. C. C.; Davidson, S. A.; Brown, J. M.; Leopold, K. R.; Evenson, K. M. (J. Mol. Spectrosc. **122** [1987] 242/5).
[7] Smith, J. H.; Robinson, D. W. (J. Chem. Phys. **71** [1979] 271/80).
[8] Dixon, R. N. (Can. J. Phys. **37** [1959] 1171/86).
[9] Litvak, M. M.; Rodriguez Kuiper, E. N. (Astrophys. J. **253** [1982] 622/33).

2.1.1.3.3 Infrared Spectrum

Absorption Spectrum. The fundamental vibration-rotation bands of the **free radical** in its electronic ground state X $^3\Sigma^-$ and lowest excited state a $^1\Delta$ were measured at high resolution.

Using a difference frequency laser and a long-path, Zeeman-modulated discharge cell with flowing NH_3, twenty-nine lines between 2949.9649 and 3293.1478 cm^{-1} were measured with an accuracy of 0.003 cm^{-1} for NH(X $^3\Sigma^-$, v=1←0). The lines were identified according to the coupling scheme for Hund's case (b) to be P-branch (N=1 to 5) and R-branch (N=0 to 5) transitions, exhibiting fine-structure splitting due to $\Delta J = \Delta N$ transitions between the rotational substates J=N−1, N, N+1; six satellite lines ($\Delta J \neq \Delta N$) of ^PQ-, ^PR-, ^RP-, and ^RQ-type were also identified. Analysis of the spectrum using a Hamiltonian [1] that includes vibrational-rotational energy and spin interactions yielded the band origin

$$\nu_0(X\ ^3\Sigma^-, v=1 \leftarrow 0) = 3125.5724 \pm 0.0020\ cm^{-1}$$

and the rotational constants given on p. 47 [2].

The fundamental absorption band of NH(a $^1\Delta$) formed by 193-nm excimer-laser photolysis of gaseous HN_3 was recorded between 3177 and 3535 cm^{-1} using an IR color-center laser. Due to the high degree of rotational excitation resulting from the photolysis, thirty-three R-branch lines (J=2 to 34) and two Q-branch lines (J=2 and 3) were observed. A least-squares fit of twenty-one R-branch wavenumbers that have been measured with an accuracy of 0.003 cm^{-1} allowed the determination of high-order centrifugal distortion constants (cf. p. 47) and gave the band origin

$$\nu_0(a\ ^1\Delta, v=1 \leftarrow 0) = 3182.7768 \pm 0.0037\ cm^{-1}\ [3].$$

Matrix-isolated ^{14}NH, ^{14}ND, and ^{15}NH molecules formed by photolysis of normal and isotopically substituted HN_3 in solid nitrogen or argon (T=4 to 20 K) were identified by their fundamental vibrations and isotopic shifts observed in the IR spectrum. Since the observed IR frequencies correspond reasonably well with the gas-phase values for the electronic ground state and spin-conservation rules, on the other hand, require NH to be initially produced in an upper singlet state, a collisional deactivation in the matrix has been suspected. The following wavenumbers (in cm^{-1}; T=15 K) were assigned to the fundamental vibrations in the electronic ground state X $^3\Sigma^-$:

species	N_2 matrix [4]	Ar matrix [4]	Ar and N_2 matrices [5]*)
^{14}NH	3122.2±0.6	3131.6±0.4	3133±2
^{14}ND	2299±2	2318.0±0.3	2323±2
^{15}NH	3115±1	3125.3±0.4	3126±2

*) Values reported for both Ar and N_2 matrices [5]; values 4 cm^{-1} higher are quoted for ^{14}NH and ^{15}NH in [6]. The different results of [4] and [5] for the N_2 matrices have been discussed by the authors; different deposition temperatures of the matrices, 4 K [5] and 15 and 20 K [4], were used and site effects may have become operative.

Emission Spectrum. High-resolution Fourier transform (FT) spectroscopy was applied to observe the vibrational-rotational emission bands $v=1\rightarrow0$, $2\rightarrow1$, $3\rightarrow2$, $4\rightarrow3$, $5\rightarrow4$ (and some $6\rightarrow5$ lines) of ground-state $NH(X\ ^3\Sigma^-)$ in the range 2275 to 3460 cm^{-1}. The NH radicals were generated in an N_2+H_2 plasma reactor (see [7]) or by electrodeless rf discharges through flowing NH_3, and the FT spectra were recorded at resolutions of 0.020 and 0.009 cm^{-1} or 0.040 cm^{-1}, respectively. Altogether 310 rotational lines have been observed and identified mostly as the R- and P-branch transitions ($\Delta N=\Delta J=\pm1$), split into the fine-structure components P_1, P_2, P_3 and R_1, R_2, R_3 with $J=N+1$, N, $N-1$; for lower rotational quantum numbers N, some of the expected satellite lines ($\Delta N \neq \Delta J$), $^PQ_{21}$, $^PQ_{32}$, and $^PR_{31}$ in the P branch and $^RQ_{12}$, $^RQ_{23}$, and $^RP_{13}$ in the R branch, were also observed. A nonlinear least-squares fit of all measured wavenumbers using a Hamiltonian that includes centrifugal distortion terms up to third order and spin-spin and spin-rotation interactions [1] gave a set of 47 spectroscopic constants, which are the five band origins and the rotational and fine-structure constants for the vibrational levels $v=0$ to 5. These parameters in turn were used to calculate more than 700 wavenumbers for all six possible components (three main-branch lines and three satellites) of each of 122 rotational-vibrational transitions in the spectral region covered by the experiments. The calculation reproduced the experimental wavenumbers within 0.01 cm^{-1}. The band origins (ν_0), the number of observed and calculated lines (No.), the highest rotational quantum numbers in the P and R branches ($N''_{max}(P)$, $N''_{max}(R)$), and the spectral regions (ν region) for each of the vibrational bands are as follows [8]:

$v'\rightarrow v''$	ν_0 in cm^{-1}	No.[a)]	$N''_{max}(P)$ [a)]	$N''_{max}(R)$ [a)]	ν region [a)] in cm^{-1}
$1\rightarrow0$	3125.57291(16)	72 (228)	10 (19)	17 (19)	2747 to 3457 (2328 to 3461)
$2\rightarrow1$	2969.30211(12)	75 (168)	10 (14)	14 (14)	2561 to 3264 (2429 to 3265)
$3\rightarrow2$	2812.72331(15)	72 (144)	11 (12)	12 (12)	2419 to 3072 (2376 to 3073)
$4\rightarrow3$	2654.71601(34)	58 (108)	8 (9)	9 (9)	2390 to 2864 (2351 to 2865)
$5\rightarrow4$	2493.83494(130)	33 (84)	6 (7) [b)]	7 (7) [b)]	2310 to 2662 (2274 to 2663)

[a)] Results from calculation in parentheses. – [b)] P-branch transitions with $N''=2$, 1 and R-branch transitions with $N''=0$ not observed.

The $\Delta v=1$ bands of NH were also observed and analyzed in the high-resolution (0.005 cm^{-1}) FT IR emission spectrum of a radiofrequency-excited N_2+SiH_4 plasma; the measurements are an improvement over the data of [8] and give new information, especially for the region above 3500 cm^{-1}; results have not yet been published [9].

The rotational line intensities for the five lowest $\Delta v=1$ bands were measured in the FT emission spectrum of an N_2+H_2 plasma source, and the vibrational transition dipole moment for the fundamental band, $\langle 0|M(\xi)|1\rangle=-0.0648\pm0.008$ D, was derived [10]; high-quality ab initio calculations of the dipole moment function $M(\xi)$ (cf. p. 39) give $\langle 0|M(\xi)|1\rangle = -0.0625$ D (CEPA) [11] or -0.0594 D (full CI) [12].

FT IR spectroscopy was used to observe the emission spectra of glow discharges in moist air (0.1 Torr) and air-H_2 (3:1) mixtures (and in NH_3 for comparison). The fundamental vibrational-rotational band of $NH(X\ ^3\Sigma^-)$ was indeed observed in these laboratory studies which simulate the conditions in the upper atmosphere at an altitude of 60 km, where NH is assumed to be detectable by IR emission [13, 14].

Using time-resolved FT spectroscopy, the IR chemiluminescence corresponding to the $v=3\rightarrow2$, $2\rightarrow1$, and $1\rightarrow0$ bands of $NH(X\ ^3\Sigma^-)$ was observed in N_2-H_2 and Ar- N_2-H_2 mixtures, irradiated by a pulsed electron beam. The dynamics of NH formation and vibrational excitation via the reaction $N(^2D) + H_2 \rightarrow NH(X\,^3\Sigma^-, v = 1, 2, 3) + H$ (cf. p. 16) and the vibrational relaxation (cf. p. 88) were investigated [15, 16].

Einstein Coefficients. Oscillator Strengths. Einstein coefficients of spontaneous emission of $A_{v,v-1}$ (v = 1, 2, and 3) = 51.7, 92.3, and 144.4 s^{-1} for the fundamental band IR emission in $NH(X\,^3\Sigma^-)$ were calculated [16] using an RKR potential energy function based on the spectroscopic constants given in [17] and on a theoretical (ab initio CI calculation) dipole moment function [18, 19]. Theoretical (CEPA) potential energy and dipole moment functions gave $A_{1,0}$ = 34.9 and 9.7 s^{-1} for $NH(X\,^3\Sigma^-)$ and $ND(X\,^3\Sigma^-)$, respectively [20]. Transition moments and band oscillator strengths for the rotationless vibrational transitions in $NH(X\,^3\Sigma^-)$ with $v' \leftarrow v'' = 1 \leftarrow 0$ to $4 \leftarrow 0$, $2 \leftarrow 1$ to $4 \leftarrow 1$, $3 \leftarrow 2$, $4 \leftarrow 2$, and $4 \leftarrow 3$ were calculated using theoretical (CI) potential energy and dipole moment functions [21].

References:

[1] Amano, T.; Hirota, E. (J. Mol. Spectrosc. **53** [1974] 346/63).
[2] Bernath, P. F.; Amano, T. (J. Mol. Spectrosc. **95** [1982] 359/64).
[3] Hall, J. L.; Adams, H.; Kasper, J. V. V.; Curl, R. F.; Tittel, F. K. (J. Opt. Soc. Am. B Opt. Phys. **2** [1985] 781/5).
[4] Rosengren, K.; Pimentel, G. C. (J. Chem. Phys. **43** [1965] 507/16).
[5] Milligan, D. E.; Jacox, M. E. (J. Chem. Phys. **41** [1964] 2838/41).
[6] Jacox, M. E.; Milligan, D. E. (Appl. Opt. **3** [1964] 873/6).
[7] Chollet, P.; Guelachvili, G.; Morillon-Chapey, M.; Gressier, P.; Schmitt, J. P. M. (J. Opt. Soc. Am. B Opt. Phys. **3** [1986] 687/95).
[8] Boudjaadar, D.; Brion, J.; Chollet, P.; Guelachvili, G.; Vervloet, M. (J. Mol. Spectrosc. **119** [1986] 352/66).
[9] Elhanine, M.; Farrenq, R.; Guelachvili, G. (Proc. SPIE-Int. Soc. Opt. Eng. **1575** [1992] 314/5).
[10] Chackerian, C., Jr.; Guelachvili, G.; Lopez-Pineiro, A.; Tipping, R. H. (J. Chem. Phys. **90** [1989] 641/9).

[11] Meyer, W.; Rosmus, P. (J. Chem. Phys. **63** [1975] 2356/75).
[12] Chackerian, C., Jr.; Bauschlicher, C. W.; Langhoff, S. R.; Lopez-Pineiro, A.; Tipping, R. H. (to be published in Astrophys. J. from [10]).
[13] Hansen, P.; Sakai, H.; Esplin, M. (Proc. Soc. Photo.-Opt. Instrum. Eng. **191** [1979] 15/20).
[14] Sakai, H.; Hansen, P.; Esplin, M.; Johansson, R.; Peltola, M.; Strong, J. (Appl. Opt. **21** [1982] 228/34).
[15] Green, B. D.; Caledonia, G. E. (J. Chem. Phys. **77** [1982] 3821/3).
[16] Dodd, J. A.; Lipson, S. J.; Flanagan, D. J.; Blumberg, W. A. M.; Person, J. C.; Green, B. D. (J. Chem. Phys. **94** [1991] 4301/10).
[17] Huber, K. P.; Herzberg, G. (Molecular Spectra and Molecular Structure, Vol. 4, Constants of Diatomic Molecules, Van Nostrand Reinhold, New York 1979, p. 456).
[18] Goldfield, E. M.; Kirby, K. P. (J. Chem. Phys. **87** [1987] 3986/94).
[19] Goldfield, E. M. (private communication to [16]).
[20] Rosmus, P.; Werner, H. J. (J. Mol. Struct. **60** [1980] 405/8).

[21] Das, G.; Wahl, A. C.; Stevens, W. J. (J. Chem. Phys. **61** [1974] 433/4).

2.1.1.3.4 Near-Infrared, Visible, and Ultraviolet Spectra

2.1.1.3.4.1 General

Eight absorption and/or emission systems have been found in the range from near IR at ~1.2 μm to vacuum UV at 160 nm due to transitions between the known six lowest valence states of NH and ND, X $^3\Sigma^-$, a $^1\Delta$, b $^1\Sigma^+$, A $^3\Pi_i$, c $^1\Pi$, and d $^1\Sigma^+$. An energy level diagram with the spectroscopically observed transitions is shown in **Fig. 2**, which was taken from the reference and completed by the a↔X and b→a systems observed later.

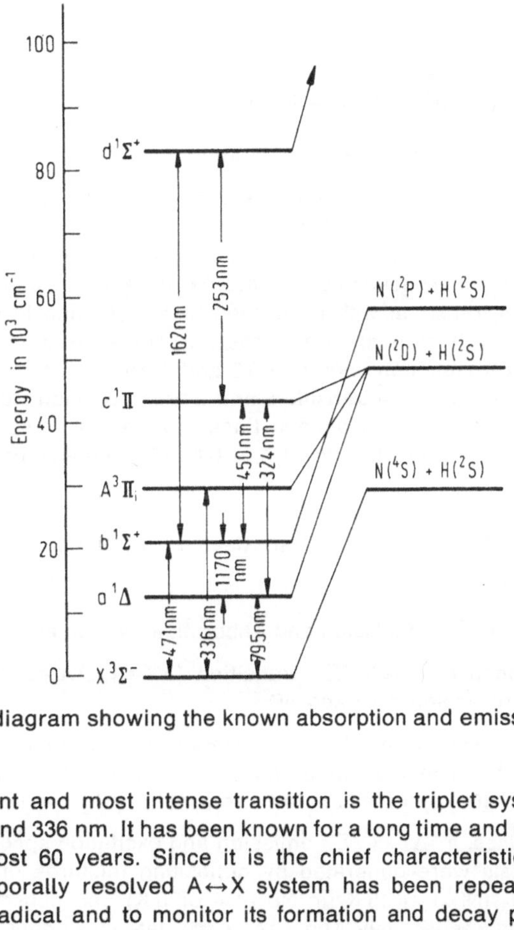

Fig. 2. Energy level diagram showing the known absorption and emission systems of NH.

The most prominent and most intense transition is the triplet system A $^3\Pi_i$↔X $^3\Sigma^-$ with its maximum around 336 nm. It has been known for a long time and extensively analyzed over a period of almost 60 years. Since it is the chief characteristic of the NH radical, the spectrally or temporally resolved A↔X system has been repeatedly used to prove the existence of the radical and to monitor its formation and decay processes in various environments, such as N- and H-containing gas molecules in electric discharges or during photolysis or radiolysis, molecular beams, flames, or noble-gas matrices. The A↔X spectrum even proved the existence of NH in the sun, in stellar atmospheres, and in comets.

The prominent singlet system c $^1\Pi$↔a $^1\Delta$ appears at 324 nm just at the short-wavelength end of the triplet system. Emission, absorption, and excitation spectra have been recorded in gaseous NH. Like in the case of the triplet system, the c↔a transition was used as an indicator for NH formation, specifically for the formation of the chemically reactive, metastable NH(a $^1\Delta$) radical.

The singlet systems, c $^1\Pi \leftrightarrow$ b $^1\Sigma^+$ at 450 nm, d $^1\Sigma^+ \rightarrow$ c $^1\Pi$ at 253 nm, and d $^1\Sigma^+ \rightarrow$ b $^1\Sigma^+$ at 162 nm, have been observed in gas-phase absorption and emission spectra.

The weak, forbidden transitions a $^1\Delta \leftrightarrow$ X $^3\Sigma^-$ at 795 nm and b $^1\Sigma^+ \leftrightarrow$ X $^3\Sigma^-$ at 471 nm were observed in the emission spectra of gaseous NH and in the emission and excitation spectra of NH in noble-gas matrices.

Recently, the transition between the two lowest metastable states, b $^1\Sigma^+ \rightarrow$ a $^1\Delta$, was detected at 1.17 μm in the emission spectrum of matrix-isolated NH.

Detailed descriptions of the absorption and emission systems of NH and ND are presented in the following sections in the order of increasing energy or decreasing wavelength.

Reference:

Graham, W. R. M.; Lew, H. (Can. J. Phys. **56** [1978] 85/99).

2.1.1.3.4.2 The b $^1\Sigma^+ \rightarrow$ a $^1\Delta$ Emission at 1.17 μm

The b $^1\Sigma^+ \rightarrow$ a $^1\Delta$ transition has been observed at 5 K in matrix-isolated NH and ND, generated by 121.6-nm photolysis of NH_3-Ar or ND_3-Ar (1:5000 to 1:20000) condensates. The b $^1\Sigma^+$, v′ = 0 level was populated either directly via the forbidden b $^1\Sigma^+ \leftarrow$ X $^3\Sigma^-$ transition with a dye laser (tuned selectively to the 0 ← 0 band) or indirectly by pumping the A $^3\Pi_i \leftarrow$ X $^3\Sigma^-$ (0 ← 0) transition with a nitrogen laser resulting in populating the b-state v′ = 0 and also v′ = 1 levels by intersystem crossing. The emission spectrum between 1169 and 1185 nm shows zero-phonon lines at 1173 and 1170 nm, which were identified as the 0 → 0 and 1 → 1 bands of the b → a transition, and a broad phonon sideband at higher wavelengths. The positions of the zero-phonon lines are (in cm^{-1}) 8520.9 (0 → 0) and 8543.6 (1 → 1) for NH, 8523.5 (0 → 0) for ND. A weak 1 → 0 band could also be detected at 854 nm.

Reference:

Ramsthaler-Sommer, A.; Becker, A. C.; Van Riesenbeck, N.; Lodemann, K.-P.; Schurath, U. (Chem. Phys. **140** [1990] 331/8).

2.1.1.3.4.3 The a $^1\Delta \leftrightarrow$ X $^3\Sigma^-$ Emission and Absorption System at 795 nm

The highly forbidden a $^1\Delta \leftrightarrow$ X $^3\Sigma^-$ transition ($\Delta S = 1$, $\Delta\Lambda = 2$) has been observed in gaseous NH and matrix-isolated NH and ND.

During ArF laser (193 nm) photolysis of **gaseous** HN_3, an emission spectrum recorded at low resolution (0.8 nm) in the range 785 to 805 nm was identified as the rotationally resolved a $^1\Delta \rightarrow$ X $^3\Sigma^-$ (0 → 0) transition; the strong Q branch is at ~794.5 nm [1].

In inert-gas **matrices**, a $^1\Delta \leftrightarrow$ X $^3\Sigma^-$ emission and excitation spectra of NH and ND were observed. The radicals were generated by depositing mixtures of inert gases with HN_3 (DN_3), HNCO, or NH_3 (ND_3) at 15 K (in Ne at 4 or 5 K) and subsequently photolyzing at 206.2, 165, or 121.6 nm, respectively. The a $^1\Delta$, v′ = 0 state was populated either immediately during the photolysis of HN_3 and HNCO or after the photolysis of NH_3 by pumping the transitions a $^1\Delta \leftarrow$ X $^3\Sigma^-$ or b $^1\Sigma^+ \leftarrow$ X $^3\Sigma^-$ (subsequent radiationless b $^1\Sigma^+ \rightarrow$ a $^1\Delta$ relaxation) with a tunable dye laser or pumping the transition A $^3\Pi_i \leftarrow$ X $^3\Sigma^-$ (subsequent A $^3\Pi_i$ − a $^1\Delta$ intersystem crossing) with a nitrogen laser. Emission spectra recorded at a low spectral resolution of 1.0 to 1.8 nm exhibit in the 800-nm region a strong zero-phonon line corresponding to the 0 → 0 band of the a → X transition and a weakly structured phonon sideband of low intensity at lower energies; the weaker 0 → 1 band appears at ~1060 nm. The following wavenumbers ν_{00}, vibrational spacings of the ground state $\Delta G_{1/2}$, and Franck-Condon factor ratios $q_{0,0}/q_{0,1}$ were measured in the various matrices upon HN_3 and HNCO photolysis [2]:

matrix	T in K	ν_{00} in cm^{-1}	$\Delta G_{1/2}$ in cm^{-1}	$q_{0,0}/q_{0,1}$
Ne	5	12686±2	3156±6	66
N_2	7	12622±3	3151±10	73
Ar	7	12606±2 [a)]	3134±2	59
Kr	7	12597±2 [b)]	3129±3	48
Xe	7	12559±2 [b)]	3109±4	48

[a)] ν_{00}(^{S}R(0) line) = 12625.5 cm^{-1} measured upon NH_3 photolysis and b $^1\Sigma^+$ or A $^3\Pi_i$ excitation of NH [3]. — [b)] High-energy satellite lines at 12605±5 (Kr) and 12583±4 (Xe) cm^{-1} [2].

Rotationally resolved a→X emission [4, 5] and a←X excitation spectra [5] were recorded with a spectral resolution of ~0.1 nm following HNCO photolysis [4] or NH_3 (ND_3) photolysis and subsequent b $^1\Sigma^+$ or A $^3\Pi_i$ excitation [4, 5]. Analysis of the rotational structure (and comparison with that expected for a gas-phase $^1\Delta \leftrightarrow {}^3\Sigma^-$ transition), analysis of the splitting of rotational lines in the different matrix environments, and measurement of a $^1\Delta$-state lifetimes (cf. p. 89) and their temperature dependencies provided insight into the interactions between the NH (ND) guest and the host lattice; models for crystal field splitting and rotation-translation coupling have been considered.

Franck-Condon factors for a $^1\Delta$ (v′=0)→X $^3\Sigma^-$ (v″) bands of NH with v″=0 to 4 and of ND with v″=0 to 6 were calculated using gas-phase potential energy functions and spectroscopic constants (literature source not given) for the a $^1\Delta$ and X $^3\Sigma^-$ states [4]. Spin-forbidden, dipole-allowed transition moments were obtained from a quantum-chemical ab initio (CI) calculation [6].

References:

[1] Rohrer, F.; Stuhl, F. (Chem. Phys. Lett. **111** [1984] 234/7).

[2] Esser, H.; Langen, J.; Schurath, U. (Ber. Bunsenges. Phys. Chem. **87** [1983] 636/43).

[3] Ramsthaler-Sommer, A.; Becker, A. C.; Van Riesenbeck, N.; Lodemann, K.-P.; Schurath, U. (Chem. Phys. **140** [1990] 331/8).

[4] Ramsthaler-Sommer, A.; Eberhardt, K. E.; Schurath, U. (J. Chem. Phys. **85** [1986] 3760/9).

[5] Blindauer, C.; Van Riesenbeck, N.; Seranski, K.; Winter, M.; Becker, A. C.; Schurath, U. (Chem. Phys. **150** [1991] 93/108).

[6] Yarkony, D. R. (J. Chem. Phys. **91** [1989] 4745/57).

2.1.1.3.4.4 The b $^1\Sigma^+ \leftrightarrow$ X $^3\Sigma^-$ Emission and Absorption System at 471 nm

The spin- ($\Delta S=1$) and parity- (+↔−) forbidden b $^1\Sigma^+ \leftrightarrow$ X $^3\Sigma^-$ transition has been observed for gaseous and matrix-isolated NH and ND.

The b $^1\Sigma^+ \rightarrow$ X $^3\Sigma^-$ emission spectrum was observed for the first time after photolysis of gaseous NH_3 or ND_3 with the Ar resonance lines at 104.8 and 106.7 nm and the Kr resonance lines at 116.5 and 123.6 nm. The spectrum recorded between 480 and 465 nm at a dispersion of 1.3 nm/mm consisted of an intense central line at 471 nm, which is the overlap of the three ^{Q}P, ^{Q}Q, and ^{Q}R branches ($\Delta N=0$), and an ^{S}R and an ^{O}P branch ($\Delta N=\pm 2$) on either side. Rotational lines with N″=0 to 2 and 0 to 6 were recorded in the ^{S}R branches and with $N''=2$ to 6 and 2 to 10 in the ^{O}P branches of NH_3 and ND_3, respectively [1 to 3] (for a schematic diagram of the rotational transitions between $^1\Sigma^+$ ($N'=J'$) and $^3\Sigma^-$ ($J''=N''-1$, N'', $N''+1$) states, see [1]).

Two alternative emission sources were used to enable the b→X emission spectrum to be recorded with a dispersion of 0.11 nm/mm and the rotational lines to be measured with an accuracy of ±0.05 or 0.02 cm^{-1}. The first was a low-pressure discharge in NH or ND to which was applied a magnetic field crossed with an electric field; ^{O}P-branch lines with J'' ($=N''-1$) $=1$ to 8 and 1 to 10 for NH and ND, respectively, and ^{S}R-branch lines with J'' ($=N''+1$) $=1$ to 7 for both were observed [4]. The second source involved the energy exchange between metastable Ar atoms from a microwave discharge and NH_3 or ND_3 to produce NH (ND) molecules in vibronic b $^1\Sigma^+$, v' levels. The emission spectrum recorded between 484 and 447 nm exhibits the Q branches (^{Q}P, ^{Q}Q, and ^{Q}R overlap) of the $\Delta v=0$ sequence up to $v=3$ (4) for NH (ND) as well as rotational lines with $N''=2$ to 9 (^{O}P) and 0 to 6 (^{S}R) in the 0→0 band of NH, and with $N''=2$ to 13 and 3 to 8 (^{O}P), 3 to 8 and 0 to 6 (^{S}R) in the 0→0 and 1→1 bands respectively, of ND [5]. Band origins ν_{00} and ν_{11} obtained from the high-resolution spectra are as follows:

molecule	ν_{00} in cm^{-1} [4]	ν_{00} in cm^{-1} [5]	ν_{11} in cm^{-1} [5]
NH	21238.26(7)	21238.5±2	–
ND	21224.87(4)	21224.7±2	21280.9±4

The b→X emission after NH_3 or ND_3 photolysis with an ArF excimer laser at 193 nm was attributed to NH, ND(b $^1\Sigma^+$) formation in a two-photon resonance process [6, 7]. Chemiluminescence due to b→X transitions resulted also from the reaction between H and HN_3; NH(b $^1\Sigma^+$) formation probably occurred in the reaction $H+N_3 \rightarrow NH+N_2$ (where N_3 originates from $H+HN_3 \rightarrow H_2+N_3$) [8].

The b $^1\Sigma^+ \rightarrow$ X $^3\Sigma^-$ emission has also been observed at 5 K in matrix-isolated NH and ND, generated by 121.6-nm photolysis of NH_3-Ar or ND_3-Ar (1:5000 to 1:20000) condensates. The b $^1\Sigma^+$, $v'=0$ level was populated either directly via the b $^1\Sigma^+ \leftarrow$ X $^3\Sigma^-$ transition with a dye laser (tuned selectively to the 0←0 band) or indirectly by pumping the A $^3\Pi_i \leftarrow$ X $^3\Sigma^-$ (0←0) transition with a nitrogen laser which results in the population of the b-state $v'=0$ and also $v'=1$ levels by intersystem crossing. The b→X bands consist of a sharp $^{Q}P(0)$ line and a broadened, crystal-field-split $^{O}P(2)$ line. The $^{Q}P(0)$ lines of the b $^1\Sigma^+ \rightarrow$ X $^3\Sigma^-$, v'→v'' bands of NH and ND have the following wavenumbers (in cm^{-1}) [9]:

molecule	v'	v''=0	v''=1	v''=2	v''=3
NH	0	21142.2	18018.9	15042.6	–
	1	24344.6	21218.7	18248.5	15426.4
ND	0	21130.2	18819.4	–	–
	1	–	21184.8	18958.2	–

The rotationally resolved excitation spectrum, recorded on the $^{S}R(1)$ line of the a $^1\Delta \rightarrow$ X $^3\Sigma^-$ emission while scanning the laser across the 0–0 band in the range 473 to 470 nm, is that of a nearly free rotor; it consists of a sharp $^{Q}P(0)$ line at 472.72 nm, an electric-dipole-forbidden but electric-quadrupole-allowed $^{R}Q(0)$ line at 472.05 nm, and a crystal-field-split $^{S}R(0)$ line around 470.5 nm [9]. The sharp $^{Q}P(0)$ line in the excitation spectra of NH and ND in Ar or Kr matrices has been the subject of a systematic study on the temperature dependency (6 to 35 K) of line position, width, and asymmetry; these allowed conclusions to be drawn and models to be tested on the electron-phonon coupling between the impurity center (NH) and the host lattice [10].

Spin-forbidden, dipole-allowed transition moments were obtained from a quantum-chemical ab initio (CI) calculation [11].

References:

[1] Masanet, J.; Gilles, A.; Vermeil, C. (J. Photochem. **3** [1974/75] 417/29).
[2] Gilles, A.; Masanet, J.; Vermeil, C. (Chem. Phys. Lett. **25** [1974] 346/7).
[3] Vermeil, C.; Masanet, J.; Gilles, A. (Int. J. Radiat. Phys. Chem. **7** [1975] 275/80).
[4] Cossart, D. (J. Chim. Phys. Phys.-Chim. Biol. **76** [1979] 1045/50).
[5] Zetzsch, C.; Stuhl, F. (Ber. Bunsenges. Phys. Chem. **80** [1976] 1348/54).
[6] Donnelly, V. M.; Baronavski, A. P.; McDonald, J. R. (Chem. Phys. **43** [1979] 271/81).
[7] Haak, H. K.; Stuhl, F. (J. Phys. Chem. **88** [1984] 2201/4).
[8] Kajimoto, O.; Kawajiri, T.; Fueno, T. (Chem. Phys. Lett. **76** [1980] 315/8).
[9] Ramsthaler-Sommer, A.; Becker, A. C.; Van Riesenbeck, N.; Lodemann, K.-P.; Schurath, U. (Chem. Phys. **140** [1990] 331/8).
[10] Hizhnyakov, V.; Seranski, K.; Schurath, U. (Chem. Phys. **162** [1992] 249/56).

[11] Yarkony, D. R. (J. Chem. Phys. **91** [1989] 4745/57).

2.1.1.3.4.5 The c $^1\Pi \leftrightarrow$ b $^1\Sigma^+$ Emission and Absorption System at 450 nm

The 0→0 band of the c $^1\Pi \rightarrow$ b $^1\Sigma^+$ transition of NH was detected at 450.2 nm and photographed at moderate dispersion in a hollow-cathode discharge in rapidly streaming NH_3; an initial rotational analysis revealed single P, Q, and R branches where the Q branch is the strongest one and the P and R branches are almost equally intense [1]. In Schüler-type discharges through NH_3 or ND_3, the 450.2-nm band of NH and its ND counterpart at 448.5 nm were observed [2]. High-resolution spectra of the c $^1\Pi \rightarrow$ b $^1\Sigma^+$, 0→0 and 0→1 bands (accuracy for strong lines: ±0.01 and ±0.02 cm^{-1}, respectively) of NH and ND were obtained in discharges through He-N_2-H_2(D_2) mixtures. The rotational structure (P, Q, R) up to J=16 and 13 was resolved in the 0→0 and 0→1 bands of NH and up to J=22 and 17 in the corresponding ND bands (P and Q branches only were resolved in the 0→1 band of ND). All bands are degraded to the red and overlapped by various emission systems of the NH^+ and ND^+ ions. The observed wavelengths of the R heads λ(R head) and the derived band origins ν_{00} and ν_{01} are as follows [3]:

molecule	λ(R head)$_{00}$ in nm	λ(R head)$_{01}$ in nm	ν_{00} in cm^{-1}	ν_{01} in cm^{-1}
NH	450.1977	525.398	22120.78(1)	18914.87(1)
ND	448.4463	501.55	22244.94(1)	19873.3(2)

Vacuum-UV photolysis of NH_3 or ND_3 with the Ar resonance lines at 104.8 and 106.7 nm resulted also in c $^1\Pi \rightarrow$ b $^1\Sigma^+$ fluorescence; in addition to the 0→0 and 0→1 bands (wrong citation of the above results), the 1→0 bands at 24231.3 and 23859.9 cm^{-1} and the 1→1 bands at 21024.0 and 21487.7 cm^{-1} for NH and ND, respectively, were observed [4]. Upon ArF laser (193 nm) photolysis of HN_3, the 0→0 band of NH was observed at 452.4 nm [5]. The rotationally resolved 0→0 bands were also recorded at the low-wavelength side of the b $^1\Sigma^+ \rightarrow$ X $^3\Sigma^-$ emission (Ar* + NH_3, ND_3; cf. p. 70) [6, figure 2]. Most recently, the c←b, 1←1 excitation spectrum, superimposed on the c←a, 1←4 band, was observed upon 248-nm laser photolysis of HN_3, the c→b, 0→0, 1 and 1→0, 1, 2 fluorescence (410 to 540 nm) was observed after c←a, 0, 1←0, 1 pumping [12].

Based on lifetime measurements in the c→b spectrum excited by electron impact on NH_3 [7], Franck-Condon factors and r-centroids were calculated for c−b transitions with v′=0 to 3 and v″=0 to 3 [8]. Intensity measurements in the c→b spectra gave transition probabilities $A_{v',v''}$ with v′, v″=0, 0; 0, 1; 1, 1; 1, 2 (HN_3 photolysis) [12] and $A_{0,0}$ (Ar^*+NH_3 collisions) [9, 10]. A quantum-chemical ab initio (CI) calculation gave the transition moment, Einstein coefficient, and Franck-Condon factor for the 0−0 transition [11].

References:

[1] Lunt, W.; Pearse, R. W. B.; Smith, E. C. W. (Proc. R. Soc. [London] A **151** [1935] 602/9).
[2] Chauvin, H.; Leach, S. (Compt. Rend. **231** [1950] 1482/4).
[3] Whittaker, F. L. (J. Phys. B **1** [1968] 977/82).
[4] Zetzsch, C. (J. Photochem. **9** [1978] 151/3).
[5] Haak, H. K.; Stuhl, F. (J. Phys. Chem. **88** [1984] 3627/33).
[6] Zetzsch, C.; Stuhl, F. (Ber. Bunsenges. Phys. Chem. **80** [1976] 1348/54).
[7] Smith, W. H. (J. Chem. Phys. **51** [1969] 520/4).
[8] Smith, W. H.; Liszt, H. S. (J. Quant. Spectrosc. Radiat. Transfer **11** [1971] 45/54).
[9] Lents, J. M., Jr. (Diss. Univ. Tennessee 1970, 277 pp.; Diss. Abstr. Int. B **32** [1971] 482).
[10] Lents, J. M. (J. Quant. Spectrosc. Radiat. Transfer **13** [1973] 297/310).

[11] Yarkony, D. R. (J. Chem. Phys. **91** [1989] 4745/57).
[12] Hack, W.; Mill, T. (J. Phys. Chem. **97** [1993] 5599/606).

2.1.1.3.4.6 The A $^3\Pi_i \leftrightarrow$ X $^3\Sigma^-$ Absorption and Emission System at 336 nm

2.1.1.3.4.6.1 Gaseous NH and ND

The A $^3\Pi_i \leftrightarrow$ X $^3\Sigma^-$ system of NH, first attributed to the NH_3 molecule, has been known since the last century (see "Stickstoff" 1936, pp. 304/5) and has been the subject of many later investigations. The spectrum can be observed upon decomposition of H- and N-containing molecules, such as NH_3, N_2H_4, HNCO, HN_3, CH_3NH_2, or of N_2-H_2 mixtures, involving decomposition by photolysis, by electron impact or collisions with metastable noble-gas atoms, in electric discharges or shock waves. The emission is generated also in flames containing NO_2 or N_2O and H_2, NH_3, or various hydrocarbons, in $H+NF_2$, $H+N_3$, $H+HN_3$ reactions, and in reactions of N^+ ions with H_2 or hydrocarbons. First, there are numerous studies on NH and/or ND spectra intended to assign the rovibrational lines, to resolve the fine- and hyperfine structure, to characterize the electronic states involved, and to derive their spectroscopic constants. Since the spectrum is well known, the A↔X emission and excitation (LIF) spectra have also been used in a great deal of investigations to monitor the production and concentration of NH radicals in their A $^3\Pi_i$ or X $^3\Sigma^-$ states and to probe and compare the internal energy distributions, i.e., populations of their vibrational, rotational, or Λ-doublet levels in various formation processes, which in turn gave insight into the dissociation kinetics of the parent molecules. In a great number of investigations, time-resolved fluorescence was used to study excited-state dynamics, such as radiative decay and collisional quenching of rovibrational levels populated initially during the formation of the radical or afterwards by state-selected excitation with a tunable laser.

The following section only refers to purely spectroscopic studies of NH and ND, in other words, analysis and characterization of the A $^3\Pi_i \leftrightarrow$ X $^3\Sigma^-$ spectrum. References for many publications that deal with the various formation processes of NH(A $^3\Pi_i$) or NH(X $^3\Sigma^-$) radicals, their identification by means of the A $^3\Pi_i \leftrightarrow$X $^3\Sigma^-$ spectrum, and with relaxation processes of A-state rovibrational levels may be found in the Sections 2.1.1.1 and 2.1.1.4.4.

The first rotational analyses of the spectra in the region of the 0→0 and 1→1 emission and 0←0 absorption bands, observed in electric discharges through NH_3 [1] and in lighting gas-NH_3 flames [2], already revealed that the A $^3\Pi_i \leftrightarrow$X $^3\Sigma^-$ system is an example of a Hund's case (a)↔Hund's case (b) transition for low quantum numbers J changing rapidly to a pure case (b)↔case (b) transition with increasing J. This means (see [3], [4, figure 1]) that each vibrational band shows spin tripling into three subbands, A $^3\Pi(\Omega)\leftrightarrow$X $^3\Sigma^-$, and each subband has three main branches, namely P_1, Q_1, R_1 for $\Omega=2$, P_2, Q_2, R_2 for $\Omega=1$, and P_3, Q_3, R_3 for $\Omega=0$, corresponding to $\Delta J=\Delta K$ $(=0, \pm1)$ transitions. If A $^3\Pi_i$ approaches case (a), additionally 3×6 satellite branches ($\Delta K=\Delta J\pm1$ and $\Delta J\pm2$) may appear which are of type N, O, P, Q, R, S, T, corresponding to transitions with $\Delta K=0, \pm1, \pm2, \pm3$. Λ-type doubling due to the A $^3\Pi_i$-state levels ($\Lambda=1$) is expected. Since the internuclear distances in the X and A states of NH are nearly equal, the Q branches pile up in a narrow region with numerous blends, whereas the P and R branches are well separated and lie on both sides of the Q heads.

Electric discharges through NH_3 and ND_3 revealed, in addition to the known 0→0 and 1→1 emission bands of NH, the corresponding bands of ND [5]. An absorption spectrum of the 0←0 band of NH with resolved P and R branches could be recorded during flash photolysis of N_2H_4 (but not of NH_3) [6]. In the emission spectra of NH and ND excited either by explosive decomposition (in a spark) of HN_3 and DN_3 or in "atomic flames" of reactive H-HN_3 and H-DN_3 mixtures, P, Q, and R branches of the 0→0, 0→1, 1→0, 1→1, 1→2, 2→1, and 2→2 bands of NH and ND have been observed [7 to 13], and rotational analyses for the 0→1 and 1→0 bands of NH [13 to 15] and for the 0→0 band of ND [16] (all at somewhat moderate resolution) were carried out. In shock-wave-excited HN_3 (DN_3), N_2H_4, or H_2-N_2 mixtures (not in NH_3), the $\Delta v=0$ sequences with v=0 to 5 for NH and v=0 to 3 for ND and extended rotational structure were observed [17 to 19].

Rotational analyses of high-resolution spectra of NH and ND in the region of the $\Delta v=0$ ($v''=v'=0$, 1, 2), $\Delta v=-1$ ($v''=0$, 1), and $\Delta v=+1$ ($v''=1$, 2; for ND also $v''=3$) bands (giving reliable spectroscopic constants, cf. pp. 41, 43/4, 49) are referred to in the following paragraphs.

NH Radical. The 0←0 and 1←0 bands of NH have been observed in absorption spectra at room temperature upon flash photolysis of HNCO vapor; the nine main branches expected could be resolved in both bands up to K=10 and 7, sixteen satellite branches for $\Delta K=0$, ±1, ±2 transitions in the 0←0 band and ten satellite branches, $\Delta K=0$, ±1 transitions in the 1←0 band. The relative accuracy of the wavenumbers of strong rotational lines was 0.03 cm^{-1} [4]. The 0→0 and 1→1 bands (main branches and a few P- and Q-type satellite lines) were observed between 319 and 347 nm (dispersion 0.06 nm/mm) in the emission spectrum from a hollow-cathode discharge through N_2-H_2-He mixtures [20]. The weak 0→1, 1→0 and, for the first time, the 1→2 and 2→1 bands were recorded in the ranges 380 to 365 nm ($\Delta v=-1$) and 315 to 302 nm ($\Delta v=+1$) with dispersions of 8 or 25 nm/mm during explosive decomposition of HN_3; only the main branches (up to K=22, 23, 18, and 15, respectively) could be observed [21]. Laser-induced fluorescence (LIF) of the A←X system was observed in a discharge (MW) flow system with pure NH_3 or NH_3-Ar and NH_3-N_2 mixtures; tuning a dye laser (band width ~0.3 cm^{-1}) on the Q- and R-branch regions of the 0←0 band and on the R-branch region of the 1←0 band, almost 100 excitations could be identified (using the assignment of [4]) with N″ up to 8, 6 and 7, respectively [22 to 24].

Using a Fourier transform spectrometer to observe the emission from a hollow-cathode discharge through a continuous flow of He (4.5 Torr) with small amounts of N_2 (40 mTorr) and H_2 (120 mTorr), the precision could be improved by more than two orders of magnitude over that of previous measurements; for strong unblended lines in the 0→0 band, for example, the precision was ±0.0002 cm^{-1}. Between 420 and 280 nm covering the seven $\Delta v=0$, ±1 bands with $v' \leq 2$, a total of 1240 lines were measured and assigned as given below; the analysis, which included also high-resolution IR data of NH(X $^3\Sigma^-$), yielded the best spectroscopic constants of NH so far. Unfortunately, the band origins were not published; they may, however, be recalculated using the term values (in cm^{-1}) $T_0=29761.1829(1)$, $T_1=32795.9283(3)$, $T_2=35633.7219(6)$ for the A $^3\Pi_i$ state and $T_0=0.0$, $T_1=3125.57292(25)$, $T_2=5094.87617(57)$ for the X $^3\Sigma^-$ state [25]. The Q-band heads λ_Q (in nm) and the assigned rotational branches with their highest quantum numbers of the total angular momentum J_{max} as read from [25, tables 1 to 7] are as follows:

band	λ_Q	J_{max} in main branches P_i; Q_i; R_i	satellite branches; J_{max}
0→0	336	34, 33, 32; 31, 29, 27; 32, 30, 32	all but $^PR_{12}$, $^SQ_{31}$, $^NP_{13}$, $^TR_{31}$; 19
1→1	337	25, 24, 23; 25, 25, 20; 23, 22, 21	all of 0→0 but $^SR_{32}$, $^RP_{31}$, $^OQ_{13}$; 12
2→2	339	16, 15, 15; 13, 15, 14; 12, 10, 11	all of 1→1 but $^OP_{23}$, $^SR_{21}$; 8
1→0	305	18, 16, 16; 19, 18, 18; 19, 18, 17	*)
2→1	308	10, 10, 9; 10, 10, 9; 9, 7, 6	*)
0→1	376	17, 19, 16; 19, 21, 20; 19, 18, 15	*)
1→2	375	17, 17, 13; 16, 14, 12; 13, 12, 11	*)

*) No satellite branches were measured in the $\Delta v=\pm1$ bands.

An LIF study of an NH molecular beam from a MW discharge in NH_3 enabled at ultrahigh resolution the hyperfine splitting (hfs) of both nuclei in 36 rotational lines of the 0→0 band to be resolved; these are the main-branch lines (N″ values in parentheses) Q_1(1 to 3), R_1(0, 2 to 7), Q_2(1), R_2(3), Q_3(4 to 7), R_3(1 to 7), P_3(3) and the satellite lines $^QR_{12}$(3), $^QR_{23}$(1, 2), $^RQ_{21}$(4 to 6), $^QP_{21}$(1, 2), $^QP_{32}$(3 to 6); more than 300 resolved hf lines have been listed with splittings ranging from 20 to 1000 MHz ($\sim1\cdot10^{-5}$ to $4\cdot10^{-4}$ nm) relative to the strongest hfs component, $F'=J'+3/2 \leftarrow F''=J''+3/2$ ($J+I_N=F_1$, $F_1+I_H=F$; $I_N=1$, $I_H=1/2$), in each rotational transition together with the hfs parameters for the X $^3\Sigma^-$ and, for the first time, the A $^3\Pi_i$ states (cf. p. 41) [26].

Collision broadening of absorption lines of NH, generated by shock heating of NH_3 (3%) in Ar, was measured on five R_1-branch lines of the 0←0 band over a temperature range 2000 to 2600 K [27].

ND Radical. High-resolution emission spectra of ND have been observed in the 0→0 and 1→1 bands (dispersion 0.04 nm/mm; main branches observed up to K=31 and 29) in hollow-cathode discharges through DN_3 [28], and in the 0→0, 1→1, and 2→2 bands (dispersion 0.06 nm/mm; main branches observed up to K=43, 39, and 31 and a few satellite lines recognized) in hollow-cathode discharges through flowing He-N_2-D_2 mixtures [29, 30]. The 0→0 band was recorded at low rotational temperatures which permitted the dense Q-branch lines at low rotational quantum numbers and numerous satellite lines to be resolved: The nine main branches and lines in all satellite branches except for two ($^NP_{13}$ and $^TR_{31}$) were observed using electrodeless microwave discharges through Ne-N_2-D_2 mixtures (dispersion 0.056 nm/mm; $K_{max}=9$) [31, 32] and upon flash photolysis of DN_3 (dispersion 0.05 nm/mm; $J_{max}=14$ in the main branches, $J_{max}=11$ in the satellite branches) [33].

The LIF excitation spectrum in the region of the $\Delta v = +1$ sequence with $v'' = 0$, 1, and 2 was measured, using a molecular ND beam produced by 193-nm photolysis of ND_3 and a tunable dye laser operating in the range 316 to 300 nm. Because of the cold rotational distribution in the beam, only transitions involving low J values were recorded; these are the main-branch lines R_1(1 to 4), Q_1(3), R_2(1, 2), Q_2(1), R_3(1), Q_3(1) in the 1←0 and 2←1 bands, R_1(1 to 3), Q_1(3), R_2(1, 2), Q_2(1), Q_3(1) in the 3←2 band, and 20, 16, and 11 lines (J=0 to 3) in 13, 10, and 8 satellite branches, respectively; wavenumbers are accurate within $\pm 0.3\ cm^{-1}$ [34].

Wavenumbers of the band origins (in cm^{-1}) obtained from high-resolution spectra of ND are as follows:

band	origin	Ref.	band	origin	Ref.
0→0	29798.72±0.03*)	[33]	1→0	32048.11(7)	[34]
1→1	29738.42±0.05	[29]	2→1	31886.17(7)	[34]
2→2	29658.19±0.02	[30]	3→2	31699.74(14)	[34]

*) 29798.75±0.03 [29], 29799.5 [31, 32].

References:

[1] Funke, G. W. (Z. Physik **96** [1935] 787/98).
[2] Funke, G. W. (Z. Physik **101** [1936] 104/12).
[3] Herzberg, G. (Molecular Spectra and Molecular Structure, Vol. 1, Spectra of Diatomic Molecules, Van Nostrand, Princeton, N. J., 1950, pp. 264/5).
[4] Dixon, R. N. (Can. J. Phys. **37** [1959] 1171/86).
[5] Chauvin, H.; Leach, S. (Compt. Rend. **231** [1950] 1482/4).
[6] Ramsay, D. A. (J. Phys. Chem. **57** [1953] 415/7).
[7] Pannetier, G. (Compt. Rend. **232** [1951] 817/8).
[8] Pannetier, G.; Gaydon, A. G. (J. Chim. Phys. **48** [1951] 221/4).
[9] Pannetier, G.; Guenebaut, H.; Gaydon, A. G. (Compt. Rend. **240** [1955] 958/60).
[10] Pannetier, G.; Guenebaut, H. (Compt. Rend. **245** [1957] 929/31).

[11] Pannetier, G.; Guenebaut, H.; Gaydon, A. G. (Compt. Rend. **246** [1958] 88/90).
[12] Pannetier, G.; Guenebaut, H. (Acta Chim. Acad. Sci. Hung. **18** [1959] 347/64).
[13] Guenebaut, H. (Bull. Soc. Chim. Fr. **1959** 962/1018, 1000/11).
[14] Pannetier, G.; Guenebaut, H. (Bull. Soc. Chim. Fr. **1958** 1463/9).
[15] Guenebaut, H.; Pannetier, G. (Compt. Rend. **250** [1960] 3613/5).
[16] Pannetier, G.; Guenebaut, H.; Hajal, I. (Bull. Soc. Chim. Fr. **1959** 1159/60).
[17] Guenebaut, H.; Pannetier, G.; Goudmand, P. (Compt. Rend. **251** [1960] 1166/8).
[18] Guenebaut, H.; Pannetier, G.; Goudmand, P. (Bull. Soc. Chim. Fr. **1962** 80/6).
[19] Pannetier, G.; Goudmand, P.; Dessaux, O.; Guenebaut, H. (Compt. Rend. **256** [1963] 3082/5).
[20] Murai, T.; Shimauchi, M. (Sci. Light [Tokyo] **15** [1966] 48/67).

[21] Malicet, J.; Brion, J.; Guenebaut, H. (J. Chim. Phys. Phys. Chim. Biol. **67** [1970] 25/30).
[22] Anderson, W. R.; Crosley, D. R. (Chem. Phys. Lett. **62** [1979] 275/8).
[23] Anderson, W. R.; Crosley, D. R.; Jones, J. E.; Allen, J. E., Jr. (Chem. Phys. Processes Combust. **1978** 58/1-58/3).
[24] Crosley, D. R.; Anderson, W. R. (AD-A091791 [1980] 42 pp. from Gov. Rep. Announce. Index [U. S.] **81** No. 6 [1981] 1075).

[25] Brazier, C. R.; Ram, R. S.; Bernath, P. F. (J. Mol. Spectrosc. **120** [1986] 381/402).
[26] Ubachs, W.; Ter Meulen, J. J.; Dymanus, A. (Can. J. Phys. **62** [1984] 1374/91).
[27] Chang, A. Y.; Hanson, R. K. (J. Quant. Spectrosc. Radiat. Transfer **42** [1989] 207/17).
[28] Kopp, I.; Kronekvist, M.; Aslund, N. (Arkiv Fys. **30** [1965] 9/17).
[29] Shimauchi, M. (Sci. Light [Tokyo] **15** [1966] 161/5).
[30] Shimauchi, M. (Sci. Light [Tokyo] **16** [1967] 185/90).

[31] Krishnamurty, G.; Narasimham, N. A. (Proc. Indian Acad. Sci. A **67** [1968] 50/60).
[32] Krishnamurty, G.; Narasimham, N. A. (Proc. 1st Int. Conf. Spectrosc., Bombay 1967, Vol. 1, pp. 142/4).
[33] Bollmark, P.; Kopp, I.; Rydh, B. (J. Mol. Spectrosc. **34** [1970] 487/99).
[34] Patel-Misra, D.; Sauder, D. G.; Dagdigian, P. J. (Chem. Phys. Lett. **174** [1990] 113/8).

2.1.1.3.4.6.2 NH and ND in Liquids and Solids

A transient absorption maximum observed at 350 nm upon pulse radiolysis (600-keV electrons) of **liquid** NH_3 at 223 K was identified as the A $^3\Pi_i \leftarrow X\ ^3\Sigma^-$ transition of NH, and an extinction coefficient $\varepsilon = 2.0 \times 10^4$ L·mol^{-1}·cm^{-1} was measured [1, 2].

The A $^3\Pi_i \rightarrow X\ ^3\Sigma^-$, $1 \rightarrow 0$, $0 \rightarrow 0$, and $0 \rightarrow 1$ emission bands at 304.8, 336.0, and 376.6 nm were observed immediately after laser-induced photolysis (KrF laser at 248 nm) of **solid** NH_4ClO_4 [3].

NH and ND radicals in **inert-gas matrices** were identified by their A $^3\Pi_i \leftrightarrow X\ ^3\Sigma^-$ absorption and emission spectra. HN_3 (DN_3), NH_3 (ND_3), or mixtures of H_2 with N-containing molecules and a thirtyfold excess of a noble gas were passed through a discharge tube and then condensed at 4.2 K. The absorption spectra showed the $0 \leftarrow 0$ and $1 \leftarrow 0$ bands of NH (ND) with red shifts of a few hundred cm^{-1} compared to those of the gas-phase spectra (wavenumbers ν in cm^{-1}):

species	ν(0←0)	ν(1←0)	species	ν(0←0)	ν(1←0)
NH in Ar	29581±3	32560±3	ND in Ar	29608±6	31819±6
NH in Kr	29509±6	–	ND in Kr	29522±6	31711±6
NH in Xe	29403±30	–	ND in Xe	29476±50	31610±60

Sharp emission lines induced by the light source appeared to the red of the $0 \leftarrow 0$ absorption bands and were assumed to result from resonance fluorescence between low-lying rotational levels in the A $^3\Pi_i$ state and X $^3\Sigma^-$-state rotational levels. The matrix shifts, the vibrational perturbations, and the rotational structure of the absorption and emission bands were used to develope models that describe the interaction between the trapped NH (ND) molecules and the host-lattice atoms [4 to 9].

Low concentrations of NH in noble-gas matrices have been detected by measuring the absorption and the magnetic circular dichroism (MCD) spectrum in the region of the $0 \leftarrow 0$ band at 4.5 to 8 K after discharges through 1% mixtures of NH_3 in Ar or Xe and condensation at low temperatures [10].

UV photolysis during or after deposition of diluted mixtures of NH_3 (ND_3) or HN_3 (DN_3) in inert gases also produces NH, ND(X $^3\Sigma^-$) radicals. The $0 \leftarrow 0$ absorption bands were observed at 338 nm ($\varepsilon = 4.0 \times 10^4$ L·mol^{-1}·cm^{-1}) in NH_3-Ar (mole ratio 3:1000) deposits between 4.2 and 36 K [11] and at 336 nm in $HN_3(DN_3)$-Ar and $HN_3(DN_3)$-N_2 (1:100 to 400) deposits between 4 and 20 K [12, 13]. The $R_1(1)$ lines of the 0←0 and 1←0 bands at 338.90±0.11 and 308.07±0.14 nm were found after photodecomposition of HN_3 trapped in Kr (1:60±20; also Xe matrices) at 4.2 K [14]. Weak emission bands at 335.92 and 337.06 nm in electrical glow discharges from a solid nitrogen condensate (4 K) were identified to be the Q maxima of the 0→0 and 1→1 bands [15]. An intense line at 337.9 nm and a much weaker one at 377.5 nm appearing at 25 to 31 K were identified to be the 0→0 and 0→1 bands in the luminescence spectrum recorded during UV photolysis and simultaneously warming up an NH_3-Ar (1:100) condensate of 10 K [16].

Time-resolved fluorescence was used to study radiative decay and vibrational relaxation of A- and X-state levels with v≤2 and interactions with the host lattice in Ne, Ar, and Kr matrices at 4 to 21 K [17 to 21].

References:

[1] Belloni, J.; Cordier, P.; Delaire, J. (Chem. Phys. Lett. **27** [1974] 241/4).
[2] Belloni, J.; Billiau, F.; Cordier, P.; Delaire, J. A.; Delcourt, M. O. (J. Phys. Chem. **82** [1978] 532/6).
[3] Chen, K.; Wang, G.; Kuo, C.; Shyy, I.; Chang, Y. (Chem. Phys. Lett. **167** [1990] 351/5).
[4] Robinson, G. W.; McCarty, M., Jr. (J. Chem. Phys. **28** [1958] 350).
[5] Robinson, G. W.; McCarty, M., Jr. (Can. J. Phys. **36** [1958] 1590/1).
[6] McCarty, M., Jr.; Robinson, G. W. (J. Am. Chem. Soc. **81** [1959] 4472/6).
[7] McCarty, M., Jr.; Robinson, G. W. (J. Chim. Phys. **56** [1959] 723/31).
[8] McCarty, M., Jr.; Robinson, G. W. (4th Int. Symp. Free Radical Stab. Trapped Radicals Low Temp., Washington, D. C., 1959, pp. F-III-1/F-III-20).
[9] McCarty, M., Jr.; Robinson, G. W. (Mol. Phys. **2** [1959] 415/30).
[10] Lund, P. A.; Hasan, Z.; Schatz, P. N.; Miller, J. H.; Andrews, L. (Chem. Phys. Lett. **91** [1982] 437/9).
[11] Schnepp, O.; Dressler, K. (J. Chem. Phys. **32** [1960] 1682/6).
[12] Milligan, D. E.; Jacox, M. E. (J. Chem. Phys. **41** [1964] 2838/41).
[13] Jacox, M. E.; Milligan, D. E. (Appl. Opt. **3** [1964] 873/6).
[14] Keyser, L. F.; Robinson, G. W. (J. Am. Chem. Soc. **82** [1960] 5245/6).
[15] Bass, A. M.; Broida, H. P. (Phys. Rev. [2] **101** [1956] 1740/7).
[16] Van de Bult, C. E. P. M.; Allamandola, L. J.; Baas, F.; Van Ijzendoorn, L.; Greenberg, J. M. (J. Mol. Struct. **61** [1980] 235/8).
[17] Bondybey, V. E.; Brus, L. E. (J. Chem. Phys. **63** [1975] 794/804).
[18] Goodman, J.; Brus, L. E. (J. Chem. Phys. **65** [1976] 1156/64).
[19] Goodman, J.; Brus, L. E. (J. Chem. Phys. **65** [1976] 3146/52).
[20] Bondybey, V. E. (J. Chem. Phys. **65** [1976] 5138/40).
[21] Bondybey, V. E.; English, J. H. (J. Chem. Phys. **73** [1980] 87/92).

2.1.1.3.4.6.3 Franck-Condon Factors. r-Centroids. Transition Probabilities. Transition Moments

Based on lifetime measurements in various A-state, v′=0 and 1, rotational levels (NH_3 excitation by electron beam) [1], Franck-Condon (FC) factors $q_{v',v''}$, r-centroids $\bar{r}$, oscillator strengths f, and electronic transition moments R_e were calculated for Q-branch transitions with N=0, 5, 10, and 15 in the 0→0 and 1→1 bands [2]. Earlier lifetime measurements in the (not rotationally resolved) 0→0 and 1→1 bands [3] gave FC factors, r-centroids, transition probabilities (Einstein coefficients) $A_{v',v''}$ for A↔X transitions with v′=0, 1, 2 and $v''=0$, 1, 2, and $f_{0,0}$ [4]. Intensity measurements of the $0\rightarrow1$, $0\rightarrow0$, and $1\rightarrow0$ bands upon collisions of NH_3 with metastable Ar atoms were used to derive $A_{0,1}$, $A_{1,0}$, and $A_{1,2}$ [5, 6]. The ratio $A_{0,1}/A_{0,0}$ was obtained from LIF measurements in an NH_3 discharge flow system [7 to 9]. Intensity measurements on the entire Q branches of the $0\rightarrow0$, $1\rightarrow1$, and $2\rightarrow2$ bands in shock-wave-excited NH_3-Ar mixtures were used to derive the oscillator strengths $f_{0,0}$, $f_{1,1}$, and $f_{2,2}$ and the corresponding transition moments R_e; the $1\rightarrow1$ and $2\rightarrow2$ transitions were found to be stronger than the $0\rightarrow0$ transition which according to [6] results from the overlap with P and R branches of other bands [10]. FC factors and r-centroids for transitions with $\Delta v=0$, ±1 and $v=0$, 1, 2 are based on earlier data for band-head intensities [11, 12].

Absorption measurements in the Q-branch region of the $0\leftarrow0$ band (reflected shock-waves in NH_3-Ar and N_2-H_2-Ar mixtures) combined with FC factors [4] and a partition function based on spectroscopic constants of NH resulted in the transition probability $|R_e|^2=0.086\pm0.024$ $(ea_0)^2$ for the $A\leftarrow X$ transition [13] (in a review, the authors quote only half the value, 0.043 ± 0.024 $(ea_0)^2$ [14]).

Using spectroscopic constants for the X and A states of NH, rotationally dependent FC factors have been calculated for P, Q, R transitions with J=1, 21, 41 in the 0−0, 1−1, 1−0, 0−1 bands [15], for P and R transitions with J=1, 11, 21, 31 in the 0−1 and 1−0 bands [16], and Einstein coefficients for 89 rovibronic transitions between the X $^3\Sigma^-$, $v=0$, N=0 to 5 and the A $^3\Pi_i$, $v=0$, N=1 to 5 (and Λ-doublets) levels [17].

Recent quantum-chemical ab initio (CI) calculations gave transition dipole moments between the X $^3\Sigma^-$ and A $^3\Pi_i$ states and Einstein coefficients for the $v'-v''=0-0$, 1 and 1−0, 1, 2 transitions [18], transition moments, Einstein coefficients, and Franck-Condon factors for the $v'-v''=0-0$, 1, 2 and 1−0, 1, 2 transitions [19].

References:

[1] Smith, W. H.; Brzozowski, J.; Erman, P. (J. Chem. Phys. **64** [1976] 4628/33).
[2] Smith, W. H.; Hsu, D. K. (J. Quant. Spectrosc. Radiat. Transfer **22** [1979] 223/9).
[3] Smith, W. H. (J. Chem. Phys. **51** [1969] 520/4).
[4] Smith, W. H.; Liszt, H. S. (J. Quant. Spectrosc. Radiat. Transfer **11** [1971] 45/54).
[5] Lents, J. M., Jr. (Diss. Univ. Tennessee 1970, 277 pp.; Diss. Abstr. Int. B **32** [1971] 482).
[6] Lents, J. M. (J. Quant. Spectrosc. Radiat. Transfer **13** [1973] 297/310).
[7] Anderson, W. R.; Crosley, D. R. (Chem. Phys. Lett. **62** [1979] 275/8).
[8] Crosley, D. R.; Anderson, W. R. (AD-A091791 [1980] 42 pp. from Gov. Rep. Announce. Index [U. S.] **81** No. 6 [1981] 1075).
[9] Anderson, W. R.; Crosley, D. R.; Jones, J. E.; Allen, J. E., Jr. (Chem. Phys. Processes Combust. **1978** 58/1-58/3).
[10] Harrington, J. A.; Modica, A. P.; Libby, D. R. (J. Quant. Spectrosc. Radiat. Transfer **6** [1966] 799/805).

[11] Singh, M.; Chaturvedi, J. P. (J. Quant. Spectrosc. Radiat. Transfer **37** [1987] 81/4).
[12] Singh, M.; Chaturvedi, J. P. (Earth, Moon, Planets **38** [1987] 253/61).
[13] Monyakin, A. P.; Ovsyannikova, N. G.; Kuznetsova, L. A. (Vestn. Mosk. Univ. Khim. **30** [1975] 235/6; Moscow Univ. Chem. Bull. **30** No. 2 [1975] 82/3).
[14] Kuznetsova, L. A.; Kuz'menko, N. E.; Kuzyakov, Y. Y.; Monyakin, A. P.; Plastinin, Y. A.; Smirnov, A. D. (Teor. Spektrosk. **1977** 160/2).
[15] Bell, R. A.; Branch, D.; Upson, W. L., II (J. Quant. Spectrosc. Radiat. Transfer **16** [1976] 177/84).
[16] Elsum, I. R. (Chem. Phys. **90** [1984] 83/6).

[17] Litvak, M. M.; Rodriguez Kuiper, E. N. (Astrophys. J. **253** [1982] 622/33).
[18] Kirby, K. P.; Goldfield, E. M. (J. Chem. Phys. **94** [1991] 1271/6).
[19] Yarkony, D. R. (J. Chem. Phys. **91** [1989] 4745/57).

2.1.1.3.4.7 The c $^1\Pi \leftrightarrow$ a $^1\Delta$ Emission and Absorption System at 324 nm

Besides the prominent A $^3\Pi_i \rightarrow$ X $^3\Sigma^-$ emission system at 336 nm, electric discharges through NH_3 (ND_3) or N_2-H_2 (D_2) mixtures revealed at lower wavelengths a system of red-degraded emission bands with simple P, Q, R structure, i.e., intense single P and Q branches and a weaker R branch forming the head, and obvious Λ-type doubling due to the upper state in all but the lowest J-value lines. The rotational structure and the missing lines were recognized to be characteristic of a $^1\Pi \rightarrow {}^1\Delta$ transition already in the earliest studies which dealt with the 0→0, 1→0, and 0→1 bands of NH (ND) at 324 (323.5), 303.5 (307.5), and 361.0 (350.1) nm, respectively, and photographed at moderate resolution [1 to 5]. These investigations were followed by analyses of high-resolution spectra of the 0→0 [6, 7] and 0→1 [7] bands of NH and of the 0→0, 1→0, and 0→1 bands of ND [8]. A high-resolution Fourier transform emission spectrum of the 0→0, 1→0, and 0→1 bands of NH (hollow-cathode discharges through flowing He(4.5 Torr)-N_2(0.12 Torr)-H_2(0.04 Torr) mixtures) enabled the Λ-type doubling due to the c $^1\Pi$ and the a $^1\Delta$ states to be resolved [9]. Based on the 0→0-band analyses of [7, 9], the ν_{00} value for the c→a system of NH as quoted by Huber and Herzberg [10] from the analysis of [6] has to be revised (see table on p. 80).

High-resolution spectra of the 0→2 and 1→2 bands of ND at 381.5 and 359.2 nm excited by MW discharges through mixtures of 0.2% ND_3 in Ar have been analyzed [11].

Upon vacuum-UV photolysis of NH_3 and ND_3 with Ar or Kr resonance lines, the rotationally resolved 0→0, 1→0 bands of NH and ND [12, 13], the 1→1, 1→2, and 0→2 bands of NH, and the 2→0, 2→1, 1→1, 1→2, 1→3, and 0→2 bands of ND [13] were recorded at moderate resolution. Rotational predissociation of NH(c $^1\Pi$, v'=0, 1) is obvious because of the significant decrease in intensity for rotational lines with J'≥18 and 11, respectively [13]. Rotationally resolved 1→0, 0→0, and 0→1 bands at 304.2, 325.1, and 362.7 nm were recorded at resolutions of 0.2 and 0.08 nm after ArF laser (193 nm) photolysis of HN_3 [14].

The c←a absorption spectrum in the region of the 0←0, 1←0, and 2←0 bands has been observed in a flash discharge through NH_3; a rotational analysis of the 2←0 band is presented [15].

More recently, the technique of laser-induced fluorescence (LIF) has been applied to study highly resolved c←a excitation spectra; the excited NH(a $^1\Delta$, v'') radicals were generated by UV photolysis of HN_3 at 266 or 248 nm (v'' up to 3 or 4, respectively) and the c $^1\Pi$, v', J'←a $^1\Delta$, v'', J'' rovibronic transitions probed with a tunable dye laser. Excitation spectra of all vibronic bands c, v'=0←a, v''=0 to 3 in the range 324 to 466 nm and c, v'=1←a, v''=1 to 4 in the range 303 to 478 nm and rotational analyses of the 0←2, 3 and 1←4 bands are presented [16] (results for the 0←2 and 0←3 bands have been published in [17], for the 1←4 band in [29]). Excitation and dispersed emission spectra of v'=0, 1↔v''=0 to 3 transitions in NH have also been studied by [18], excitation spectra of the 0←1 and 1←2 and emission spectra of the 0→0 and 1→0 transitions of NH and ND by [19].

Excitation spectra of the c $^1\Pi$, v'=0, 1←a $^1\Delta$, v''=0, 1 transitions of NH and ND were recorded after photolysis of NH_3 or ND_3 at 193 nm [20] and of the 0←0 transition of NH after photolysis of HNCO at 193 nm [21].

An LIF study under ultrahigh resolution using an NH molecular beam from MW discharges in NH_3 enabled the hyperfine splittings (hfs) of both nuclei in each of the Λ-doubling components of the 0→0 band lines Q(2) to Q(7) and P(2), P(5), and P(6) to be resolved; some 120 splittings ranging from 25 to 390 MHz ($\sim 1 \cdot 10^{-5}$ to $1.4 \cdot 10^{-4}$ nm) relative to the strongest hfs component, $F' = J' + 3/2 \leftarrow F'' = J'' + 3/2$, in each rotational transition have been listed, and hfs parameters (cf. p. 42) and absolute wavenumbers for the c $^1\Pi$, v' = 0, J' ← a $^1\Delta$, v'' = 0, J'' transitions derived [22].

The following tables list the band origins and the highest J values for the rotational structure of NH and ND, derived from high-resolution LIF spectra [16, 17, 22, 29], the Fourier transform emission spectrum [9], and the absorption spectrum [15] of NH, and from the emission spectra recorded at high resolution for NH [7] and ND [8, 11] and at moderate resolution for NH and ND [13].

Band origins $\nu_{v',v''}$ (in cm^{-1}) and J_{max} values for NH:

v' → / v'' ↓	0: $\nu_{v',v''}$	J_{max}	Ref.	1: $\nu_{v',v''}$	J_{max}	Ref.	2: $\nu_{v',v''}$	J_{max}	Ref.
0	30704.0743(32)	7	[22]	32825.5375(67)	12	[9]	34572.3	7	[15]
	30704.0818(16)	19	[9]						
	30704.101(26)	–	[7]						
1	27521.2939(16)	18	[9]	29690.2 ± 3	–	[13]			
	27521.343(13)	15	[7]						
2	24484.61 ± 1.1	13	[17,	26652.0 ± 3	–	[13]			
3	21591.60 ± 0.07	7	16]						
4				20962.66 ± 0.12	5	[16, 29]			

Band origins $\nu_{v',v''}$ (in cm^{-1}) and J_{max} values for ND:

v' → / v'' ↓	0: $\nu_{v',v''}$	J_{max}	Ref.	1: $\nu_{v',v''}$	J_{max}	Ref.	2: $\nu_{v',v''}$	J_{max}	Ref.
0	30849.06 ± 0.03	30	[8]	32470.79 ± 0.04	15	[8]	33883.8 ± 3	–	[13]
1	28492.89 ± 0.02	16	[8]	30113.9 ± 3	–	[13]	31531.7 ± 3	–	[13]
2	26215.53 ± 0.08	10	[11]	27837.0 ± 0.05	10	[11]			
	26216.4 ± 3	–	[13]	27834.3 ± 3	–	[13]			
3				25636.4 ± 3	–	[13]			

The c $^1\Pi$ → a $^1\Delta$ fluorescence or c $^1\Pi$ ← a $^1\Delta$ excitation (LIF) spectra of NH or ND are presented in numerous publications which report on various formation processes of NH (ND) in the c $^1\Pi$ or a $^1\Delta$ states, such as formation by VUV photolysis of NH_3, HN_3, HNCO, CH_3NH_2, or CH_2NHCH_2, by electron impact on NH_3, HN_3, HNCO, N_2H_4, CH_3NH_2, or CH_2NHCH_2, in collisions of metastable noble-gas atoms with NH_3, HN_3, or HNCO, and in chemical reactions of H atoms with NF_2 or N_3. The rotationally resolved c $^1\Pi$ ↔ a $^1\Delta$ spectra supply information on the branching ratios of electronic states and vibrational, rotational, and even Λ-doublet levels populated in a definite formation process; time-resolved spectra give insight into the subsequent relaxation processes. For details and literature see Sections 2.1.1.1 and 2.1.1.4.5.

Franck-Condon factors and r-centroids for c $^1\Pi$ (v' = 0 to 3) ↔ a $^1\Delta$ (v'' = 0 to 3) transitions of NH and ND were calculated using the own spectroscopic constants and literature data and assuming a Morse potential function for the molecules [11]. Based on lifetime measurements in the c → a spectrum excited by electron impact on NH_3 [23], Franck-Condon factors and r-centroids were calculated for c−a transitions with v′=0 to 3 and v″=0 to 4 [24]. Intensity measurements on the 0→0, 1, 2, 3 and 1→0, 1, 2, 3, 4 bands upon HN_3 photolysis and NH(c←a) excitation were used to derive the transition probabilities $A_{v',v''}$ [29]; intensity measurements on the 0→0 and 0→1 bands upon collisions of NH_3 with metastable Ar atoms gave $A_{0,1}$ [25, 26]. Intensity measurements on the 0↔1, 2 and 1↔1, 2, 3 emission and LIF bands yielded the relative (with respect to the 0−0 and the 1−0 bands, respectively) Einstein transition probabilities $A_{v',v''}$ for emission and $B_{v',v''}$ for absorption [18]. A quantum-chemical ab initio (CI) calculation gave the transition moment, Einstein coefficient, and Franck-Condon factor for the 0−0 transition [27].

References:

[1] Pearse, R. W. B. (Proc. R. Soc. [London] A **143** [1934] 112/23).
[2] Dieke, G. H.; Blue, R. W. (Phys. Rev. [2] **45** [1934] 395/400).
[3] Nakamura, G.; Shidei, T. (Jpn. J. Phys. **10** [1934] 5/10).
[4] Chauvin, H.; Leach, S. (Compt. Rend. **231** [1950] 1482/4).
[5] Florent, R.; Leach, S. (J. Phys. Radium [8] **13** [1952] 377/85).
[6] Shimauchi, M. (Sci. Light [Tokyo] **13** [1965] 53/63).
[7] Ramsay, D. A.; Sarre, P. J. (J. Mol. Spectrosc. **93** [1982] 445/6).
[8] Hanson, H.; Kopp, I.; Kronekvist, M.; Aslund, N. (Arkiv Fys. **30** [1965] 1/8).
[9] Ram, R. S.; Bernath, P. F. (J. Opt. Soc. Am. B Opt. Phys. **3** [1986] 1170/4).
[10] Huber, K. P.; Herzberg, G. (Molecular Spectra and Molecular Structure, Vol. 4, Constants of Diatomic Molecules, Van Nostrand Reinhold, New York 1979, pp. 456/61).

[11] Cheung, W. Y.; Gelernt, B.; Carrington, T. (Chem. Phys. Lett. **66** [1979] 287/90).
[12] Masanet, J.; Gilles, A.; Vermeil, C. (J. Photochem. **3** [1974/75] 417/29).
[13] Zetzsch, C. (J. Photochem. **9** [1978] 151/3).
[14] Haak, H. K.; Stuhl, F. (J. Phys. Chem. **88** [1984] 3627/33).
[15] Dabrovski, I.; Herzberg, G. (unpublished results, private communication to [28]).
[16] Mill, T. (Ber.-Max-Planck-Inst. Strömungsforsch. **14** [1990] 1/101, 12/30).
[17] Hack, W.; Mill, T. (J. Mol. Spectrosc. **144** [1990] 358/65).
[18] Nelson, H. H.; McDonald, J. R. (J. Chem. Phys. **93** [1990] 8777/83).
[19] Bohn, B.; Stuhl, F.; Parlant, G.; Dagdigian, P. J.; Yarkony, D. R. (J. Chem. Phys. **96** [1992] 5059/68).
[20] Kenner, R. D.; Rohrer, F.; Stuhl, F. (J. Chem. Phys. **86** [1987] 2036/43).

[21] Drozdoski, W. S.; Baronavski, A. P.; McDonald, J. R. (Chem. Phys. Lett. **64** [1979] 421/5).
[22] Ubachs, W.; Meyer, G.; Ter Meulen, J. J.; Dymanus, A. (J. Mol. Spectrosc. **115** [1986] 88/104).
[23] Smith, W. H. (J. Chem. Phys. **51** [1969] 520/4).
[24] Smith, W. H.; Liszt, H. S. (J. Quant. Spectrosc. Radiat. Transfer **11** [1971] 45/54).
[25] Lents, J. M., Jr. (Diss. Univ. Tennessee 1970, 277 pp.; Diss. Abstr. Int. B **32** [1971] 482).
[26] Lents, J. M. (J. Quant. Spectrosc. Radiat. Transfer **13** [1973] 297/310).
[27] Yarkony, D. R. (J. Chem. Phys. **91** [1989] 4745/57).
[28] Graham, W. R. M.; Lew, H. (Can. J. Phys. **56** [1978] 85/99).
[29] Hack, W.; Mill, T. (J. Phys. Chem. **97** [1993] 5599/606).

2.1.1.3.4.8 The d $^1\Sigma^+ \rightarrow$ c $^1\Pi$ Emission System at 253 nm

The d $^1\Sigma^+ \rightarrow$ c $^1\Pi$ emission spectrum was excited in hollow-cathode discharges through flowing NH_3 (0→0 and weak 1→1 bands at 255.7 to 247.5 nm) [1, 2], in mildly condensed transformer discharges through flowing NH_3 (0→0, 0→1, and 1→1 bands at 268.4 to 246.6 nm) [3], in positive column discharges through flowing He-N_2-H_2 (D_2) mixtures (0→0 and 1→1 bands of NH and ND, 2→2 band of ND, at 246 to 256 nm) [4], and by impact of an electron beam on NH_3 or ND_3 from supersonic gas jets (six bands of NH and eight bands of ND (see below) in the range 268.4 to 235.4 nm) [5]. The rotational structure consists of single P, Q, and R branches characteristic of a $^1\Sigma \rightarrow {}^1\Pi$ transition. For the 0→0, 1→1, and 0→1 bands (J_{max}=23, 16, and 16) of NH and for the 0→0, 1→1, and 2→2 bands (J_{max}=31, 26, and 18) of ND, the wavenumbers listed for the rotational lines from the high-resolution studies (accuracy better than 0.1 cm^{-1}) are in good agreement [3 to 5]. A rotational analysis yielded the following wavenumbers (in cm^{-1}) for the band origins [5]:

NH				ND					
0→0	39512.26	0→1	37389.72	0→0	39484.22	0→1	37863.33		
1→0	42042.73	1→1	39920.00	1→0	41362.43	1→1	39740.59	1→2	38314.93
2→1	42306.49	2→2	40612.41	2→2	40116.22	2→3	38948.03	3→3	40672.45

Predissociation by rotation of the c $^1\Pi$ state is evident from the abrupt disappearance or weakening of rotational structure simultaneously in all three branches at J″≥24 or 32 in the 0→0 bands of NH or ND, at J″≥17 in the 0→1 band of NH, and at J″≥26, 19, and 6 in the 1→1, the 2→1 and 2→2, and in the 2→3 and 3→3 bands of ND [3, 5, 6]. Predissociation of the d $^1\Sigma^+$ state as concluded from the cutoff in rotational structure at J′≥16 in the 1→1 band of NH [3, 6] most likely does not occur, since rotational levels up to J′=23 could be observed in the intense 1→2 and 1→3 bands of the d $^1\Sigma^+ \rightarrow$ b $^1\Sigma^+$ system (see next section) [5].

Franck-Condon factors and r-centroids were calculated for d−c transitions with v′=0 to 4 and v″=0 to 3 [7] using lifetime data for the d→c spectrum excited by electron impact on NH_3 [8].

References:

[1] Lunt, W.; Pearse, R. W. B.; Smith, E. C. W. (Nature **136** [1935] 32).
[2] Lunt, W.; Pearse, R. W. B.; Smith, E. C. W. (Proc. R. Soc. [London] A **155** [1936] 173/82).
[3] Narasimham, N. A.; Krishnamurty, G. (Proc. Indian Acad. Sci. A **64** [1966] 97/110).
[4] Whittaker, F. L. (Proc. Phys. Soc. [London] **90** [1967] 535/41).
[5] Graham, W. R. M.; Lew, H. (Can. J. Phys. **56** [1978] 85/99).
[6] Krishnamurty, G.; Narasimham, N. A. (J. Mol. Spectrosc. **29** [1969] 410/4).
[7] Smith, W. H.; Liszt, H. S. (J. Quant. Spectrosc. Radiat. Transfer **11** [1971] 45/54).
[8] Smith, W. H. (J. Chem. Phys. **51** [1969] 520/4).

2.1.1.3.4.9 The d $^1\Sigma^+ \rightarrow$ b $^1\Sigma^+$ Emission System at 162 nm

Two of the emission sources for d $^1\Sigma^+ \rightarrow$ c $^1\Pi$ excitation, electric discharges in flowing He-N_2-H_2 (D_2) mixtures and electron bombardment of NH_3, also excite the d $^1\Sigma^+ \rightarrow$ b $^1\Sigma^+$ system, the vibrational bands of which show single P and R branches characteristic of a $^1\Sigma \rightarrow {}^1\Sigma$ transition. The spectra of NH and ND generated by electric discharges were too weak for rotational analysis; R-branch heads observed for NH at 170.846, 180.101, and 181.440 nm were attributed to the 0→1, 0→2, and 1→3 bands, those observed for ND at 168.315, 175.036, 169.517, 176.103, and 177.192 nm to the 0→1, 0→2, 1→2, 1→3, and 2→4 bands; wavenumbers for the missing 0→0 and 1→2 band heads of NH and the missing 0→0 band head of ND (unobservable due to overlap) and wavenumbers for all band origins were calculated using the known spectroscopic data [1, 2] for the states involved; ν_{00} = 61619.23 cm^{-1} for NH and ν_{00} = 61721.63 cm^{-1} for ND [3]. Electron bombardment on NH_3 enabled the observation and analysis of 15 rotational-vibrational bands (wavenumber accuracy ± 0.06 cm^{-1}) that involve vibrational levels v′ = 0 to 6 (except 4) in the d state and v″ = 1 to 9 in the b state and rotational levels with J_{max} = 24 to 26 in the 0→1, 0→2, 1→2, and 1→3 and J_{max} = 8 to 12 in the remaining bands; the 0→0 band could not be analyzed because of overlap by H_2 bands. Wavenumbers (in cm^{-1}) of the band origins are as follows [4]:

0→0	61619.60*⁾	0→1	58413.75	0→2	55350.61
1→2	57880.82	1→3	54958.51		
2→3	57345.28	2→4	54562.63	2→5	51917.61
3→4	56807.98	3→5	54163.48	3→6	51657.29
5→6	55723.73	5→7	53356.46	5→8	51129.90
6→8	52958.53	6→9	50876.07		

*⁾ Calculated value.

Using the molecular constants from [3] and modified RKR potential energy functions for the b $^1\Sigma^+$ and d $^1\Sigma^+$ states, Franck-Condon factors and r-centroids for d − b transitions with v′ = 0 to 3 and v″ = 0 to 3 were calculated [5].

References:

[1] Whittaker, F. L. (Proc. Phys. Soc. [London] **90** [1967] 535/41).
[2] Whittaker, F. L. (J. Phys. B **1** [1968] 977/82).
[3] Whittaker, F. L. (Can. J. Phys. **47** [1969] 1291/3).
[4] Graham, W. R. M.; Lew, H. (Can. J. Phys. **56** [1978] 85/99).
[5] Rao, T. V. R.; Lakshman, S. V. J. (Indian J. Pure Appl. Phys. **11** [1973] 539/40).

2.1.1.3.5 Resonance-Enhanced Multiphoton Ionization (REMPI) Spectra

Multiphoton ionization by UV radiation, (2+1) REMPI of NH(X $^3\Sigma^-$) and NH(a $^1\Delta$), with the high-lying valence state d $^1\Sigma^+$ or a number of singlet and triplet Rydberg states as the intermediate states responsible for the resonance enhancement has been comprehensively investigated most recently (further studies obviously are in progress). Laser excitation wavelengths (tunable dye lasers) between 288 and 220 nm were used which give two-photon energies of 69000 to 91000 cm^{-1}. The NH, ND(a $^1\Delta$) radicals, which enable REMPI via resonant singlet states, were produced by HN_3 (DN_3) photodissociation with the same laser radiation as used for the (2+1) excitation and ionization processes. The NH, ND(X $^3\Sigma^-$) radicals, which enable REMPI via resonant triplet states, were obtained by reacting excess F atoms with NH_3 (ND_3) in a flow reactor. REMPI spectra were obtained by recording either the fraction of NH^+ or ND^+ ions (with a time-of-flight mass spectrometer) or the photoelectron yield (REMPI-PES) both as a function of the excitation wavelength. These spectra are consistent with the excitation or absorption spectra of the NH (ND) radicals for the two-photon transitions between the lower X $^3\Sigma^-$ or a $^1\Delta$ states and the resonant intermediate state;

they allow the latter to be characterized by analyzing the rotational-vibrational structure which, in addition to the P, Q, R branches appearing in one-photon spectra, exhibits S and O branches due to ΔJ or $\Delta N = \pm 2$ transitions. Moreover, electronic transitions with $\Delta\Lambda = \pm 2$ (e.g. $^1\Delta \leftrightarrow {}^1\Sigma$) occur. Vibrationally and rotationally resolved REMPI-PE spectra were obtained by measurements of the photoelectron kinetic energy at selected excitation wavelengths which show ionizations of selected intermediate state levels (v', J', N') into definite (v^+, J^+, N^+) levels of the ion. Besides REMPI via the two-photon d $^1\Sigma^+ \leftarrow\leftarrow$ a $^1\Delta$ transition, REMPI via a number of hitherto undetected Rydberg states, identified as the first members of $\{NH^+\ X\ ^2\Pi_r$, a $^4\Sigma^-$, or B $^2\Delta\}$ nsσ, npσ, npπ, ndσ, or ndπ series, have been investigated. Five singlet Rydberg states, f $^1\Pi$, g $^1\Delta$, h $^1\Sigma^+$, i $^1\Pi$, and j $^1\Delta$ for NH and ND, three triplet Rydberg states, B $^3\Pi$, D $^3\Pi$, and F $^3\Sigma^-$ for NH, and five triplet Rydberg states, B $^3\Pi$, C $^3\Sigma^-$, D $^3\Pi$, E $^3\Sigma^-$, and F $^3\Sigma^-$ for ND have been detected and characterized. Details for the individual REMPI processes are given below.

The d $^1\Sigma^+ \leftarrow\leftarrow$ a $^1\Delta$ Transition. This resonance enhancement gives rise to band systems of NH and ND between 287 and 273 nm with well-resolved O-, P-, Q-, R-, and S-branch lines in the 0−0 band centered around 285 nm [1 to 5], in the 1−0 band around 274 nm [1, 3], and in the 2−1 band in the range 278 to 279 nm [1]. The assignment is supported by the so-called J_{min} pattern, i.e., by the five J''_{min} values of the rotational branches correlating in a definite manner with the upper- and lower-state symmetries, and by rotational and vibrational analyses which yielded rotational constants and vibrational spacings in good agreement with those known from one-photon transitions involving the d and a states. The following band origins ν_0 (in cm^{-1}) were determined [1, 3] (according to [1], the discrepancies of the ν_0 values are presumably due to neglecting the air-to-vacuum correction of the wavenumbers by [3]):

band	ν_0(NH) a) [1]	ν_0(ND) a)	ν_0(NH) b) [3]	ν_0(ND) b)
$v=0\leftarrow0$	70202.8(5)	70297.6(5)	70211.39(13)	70306.91(12)
$v=1\leftarrow0$	72732.1(5)	72175.8(5)	72740.89(16)	72185.13(19)
$v=2\leftarrow1$	71935.3(5)	71620.3(5)	—	—

a) Values in parentheses represent the likely uncertainty in interpolating between the Ne calibration lines. — b) One standard deviation in parentheses.

It may be noted that early REMPI studies of NH_3 in the range 380 to 500 nm identified an absorption band at 396 nm as the three-photon d $^1\Sigma^+$, $v'=2\leftarrow\leftarrow\leftarrow$ a $^1\Delta$, $v''=0$ transition (75780 cm^{-1}) in NH and subsequent ionization by a fourth photon. NH(a $^1\Delta$) obviously results from collisional quenching of NH(c $^1\Pi$) (a three-photon photolysis product of NH_3), because the (3+1) process was observed only in static NH_3 gas but not in an expansion-cooled NH_3 beam where the environment is essentially collision-free [6, 7].

The f $^1\Pi$, g $^1\Delta$, h $^1\Sigma$, i $^1\Pi$, j $^1\Delta\leftarrow\leftarrow$ a $^1\Delta$ Transitions. Resonance enhancements due to the three Rydberg $\{NH^+\ X\ ^2\Pi_r\}$ 3p states, f $^1\Pi$, g $^1\Delta$, and h $^1\Sigma$, were identified at excitation wavelengths of 283 to 245 nm [3, 8, 9]. At shorter wavelengths down to 235 nm, the two higher Rydberg states i $^1\Pi$ and j $^1\Delta$ were found; their characterization was complicated by perturbations observed in the spectra. Based on the results of REMPI-PES following excitation via selected (v', J') levels of the i and j states, the following conclusions were drawn: The i $^1\Pi$ state has the ground-state ionic core $\{NH^+\ X\ ^2\Pi_r\}$ and a mixed 3pσ/3dσ character; a close lying $^1\Sigma^+$ or $^1\Sigma^-$ state, most likely arising from the $\{NH^+\ X\ ^2\Pi_r\}$ 3dπ configuration, was assumed to be the perturbing state of ND(i $^1\Pi$, v=0). The j $^1\Delta$ state exhibits significant core mixing with the configurations $\{NH^+\ B\ ^2\Delta\}$ 3sσ and $\{NH^+\ X\ ^2\Pi_r\}$ 3dπ involved; states perturbing the lowest vibrational levels of j $^1\Delta$ were found to have $^1\Delta$ symmetry in the case of NH, and $^1\Sigma$, $^1\Pi$, and $^1\Delta$ symmetry in the case of ND; the $^1\Pi$ perturber most likely involves $\{NH^+\ X\ ^2\Pi_r\}$ 3dσ or 3dδ contributions; the identity of the $^1\Sigma$ and $^1\Delta$ perturbers is unknown [10]. The analysis of a number of rovibrational REMPI bands (including hot bands from a $^1\Delta$, v=1 and 2) gave the following assignments and band origins ν_0 (in cm^{-1}; as in the case of the d←←a transition, there is a systematic shift in the wavenumbers reported by [3] and [8]; see preceding table and remark):

transition	v′−v″	ν_0(NH) [a), c)] [8] (f, g, h); [10] (i, j)	ν_0(ND) [a), c)]	ν_0(NH) [b)] [3]	ν_0(ND) [b)]
f $^1\Pi$←←a $^1\Delta$	0−0	73777.8(10)	73834.1(8)	73789.13(13)	73843.86(13)
	1−0	−	75637.8(7)	−	−
	2−0	−	77421.4(7)	−	−
g $^1\Delta$←←a $^1\Delta$	0−0	75542.0(10)	75602.9(6)	75551.01(14)	75585.86(33)
	1−0	78005.0(10)	77479.4(7)	[78004.2(2)] [d)]	−
	0−1	−	−	72368.42(29)	−
	1−1	−	−	74830.75(15)	75133.75(17)
	2−0	80036.8(6)	79141.8(6)	−	−
	3−0	−	80580.7(7)	−	−
h $^1\Sigma$←←a $^1\Delta$	0−0	76870.3(6)	76911.2(6)	76942.25(23)	76955.60(22)
	2−1 [e)]	−	78610.0(10)	−	−
i $^1\Pi$←←a $^1\Delta$	0−0	79532.1(7)	79557.2(9)	−	−
j $^1\Delta$←←a $^1\Delta$	0−0	82994.0(8)	83089.7(6)	−	−
	1−0	85340.4(6)	84818.3(5)	−	−
	2−1 [e)]	84509.0(6)	−	−	−
	2−0	−	86371.0(5)	−	−
$^1\Pi$ perturber		−	83022(2)	−	−
$^1\Sigma$ perturber		−	83121(6)	−	−
$^1\Delta$ perturber		83261(3)	85101.6(9)	−	−

a) Values in parentheses represent the probable uncertainty in interpolating between the Ne calibration lines. — b) One standard deviation in parentheses. — c) Hot bands from a $^1\Delta$, v=1 (and v=2 for j $^1\Delta$ of ND) were also observed. — d) From a wavelength scan of the total electron current in REMPI-PES study [9]. — e) From analysis of hot bands.

The B $^3\Pi$, C $^3\Sigma^-$, D $^3\Pi$, E $^3\Sigma^-$, F $^3\Sigma^-$←←X $^3\Sigma^-$ Transitions. Using NH radicals in their electronic ground state X $^3\Sigma^-$ and laser excitation wavelengths of 235 to 220 nm, REMPI via triplet Rydberg states was observed [11, 12]. The spectra exhibit the O_i, P_i, Q_i, R_i, and S_i (i=1, 2, 3) branches expected for ΔJ=ΔN=0, ±1, ±2 transitions and fine-structure splitting; because of limited resolution (~0.8 cm^{-1} at the two-photon energy), the rotational structure was incompletely resolved. Analysis based on the rotational line positions and line strengths led only to a partial characterization of the detected Rydberg states. The B $^3\Pi$, D $^3\Pi$, and E $^3\Sigma^-$ states were viewed as "regular" Rydberg states; B $^3\Pi$ was identified as the triplet counterpart of f $^1\Pi$, $\{NH^+\ X\ ^2\Pi_r\}$ 3pσ, and D $^3\Pi$ was found to arise mainly from the $\{NH^+\ a\ ^4\Sigma^-\}$ 3pπ configuration, and E $^3\Sigma^-$ probably from the $\{NH^+\ X\}$ 3pπ configuration. The C $^3\Sigma^-$ state, which perturbs the v=0 levels of the B $^3\Pi$ state of NH and ND and provides its own resonance enhancements in the case of ND, and the F $^3\Sigma^-$ state both have significant valence character; they can be identified with the 3 $^3\Sigma^-$ and 4 $^3\Sigma^-$ states obtained from an ab initio CI calculation [13]. The following band origins ν_0 (in cm^{-1})

were determined with an uncertainty of ±1.0 cm^{-1}, allowing for a possible error in the interpolation between the Ne calibration lines [12]:

transition	v′−v″	ν_0(NH)	ν_0(ND)
B $^3\Pi \leftarrow\leftarrow$ X $^3\Sigma^-$	0−0	85538.6	85588.1
	1−0	87957.9	87387.2
C $^3\Sigma^- \leftarrow\leftarrow$ X $^3\Sigma^-$	0−0	—	85734.3
D $^3\Pi \leftarrow\leftarrow$ X $^3\Sigma^-$	0−0	89277.1	89394.3
E $^3\Sigma^- \leftarrow\leftarrow$ X $^3\Sigma^-$	0−0	—	89829.3
F $^3\Sigma^- \leftarrow\leftarrow$ X $^3\Sigma^-$	0−0	90276.8	90311.8

$NH^+ \leftarrow$ NH Transitions. The REMPI-PES technique allows selection of a single rovibrational level (v′, N′, J′) in the resonant Rydberg state so that only a few levels of the ion (v^+, N^+, J^+) are accessed; this procedure substantially simplifies the PES (compared with traditional PES). Inspection of the vibrational-rotational structure and intensity distributions among the vibrational bands or rotational lines enables conclusions to be drawn about the nature of the intermediate Rydberg orbital and about the photoelectron dynamics. Special criteria are the vibrational branching ratio, which may indicate either a Franck-Condon or a non-Franck-Condon vibrational distribution in the ion, and the propensities for certain selection rules in the Δv ($=v^+ - v'$) and ΔN ($=N^+ - N'$) transitions. Deviations from Franck-Condon behavior and violations of the expected selection rules, $\Delta v = 0$ and ΔN = odd, are explained by rapid Rydberg orbital evolution (i.e., a rapidly changing Rydberg orbital with increasing internuclear distance r already in the vicinity of r_e) and by the existence of Cooper minima for certain ionization channels (i.e., minima in the photoionization cross section vs. photon (or photoelectron kinetic) energy curves arising from zeros in the dipole matrix element for the transition from the resonant state into the continuum at definite photon energies; see [14 to 16]). Rotationally resolved REMPI-PES are presented for $NH^+ \leftarrow$ NH transitions via the Rydberg states f $^1\Pi$, v′=0, J′=13; g $^1\Delta$, v′=1, J′=14 [9]; f $^1\Pi$, v′=0, J′=11; g $^1\Delta$, v′=0, J′=16; and h $^1\Sigma$, v′=0, J′=11 [17], only vibrationally resolved REMPI-PES for ionizations via the states i $^1\Pi$, v′=0, J′=1 and j $^1\Delta$, v′=0, 1, 2, J′=2 [10].

In the case of the intermediate f, g, and h states, the spectra illustrate transitions into the ionic levels X $^2\Pi_r$, $v^+=0$ to 3, $J^+=J'$, $J'\pm1$, $J'\pm2$ and clearly demonstrate the expected $\Delta v=0$ propensity but unusually strong transitions with ΔN = even, particularly $\Delta N=0$ [9, 17]. Combining the experimental results with quantum-chemical calculations of ground-state, resonant-state, and photoelectron orbitals, a number of Cooper minima were predicted to occur in various photoelectron channels of the f, g, and h states; calculations of the photoelectron angular momentum distributions and the angular momentum compositions of photoelectron matrix elements also provided insight into the origin of these Cooper minima [17].

In the case of the intermediate j state, transitions into X $^2\Pi_r$, $v^+=0$ to 2 levels showing a clear $\Delta v=0$ propensity were also observed. For the j $^1\Delta$, v′=0, 1, and 2 intermediates however, the spectra demonstrate the excitation of NH^+(X $^2\Pi_r$, $v^+=0$ to 2, 0 to 3, and 0 to 5), respectively, in non-Franck-Condon vibrational branching ratios and also show some weak probability to form NH^+(A $^2\Sigma^-$) ions and a high probability of forming NH^+(B $^2\Delta$) ions, the latter with a clear $\Delta v=0$ preference (unpublished results of rotationally resolved REMPI-PES following excitation via higher J′ levels of the j state [18] are included in the analysis) [10].

Quantum-chemical calculations for (3+1) REMPI via the B $^3\Pi$ ($\{NH^+ X\ ^2\Pi_r\}$ 3pσ) Rydberg state predicted Cooper minima resulting in non-Franck-Condon vibrational distributions [19] and unusual rotational branching ratios [20] for various vibrational transitions.

N^+ and H^+ Fragment-Ion Formation. Wavelength scans of the d $^1\Sigma^+ \leftarrow\leftarrow$ a $^1\Delta$ (2+1) REMPI spectrum were also obtained by monitoring the $^{14}N^+$, H^+, or D^+ ion channels; the peaks in these spectra match with the features in the excitation spectra forming NH^+ or ND^+; variations in the intensities, however, are large [1]. Monitoring the $^{14}N^+$ ion channel for REMPI spectra via the triplet Rydberg states enabled the C $^3\Sigma^- \leftarrow\leftarrow$ X $^3\Sigma^-$ transition in ND and the F $^3\Sigma^- \leftarrow\leftarrow$ X $^3\Sigma^-$, 0–0 bands in NH and ND to be better observed than wavelength scans for the $^{14}NH^+$ and $^{14}ND^+$ ions [12]. Various mechanisms for the daughter-ion formation were discussed; most probably, a three-photon process leads to highly excited (autoionizing) molecular states and dissociation by a fourth photon [1, 12].

References:

[1] Ashfold, M. N. R.; Clement, S. G.; Howe, J. D.; Western, C. M. (J. Chem. Soc. Faraday Trans. **87** [1991] 2515/23).
[2] Ashfold, M. N. R.; Clement, S. G.; Howe, J. D.; Western, C. M. (Faraday Discuss. Chem. Soc. **91** [1991] 411/2).
[3] Johnson, R. D., III; Hudgens, J. W. (J. Chem. Phys. **92** [1990] 6420/5).
[4] Chu, J.-J.; Marcus, P.; Dagdigian, P. J. (J. Chem. Phys. **93** [1990] 257/67).
[5] Dagdigian, P. J. (Faraday Discuss. Chem. Soc. **91** [1991] 412/3).
[6] Nieman, G. C.; Colson, S. D. (J. Chem. Phys. **68** [1978] 5656/7).
[7] Glownia, J. H.; Riley, S. J.; Colson, S. D.; Nieman, G. C. (J. Chem. Phys. **73** [1980] 4296/309).
[8] Clement, S. G.; Ashfold, M. N. R.; Western, C. M. (J. Chem. Soc. Faraday Trans. **88** [1992] 3121/8).
[9] De Beer, E.; Born, M.; De Lange, C. A.; Westwood, N. P. C. (Chem. Phys. Lett. **186** [1991] 40/6).
[10] Clement, S. G.; Ashfold, M. N. R.; Western, C. M.; De Beer, E.; De Lange, C. A.; Westwood, N. P. C. (J. Chem. Phys. **96** [1992] 4963/73).
[11] Clement, S. G.; Ashfold, M. N. R.; Western, C. M.; Johnson, R. D., III; Hudgens, J. W. (J. Chem. Phys. **96** [1992] 5538/40).
[12] Clement, S. G.; Ashfold, M. N. R.; Western, C. M.; Johnson, R. D., III; Hudgens, J. W. (J. Chem. Phys. **97** [1992] 7064/72).
[13] Goldfield, E. M.; Kirby, K. P. (J. Chem. Phys. **87** [1987] 3986/94).
[14] Cooper, J. W. (Phys. Rev. [2] **128** [1962] 681/93); Fano, U.; Cooper, J. W. (Rev. Mod. Phys. **40** [1968] 441/507).
[15] Msezane, A. Z.; Manson, S. T. (Phys. Rev. Lett. **48** [1982] 473/5); Manson, S. T. (Phys. Rev. [3] A **31** [1985] 3698/703).
[16] Chupka, W. A. (J. Chem. Phys. **87** [1987] 1488/98).
[17] Wang, K.; Stephens, J. A.; McKoy, V.; De Beer, E.; De Lange, C. A.; Westwood, N. P. C. (J. Chem. Phys. **97** [1992] 211/21).
[18] De Beer, E.; Born, M.; De Lange, C. A.; Westwood, N. P. C. (unpublished results quoted in [10]).
[19] Wang, K.; Stephens, J. A.; McKoy, V. (J. Chem. Phys. **93** [1990] 7874/82).
[20] Wang, K.; McKoy, V. (J. Chem. Phys. **95** [1991] 4977/85).

2.1.1.4 Relaxation Processes

2.1.1.4.1 Deactivation of X $^3\Sigma^-$-State Vibrational Levels

Collision-induced vibrational relaxation was studied on vibrationally excited NH(X $^3\Sigma^-$), produced by pulsed electron impact on N_2-H_2 or N_2-H_2-Ar **gaseous** mixtures; time-resolved IR Fourier transform spectroscopy was used to observe the $v=3\rightarrow2$, $2\rightarrow1$, and $1\rightarrow0$ fundamental band emission (2500 to 3400 cm^{-1}) which allowed the time-dependent vibrational populations to be determined. The following rate constants for $v\rightarrow v-1$ transitions, $k_{v\rightarrow v-1}$, were derived at room temperature for the collision partners N_2, Ar, and H_2 [1]:

	$k_{v\rightarrow v-1}$ in 10^{-14} $cm^3\cdot molecule^{-1}\cdot s^{-1}$		
v	N_2	Ar	H_2
1	1.2 ± 0.5	0.2 ± 0.1	≤50
2	3.8 ± 1.5	0.5 ± 0.2	≤100 *)
3	7.5 ± 2.5	0.8 ± 0.3	≤150 *)

*) May include reactive quenching via the reaction $NH+H_2\rightarrow NH_2+H$.

Vibrational relaxation of the X $^3\Sigma^-$, $v''=1$ level of NH and ND in **noble-gas matrices** was studied by an optical-optical double resonance technique (using excitation of A $^3\Pi_i$, $v'=0$ by a pump laser and probing the X $^3\Sigma^-$, $v''=1$ population by another laser with variable time delays). The following lifetimes τ were obtained at 4 K in photolyzed NH_3 (ND_3)-noble gas (1:10000) samples [2]:

matrix	τ in μs for NH	τ in μs for ND
Ne	~0.5	240
Ar*)	180	24000
Kr	2300	21000

*) Preliminary results: $\tau=190$ and 31000 ($\pm10\%$) μs for NH and ND [3].

The strong dependence of τ upon the host lattice and the strong isotopic effect, which are contrary to the predictions of the so-called energy-gap law for multiphonon relaxation processes, indicate a relaxation mechanism with transfer of the vibrational energy into rotational energy of the impurity molecule rather than transfer into phonons of the host lattice. Theoretical models for vibrational relaxation of diatomic molecules in solids, which describe the experimental results for NH and ND, have been presented [4 to 9].

References:

[1] Dodd, J. A.; Lipson, S. J.; Flanagan, D. J.; Blumberg, W. A. M.; Person, J. C.; Green, B. D. (J. Chem. Phys. **94** [1991] 4301/10).
[2] Bondybey, V. E.; English, J. H. (J. Chem. Phys. **73** [1980] 87/92).
[3] Bondybey, V. E. (J. Chem. Phys. **65** [1976] 5138/40).
[4] Berkowitz, M.; Gerber, R. B. (Chem. Phys. Letters **49** [1977] 260/4).
[5] Gerber, R. B.; Berkowitz, M. (Phys. Rev. Lett. **39** [1977] 1000/4).
[6] Gerber, R. B.; Berkowitz, M. (Chem. Phys. Lett. **56** [1978] 105/8).
[7] Berkowitz, M.; Gerber, R. B. (Chem. Phys. **37** [1979] 369/88).
[8] Knittel, D.; Lin, S. H. (Mol. Phys. **36** [1978] 893/906).
[9] Diestler, D. J.; Knapp, E. W.; Ladouceur, H. D. (J. Chem. Phys. **68** [1978] 4056/65).

2.1.1.4.2 Deactivation of the a $^1\Delta$ State

NH(a $^1\Delta$) radicals for kinetic studies are usually generated by UV and VUV photolysis of HN_3 at wavelengths of 313, 308, 285, 266, 248, or 193 nm and, in some cases, by VUV photolysis (193 nm) of HNCO, NH_3, and N_2H_4. The decay of NH(a $^1\Delta$) by spontaneous radiative transitions into the ground state and by collisions with parent or foreign-gas molecules (quenching) was monitored either by observing the chemiluminescence from the forbidden a $^1\Delta \rightarrow$ X $^3\Sigma^-$ transition at 795 nm or by probing the decreasing NH(a $^1\Delta$) or increasing NH(X $^3\Sigma^-$) concentrations by laser-induced fluorescence (LIF) excitation of the c $^1\Pi \leftarrow$ a $^1\Delta$ or A $^3\Pi_i \leftarrow$ X $^3\Sigma^-$ transitions at 324 or 336 nm. Despite the long radiative lifetime of the metastable a $^1\Delta$ state, its high chemical reactivity in collisions with the parent molecules complicates the time-resolved spectroscopic measurements; thus, data for the radiative lifetime are scarce and somewhat disagreeing (only one measurement of τ_{rad} by a more direct method is available, see next paragraph). Collision-induced deactivation of NH(a $^1\Delta$) has been studied for a number of collision partners; besides the parent molecules HN_3, HNCO, NH_3, and N_2H_4, suitable collision partners include noble-gas atoms, the di- and triatomics H_2, N_2, O_2, F_2, CO, NO, HF, HCl, H_2O, CO_2, N_2O, NO_2, HCN, and a number of hydrocarbons. For some of these, only the physical quenching pathway, the a $^1\Delta \rightarrow$ X $^3\Sigma^-$ intersystem crossing (isc), is allowed; for others, physical quenching and chemical reaction (cf. pp. 119/22) occur with various intensity. Two comparative and critical reviews on recent isc experiments are given in [1] and [2].

Radiative Lifetime τ_{rad}

The Einstein coefficient of spontaneous emission, $A = 3.7 \pm 0.6\ s^{-1}$, thus $\tau_{rad} = 0.27 \pm 0.04$ s, was determined by measuring the absolute a $^1\Delta \rightarrow$ X $^3\Sigma^-$ emission rate using a calibrated (emission intensity from the O+NO reaction) optical detection system and the absolute NH(a $^1\Delta$) concentration using an ESR spectrometer; the NH(a $^1\Delta$) molecules were generated by the reaction $F + NH_2 \rightarrow NH + HF$ in a fast-flow reaction chamber equipped with these two detection systems [3].

By observing the disappearance of the a $^1\Delta \rightarrow$ X $^3\Sigma^-$ emission during ArF (193 nm) photolysis of HN_3 in dilute mixtures with Q = N_2, He, or Ar (semilogarithmic plots of the intensity at 789 nm vs. time (up to ~5 ms) and Stern-Volmer plots of the decay rates τ^{-1} vs. total pressure at very small HN_3:Q mixing ratios), the NH(a $^1\Delta$) radicals were found to decay with rates faster than 33 s^{-1}, thus $\tau_{rad} > 0.03$ s [4]. LIF due to A $^3\Pi_i \leftarrow$ X $^3\Sigma^-$ excitation following 193-nm photolysis of HNCO and collision-induced a $^1\Delta \rightarrow$ X $^3\Sigma^-$ quenching by O_2 yielded $\tau_{rad} \approx 0.05$ s [5]. (The disappearance of NH(a $^1\Delta$) within less than 0.07 ms, observed by absorption [6], chemiluminescence [7, 8], and LIF [9] is due to chemical reactions with the parent molecules, see p. 121.)

Considerably longer lifetimes of more than 1 s have been observed during the a $^1\Delta \rightarrow$ X $^3\Sigma^-$ emission in noble-gas matrices (in situ photolysis of HN_3 or HNCO) [10 to 12], and gas-phase values of $\tau_{rad} = 3.3$ s [10] and 1.9 s [11] were extrapolated. These compare favorably (fortuitously ?) with ab initio (CI) results, $\tau_{rad} = 1.7$ s [13] or 2.18 s [14].

Rate Constants for Collision-Induced a $^1\Delta \rightarrow$ X $^3\Sigma^-$ Intersystem Crossing (k_q in $cm^3 \cdot molecule^{-1} \cdot s^{-1}$ at room temperature, unless otherwise stated).

Noble-Gas Atoms. There are no reactive channels for collisions of NH(a $^1\Delta$) with noble-gas atoms. Electronic quenching by He, Ar, and Kr is inefficient, whereas rapid isc has been observed for Xe.

Rate constants of $10^{16} \cdot k_{He} = 8.6 \pm 1.7$, 10.3 ± 1.7, and 9.9 ± 3.3 for NH(a $^1\Delta$) in the vibrational levels v″ = 1, 2, and 3 were derived from c ← a LIF [15], [34, p. 35], which confirms the earlier findings $k_{He}<10^{-15}$ (c←a LIF) [16] and $k_{He}=(1\ \text{to}\ 10)\cdot 10^{-16}$ (a→X fluorescence) [4].

When quenching with Ar, the a→X fluorescence also yielded $k_{Ar}=(1\ \text{to}\ 10)\cdot 10^{-16}$ [4] which is more than an order of magnitude smaller than $k_{Ar}=1.2\cdot 10^{-14}$ derived from a Stern-Volmer plot for the quenching of NH(a $^1\Delta$) by HN_3 in the presence of Ar (c←a LIF) [9]; the reason for the discrepancy could not be found [4].

$k_{Kr}\leq 1\cdot 10^{-14}$ resulted from c←a LIF following HN_3 photolysis [17, pp. 23, 32].

When quenching with Xe, both the decay of NH(a $^1\Delta$, v''=0, 1) and the growth of NH(X $^3\Sigma^-$, v''=0, 1, 2) were probed by c←a and A←X LIF excitation, and the following rate constants were obtained:

$10^{11}\cdot k_{Xe}$ for NH(a $^1\Delta$, v''=0)	Ref.	$10^{11}\cdot k_{Xe}$ for NH(a $^1\Delta$, v''=1)	Ref.
1.09 ± 0.04 a)	[18]	1.06 ± 0.02	[1]
1.2 ± 0.1	[19], [17, pp. 23, 32]	1.3 ± 0.1	[20]
1.7 ± 0.2 b)	[21]		
2.3 c)	[22]		

a) Result for T=298 K; $10^{11}\cdot k_{Xe}=1.30\pm 0.03$ for T=476 K [18]. — b) An error limit of 0.02 (presumably misprinted) is given in [21]. — c) $k_{Xe}/k_{react}=0.187$ by measuring the quantum yields of N_2, H_2, and NH_4N_3 upon UV photolysis of HN_3-Xe mixtures at 303 K [22]; the rate constant for reactive quenching by HN_3, $k_{react}=1.21\cdot 10^{-10}\ cm^3\cdot molecule^{-1}\cdot s^{-1}$, was used to derive k_{Xe} [19, 21]. $k_{react}=1.7\cdot 10^{-10}\ cm^3\cdot molecule^{-1}\cdot s^{-1}$ [9] gave $k_{Xe}=3.3\cdot 10^{-11}$ [23].

Relative cross sections for the formation of NH(X $^3\Sigma^-$, v''=0 to 4) levels during quenching of NH(a $^1\Delta$) were derived from A←X LIF, excited in crossed NH(a $^1\Delta$) and Xe beams [2].

Diatomic and Triatomic Molecules. Quenching by $\mathbf{H_2}$ proceeds preferably via the chemically reactive channel yielding rate constants of $\sim(3\ \text{to}\ 5)\cdot 10^{-12}$ using c←a LIF [16, 24] and a→X emission [25]. A←X LIF experiments gave $k_{H_2}\leq 5\cdot 10^{-14}$ for the isc [24].

Quenching of NH(a $^1\Delta$) by $\mathbf{N_2}$ is solely a physical process. Rate constants for the lowest vibrational level, v=0, are in reasonable agreement within the given error limits: $10^{14}\cdot k_{N_2}=7.93\pm 0.66$ at 306 K [18], 7.5 ± 0.6 [16], 6.8 ± 0.5 [25], 6.6 ± 2.4 at 291 K [20], 8.3 ± 0.8 [19], [17, pp. 25, 32]. Measurements of the temperature dependence of k_{N_2} for NH(a $^1\Delta$, v=0) (T=306 to 596 K) [18] and for NH, ND(a $^1\Delta$, v=0) and NH(a $^1\Delta$, v=1) (T=290 to 527 K) [20] showed that the reaction NH(a $^1\Delta$)+N_2(X $^1\Sigma_g^+$) has an activation energy of $\sim$450 cm^{-1} and that the process probably takes place via a long-lived HN_3 complex (the existence of an entrance channel barrier to NH(a $^1\Delta$)+N_2 isc is consistent with experimental and theoretical work that indicated the presence of an exit channel barrier of the HN_3→NH(a $^1\Delta$)+N_2 potential surface, cf. p. 120). Quenching of NH(a $^1\Delta$) in the vibrational levels v=1, 2, and 3 is about three to four times faster than in the v=0 level. In addition to electronic quenching, vibrational relaxation takes place which is comparable in rate with the isc. Rate constants are: $10^{13}\cdot k_{N_2}=2.47\pm 0.03$ for v=1 [1], 2.2 ± 0.3, 2.3 ± 0.2, 2.5 ± 0.2 for v=1, 2, 3 [34, pp. 38/43], 2.2 and 2.5 for v=1 and 2 [15]; the latter two have been corrected for vibrational deactivation of higher vibrational levels to give $k_{N_2}=(2.8\pm 0.3)\cdot 10^{-13}$ for both the v=1 and 2 levels [15].

For removing NH(a $^1\Delta$) by collisions with $\mathbf{O_2}$, there are a number of exothermic reaction channels (yielding NO+OH, NO_2+H, HNO+O) in addition to the somewhat dominant

NH(a $^1\Delta \rightarrow X\,^3\Sigma^-$) isc which was found to proceed via NH(a $^1\Delta$) + O_2(X $^3\Sigma_g^-$) $\rightarrow$ NH(X $^3\Sigma^-$) + O_2(b $^1\Sigma_g^+$). However, there is considerable disagreement among the reported rate constant measurements. The large value $k_{O_2} = 1.55 \cdot 10^{-11}$ was obtained using A$\leftarrow$X LIF for NH(X $^3\Sigma^-$) detection, and 193-nm photolysis of HNCO [5] and 266-nm photolysis of HN_3 [24] (see comments in [1]). On the other hand, values lower by a factor of ~300, namely, $10^{14} \cdot k_{O_2} = 4.5 \pm 0.5$, 6.2 ± 0.8, and 5.65 ± 0.15, were derived for NH(a $^1\Delta$, v=0) using UV (193, 248, 266, 308 nm) photolysis of HN_3 and NH(c$\leftarrow$a and A$\leftarrow$X) LIF [19], [17, pp. 26, 32], NH(a$\rightarrow$X) and O_2(b$\rightarrow$X) emission [25], and NH(c$\leftarrow$a) LIF [18], respectively. The ratio of NH(a $^1\Delta$) isc to total depletion of NH(a $^1\Delta$), $k_{O_2}/k \geq 0.6$, was obtained by comparing the NH(X $^3\Sigma^-$) concentrations produced by collisions of NH(a $^1\Delta$) with O_2 and with Xe [19], [17, pp. 36/7]. The removal of NH(a $^1\Delta$, v=2 and 1) levels with the higher rates $10^{14} \cdot k_{O_2} = 12.4 \pm 1.0$ and 8.6 ± 0.4 and the appearance of NH(X $^3\Sigma^-$, v=2, 1, 0) with rates (in 10^{14}) of 18.8 ± 0.9, 11.0 ± 1.1, 4.6 ± 0.5 was probed by c$\leftarrow$a and A$\leftarrow$X LIF following 266-nm photolysis of HN_3; the increase in the rate of NH(a $^1\Delta$, v>0) decay can be explained by populating of NH(X $^3\Sigma^-$, v>0) levels (rather than vibrational relaxation) [1] and possibly accounts for the O_2(b $^1\Sigma_g^+$) rise which was faster by a factor of 3 than the NH(a $^1\Delta$) decay observed by [25]. Measurements of the temperature dependence of k_{O_2} for NH(a $^1\Delta$, v=0) (T=298 to 476 K) showed that the NH(a $^1\Delta$)+O_2 quenching has an activation energy of 8.66 kJ/mol which was attributed, by analogy to the NH(a $^1\Delta$)+N_2 system, to a barrier in the entrance channel for the formation of a long-lived complex that dissociates to O_2(b $^1\Sigma_g^+$) [18].

In collisions of NH(a $^1\Delta$) with **CO** molecules, the reactive channel (giving NCO+H) dominates; only 12% of the observed depletion occurred by a$\rightarrow$X isc [26]. The measured overall rate constants, $10^{11} \cdot k = 1.350 \pm 0.070$ [25], 2.0 ± 0.7 (independent of temperature between 293 and 459 K) [26], 1.64 ± 0.08 (at 301 K and increasing to 2.24 ± 0.03 at 619 K) [18] for NH(a $^1\Delta$, v=0) and $10^{11} \cdot k = 1.82 \pm 0.03$ for NH(a $^1\Delta$, v=1) [1], then give values of $10^{12} \cdot k_{CO} = 1.62$, 2.4, 1.97, and 2.28 for the isc, respectively. Relative cross sections for the formation of NH(X $^3\Sigma^-$, v=0, 1, 2, 3) levels in the quenching of NH(a $^1\Delta$) were derived from A$\leftarrow$X LIF, excited in crossed NH(a $^1\Delta$) and CO beams [2].

The rate constant for isc by **NO** is not easy to measure, because the NH(X $^3\Sigma^-$) is consumed by fast subsequent reactions with NO [26]. Measurements of the NH(a $^1\Delta$) decay and NH(X $^3\Sigma^-$) growth profiles by c$\leftarrow$a and A$\leftarrow$X LIF showed that about half of NH(a $^1\Delta$) was removed due to isc. $k_{NO} = 2.5 \cdot 10^{-11}$ with the overall rate constant $k = (4.8 \pm 0.1) \cdot 10^{-11}$ and $k_{NO}/k = 0.53 \pm 0.1$ was obtained by [21], whereas $k_{NO} = 1.1 \cdot 10^{-11}$ with $k = 2.8 \cdot 10^{-11}$ (preliminary results in [17, 19]) and $k_{NO}/k = 0.40$ was obtained by [27].

For quenching NH(a $^1\Delta$) with **HF** and **F_2**, the overall rate constants $10^{13} \cdot k = 7.3 \pm 2.6$ and 6.3 ± 1.6 were reported without discussing the reaction mechanism [16]. Quenching of NH(a $^1\Delta$) by **HCl** is a chemically reactive process ($k = 7.9 \cdot 10^{-11}$) [8].

For **CO_2**, **N_2O**, **H_2O**, and **HCN**, the following overall rate constants k, branching ratios k_q/k, and electronic quenching rate constants k_q were obtained:

quencher Q	$10^{11} \cdot k$	k_q/k	$10^{12} \cdot k_q$	Ref.
CO_2*)	0.023 ± 0.005	0.24 ± 0.08	0.055	[28]
N_2O*)	0.17 ± 0.02	0.06 ± 0.03	0.10	[28]
H_2O	4.8 ± 1.0	≤ 0.02	<0.96	[28]
HCN	3.5 ± 1.0	0.04 ± 0.01	1.4	[27]

*) $10^{11} \cdot k = 0.025 \pm 0.002$ and 0.158 ± 0.011 for CO_2 and N_2O, respectively, were reported without further analyzing the products [25].

The removal of NH(a $^1\Delta$) by the $\mathbf{NO_2}$ radical ($k=3.7\cdot10^{-11}$) is solely a chemical process [28].

Hydrocarbons. In reactions of NH(a $^1\Delta$) with C_6H_6 [28] and CH_4, C_2H_4, C_3H_8, *cis*-2-butene, and methyl acetylene [24], no ground-state NH(X $^3\Sigma^-$) could be observed by A←X LIF. The latter authors [24] showed that this can be attributed to a rapid reaction of NH(X $^3\Sigma^-$), and they derived upper limits for the rate constants of isc by CH_4 and C_3H_8, $k_q<4\cdot10^{-14}$ and $<4\cdot10^{-13}$ [24] (overall rate constants for CH_4 and C_2H_4, considerably different from those of [24], were also reported by [8]). Photolysis of HN_3 in the presence of CH_4 [29], C_2H_6 [30], C_2H_4 [31], C_3H_8 [32], and C_2H_2 [23] and chemical analysis of the reaction products yielded the ratios $k_q/k_{react}=0.177$, 0.217, 0.0, 0.262, and 0.0, where k_{react} is the rate constant for the reaction of NH(a $^1\Delta$) with the parent molecule HN_3; with $k_{react}=(1\text{ to }2)\cdot10^{-10}$ (see below), these ratios would give k_q values of $\sim(2\text{ to }5)\cdot10^{-11}$ for CH_4, C_2H_6, and C_3H_8 which are considerably higher than those derived from LIF studies.

Parent Molecules $\mathbf{HN_3}$, HNCO, $\mathbf{NH_3}$, and $\mathbf{N_2H_4}$. Quenching of NH(a $^1\Delta$) by its parent molecules is a predominantly chemically reactive process, and rate constants of $k_{react}=(1\text{ to }2)\cdot10^{-10}$ have been obtained by various methods for HN_3 [4, 6 to 9, 23 to 25, 33], NH_3, N_2H_4 [16], and HNCO [5] (but $k_{HNCO}=1.65\cdot10^{-11}$ [16]). Recently, quenching of NH(a $^1\Delta$) in its v=0 to 4 levels by HN_3 (248-nm photolysis of HN_3; c←a LIF) was considered to include vibrational and/or electronic deactivation and chemical reaction; the very similar rate constants, $10^{10}\cdot k_{HN_3}=1.21\pm0.18$ for the v=0 level [28] and 1.26 ± 0.13, 1.16 ± 0.13, 0.98 ± 0.18, and 1.13 ± 0.30 for the v=1, 2, 3, and 4 levels [15], [34, pp. 36/8] show that vibrational relaxation is negligible; unfortunately, the contributions of electronic quenching (e.g. by probing the NH(X $^3\Sigma^-$) concentrations by A←X LIF as for other quenchers, see above) have not been determined so far.

References:

[1] Adams, J. S.; Pasternack, L. (J. Phys. Chem. **95** [1991] 2975/82).
[2] Patel-Misra, D.; Dagdigian, P. J. (J. Chem. Phys. **97** [1992] 4871/80).
[3] Bradburn, G. R.; Lilenfeld, H. V. (J. Phys. Chem. **95** [1991] 555/8).
[4] Rohrer, F.; Stuhl, F. (Chem. Phys. Lett. **111** [1984] 234/7).
[5] Drozdoski, W. S.; Baronavski, A. P.; McDonald, J. R. (Chem. Phys. Lett. **64** [1979] 421/5).
[6] Paur, R. J.; Bair, E. J. (Int. J. Chem. Kinet. **8** [1976] 139/52).
[7] Baronavski, A. P.; Miller, R. G.; McDonald, J. R. (Chem. Phys. **30** [1978] 119/31).
[8] McDonald, J. R.; Miller, R. G.; Baronavski, A. P. (Chem. Phys. **30** [1978] 133/45).
[9] Piper, L. G.; Krech, R. H.; Taylor, R. L. (J. Chem. Phys. **73** [1980] 791/800).
[10] Esser, H.; Langen, J.; Schurath, U. (Ber. Bunsenges. Phys. Chem. **87** [1983] 636/43).
[11] Ramsthaler-Sommer, A.; Eberhardt, K. E.; Schurath, U. (J. Chem. Phys. **85** [1986] 3760/9).
[12] Ramsthaler-Sommer, A.; Becker, A. C.; Van Riesenbeck, N.; Lodemann, K.-P.; Schurath, U. (Chem. Phys. **140** [1990] 331/8).
[13] Marian, C. M.; Klotz, R. (Chem. Phys. **95** [1985] 213/23).
[14] Yarkony, D. R. (J. Chem. Phys. **91** [1989] 4745/57).
[15] Hack, W.; Mill, T. (J. Phys. Chem. **95** [1991] 4712/8).
[16] Bower, R. D.; Jacoby, M. T.; Blauer, J. A. (J. Chem. Phys. **86** [1987] 1954/6).
[17] Hack, W.; Wilms, A. (Ber. Max-Planck-Inst. Strömungsforsch. **1987** 82 pp.).
[18] Nelson, H. H.; McDonald, J. R.; Alexander, M. H. (J. Phys. Chem. **94** [1990] 3291/4).
[19] Hack, W.; Wilms, A. (J. Phys. Chem. **93** [1989] 3540/6).
[20] Hack, W.; Rathmann, K. (J. Phys. Chem. **96** [1992] 47/52).

[21] Yamasaki, K.; Okada, S.; Koshi, M.; Matsui, H. (J. Chem. Phys. **95** [1991] 5087/96).
[22] Kodama, S. (Bull. Chem. Soc. Jpn. **56** [1983] 2348/54).
[23] Kodama, S. (J. Phys. Chem. **92** [1988] 5019/24).
[24] Cox, J. W.; Nelson, H. H.; McDonald, J. R. (Chem. Phys. **96** [1985] 175/82).
[25] Freitag, F.; Rohrer, F.; Stuhl, F. (J. Phys. Chem. **93** [1989] 3170/4).
[26] Hack, W.; Rathmann, K. (J. Phys. Chem. **94** [1990] 3636/9).
[27] Hack, W.; Rathmann, K. (J. Phys. Chem. **94** [1990] 4155/61).
[28] Hack, W.; Wilms, A. (Z. Phys. Chem. [Munich] **161** [1989] 107/21).
[29] Kodama, S. (Bull. Chem. Soc. Jpn. **58** [1985] 2891/9).
[30] Kodama, S. (Bull. Chem. Soc. Jpn. **56** [1983] 2355/62).

[31] Kodama, S. (Bull. Chem. Soc. Jpn. **56** [1983] 2363/70).
[32] Kodama, S. (Bull. Chem. Soc. Jpn. **58** [1985] 2900/10).
[33] McDonald, J. R.; Miller, R. G.; Baronavski, A. P. (Chem. Phys. Lett. **51** [1977] 57/60).
[34] Mill, T. (Ber. Max-Planck-Inst. Strömungsforsch. **14** [1990] 104 pp.).

2.1.1.4.3 Deactivation of the b $^1\Sigma^+$ State

NH, ND(b $^1\Sigma^+$) radicals for kinetic studies are generated by VUV photolysis of NH_3, ND_3 with Ar or Kr resonance lines of 104.8 and 106.7 or 116.5 and 123.6 nm or with VUV pulses at $\lambda > 105$ nm from capacitor discharges; VUV flash photolysis of HNCO and two-photon photolysis of NH_3 or one-photon photolysis of HN_3 with 193-nm radiation from an ArF laser have also been applied. The decay of NH(b $^1\Sigma^+$) by spontaneous radiative transitions or by collisions with parent or foreign gas molecules was monitored by observing the forbidden b $^1\Sigma^+ \rightarrow$ X $^3\Sigma^-$ transition at 471 nm or by pumping the c $^1\Pi \leftarrow$ b $^1\Sigma^+$ transition at 450 nm and observing the c $^1\Pi \rightarrow$ a $^1\Delta$ fluorescence at 324 nm. Intersystem crossing A $^3\Pi_i \leftarrow$ b $^1\Sigma^+$ has been demonstrated by observing the A $^3\Pi_i \rightarrow$ X $^3\Sigma^-$ fluorescence at 338 nm. Collision-induced deactivation of NH, ND(b $^1\Sigma^+$) has been studied on a number of collision partners; besides the parent molecules NH_3, ND_3, and HN_3, noble-gases, N and O atoms, the di- and triatomics H_2, HD, D_2, N_2, O_2, CO, NO, H_2O, D_2O, CO_2, and N_2O, a number of hydrocarbons, methanol, methylamine, and hydrazine have been used. Some of these collision partners only take the physical quenching pathway, b $^1\Sigma^+ \rightarrow$ X $^3\Sigma^-$ and/or b $^1\Sigma^+ \rightarrow$ a $^1\Delta$ deactivations or, in the case of O_2, intersystem crossing NH(b $^1\Sigma^+$) + O_2(a $^1\Delta_g$ or b $^1\Sigma_g^-$) $\rightarrow$ NH(A $^3\Pi_i$) + O_2(X $^3\Sigma_g^-$). For others, chemical reactions contribute to the quenching process. The various quenching channels of the NH(b $^1\Sigma^+$) species, unlike the very reactive NH(a $^1\Delta$) radical, generally have not been investigated. Although it is more energetic than NH(a $^1\Delta$), it reacts less rapidly. The rate constants k_q reported in various publications therefore are deactivation/reaction rate constants for the processes NH(b $^1\Sigma^+$) + Q $\rightarrow$ products, and possible reaction products appearing besides deactivated NH species are occasionally mentioned.

Radiative Lifetime τ_{rad}

The most recent result, $\tau_{rad} = 53^{+17}_{-13}$ ms, was obtained from the b $^1\Sigma^+ \rightarrow$ X $^3\Sigma^-$ emission, excited by VUV flash photolysis of NH_3 or ArF laser photolysis of HN_3 [1]. This value is larger than previous ones, obtained by VUV photolysis of NH_3, and differing in turn by two orders of magnitude: τ_{rad}(in ms) = $17.8^{+4.7}_{-3}$ [2], ≥ 5 [3], ≥ 2 (probably around 6) [4], and 0.23 ± 0.08 (1.6 ± 0.2 or 2.86 ± 0.41 for ND) [5, 6]. Ab initio CI calculations gave $\tau_{rad} = 72$ ms in the Franck-Condon approximation and 97 ms if experimental excitation energies for the low-lying valence states are employed [7], or $\tau_{rad} = 100$ ms [8].

Lifetimes have been measured on the b $^1\Sigma^+$, v'=0, 1→X $^3\Sigma^-$, v''=0, 1, 2, 3 transitions between 5 and 30 K for NH(b $^1\Sigma^+$) in Ar matrices, produced by VUV photolysis of NH_3-Ar (1:5000 to 1:20000) deposits at 15 K and excited directly via b $^1\Sigma^+$←X $^3\Sigma^-$ absorption or indirectly by intersystem crossing via A $^3\Pi_i$←X $^3\Sigma^-$ pumping. $\tau_{rad,0}=1.55\pm0.1$ ms was obtained at 5 K for the v'=0 level and $\tau_{rad,1}=520\pm20$ µs for the v'=1 level; the difference, $\tau_{rad,1}^{-1}-\tau_{rad,0}^{-1}=1.3\pm0.1\ ms^{-1}$, is the vibrational relaxation rate constant k_{vib}. The considerably shorter lifetime in the Ar matrix compared to the most recent gas-phase value is attributed to matrix-induced b $^1\Sigma^+$→a $^1\Delta$ transitions which are four orders of magnitude faster than in the gas phase [9].

Quenching Rate Constants k_q (in $cm^3\cdot molecule^{-1}\cdot s^{-1}$ at room temperature, unless otherwise stated)

The quenching of NH(b $^1\Sigma^+$) was studied in the late seventies by three teams that used VUV photolysis of NH_3 and, as quenchers, He, Ar, O, and N atoms and H_2, N_2, O_2, CO, NH_3, ND_3, and CH_4 molecules [2, 10], NH_3, ND_3 [5, 6], He, Ar, Xe atoms, H_2 (HD, D_2), N_2, O_2, CO, NO, H_2O (D_2O), CO_2, N_2O, ND_3, N_2H_4 molecules, methylamine, methanol, and a number of normal and deuterated hydrocarbons [3, 11 to 14]. Subsequent investigations made use of VUV photolysis of NH_3 and Q=NH_3 [4], ArF laser photolysis of HN_3 and Q=He, Ar, N_2, HN_3 [1], ArF laser two-photon photolysis of NH_3 and Q=Ar, H_2, N_2, O_2, H_2O, CO_2, CH_4, C_2H_6 [15], and VUV flash photolysis of HNCO and Q=HNCO, He, O_2 [16]. The following tables show, that there is moderate to excellent agreement between the rate constants from studies where the same quenchers were used. Details on experimental conditions and discussions concerning the quenching processes may be found in the original papers.

Quenching by **noble-gas, N,** and **O** atoms:

Q	k_q for NH	k_q for ND	Ref.	k_q for NH	Ref.
He	$(4.2\pm3)\cdot10^{-17}$	$(1.6\pm0.5)\cdot10^{-17}$	[12]	$(7.7\pm0.3)\cdot10^{-17}$	[1]
	$(7.04\pm0.15)\cdot10^{-17}$	$<5\cdot10^{-18}$	[10]	$(9.0\pm2.0)\cdot10^{-17}$	[16]
Ar	$(1.8\pm0.3)\cdot10^{-16}$	$(4.5\pm0.5)\cdot10^{-17}$	[12]	$(7.1\pm0.6)\cdot10^{-17}$	[15]
	$(1.27\pm0.04)\cdot10^{-16}$	$<5\cdot10^{-18}$	[10]	$(1.0\pm0.05)\cdot10^{-16}$	[1]
	$(1.61\pm0.06)\cdot10^{-16}$	–	[2]	$(3.6\pm0.7)\cdot10^{-16}$	[3]
Xe	$(2.8\pm0.6)\cdot10^{-15}$	$(1.3\pm0.4)\cdot10^{-15}$	[14]		
N	$(3.38\pm0.07)\cdot10^{-11}$	–	[10]		
O	$(1.78\pm0.09)\cdot10^{-11}$	–	[10]		

Quenching by the diatomic molecules **H_2, D_2, HD, N_2, O_2, CO,** and **NO**:

Q	k_q for NH	k_q for ND	Ref.	k_q for NH	Ref.
H_2	$(8.6\pm1.5)\cdot10^{-13}$	$(8.1\pm1.3)\cdot10^{-13}$	[12]	$(4.5\pm2.1)\cdot10^{-13}$	[15]
	$(10.0\pm0.8)\cdot10^{-13}$	$(8.96\pm0.19)\cdot10^{-13}$	[10]		
D_2	$(1.7\pm0.1)\cdot10^{-14}$	$(1.8\pm0.3)\cdot10^{-14}$	[13]		
HD	$(1.2\pm0.24)\cdot10^{-14}$	$(1.25\pm0.07)\cdot10^{-14}$	[13]		
N_2	$(6.0\pm0.6)\cdot10^{-16}$	$(4.8\pm0.8)\cdot10^{-17}$	[12]	$(5.3\pm1.2)\cdot10^{-16}$	[15]
	$(4.48\pm0.20)\cdot10^{-16}$	$(3.83\pm0.25)\cdot10^{-17}$	[10]	$(5.0\pm0.2)\cdot10^{-16}$	[1]
O_2	$(2.4\pm0.3)\cdot10^{-15}$	$(9.1\pm2)\cdot10^{-16}$	[12]	$(2.44\pm0.84)\cdot10^{-15}$	[2]
				$(2.0\pm0.6)\cdot10^{-15}$	[15]
CO	$<2\cdot10^{-14}$	$<2\cdot10^{-14}$	[12]	$(1.29\pm0.11)\cdot10^{-15}$	[2]
NO	$(4.6\pm0.6)\cdot10^{-12}$	$(3.9\pm0.6)\cdot10^{-12}$	[14]		

Quenching by the triatomic molecules **H_2O, D_2O, CO_2,** and **N_2O**:

Q	k_q for NH	k_q for ND	Ref.	k_q for NH	Ref.
H_2O	$(4.9\pm1)\cdot10^{-13}$	$(4.4\pm0.8)\cdot10^{-13}$	[12]	$(1.0\pm0.21)\cdot10^{-12}$	[15]
D_2O	$(2.2\pm1.0)\cdot10^{-13}$	$(4.9\pm1.2)\cdot10^{-14}$	[13]		
CO_2	$<1\cdot10^{-14}$	$<1\cdot10^{-14}$	[12]	$(1.5\pm0.6)\cdot10^{-14}$	[15]
N_2O	$<1\cdot10^{-14}$	$<1\cdot10^{-14}$	[12]		

Quenching by the parent molecules **NH_3, ND_3, HN_3, HNCO,** and by **N_2H_4**:

Q	k_q for NH	Ref.	Q	k_q for ND	Ref.
NH_3	$(3.90\pm0.19)\cdot10^{-13}$	[2]	ND_3	$(5.22\pm0.20)\cdot10^{-14}$	[10]
NH_3	$(1.801\pm0.071)\cdot10^{-13}$	[5]	ND_3	$(0.38\pm0.3)\cdot10^{-14}$	
NH_3	$(4.1\pm0.8)\cdot10^{-13}$	[3]		or $(1.49\pm0.15)\cdot10^{-14}$	[5]
NH_3	$(3.7\pm0.3)\cdot10^{-13}$	[4]	ND_3	$1.9\cdot10^{-14}$	[12]
HN_3	$(7\pm1)\cdot10^{-11}$	[1]			
HNCO	$(5.6\pm0.3)\cdot10^{-13}$	[16]			
N_2H_4	$(3.6\pm0.7)\cdot10^{-11}$	[14]			

Quenching by **hydrocarbons, methanol,** and **methylamine**:

Q	k_q for NH	k_q for ND	Ref.	k_q for NH	Ref.
CH_4	$(1.8\pm0.3)\cdot10^{-13}$	$(1.8\pm0.2)\cdot10^{-13}$	[12]	$(1.90\pm0.08)\cdot10^{-13}$	[2]
				$(8.5\pm1.5)\cdot10^{-14}$	[15]
CD_4	$(3.1\pm1)\cdot10^{-14}$	$(3.2\pm0.4)\cdot10^{-14}$	[13]		
C_2H_2	$(5.5\pm1.5)\cdot10^{-14}$	$(4.7\pm2)\cdot10^{-14}$	[12]		
C_2D_2	$(1.2\pm0.25)\cdot10^{-14}$	$(9.8\pm1.6)\cdot10^{-15}$	[13]		
C_2H_4	$(1.4\pm0.6)\cdot10^{-13}$	$(1.6\pm0.15)\cdot10^{-13}$	[12]		
C_2D_4	$(1.3\pm0.5)\cdot10^{-14}$	$(1.1\pm0.25)\cdot10^{-14}$	[13]		
C_2H_6				$(2.6\pm0.6)\cdot10^{-13}$	[15]
C_3H_6		$(5.7\pm1.3)\cdot10^{-13}$	[14]	$(4.7\pm0.7)\cdot10^{-13}$	[12]
1-C_4H_8	$(6.3\pm1.4)\cdot10^{-13}$	$(7.4\pm0.8)\cdot10^{-13}$	[14]		
1,3-C_4H_6	$\sim2\cdot10^{-10}$		[14]		
C_6H_6	$(3.0\pm1.3)\cdot10^{-13}$		[14]		
CH_3OH	$(10.90\pm0.90)\cdot10^{-12}$	$(6.00\pm0.90)\cdot10^{-12}$	[14]		
CH_3NH_2	$(10.40\pm2.00)\cdot10^{-12}$	$(8.60\pm0.50)\cdot10^{-12}$	[14]		

References:

[1] Blumenstein, U.; Rohrer, F.; Stuhl, F. (Chem. Phys. Lett. **107** [1984] 347/50).
[2] Gelernt, B.; Filseth, S. V.; Carrington, T. (Chem. Phys. Lett. **36** [1975] 238/41).
[3] Zetzsch, C.; Stuhl, F. (Chem. Phys. Lett. **33** [1975] 375/7).
[4] Nguyen Xuan, C.; Di Stefano, G.; Lenzi, M.; Margani, A. (J. Chem. Phys. **74** [1981] 6219/23).
[5] Masanet, J.; Lalo, C.; Durand, G.; Vermeil, C. (J. Photochem. **9** [1978] 171/3).
[6] Masanet, J.; Lalo, C.; Durand, G.; Vermeil, C. (Chem. Phys. **33** [1978] 123/30).
[7] Marian, C. M.; Klotz, R. (Chem. Phys. **95** [1985] 213/23).

[8] Yarkony, D. R. (J. Chem. Phys. **91** [1989] 4745/57).
[9] Ramsthaler-Sommer, A.; Becker, A. C.; Van Riesenbeck, N.; Lodemann, K.-P.; Schurath, U. (Chem. Phys. **140** [1990] 331/8).
[10] Gelernt, B.; Filseth, S. V.; Carrington, T. (J. Chem. Phys. **65** [1976] 4940/4).

[11] Zetzsch, C.; Stuhl, F. (Ber. Bunsenges. Phys. Chem. **79** [1975] 1156).
[12] Zetzsch, C.; Stuhl, F. (Ber. Bunsenges. Phys. Chem. **80** [1976] 1354/64).
[13] Zetzsch, C.; Stuhl, F. (J. Chem. Phys. **66** [1977] 3107/11).
[14] Zetzsch, C. (Ber. Bunsenges. Phys. Chem. **82** [1978] 1098/102).
[15] Van Dijk, C. A.; Sandholm, S. T.; Davis, D. D.; Bradshaw, J. D. (J. Phys. Chem. **93** [1989] 6363/7).
[16] Presser, N.; Zhu, Y.-F.; Gordon, R. J. (J. Phys. Chem. **91** [1987] 4383/8).

2.1.1.4.4 Deactivation of the A $^3\Pi_i$ State

Gaseous NH, ND(A $^3\Pi_i$) radicals for kinetic studies are generated by VUV photolysis of NH_3, ND_3 and, in a few cases, N_2H_4 or CH_3NH_2, mostly using 193-nm radiation from an ArF laser which resulted in a "hot" rotational distribution of the ground vibrational level. VUV flash photolysis of HN_3 and HNCO, multiphoton IR photolysis of DN_3, MW discharges through NH_3-noble-gas mixtures, and, for the earlier studies, electron impact on NH_3 have been applied. The decay of NH(A $^3\Pi_i$) by spontaneous radiative transitions or by collisions with parent or foreign-gas molecules was monitored by observing the A $^3\Pi_i \rightarrow$ X $^3\Sigma^-$ fluorescence spectrum in the range 336 to 338 nm. Selected excitation of A-state rovibrational and fine-structure levels was achieved by pumping selected A $^3\Pi_i \leftarrow$ X $^3\Sigma^-$ lines with a tunable dye laser after a time delay long enough to enable A → X relaxation to be completed and LIF to be observed. Collision-induced deactivation of the A state was extensively studied for many collision partners, such as noble-gas atoms and SF_6, diatomic molecules H_2, D_2, N_2, O_2, CO, NO, triatomic molecules H_2O, D_2O, CO_2, N_2O, a number of hydrocarbons and methanol as well as the parent molecules NH_3, ND_3, HN_3, DN_3, HNCO, CH_3NH_2, and N_2H_4. A large number of data for the radiative lifetime and the quenching rate constants, not always in satisfactory agreement, have accumulated. Most recently, subtle methods of laser spectroscopy allowed collision-induced relaxation between fine-structure levels, i.e., within the spin-orbit triplets or the Λ doublets, to be measured.

For studies of radiative decay and vibrational relaxation of matrix-isolated NH (ND), mixtures of NH_3 (ND_3) and noble gases were photolyzed during deposition on a cold substrate and time-resolved A $^3\Pi_i \leftarrow$ X $^3\Sigma^-$ LIF transitions were measured.

Radiative Lifetime τ_{rad}

Lifetimes of NH(A $^3\Pi_i$) obtained by various methods are, with a few exceptions, in the range from 400 to 500 ns. Recent laser-spectroscopic studies indicate that the most reliable values for the ground-vibrational level v′ = 0 are around 420 ns. A recent quantum-chemical ab initio (CI) calculation resulted in τ_{rad} = 407 and 478 ns for the v′ = 0 and 1 levels [1]. The following table gives a compilation of the various experimental results for τ_{rad} (≈ zero-pressure lifetime τ_0) of NH and, in braces, for ND beginning with the most recent studies (some information on the experimental conditions, comments on individual results, and possible corrections are given in the remarks below the table):

τ_{rad} in ns	method	remark	Ref.
422 ± 55 420 ± 48	NH_3 laser photolysis at 193 nm; A $^3\Pi_i$, Ω = 2, 1, 0, v′ = 0, N′ = 3 to 8 → X $^3\Sigma^-$ LIF	a)	[2]
423 ± 4	NH_3 laser photolysis at 193 nm; A $^3\Pi_i$, v′ = 0, N′ ≤ 24 → X $^3\Sigma^-$ LIF	b)	[3]
{441 ± 10}	ND_3 laser photolysis at 193 nm; A $^3\Pi_i$, v′ = 1, 2, 3, J′ ≤ 4 → X $^3\Sigma^-$ LIF	c)	[4]
423 ± 20	NH_3 laser photolysis at 193 nm; A $^3\Pi_i$, v′ = 0, N′ = 1 to 4 → X $^3\Sigma^-$ emission		
424 ± 8	A $^3\Pi_i$, v′ = 0, N′ = 8 to 12 → X $^3\Sigma^-$ emission		
449 ± 20	A $^3\Pi_i$, v′ = 1, N′ = 1 to 4 → X $^3\Sigma^-$ emission		
461 ± 10	A $^3\Pi_i$, v′ = 1, N′ = 4 to 6 → X $^3\Sigma^-$ emission	d)	[5]
{407 ± 20}	ND_3 laser photolysis at 193 nm; A $^3\Pi_i$, v′ = 0, N′ = 1 to 6 → X $^3\Sigma^-$ emission		
{415 ± 5}	A $^3\Pi_i$, v′ = 0, N′ = 8 to 15 → X $^3\Sigma^-$ emission		
{447 ± 15}	A $^3\Pi_i$, v′ = 1, N′ = 4 to 6 → X $^3\Sigma^-$ emission		
500 ± 40	HN_3 and HNCO flash photolysis at 121.6 nm; A $^3\Pi_i$, v′ = 0 → X $^3\Sigma^-$ emission		[6]
463 ± 45	HN_3 photolysis by Xe-resonance lamp; A $^3\Pi_i$, v′ = 0 → X $^3\Sigma^-$ emission	e)	[7]
418 ± 8	NH_3-Ar, NH_3-He, MW discharge; A $^3\Pi_i$, v′ = 0, N′ = 1 to 7 ← X $^3\Sigma^-$ LIF	f)	[8]
453 ± 10	NH_3, electron impact; high-frequency deflection technique; A → X emission	g)	[9]
470 ± 40	NH_3 laser photolysis at 193 nm; A $^3\Pi_i$, v′ = 0, N′ ≤ 12 → X $^3\Sigma^-$ emission		[10]
400 ± 60	NH_3 laser photolysis at 172 nm; A $^3\Pi_i$, v′ = 0, 1, J′ ≤ 25 → X $^3\Sigma^-$ emission	h)	[11]
424 ± 6	NH_3, pulsed electron impact; A $^3\Pi_i$, v′ = 0, 1(unresolved) → X $^3\Sigma^-$ emission		[12]
{~500 ± 50}	DN_3, laser IR multiphoton excitation and dissociation; A $^3\Pi_i$ → X $^3\Sigma^-$ emission		[13]
490 ± 50	N_2H_4 laser photolysis at 193 nm; A $^3\Pi_i$, v′ = 0, 1 → X $^3\Sigma^-$ emission	i)	[14]
545 ± 65	CH_3NH_2 laser photolysis at 193 nm; A $^3\Pi_i$, v′ = 0 → X $^3\Sigma^-$ emission		[15]
347 ± 5	NH_3, electron impact; electron-photon delayed coincidence, A $^3\Pi_i$ → X $^3\Sigma^-$ emission	j)	[16]
404 ± 5	NH_3, electron impact; high-frequency deflection technique, A $^3\Pi_i$ → X $^3\Sigma^-$ emission	k)	[17]
500 ± 60	NH_3, electron impact; A $^3\Pi_i$, v′ = 0, 1 → X $^3\Sigma^-$ intensity vs. NH_3 pressure		[18]
410 ± 20	NH_3, electron impact; time-sampling technique, A $^3\Pi_i$, v′ = 0, 1 → X $^3\Sigma^-$ emission		[19]
455 ± 90	NH_3, electron impact; phase-shift technique, A $^3\Pi_i$, v′ = 0, 1 → X $^3\Sigma^-$ emission		[20]
460 ± 80	NH_3, electron impact; phase-shift technique, A $^3\Pi_i$, v′ = 0 → X $^3\Sigma^-$ emission		[21]
425 ± 60	NH_3, electron impact; intensity decay of A $^3\Pi_i$, v′ = 0, 1 → X $^3\Sigma^-$ emission		[22]

a) Selected excitation of A $^3\Pi_i$, $\Omega=2$, 1, 0, $v'=0$, $N'=3$ to 8 rotational and fine-structure levels; the lifetimes of the lowest levels with $N'=3$ and $\Omega=2$ and 1 (pumping of the $P_1(4)$ and $P_2(4)$ lines) are given above; pumping of the $Q_2(3)$ line gave 442 ± 35 ns; values for levels with $N'>3$ are between 402 ± 21 and 423 ± 24 [2]. Preliminary result for the A $^3\Pi_i$, $\Omega=2$, $v'=0$, $N'=3$ level: 422 ± 45 ns [23].

b) Selected excitation of A $^3\Pi_i$, $v'=0$ rotational levels $N'=4$ (τ_{rad} given above) and $N'=10$, 14, 19, 24 ($\tau_{rad}=435\pm5$, 453 ± 6, 491 ± 8, 525 ± 8 ns) in flowing NH_3-Ar mixtures; the increase of the lifetime with increasing N′ is due to an electronic transition moment which decreases with increasing internuclear distance, coupled with a significant degree of centrifugal distortion in the light NH molecule [3].

c) Selected excitation of ND(A $^3\Pi_i$, $v'=1$, 2, 3) rotational and fine-structure levels in a pulsed supersonic beam; average values for the $J'=1$ to 4 levels for $v'=1$(given above) and $v'=2$ and 3 (428 ± 16 and 434 ± 20 ns). For the NH(A$^3\Pi_i$, $v'=2$, $J'=1$ to 3) fine-structure levels, lifetimes as short as 335 ± 30 ($J'=1$, F_{3f}) to 210 ± 15 ns ($J'=3$, F_{1e}) have been measured as a result of predissociation induced by the repulsive $^5\Sigma^-$ state; by using a high-quality ab initio calculation [24] and the most reliable (at that time) available experimental results for the NH(A $^3\Pi_i$, $v'=0$, 1) levels [5, 25] (see above and remarks d) and f)), predissociation rates and a pure radiative lifetime of 478.5 ns have been derived for NH(A $^3\Pi_i$, $v'=2$) [4].

d) Zero-pressure lifetimes for higher rotational levels, $N'\geq4$, were obtained in neat NH_3 and ND_3 and belong to the nascent rotational population after photolysis, whereas the lifetimes for the lowest rotational levels (relaxed rotational population) were obtained in NH_3-N_2 and ND_3-N_2 mixtures [5]. These results confirm those of [23] (see remark a)). They supersede earlier results from the same laboratory, after an uncorrected systematic error (nonlinear response of the photomultiplier) was detected [5]: 465 ± 14 ns for A $^3\Pi_i$, $v'=0$, $N'=8$ to 13 [26], 480 ± 40 $\{435\pm40\}$ ns for A $^3\Pi_i$, $v'=0$, 1 [27] were derived from A→X fluorescence decay after 193-nm photolysis of NH_3 $\{ND_3\}$ and 470 ± 30 ns for A $^3\Pi_i$, $v'=0$ derived from A→X fluorescence decay after 193-nm photolysis of CH_3NH_2 [28]. For the $v'=2$ level, a lifetime of 406 ± 35 ns was measured from A→X fluorescence decay after 193-nm photolysis of HN_3 [29].

e) From a Stern-Volmer plot for the quenching of NH(A $^3\Pi_i$) by NO at $p(HN_3)=0.123$ Torr. At higher HN_3 pressures of 0.20, 0.30, and 0.40 Torr, lifetimes of 455 ± 2, 441 ± 25, and 412 ± 25 ns were measured [7].

f) Prior LIF measurements with NH, produced in the NH_3+F reaction (MW discharge through a flowing NH_3-CF_4-He mixture), yielded lifetimes of 440 ± 15 and 420 ± 35 ns for the A $^3\Pi_i$, $v'=0$ and 1 levels [25].

g) Fluorescence decay was measured for the F_{1e} fine-structure components of the A $^3\Pi_i$, $v'=0$, $N'=1$ to 17 and $v'=1$, $N'=3$ and 5 levels. The lifetimes of the $v'=0$, N′ levels vary unsystematically between 439 ± 10 and 483 ± 10 ns; the value for $N'=1$ is given above. Values of 488 ± 10 and 486 ± 10 ns were measured for the $v'=1$, $N'=3$ and 5 levels. Purely radiative decay and possible predissociation of the A $^3\Pi_i$ rovibronic levels into the X $^3\Sigma^-$ continuum were investigated with the aid of an ab initio (CASSCF) calculation [9].

h) Average of the results for $v'=0$ and 1 [11].

i) The value given above is for the $v'=0$ level; 390 ± 40 ns was measured for the $v'=1$ level [14].

j) Presumably erroneous result, cf. remark h) on p. 108.

k) Value for the A $^3\Pi_i$, v′=0, N′=4 level; measurements for the individual rotational levels v′=0, N′=4 to 31 yielded values between 399 and 408 (±4) ns for N′=4 to 12, increased to 453±4 ns for N′=25 (decrease of the transition moment with increasing centrifugal distortion) followed by a decrease to 96±2 ns for N′=31 (predissociation); for the A $^3\Pi_i$, v′=1, N′=5 to 24 levels, lifetimes between 402 and 426 (±6) ns were observed for N′=5 to 15 followed by a decrease to 41±3 ns for N′=24 [17]; cf. however remark i) on p. 108.

Quenching Rate Constants k_q (in $cm^3 \cdot molecule^{-1} \cdot s^{-1}$) **and Cross Sections σ_q** (in $Å^2$)

$\sigma_q = k_q \cdot (8\,kT/\pi\mu)^{-1/2}$, where k = Boltzmann constant, T = absolute temperature, μ = reduced mass of the colliding pair.

Noble-Gas Atoms and SF_6. The rate constant and cross section for quenching rotationally "hot" NH(A $^3\Pi_i$, v′=0, N′=8 to 12) by **Xe** are $k_q = (4.1 \pm 0.3) \cdot 10^{-11}$ and $\sigma_q = 6.0$ [26, 29] (see however remark d) on p. 98). Values measured (presumably) for lower rotational levels are $\sigma_q = 5.1$ (NH) and 1.9 (ND) [30], $k_q = 3.1 \cdot 10^{-11}$ (NH) [31].

For quenching NH, ND(A $^3\Pi_i$) with **Kr**, a cross section of $\sigma_q = 0.10$ was reported [30].

Results for **Ar** differ by up to four orders of magnitude: $10^{14} \cdot k_q = 170$ ($\sigma_q = 0.07$) [15], <8.6 ($\sigma_q < 0.011$) [26], ≈10 [10], <10 (ND) [5], <20 [32], =0.016±0.003 [33].

The remaining species are ineffective quenchers: $10^{14} \cdot k_q < 6.4$ ($\sigma_q < 0.0046$) [26], =20 [32], =0.0050±0.0009 [33] for **He**, ≤20 (NH) and <30 (ND) [5] for **SF_6**.

Diatomic Molecules H_2, D_2, N_2, O_2, CO, NO. The following table gives the rate constants and cross sections generally measured at room temperature (except those falling under remarks e) to h)). These are either for a selected rovibrational level v′=0, N′, J′ of the A $^3\Pi_i$ state (mostly the lowest one) or for a manifold of N′, J′ levels which may exhibit the normal (Boltzmann) distribution or some other "hot" distribution. Explanations, some other details, and additional results are given in the remarks following the table. Results for ND are given in braces.

Q	$10^{11} \cdot k_q$	σ_q	remark	Ref.	$10^{11} \cdot k_q$	σ_q	remark	Ref.
H_2	13±2	7.1±1.2	a)	[8]	8.4±0.5	–	e)	[34]
	11±1.5	5.9±1.8	b)	[3]	8.5±0.3	–	f)	[34]
	8.4±0.4	–	c)	[5]	18.8±2.1	4.5±0.5	g)	[35, 36]
	5.5±0.5	2.9	d)	[37]	7 to 13	–	h)	[38]
	4.8±0.5	–	d)	[10]				
D_2	3.3±0.5	2.3±0.4	i)	[3]	3.4±0.4	2.4	d)	[37]
	{5.8±0.3}	–	c)	[5]	12.6±1.4	4.2±0.5	g)	[35, 36]
N_2	0.003±0.005 {≤0.004}	–	j)	[5]	<0.2	<0.1	g)	[35, 36]
	≤0.008	–	e)	[34]	≤0.005	–	f)	[34]
O_2	8.7±1.7	11±2	k)	[8]	5.56±0.6	–	n)	[34]
	4.4±0.5	–	l)	[5]	3.4±0.3	4.3	m)	[26]
	2.9±0.2	–	d)	[10]	8.0±0.9	4.7±0.6	g)	[35]
CO	19±2	23±3	o)	[8]	12.2±0.7	–	q)	[34]
	13±2	17±3	p)	[3]	12.3±1.4	6.9±0.8	g)	[35, 36]
	5.2±0.5	6.4	m)	[26]				
NO	13.4±0.9	17	m)	[26]	7.1	2.8	d)	[15]
	7.3±1.0	–	d)	[10]	47.8±8.8	60±11	r)	[7]

a) Selective excitation of the rotational levels N′=1 (results given above) and N′=2 to 6 ($10^{11}\cdot k_q$ decreased from 11±2 to 3.4±1.5, σ_q from 5.8±1.0 to 1.8±0.8) [8].

b) Selective excitation of the rotational levels N′=1 (results given above) and N′=4, 10, 12, 14 19, 24 ($10^{11}\cdot k_q$ decreased from 8±1 to 0.9±0.4, σ_q from 4.3±0.5 to 0.5±0.3) [3].

c) The values given above are for a thermalized (N_2 buffer gas) rotational distribution N′=1 to 4 for NH and N′=1 to 6 for ND; corresponding values for a rotationally "hot" distribution (absence of buffer gas) are 6.6±0.4 (N′=8 to 12) for NH and 5.2±0.15 (N′=8 to 15) for ND. These all supersede the previous results [26] which suffer from a systematic error (cf. remark d) on p. 98). – For the v′=1 level of NH in thermalized (N′=1 to 4) and "hot" (N′=4 to 6) rotational distributions, $10^{11}\cdot k_q=8.5\pm0.5$ and 6.7±0.6 were measured and the rate constants for vibrational relaxation, $10^{11}\cdot k_{v'=1\to0}=10^{11}\cdot(k_{q,v'=1}-k_{q,v'=0})\leq0.8$ and ≤0.9 derived, respectively [5].

d) High rotational excitation of the A $^3\Pi_i$, v′=0 level [10, 15, 37].

e) Thermalized rotational distribution at 243±3 K [34].

f) Thermalized rotational distribution at 415±5 K [34].

g) Thermalized rotational distribution at 1400 K [35]. The rotational level dependence and the temperature dependence of quenching in NH are discussed in [39].

h) Selective excitation of the A $^3\Pi_i$, v′=0, J′=8, F_3 level in an H_2-N_2O flame at 900 to 2700 K [38] (preliminary publications [40, 41]).

i) Selective excitation of the rotational levels N′=10 (results given above) and N′=14, 19, 24 ($10^{11}\cdot k_q$ decreased from 1.6±0.5 to 1.2±0.5, σ_q from 1.1±0.3 to 0.8±0.3) [3].

j) "Hot" rotational distribution N′=8 to 12 {8 to 15}. For the v′=1 level, $10^{11}\cdot k_q=$ 0.097±0.003 and $10^{11}\cdot k_{v'=1\to0}=0.094\pm0.006$ were measured [5]. Previous results for low N′ were $10^{11}\cdot k_q<0.0049$ and $\sigma_q<0.0061$ [26], $10^{11}\cdot k_q<0.2$ [32].

k) Selective excitation of the rotational levels N′=1 (results given above) and N′=2 to 5 ($10^{11}\cdot k_q$ decreased from 7.7±1.3 to 4.2±1.6, σ_q from 9.8±1.6 to 5.4±2.0) [8].

l) Thermalized rotational distribution. For the v′=1 level in thermalized (N′=1 to 4) and "hot" (N′=4 to 6) rotational distributions, $10^{11}\cdot k_q=8.9\pm0.75$ and 6.2±0.3 and $10^{11}\cdot k_{v'=1\to0}=4.5\pm1.2$ and 2.3±0.6 were measured, respectively [5].

m) Rotationally "hot" (N′=8 to 12) NH [26] (see however remark d) on p. 98).

n) Thermalized rotational distribution at 296±3 K; $10^{11}\cdot k_q=5.91\pm0.27$ or 6.15±0.2 at 243±3 K and 6.09±0.28 or 5.86±0.17 at 273±3 K [34].

o) Selective excitation of the rotational levels N′=1 (results given above) and N′=2 to 6 ($10^{11}\cdot k_q$ decreased from 17±1.5 to 5.1±1.6, σ_q from 21±2 to 6.3±2.0) [8].

p) Selective excitation of the rotational levels N′=1 (results given above), N′=4, 10, 12, 14, 19 ($10^{11}\cdot k_q$ decreased from 10.9±1.4 to 1.4±0.4, σ_q from 14±2 to 1.7±0.5), and N′=24 ($10^{11}\cdot k_q=2.4\pm0.4$, $\sigma_q=3.0\pm0.5$) [3].

q) Or 12.4±0.8 or 11.9±0.8, all for a thermalized rotational distribution at 296±3 K; $10^{11}\cdot k_q=12\pm0.9$, 12.0±0.2, or 10.0±0.2 at 243±3 K and 9.7±0.5 at 415±5 K [34].

r) Estimated from NH(c $^1\Pi\to$A $^3\Pi_i$) quenching by NO [7].

Triatomic Molecules H_2O, D_2O, CO_2, N_2O. Room-temperature values (except those under remark h)) are as follows (cf. introduction of preceding table):

Q	$10^{11}\cdot k_q$	σ_q	remark	Ref.	$10^{11}\cdot k_q$	σ_q	remark	Ref.
H_2O	46±4	52±5	a)	[8]	36.7±1.5	42	c)	[42]
	35.4±1.5	–	b)	[34]				
D_2O	44±5	52±6	d)	[8]				
CO_2	0.8±0.3	1.0±0.4	e)	[8]	0.87±0.07	–	g)	[34]
	0.63±0.05	–	f)	[5]	1.9±0.2	1.2±0.2	h)	[35, 36]
N_2O	1.0±0.3	1.3±0.4	e)	[8]	1.23±0.06	–	g)	[34]
	3.5±0.9	–	i)	[5]	4.6±0.5	2.8±0.3	h)	[35, 36]

a) Selective excitation of the rotational levels N′=1 (results given above) and N′=2, 3, 5, 6, and 7 ($10^{11}\cdot k_q$ decreased from 47±4 to 30±6 and 32±7, σ_q from 54±5 to 34±6 and 37±7) [8].

b) Or 37±2, both for a thermalized rotational distribution at 296±3 K; $10^{11}\cdot k_q$=39±3 at 253 K and 30±3 at 415±5 K [34].

c) Thermalized rotational distribution [42].

d) Selective excitation of the rotational levels N′=1 (results given above) and N′=2 to 7 ($10^{11}\cdot k_q$ decreased from 44±4 to 27±5, σ_q from 51±5 to 31±6) [8].

e) Selective excitation of the rotational levels N′=2 (results given above) and N′=5 ($10^{11}\cdot k_q$=1.0±0.3, σ_q=1.3±0.4 for Q=CO_2 and $10^{11}\cdot k_q$=0.6±0.6, σ_q=0.8±0.8 for Q=N_2O) [8].

f) Thermalized rotational distribution. For the v′=1 level in a thermalized (N′=1 to 4) distribution, $10^{11}\cdot k_q$=6.1±0.6 and $10^{11}\cdot k_{v'=1\to 0}$=5.5±0.65 were measured [5]. Previous results for the v′=0 level: $10^{11}\cdot k_q$=0.73±0.05 (σ_q=0.97; see however remark d) on p. 98) [26] and 0.45 [31].

g) Or four other results between 0.75±0.01 and 0.70±0.03 for a thermalized rotational distribution at 296±3 K; $10^{11}\cdot k_q$=1.02±0.03 or 0.80±0.05 at 243±3 K, 0.71±0.03 at 273±3 K, and 0.83±0.06 at 415±5 K were measured for Q=CO_2 and $10^{11}\cdot k_q$=1.43±0.08 at 243±3 K and 0.94±0.09 at 415±5 K were measured for Q=N_2O [34].

h) Thermalized rotational distribution at 1400 K [35, 36].

i) "Hot" rotational distribution N′=8 to 12. For the v′=1 level, $10^{11}\cdot k_q$=5.9±0.5 and $10^{11}\cdot k_{v'=1\to 0}$=2.4±1.4 were measured [5].

Parent Molecules NH_3, ND_3, HN_3, DN_3, HNCO, CH_3NH_2, and N_2H_4. Most of the quenching studies used NH_3 (and in two cases ND_3) as the parent molecule of NH (ND). Selected excitation of A-state rotational (N′) and fine structure F_1, F_2, and F_3 levels of NH and monitoring the time-resolved A $^3\Pi_i\to$X $^3\Sigma^-$ fluorescence (LIF) resulted in rate constants and cross sections for individual A $^3\Pi_i$, v′=0, N′ and A $^3\Pi_i$, v′=0, N′, Ω=2, 1, 0 levels. Results for the lowest vibrational levels v′=0 and/or 1 came from following the A→X emission in the 0→0 and 1→1 bands. The following rate constants and cross sections were measured for NH and, in braces, for ND at room temperature (if no other temperature is given) for the lowest rovibrational levels A $^3\Pi_i$, v′=0, N′ (additional results, comments, and possible corrections are given in the remarks below the table):

$10^{10}\cdot k_q$	σ_q	remark	Ref.	$10^{10}\cdot k_q$	σ_q	remark	Ref.
8.2 ± 0.7	92 ± 8	a)	[8]	5.97 ± 0.25	– (243±3 K)	e)	[34]
6.5 ± 0.8	73 ± 9	b)	[3]	6.0 ± 0.5	– (415±5 K)	e)	[34]
6.2 ± 0.5	–	c)	[2]	4.90 ± 0.54	26±3 (1400 K)	e)	[35]
6.41 ± 0.27	–	d)	[5]	3.5 ± 0.7	–	f)	[11]
$\{5.93\pm0.12\}$	–	d)	[5]	5.1 (2.0 to 2.2)	–	g)	[18]

a) Selective excitation of the rotational levels N′=1 (results given above) and N′=2 to 7 ($10^{10}\cdot k_q$ decreased from 8.6±0.7 to 5.8±0.8, σ_q from 96±8 to 65±9) [8].

b) Selective excitation of the rotational levels N′=4 (results given above) and N′=10, 14, 19, 24 ($10^{10}\cdot k_q$ decreased from 3.3±0.4 to 1.9±0.3, σ_q from 37±5 to 22±3) [3].

c) Selective excitation of the rotational fine-structure levels N′=3, $F_1(\Omega=2)$ (result given above) and $F_2(\Omega=1)$ and N′=4 to 8, $F_1(\Omega=2)$, $F_2(\Omega=1)$, $F_3(\Omega=0)$ ($10^{10}\cdot k_q$ decreased with increasing N′ from 5.9±0.4 to 3.5±0.3) [2] (see also [23] and pp. 103/5).

d) The values given above are for the thermalized (N_2 buffer gas) rotational distributions N′=1 to 4 for NH and N′=1 to 6 for ND; corresponding values for rotationally "hot" distributions (absence of buffer gas) are 4.50±0.15 (N′=8 to 12) for NH and 4.32±0.15 (N′=8 to 15) for ND, values for the Q branch only (for comparison with previous work) are 4.34±0.12 {3.85±0.14}. These all supersede previous results [26, 27] which were affected by a systematic error (cf. remark d) on p. 98). For the v′=1 level of NH in a thermalized (N′=1 to 4) and "hot" (N′=4 to 6) rotational distribution, $10^{10}\cdot k_q=6.6\pm1.1$ and 5.5 ± 0.25 were measured and the rate constants for vibrational relaxation, $10^{10}\cdot k_{v'=1\rightarrow0}=10^{10}\cdot(k_{q,v'=1}-k_{q,v'=0})\leq1.2$ and $=1.0\pm0.4$, respectively, derived; values for ND(A $^3\Pi_i$, v′=1, N′=4 to 6) are $10^{10}\cdot k_q=5.2\pm0.35$ and $10^{10}\cdot k_{v'=1\rightarrow0}=0.9\pm0.5$.

e) Thermalized rotational distributions.

f) For NH(A $^3\Pi_i$, v′=0, 1, J′≤25).

g) $k_q=(5.1\pm0.3)\cdot10^{-10}$ is reported, but the Stern-Volmer plot depicted in [18] gives a value of $(2.0 \text{ to } 2.2)\cdot10^{-10}$ (see [8, 26, 43]).

For the quenching of NH, ND(A $^3\Pi_i$, v′=0) by the remaining parent molecules, the following rate constants $10^{10}\cdot k_q$ have been measured:

HN_3	4.7 ± 0.1	[6]	DN_3	{2.3} [a)]	[13]
HNCO	4.3 ± 0.3	[6]	CH_3NH_2	5.5 ± 0.15 ($\sigma_q=22$)	[15]
	0.0065 ± 0.0005 [b)]	[33]		3.4 ± 0.2 [c)]	[28]

a) Result for ND; $k_q=7.5\cdot10^6$ Torr$^{-1}\cdot$s^{-1} given in [13]. – b) NH(A $^3\Pi_i$) is the product of collision-induced intersystem crossing, NH(b $^1\Sigma^+$)+O_2(a $^1\Delta_g$)→NH(A $^3\Pi_i$)+O_2(X $^3\Sigma_g^-$), observed upon VUV flash photolysis of an HNCO-O_2-He mixture. – c) Value presumably too small due to a systematic error; cf. remark d) on p. 98.

For Q=N_2H_4, a cross section $\sigma_q\approx40$ has been estimated [44].

Hydrocarbons and Methanol. Room temperature values (except those under remarks g) and i)) are as follows (cf. introduction of table on p. 99):

Q	$10^{11}\cdot k_q$	σ_q	remark	Ref.	$10^{11}\cdot k_q$	σ_q	remark	Ref.
CH_4	20±3	22±3	a)	[8]	7.3±1.4	8	d)	[26]
	20±3	22±3	b)	[3]	9.4	3.3	d, e)	[15]
	8.5±0.5	–	c)	[5]	7.5±1.7	8.3	f)	[45]
	6.6±1.5	–	d)	[10]	15.4±2.6	7.8±1.5	g)	[35, 36]
C_2H_6	37±4	47±5	a)	[8]	18.8±2.0	23.9	f)	[45]
	27±2	–	h)	[34]				
C_3H_8	48±2	–	h)	[34]	24.9±2.8	33.6	f)	[45]
c-C_3H_6	32±2	–	h)	[34]	14.6±1.5	19.5	f)	[45]
n-C_4H_{10}					32.0±3.3	44.6	f)	[45]
i-C_4H_{10}	37±2	–	h)	[34]	30.4±2.5	42.3	f)	[45]
C_2H_4	48±3	–	i)	[34]	19.8±1.5	24.9	f)	[45]
	34	13	d, e)	[15]				
C_3H_6	58±3	–	h)	[34]	30.5±1.8	40.9	f)	[45]
1-C_4H_8					35.9±2.2	49.7	f)	[45]
i-C_4H_8					58.1±3.7	80.5	f)	[45]
t-C_4H_8	63±4	–	h)	[34]	58.5±2.3	81.0	f)	[45]
c-C_4H_8					59.6±3.0	82.6	f)	[45]
CH_3OH	37	15	d, e)	[15]				

a) Selective excitation of the rotational levels N′=1 (results given above) and N′=2 to 6 for CH_4 ($10^{11}\cdot k_q$ decreased from 19±2 to 7.4±1.4, σ_q from 21±2 to 8.2±1.6) and N′=2 to 7 for C_6H_6 ($10^{11}\cdot k_q$ decreased from 37±3 to 20±3, σ_q from 46±4 to 26±4) [8].

b) Selective excitation of the rotational levels N′=1 (results given above) and N′=4, 10, 14, 19, and 24 ($10^{11}\cdot k_q$ decreased from 12±4 to 2.2±0.6 and 2.6±0.9, σ_q from 14±4 to 2.5±0.7 and 2.9±1.0) [3].

c) For a thermalized (N_2 buffer gas) rotational distribution N′=1 to 4 for NH. For the v′=1 level of NH in a thermalized rotational distribution, $10^{11}\cdot k_q=15.5\pm1.75$ was measured and the rate constant for vibrational relaxation, $10^{11}\cdot k_{v'=1\to0}=7.1\pm3.1$, derived [5].

d) High rotational excitation of the A $^3\Pi_i$, v′=0 level [10, 15, 26].

e) Considered to be reactive quenching [15].

f) Presumably thermalized rotational distribution [45]; results considered not very reliable by [34]; see also p. 111.

g) Thermalized rotational distribution at 1400 K [35, 36].

h) Thermalized rotational distribution at 296±3 K. For Q=C_2H_6, $10^{11}\cdot k_q=26.0\pm0.8$ at T=243±3 K and 24±1 at 415±5 K [34].

i) Thermalized rotational distribution at 243±3 K [34].

In low-pressure N_2O-Q flames with Q=CH_4, C_2H_2, C_2H_4, and C_3H_8, quenching of NH(A $^3\Pi_i$) was measured in the range 1600 to 2700 K; therefrom, a general value for the rate constant of $k_q\approx1\cdot10^{-10}$ (for flame diagnostics) was recommended [38] (preliminary publications [40, 41]).

Collision-Induced Rotational, Spin, and Λ-Doublet Relaxation

Most recently, selected excitation of A $^3\Pi_i$, v′=0 rotational (N′, J′) and fine-structure (F_1, F_2, F_3; e, f) levels and monitoring the A→X LIF transition have been applied to look into three additional relaxation processes (or combinations of them) induced by collisions with the parent molecule NH_3 ($NH(X\ ^3\Sigma^-)$ generation by 193-nm photolysis of NH_3) or the foreign quenchers Ar and He [2, 23, 46]. These processes involve (1) rotational relaxation within one of the three spin units F_1, F_2, or F_3 (rate constant $k_{\Delta J}$), (2) spin relaxation, i.e., fine-structure changes $F_1 \leftrightarrow F_2$, $F_1 \leftrightarrow F_3$, $F_2 \leftrightarrow F_3$ (rate constant $k_{\Delta\Omega}$), and (3) Λ-doublet relaxation e↔f, i.e., mixing of the antisymmetric and symmetric Λ-doublet states (rate constant $k_{a,s}$); for a schematic diagram of the energy levels and the possible transitions, see [46, figure 1]. Rotational relaxation was found to be fast with downward transitions being more likely than upward transitions and collisions with NH_3 being more efficient than collisions with Ar: for $\Delta J = -1$ and $\Delta J = +1$, respectively, $10^{11} \cdot k_{\Delta J}$ ($J'=4$, $N'=3$) = 36.6 ± 1.3 and 24.6 ± 1.5 for $Q = NH_3$, 4.4 ± 0.1 and 0.7 ± 0.1 for $Q = Ar$, $10^{11} \cdot k_{\Delta J}$ ($J'=4$, $N'=5$) = 28.7 ± 0.3 and 10.8 ± 0.4 for $Q = NH_3$. Spin-relaxation ($\Delta\Omega$) processes are roughly half as efficient as ΔJ processes based on the relative populations in the various sublevels $A\ ^3\Pi_i$, $v'=0$, $N'=3$ to 6, $J'=3$ to 6, $\Omega = 2, 1, 0$ which result from pumping various P_i- and Q_i-branch lines and subsequent collisional relaxation ($k_{\Delta\Omega}$ not derived); furthermore, a propensity for spin conservation was observed for collisions with NH_3, whereas in collisions with Ar, spin is not conserved [46]. Λ-doublet mixing was found to be a very fast process in collisions with NH_3, particularly for low rotational levels: the rate constant $10^{10} \cdot k_{a,s}$ had a value of 10.6 ± 0.8 for the $N'=3$, $J'=4$, $\Omega=2$ level and decreased with increasing rotational energy to 2.4 ± 0.1 for $N'=8$, $J'=7$, $\Omega=0$; the least energetic F_1 component exhibited the largest mixing efficiency and the most energetic F_3 component showed the smallest efficiency [2]; a Stern-Volmer plot depicted in [23] exhibits the intercept close to zero (slope $k_{a,s} \approx 1 \cdot 10^{-9}$) which indicates that Λ-doublet mixing is very slow without perturbing collisions. Rate constants for collisions with Ar and He are about a factor 3 to 8 smaller than those for NH_3 [2]. In the case of He, these data reveal good agreement with theoretical results based on ab initio calculations (CEPA) of NH-He potential surfaces [47].

In previous experiments, spin relaxation in $A\ ^3\Pi_i$, $v'=0$, $N'=1, 2, 3$, $\Omega = 2, 1, 0$ levels by collisions with NH_3 has been demonstrated in supersonic nozzle expansions of NH_3-He mixtures [48].

Relaxation of NH($A\ ^3\Pi_i$, v′, J′) in Noble-Gas Matrices

By recording the time-resolved $A\ ^3\Pi_i \rightarrow X\ ^3\Sigma^-$ LIF from NH and ND radicals in noble-gas matrices between 4 and 25 K, the following lifetimes for the A-state vibrational levels $v'=0$, 1, and 2 have been measured (in ns) [49]:

matrix	NH			ND		
	$v'=0$	$v'=1$	$v'=2$	$v'=0$	$v'=1$	$v'=2$
Ne [a)]	340 ± 10	340 ± 10	–	330 ± 10	335 ± 10	360 ± 10
Ar [a)]	235 ± 5	185 ± 5	96 ± 10	220 ± 10	237 ± 10	228 ± 10
Kr [b)]	210 ± 10	45 ± 5	–	200 ± 10	170 ± 10	–

[a)] Independent of temperature between 3.8 to 6.0 K in an Ne matrix and 3.8 to 25.0 K in an Ar matrix. — [b)] $\tau(v'=0)$ for NH and ND becomes much shorter near 10 and 25 K, respectively; $\tau(v'=1)$ is independent of temperature for NH and becomes shorter at $T \geq 10$ K for ND.

The $A\ ^3\Pi_i$, v′ levels exhibiting lifetimes of 200 to 360 ns are considered to be purely radiative, whereas vibrational relaxation, $v'=2 \rightarrow 1$ and $1 \rightarrow 0$, is responsible for the shorter $\tau(v'=1$ and 2) for NH in Ar and $\tau(v'=1)$ for NH and ND in Kr [49]. The studies were extended to NH and ND in mixed Ar-Ne, Ar-Kr, Ar-Xe, and Ar-H_2 matrices [50]. Exothermic, near-resonant, and anomalously fast vibrational energy transfer processes, such as NH(A $^3\Pi_i$, v' = 1) + CO(v = 0) → NH(A $^3\Pi_i$, v' = 0) + CO(v = 1), have been observed in mixed Ar-CO matrices [51].

A theoretical study of phonon-assisted librational relaxation, with NH(A $^3\Pi_i$) in solid Ar being the model case, and estimates of the relaxation rates at 4 to 24 K have been presented [52].

References:

[1] Kirby, K. P.; Goldfield, E. M. (J. Chem. Phys. **94** [1991] 1271/6).
[2] Kaes, A.; Stuhl, F. (J. Chem. Phys. **97** [1992] 7362/70).
[3] Chappell, E. L.; Jeffries, J. B.; Crosley, D. R. (J. Chem. Phys. **97** [1992] 2400/5).
[4] Patel-Misra, D.; Parlant, G.; Sauder, D. G.; Yarkony, D. R.; Dagdigian, P. J. (J. Chem. Phys. **94** [1991] 1913/22).
[5] Kenner, R. D.; Kaes, A.; Browarzik, R. K.; Stuhl, F. (J. Chem. Phys. **91** [1989] 1440/5).
[6] Hikida, T.; Maruyama, Y.; Saito, Y.; Mori, Y. (Chem. Phys. **121** [1988] 63/71).
[7] Sasaki, S.; Tsunashima, S.; Sato, S. (Bull. Chem. Soc. Jpn. **59** [1986] 1671/4).
[8] Garland, N. L.; Crosley, D. R. (J. Chem. Phys. **90** [1989] 3566/73).
[9] Gustafsson, O.; Kindvall, G.; Larsson, M.; Olsson, B. J.; Sigray, P. (Chem. Phys. Lett. **138** [1987] 185/94).
[10] Ni, T.; Yu, S.; Ma, X.; Kong, F. (Chem. Phys. Lett. **126** [1986] 413/6).

[11] Hellner, L.; Grattan, K. T. V.; Hutchinson, M. H. R. (J. Chem. Phys. **81** [1984] 4389/95).
[12] Fujita, I. (Z. Phys. Chem. [Munich] **136** [1983] 187/96).
[13] Simpson, T. B.; Mazur, E.; Lehmann, K. K.; Burak, I.; Bloembergen, N. (J. Chem. Phys. **79** [1983] 3373/81).
[14] Hawkins, W. G.; Houston, P. L. (J. Phys. Chem. **86** [1982] 704/9).
[15] Nishi, N.; Shinohara, H.; Hanazaki, I. (Reza Kenkyu **10** [1982] 394/9).
[16] Cvejanovic, D.; Adams, A.; King, G. C. (J. Phys. B **11** [1978] 1653/62).
[17] Smith, W. H.; Brzozowski, J.; Erman, P. (J. Chem. Phys. **64** [1976] 4628/33).
[18] Clerc, M.; Schmidt, M.; Hagege-Temman, J.; Belloni, J. (J. Phys. Chem. **75** [1971] 2908/14).
[19] Sawada, T.; Kamada, H. (Bull. Chem. Soc. Jpn. **43** [1970] 331/4).
[20] Smith, W. H. (J. Chem. Phys. **51** [1969] 520/4).

[21] Fink, E.; Welge, K. H. (Z. Naturforsch. **19a** [1964] 1193/201).
[22] Bennett, R. G.; Dalby, F. W. (J. Chem. Phys. **32** [1960] 1716/9).
[23] Kaes, A.; Stuhl, F. (Chem. Phys. Lett. **146** [1988] 169/73).
[24] Yarkony, D. R. (J. Chem. Phys. **91** [1989] 4745/57).
[25] Fairchild, P. W.; Smith, G. P.; Crosley, D. R.; Jeffries, J. B. (Chem. Phys. Lett. **107** [1984] 181/6).
[26] Hofzumahaus, A.; Stuhl, F. (J. Chem. Phys. **82** [1985] 3152/9).
[27] Haak, H. K.; Stuhl, F. (J. Phys. Chem. **88** [1984] 2201/4).
[28] Haak, H. K.; Stuhl, F. (J. Phys. Chem. **88** [1984] 3627/33).
[29] Rohrer, F.; Stuhl, F. (J. Chem. Phys. **86** [1987] 226/33).
[30] Kawasaki, M.; Hirata, Y.; Tanaka, I. (J. Chem. Phys. **59** [1973] 648/53).

[31] Cody, R. J.; Allen, J. E., Jr. (16th Intern. Symp. Free Radicals, Lauzelle-Ottignies, Belgium, 1983 from [26, 37]).
[32] Zetzsch, C. (Habilitationsschrift, Ruhr Univ., Bochum 1977 from [26]).
[33] Presser, N.; Zhu, Y.-F.; Gordon, R. J. (J. Phys. Chem. **91** [1987] 4383/8).

[34] Kenner, R. D.; Pfannenberg, S.; Heinrich, P.; Stuhl, F. (J. Phys. Chem. **95** [1991] 6585/93).
[35] Garland, N. L.; Jeffries, J. B.; Crosley, D. R.; Smith, G. P.; Copeland, R. A. (J. Chem. Phys. **84** [1986] 4970/5).
[36] Garland, N. L.; Crosley, D. R. (Symp. Int. Combust. Proc. **21** [1986/88] 1693/702).
[37] Cody, R. J.; Allen, J. E., Jr. (NBS Spec. Publ. [U. S.] 716 [1986] 176/82).
[38] Rensberger, K. J.; Copeland, R. A.; Wise, M. L.; Crosley, D. R. (Symp. Int. Combust. Proc. **22** [1988/89] 1867/75).
[39] Crosley, D. R. (J. Phys. Chem. **93** [1989] 6273/82).
[40] Rensberger, K. J.; Dyer, M. J.; Wise, M. L.; Copeland, R. A. (AIP Conf. Proc. No. 172 [1988] 750/2).

[41] Copeland, R. A.; Wise, M. L.; Rensberger, K. J.; Crosley, D. R. (Appl. Opt. **28** [1989] 3199/205).
[42] Heinrich, P.; Kenner, R. D.; Stuhl, F. (Chem. Phys. Lett. **147** [1988] 575/80).
[43] Umemoto, H.; Kikuma, J.; Tsunashima, S.; Sato, S. (Chem. Phys. **123** [1988] 159/64).
[44] Becker, K. H.; Welge, K. H. (Z. Naturforsch. **19a** [1964] 1006/15).
[45] Sasaki, S.; Kano, A.; Tsunashima, S.; Sato, S. (Bull. Chem. Soc. Jpn. **59** [1986] 1675/81).
[46] Kaes, A.; Stuhl, F. (J. Chem. Phys. **97** [1992] 4661/8).
[47] Alexander, M. H.; Dagdigian, P. J.; Lemoine, D. (J. Chem. Phys. **95** [1991] 5036/46).
[48] Carrick, P. G.; Engelking, P. C. (Chem. Phys. Lett. **108** [1984] 505/8).
[49] Bondybey, V. E.; Brus, L. E. (J. Chem. Phys. **63** [1975] 794/804).
[50] Goodman, J.; Brus, L. E. (J. Chem. Phys. **65** [1976] 3146/52).

[51] Goodman, J.; Brus, L. E. (J. Chem. Phys. **65** [1976] 1156/64).
[52] Gerber, R. B.; Berkowitz, M.; Yakhot, V. (Mol. Phys. **36** [1978] 355/63).

2.1.1.4.5 Deactivation of the c $^1\Pi$ State

NH(c $^1\Pi$) radicals for kinetic measurements were obtained from various sources. Earlier studies used electron impact on NH_3 to excite the c $^1\Pi \rightarrow$ a $^1\Delta$ and c $^1\Pi \rightarrow$ b $^1\Sigma^+$ fluorescence spectra of NH at 325 and 450 nm; their decay was observed by using various techniques to derive the radiative lifetime of the c $^1\Pi$ state. VUV photolysis of NH_3 (ND_3), HNCO, or HN_3 with the Lyman-α line at 121.6 nm, the Kr and Xe resonance lines at 123.6 and 147.0 nm, and, in the case of NH_3, synchrotron radiation in the 100- to 200-nm region also enabled the observation of c$\rightarrow$a fluorescence decay. In one experiment, pulsed discharges in ND_3-Ar mixtures were used to observe c$\rightarrow$a laser-induced fluorescence (LIF) following c$\leftarrow$b excitation. More recently, UV laser photolysis of HN_3 has been applied to produce the NH(c$\rightarrow$a) emission directly (ArF laser at 193 nm) or indirectly (LIF) via c$\leftarrow$a excitation of NH(a $^1\Delta$) (KrF laser at 248 nm, doubled dye laser at 285 nm, XeCl laser at 308 nm). Upon collisional quenching by foreign gas atoms and molecules, not only c$\rightarrow$a emission decay, but a simultaneous increase of the intensity in the triplet system A $^3\Pi_i \rightarrow$ X $^3\Sigma^-$ was observed for some of the quenchers; this was interpreted as collision-induced c $^1\Pi \rightarrow$ A $^3\Pi_i$ intersystem crossing (isc) or spin conversion which, for a few quenchers, dominates the ordinary c$\rightarrow$a, b quenching. Thus, quenching and/or isc rate constants (Stern-Volmer plots τ^{-1} vs. p(Q)) or cross sections and the spin conversion efficiencies have been measured for a number of quenchers, such as noble-gas atoms and SF_6, di- and triatomic molecules H_2, D_2, N_2, O_2, CO, NO, H_2O, CO_2, and N_2O, the parent molecules NH_3 and HN_3, and a number of hydrocarbons. Details on experimental conditions, discussions concerning the quenching processes, and discussions about the discrepancies between some results of various authors as well as mutual criticism may be found in the original papers.

Radiative Lifetime τ_{rad}

The lifetimes of NH(c $^1\Pi$, v'=0) obtained by various methods are with a few exceptions in the range 400 to 500 ns. Recent laser-spectroscopic studies indicate that the most reliable values are between 460 and 480 ns. The following table is a compilation of the various experimental results for τ_{rad} ($\approx$zero-pressure lifetime τ_0), beginning with the most recent studies (comments on individual results and possible corrections are given in the remarks below the table; results for ND are given in braces):

τ_{rad} in ns	method	remark	Ref.
463±11 {500±10}	HN_3 / DN_3 laser photolysis at 193 and 248 nm; c $^1\Pi$, v'=0, 1, J'←a $^1\Delta$ LIF	a)	[1]
460±20	HN_3 laser photolysis at 193 nm; c $^1\Pi$, v'=0, J'=2 to 9→a $^1\Delta$ emission	b)	[2]
480±40	HN_3 laser photolysis at 308 nm; c $^1\Pi$, v'=0, J'=2, 4, 8←a $^1\Delta$ LIF	c)	[3]
480±40	NH_3 laser photolysis at 121.6 nm; c $^1\Pi$, v'=0, J'=1 to 5→a $^1\Delta$ emission	d)	[4]
440±40	HN_3 flash photolysis at 121.6 nm; c $^1\Pi$, v'=0, J'=1 to 5→a $^1\Delta$ emission	d)	[4]
470±50	HNCO flash photolysis at 121.6 nm; c $^1\Pi$, v'=0, J'=2 to 4→a $^1\Delta$ emission	d)	[4]
455±21	HN_3 photolysis by an Xe-discharge lamp; c $^1\Pi$→a $^1\Delta$ emission	e)	[5]
480±80	HN_3 laser photolysis at 193 nm; c $^1\Pi$→a $^1\Delta$ emission	f)	[6]
470	HN_3 laser photolysis at 285 nm; c $^1\Pi$, v'=0←a $^1\Delta$ LIF (P(2), Q(5) lines)		[7]
437.5±1.3	NH_3, pulsed electron impact; c $^1\Pi$, v'=0→a $^1\Delta$, v''=0 emission		[8]
{425±90}	ND_3-Ar, pulsed discharge; c $^1\Pi$→a $^1\Delta$ LIF via c $^1\Pi$←b $^1\Sigma^+$ excitation	g)	[9]
364±5	NH_3, electron impact; electron-photon delayed coincidence, c→a emission	h)	[10]
411±4	NH_3, electron impact; high-frequency deflection technique, c→a emission	i)	[11]
500±100	NH_3, electron impact; Stern-Volmer plot of c→a intensity vs. NH_3 pressure		[12]
430±20	NH_3, electron impact; time-dependent intensity of c→a emission		[13]
480±90	NH_3, electron impact; phase-shift technique, c→a and c→b emission		[14]
435±40	NH_3, electron impact; phase-shift technique, c→a emission		[15]

a) Values for the NH(c $^1\Pi$, v'=0, J'=1, 2) and ND(c $^1\Pi$, v'=0, J'=1, 2, 3) levels. τ_{rad} varies between 467±11 and 461±11 ns for the NH, v'=0, J'=1 to 7 levels, and decreases steadily to 398±13 ns for levels with J'=7 to 12; in the case of ND, τ_{rad} increases steadily from 510±13 to 565±16 ns for J'=7 to 16. The lifetimes of the v'=1, J' levels of NH and ND are much shorter due to predissociation by the repulsive $^5\Sigma^-$ state: 72±7 to 60±8 ns for J'=1 to 6 {236±10 to 196±12 ns for J'=1 to 8} [1].

b) Average value for the $J'=1$ to 9 levels from this [2] and earlier [16, 17] work. For the levels $v'=1$, $J'=1$ and 4, $\tau_{rad}=67\pm7$ ns was measured [2]. These values are well reproduced by a recent theoretical study which combines ab initio (MC SCF) electronic structure and coupled-states dynamics calculations; $\tau_{rad}=460\pm11$ ns is the average value for NH(c $^1\Pi$, $v'=0$, $J'=1$ to 9) and $\tau_{rad}=64\pm9$ ns the average value for NH(c $^1\Pi$, $v'=1$, $J'=1$ to 4); furthermore, $\tau_{rad}=504\pm1$ ns was obtained for ND(c $^1\Pi$, $v'=0$, $J'=1$) [18].

c) Quenching of thermalized and translationally hot NH(c $^1\Pi$) by the HN_3 parent molecule [3].

d) A lifetime of 310 ± 30 ns was measured for the $J'=14$ level of NH(c $^1\Pi$) produced by NH_3 photolysis [4]. Preliminary results from the studies of HN_3 photolysis are reported in [19].

e) From a Stern-Volmer plot for quenching NH(c $^1\Pi$) by the parent molecule, $\tau_{rad}=448\pm22$, 455 ± 18, and 435 ± 13 ns for quenching by CO, NO, and CO_2, respectively. With He as a quencher, the very low value 357 ± 13 ns was obtained [5].

f) Emission at 326.0 nm with a bandwidth of 0.2 nm, corresponding to the region of the Q(6) and partly of the P(2), P(3), Q(5), and Q(7) lines [6].

g) This result was obtained by tuning the laser to the P(7) line of the c $^1\Pi$, $v'=0\leftarrow$ b $^1\Sigma^+$, $v''=0$ line; tuning to the Q(4) line gave a lifetime of only 350 ± 70 ns [9].

h) Erroneous result according to [3].

i) Lifetime of the c $^1\Pi$, $v'=0$, $J'=2$ level; measurements of the individual rotational levels $J'=2$ to 17 showed a steady decrease to 226 ± 5 ns; lifetimes between 57.1 and 41.7 (±0.5) ns were measured for the c $^1\Pi$, $v'=1$, $J'=2$ to 10 levels [11]. According to [2], the values are too low, presumably due to faulty pressure measurements; a rough correction raised the lifetime of the c $^1\Pi$, $v'=0$, $J'=2$ level to ~450 ns [2].

Quenching Rate Constants k_q (in $cm^3 \cdot molecule^{-1} \cdot s^{-1}$) **and Cross Sections σ_q** (in $Å^2$)

$\sigma_q = k_q \cdot (8\,kT/\pi\mu)^{-1/2}$, where k = Boltzmann constant, T = absolute temperature, μ = reduced mass of the colliding pair.

Rate constants for the total quenching and the intersystem crossing are denoted $k_{q,tot}$ and $k_{q,isc}$, the corresponding cross sections $\sigma_{q,tot}$ and $\sigma_{q,isc}$; the efficiency for intersystem crossing or spin conversion is $\sigma_{q,isc}/\sigma_{q,tot}=p_{isc}$.

Noble-Gas Atoms and SF_6. He and Ar atoms and the SF_6 molecule are particularly inefficient quenchers of NH(c $^1\Pi$), whereas quenching and collision-induced isc has been observed for Kr and Xe. The following rate constants, cross sections, and spin conversion efficiencies have been measured at room temperature (asterisked σ_q values were converted from the original [20] k_q values by [21, 22]; values in braces are for ND):

Q	$k_{q,tot}$	$\sigma_{q,tot}$	$k_{q,isc}$	$\sigma_{q,isc}$	p_{isc}	remark	Ref.
He	$<1.1\cdot10^{-13}$	<0.002	$<0.1\cdot10^{-13}$	<0.0003	–	a)	[20]
	$(1.8\pm0.2)\cdot10^{-13}$	0.01	–	–	–		[5]
	$\leq0.06\cdot10^{-13}$	≤0.0004	–	–	–	b)	[2]
	–	<0.001	–	–	–		[3]
Ar	–	≤0.01	–	–	–		[23]
	$<1.1\cdot10^{-13}$	<0.005	$<0.1\cdot10^{-13}$	<0.0006	–	a)	[20]

Table (continued)

Q	$k_{q,tot}$	$\sigma_{q,tot}$	$k_{q,isc}$	$\sigma_{q,isc}$	p_{isc}	remark	Ref.
Ar	$\leq 0.06 \cdot 10^{-13}$	≤ 0.0008	–	–	–	b)	[2]
	–	<0.001	–	–	–		[3]
	(3 or 2) $\cdot 10^{-12}$	–	–	–	–	c)	[4]
Kr	$1.24 \cdot 10^{-11}$	1.7*)	$0.14 \cdot 10^{-11}$	0.2*)	0.11		[20]
	$\{1.19 \cdot 10^{-11}\}$	{0.498}	$\{0.19 \cdot 10^{-11}\}$	{0.088}	{0.18}		[20]
	–	2.34 ± 0.06	–	0.06 ± 0.01	0.03		[22]
	$(1.85 \pm 0.27) \cdot 10^{-11}$	2.62 ± 0.38	–	–	–	b)	[2]
Xe	$2.07 \cdot 10^{-10}$	30*)	$0.47 \cdot 10^{-10}$	7*)	0.23		[20]
	$\{2.2 \cdot 10^{-10}\}$	{10.5}	$\{1.1 \cdot 10^{-10}\}$	{5.2}	{0.5}		[20]
	–	28.2 ± 1.6	–	9.0 ± 0.9	0.32		[22]
	$(2.29 \pm 0.11) \cdot 10^{-10}$	33.4 ± 1.6	–	<5	≤ 0.15	b)	[2, 17]
SF_6	$(2.2 \pm 0.5) \cdot 10^{-13}$	0.03 ± 0.007	–	–	–	b)	[2]

a) Values for NH and ND.

b) For NH(c $^1\Pi$, v′=0) and a thermalized (Boltzmann) J′ distribution at 298±4 K. Results for "unrelaxed" NH (in the absence of an Ar buffer gas) and Q=Kr and Xe are also given in [2, 17].

c) Values that show rotational relaxation for NH(c $^1\Pi$, v′=0, J′=1 to 5) from parent molecules NH_3 and HN_3, respectively; for the J′≈14 level, the rate constant is increased to $k_{q,tot} = (1.0 \pm 0.3) \cdot 10^{-11}$.

For the quenching of NH(c $^1\Pi$, v′=1) by Xe, $k_{q,tot} = (1.7 \pm 0.5) \cdot 10^{-10}$, and for the rotational relaxation of the J′=1, 4, and 8 levels, $k_{rot} = (5 \pm 2) \cdot 10^{-11}$ were measured [24].

Diatomic and Triatomic Molecules. N_2, H_2, and D_2 are rather inefficient quenchers, whereas cross sections greater than 20 Å^2 were observed for the remaining collision partners and as much as 90 Å^2 for the highly polar H_2O molecule. Isc has been demonstrated for O_2, NO, and N_2O as collision partners. The following rate constants, cross sections, and spin conversion efficiencies have been measured at room temperature (asterisked σ_q values have been converted from the original k_q values by [21, 22]; values in braces are for ND):

Q	$k_{q,tot}$	$\sigma_{q,tot}$	$k_{q,isc}$	$\sigma_{q,isc}$	p_{isc}	remark	Ref.
H_2	$(1.54 \pm 0.08) \cdot 10^{-10}$	8.15 ± 0.42	–	–	–	a)	[2]
	–	8.2 ± 0.3	–	–	–	b)	[21]
	$1.5 \cdot 10^{-10}$	>8*)	–	–	–		[25]
	$3.8 \cdot 10^{-10}$	20*)	$<0.001 \cdot 10^{-10}$	–	–		[20]
	$\{4.0 \cdot 10^{-10}\}$	–	$\{<0.001 \cdot 10^{-10}\}$	–	–		[20]
	–	3	–	–	–		[23]
D_2	$(1.09 \pm 0.09) \cdot 10^{-10}$	7.71 ± 0.64	–	–	–		[2]
N_2	$(1.34 \pm 0.07) \cdot 10^{-11}$	1.67 ± 0.09	–	–	–	a)	[2]
	$(1.4 \pm 0.1) \cdot 10^{-11}$	1.7*)	–	–	–		[4]
	–	1.43 ± 0.11	–	–	–	b)	[21]
	$\sim 1.0 \cdot 10^{-11}$	1.8*)	$<0.007 \cdot 10^{-11}$	–	–	c)	[20]
	$\sim 0.17 \cdot 10^{-11}$	0.5	–	–	–		[23]

Table (continued)

Q	$k_{q,tot}$	$\sigma_{q,tot}$	$k_{q,isc}$	$\sigma_{q,isc}$	p_{isc}	remark	Ref.
O_2	$(1.96\pm0.10)\cdot10^{-10}$	24.9 ± 1.3	–	–	–	a)	[2]
	–	21.4 ± 1.3	–	10.7 ± 1.2	0.50	b)	[22]
	$(1.92\pm0.15)\cdot10^{-10}$	25*)	–	23*)	0.95 ± 0.05	a)	[17]
	–	16*)	–	5*)	0.33		[26]
	$4.2\cdot10^{-10}$	>53*)	–	–	–		[25]
CO	$(3.42\pm0.25)\cdot10^{-10}$	42.6 ± 3.1	–	–	–	a)	[2]
	–	44.8 ± 2.4	–	–	–	b)	[21]
	–	27*)	–	–	–		[26]
	$(3.94\pm0.63)\cdot10^{-10}$	49	–	–	–		[5]
	$5.0\cdot10^{-10}$	>62*)	–	–	–		[25]
NO	–	62.9 ± 4.2	–	27.9 ± 2.4	0.44		[22]
	$(5.78\pm0.56)\cdot10^{-10}$	72.8 ± 7.0	–	69	0.95 ± 0.05	a)	[17]
	$(3.84\pm0.56)\cdot10^{-10}$	48	–	–	–		[5]
	$1.7\cdot10^{-10}$	>21*)	–	–	–		[25]
H_2O	$(7.94\pm1.31)\cdot10^{-10}$	90.4 ± 14.9	–	–	–	a)	[2]
CO_2	$(2.09\pm0.14)\cdot10^{-10}$	27.8 ± 1.9	–	–	–	a)	[2]
	–	25.4 ± 2.0	–	–	–	b)	[21]
	$(2.23\pm0.33)\cdot10^{-10}$	30	–	–	–		[5]
N_2O	$(4.43\pm0.20)\cdot10^{-10}$	59.0 ± 2.7	–	<6	≤0.1	a)	[2, 17]

a) For NH(c $^1\Pi$, v′=0) and in a thermalized (Boltzmann) J′ distribution at 298±4 K. Results for "unrelaxed" NH (in the absence of the Ar buffer gas) are also given in [2, 17]. $k_{q,tot}$ values at 243±3 and 415±5 K were measured for Q = H_2, N_2, CO, O_2, CO_2, and N_2O, at 253±3 and 415±5 K for H_2O; at 296±3 K, $k_{q,tot}=(1.18\pm0.07)\cdot10^{-11}$ for N_2 and $k_{q,tot}=(6.9\pm0.4)\cdot10^{-10}$, $\sigma_{q,tot}=78\pm5$ for H_2O [27].

b) Thermalized radicals; the more efficient quenchers CO, NO, and N_2O were also tested for "translational hot" NH (in the absence of the He buffer gas).

c) For NH and ND.

For completely quenching NH(c $^1\Pi$, v′=1) by N_2, $k_{q,tot}=(1.6\pm0.7)\cdot10^{-11}$, and for the contributions of the electronic and vibrational relaxation, $k_{q,el}=1.1\cdot10^{-11}$ and $k_{q,vib}=0.5\cdot10^{-11}$ were measured [24].

Parent Molecule and Foreign Quencher NH_3; Parent Molecules HN_3 and HNCO. The rate constants for NH_3 as a foreign quencher, $k_q\cdot10^{10}=6.64\pm0.15$, 7.4 ± 0.4, 8.4 ± 0.3 at T=243±3, 296±3, 415±5 K (NH(c $^1\Pi$) from 193-nm photolysis of HN_3) [27], and as the parent molecule, $k_q=(5.7\pm0.2)\cdot10^{-10}$ at room temperature (121.6-nm photolysis of NH_3) [4], are in reasonable agreement and correspond to a cross section of $\sigma_q=83$ at 298 K [2]. Slightly higher values were obtained from pulsed radiolysis of NH_3, $k_q=9.5\cdot10^{-10}$ [12], and photolysis of NH_3 using synchrotron radiation, $\tau_{rad}\cdot k_q=5\cdot10^{-16}$ cm^3/molecule, thus $k_q\approx1\cdot10^{-9}$ [28]. Obviously too low is the value $\sigma_q\approx16$ (electron impact on NH_3) [23].

Experiments with HN_3 as the photolyzed parent molecule gave the following rate constants and cross sections in reasonable agreement:

k_q	$(6.39\pm0.24)\cdot10^{-10}$ a)	$(7.4\pm0.4)\cdot10^{-10}$	$(9.8\pm0.4)\cdot10^{-10}$
σ_q	85.3 ± 3.2 a)	99 b)	133
Ref.	[3]	[6]	[4]

[a] Results for thermalized NH(c $^1\Pi$); for translationally hot NH (absence of He buffer gas) k_q and σ_q were also measured [3]. — [b] Evaluated by [3] by assuming that the NH(c $^1\Pi$) radicals were translationally thermal.

Less certain rate constants, $k_q=(1.1\pm0.5)\cdot10^{-9}$ for NH and $k_q=(1.2\pm0.4)\cdot10^{-9}$ for ND, were derived from the decay of NH, ND(c $^1\Pi$, v'=0, J'=2) [1]. Much lower values, $k_q=(8.6\pm1.7)\cdot10^{-11}$ and $\sigma_q=12$, were reported by [5]; according to [3], the pressure of the very adsorptive HN_3 must have been underestimated (rather overestimated?) during the fluorescence measurements.

For rotationally "unrelaxed" NH(c $^1\Pi$, v'=0) (in the absence of Ar buffer gas), $k_q=(8.50\pm0.32)\cdot10^{-10}$ and $\sigma_q=49.3$ were measured [2].

Electronic quenching of NH(c $^1\Pi$, v'=1) by HN_3 gave $k_q=7.5\cdot10^{-10}$, and for the rotational relaxation of the J'=1 level, $k_{rot}=8.3\cdot10^{-10}$ (vibrational relaxation negligible) [24].

The parent molecule HNCO quenches NH(c $^1\Pi$) at room temperature with $k_q=(7.2\pm0.3)\cdot10^{-10}$ [4].

Hydrocarbons. Recent measurements of c→a LIF [21] and c→a emission [2, 27] following HN_3 laser photolysis gave the following rate constants and cross sections for quenching by some simple hydrocarbons; the last column gives the cross sections recommended on the basis of these results:

Q	σ_q [21]	$10^{10}\cdot k_q$ [2]	σ_q [2]	$10^{10}\cdot k_q$ [27]	$\sigma_{q,rec}$ [2]
CH_4	23.7 ± 1.1	2.51 ± 0.21	27.8 ± 2.3	–	26
C_2H_6	65.9 ± 4.2	–	–	5.4 ± 0.2	67
C_3H_8	98.7 ± 4.7	7.14 ± 0.34	95.1 ± 4.5	–	96
C_2H_4	75.4 ± 8.8	–	–	6.8 ± 0.2	84
C_3H_6	109.0 ± 5.8	8.07 ± 0.71	106.8 ± 9.4	–	108
c-C_3H_6	–	7.53 ± 0.37	99.7 ± 4.9	–	100
C_2H_2	83.0 ± 10.1	–	–	–	83

Earlier studies had given considerably lower k_q and σ_q values for Q = CH_4, C_2H_6, C_3H_8, C_2H_4, C_3H_6, and c-C_3H_6 [29]; the cause of the discrepancy is unclear; insufficiently correcting for electric noise contributions in evaluating the rate constants has been suspected by [21]. Furthermore, k_q values around $3\cdot10^{-10}$, σ_q values around 44 for two C_4H_{10} isomers, and k_q values around $5\cdot10^{-10}$, and σ_q values around 73.5 for four C_4H_8 isomers have been reported by [29].

References:

[1] Bohn, B.; Stuhl, F.; Parlant, G.; Dagdigian, P. J.; Yarkony, D. R. (J. Chem. Phys. **96** [1992] 5059/68).
[2] Kenner, R. D.; Rohrer, F.; Stuhl, F. (J. Phys. Chem. **93** [1989] 7824/32).
[3] Umemoto, H.; Kikuma, J.; Tsunashima, S.; Sato, S. (Chem. Phys. **120** [1988] 461/7).
[4] Hikida, T.; Maruyama, Y.; Saito, Y.; Mori, Y. (Chem. Phys. **121** [1988] 63/71).
[5] Sasaki, S.; Tsunashima, S.; Sato, S. (Bull. Chem. Soc. Jpn. **59** [1986] 1671/4).
[6] Haak, H. K.; Stuhl, F. (J. Phys. Chem. **88** [1984] 3627/33).
[7] Piper, L. G.; Krech, R. H.; Taylor, R. L. (J. Chem. Phys. **73** [1980] 791/800).
[8] Fujita, I. (Z. Phys. Chem. [Munich] **136** [1983] 187/96).
[9] Gelernt, B.; Smith, A. L. (Chem. Phys. Lett. **60** [1979] 261/4).
[10] Cvejanovic, D.; Adams, A.; King, G. C. (J. Phys. B **11** [1978] 1653/62).

[11] Smith, W. H.; Brzozowski, J.; Erman, P. (J. Chem. Phys. **64** [1976] 4628/33).
[12] Clerc, M.; Schmidt, M.; Hagege-Temman, J.; Belloni, J. (J. Phys. Chem. **75** [1971] 2908/14).
[13] Sawada, T.; Kamada, H. (Bull. Chem. Soc. Jpn. **43** [1970] 331/4).
[14] Smith, W. H. (J. Chem. Phys. **51** [1969] 520/4).
[15] Fink, E.; Welge, K. H. (Z. Naturforsch. **19a** [1964] 1193/201).
[16] Rohrer, F. (Diss. Ruhr-Univ., Bochum 1987 from [2]).
[17] Rohrer, F.; Stuhl, F. (J. Chem. Phys. **86** [1987] 226/33).
[18] Parlant, G.; Dagdigian, P. J.; Yarkony, D. R. (J. Chem. Phys. **94** [1991] 2364/7).
[19] Maruyama, Y.; Hikida, T.; Mori, Y. (Chem. Phys. Lett. **116** [1985] 371/3).
[20] Kawasaki, M.; Hirata, Y.; Tanaka, I. (J. Chem. Phys. **59** [1973] 648/53).

[21] Umemoto, H.; Kikuma, J.; Tsunashima, S.; Sato, S. (Chem. Phys. **125** [1988] 397/402).
[22] Umemoto, H.; Kikuma, J.; Tsunashima, S.; Sato, S. (Chem. Phys. **123** [1988] 159/64).
[23] Becker, K. H.; Welge, K. H. (Z. Naturforsch. **19a** [1964] 1006/15).
[24] Mill, T. (Ber. Max-Planck-Inst. Strömungsforsch. **14** [1990] 1/101, 65/76).
[25] Okabe, H. (J. Chem. Phys. **49** [1968] 2726/33).
[26] Kawasaki, M. (private communication from [5, 21, 22]).
[27] Kenner, R. D.; Pfannenberg, S.; Heinrich, P.; Stuhl, F. (J. Phys. Chem. **95** [1991] 6585/93).
[28] Suto, M.; Lee, L. C. (J. Chem. Phys. **78** [1983] 4515/22).
[29] Sasaki, S.; Kano, A.; Tsunashima, S.; Sato, S. (Bull. Chem. Soc. Jpn. **59** [1986] 1675/81).

2.1.1.4.6 Deactivation of the d $^1\Sigma^+$ State

Lifetimes of the d $^1\Sigma^+$, v′=0, J′≤22 levels of NH have been determined with a phase-shift technique from P-, Q-, and R-branch lines in the d $^1\Sigma^+ \rightarrow$ b $^1\Sigma^+$, v=0→0 emission spectrum at 253 nm, excited by electron impact on NH_3; $\tau_{rad} = 46 \pm 5$ ns is the average value of the Q(1) to Q(10) lines. For ND, the P and Q branches of the 0→0 transition (rotational structure not well resolved) gave $\tau_{rad} = 62 \pm 6$ ns (lifetimes were also taken from the P and Q branches of the 1→1 band) [1]. An earlier study of the d→c emission at low spectral resolution gave only 18±3 ns for the radiative lifetime of the d $^1\Sigma^+$, v′=0 level [2].

References:

[1] Hsu, D. K.; Smith, W. H. (J. Chem. Phys. **66** [1977] 1835/6).
[2] Smith, W. H. (J. Chem. Phys. **51** [1969] 520/4).

2.1.1.5 Reactions of NH

Walter Hack
Max-Planck-Institut für Strömungsforschung
Göttingen

Elementary collision-induced processes of NH radicals can be of a chemical nature or of a physical nature (change of quantum state (excitation or quenching)). Only the product analysis gives a definite answer for the interaction pathway. Many of the investigations concerning elementary processes of NH were done in the gas phase, which is significantly different from the kinetic situation in condensed media. Therefore, the two situations are described in separate sections.

2.1.1.5.1 Elementary Reactions of NH in the Gas Phase

NH radicals in their electronic ground state are mainly studied with respect to their chemical behavior, whereas for NH in electronically excited states, the quenching processes also have to be taken into account; thus NH(X) and NH* are treated in separate paragraphs.

2.1.1.5.1.1 NH(X) Reactions

For the simplest bimolecular reaction in the N-H system

$$NH(X\,^3\Sigma^-) + H(^2S) \rightarrow H_2(X\,^1\Sigma_g^+) + N(^4S)$$

at temperatures between 1790 and 2200 K a value of the rate constant of $k = 3.0 \times 10^{13}$ $cm^3 \cdot mol^{-1} \cdot s^{-1}$ was determined from measurements in H_2-O_2-Ar flames doped with CH_3CN [1]. In good agreement with this value is a rate constant of $k(T) = 3.2 \times 10^{13}$ $\exp(-1.36\ kJ \cdot mol^{-1}/RT)$ $cm^3 \cdot mol^{-1} \cdot s^{-1}$ which was obtained from the reverse reaction [2].

The collision-induced dissociation of NH via

$$NH(X) + Ar \rightarrow N(^4S) + H(^2S) + Ar$$

has, in the temperature range of 3140 to 3320 K, a rate constant $k(T) = 2.65 \times 10^{14}$ $\exp(-31.6\ kJ \cdot mol^{-1}/RT)$ $cm^3 \cdot mol^{-1} \cdot s^{-1}$ [3].

The reaction of NH(X) with hydrogen molecules

$$NH(X) + H_2(X) \rightarrow \text{products}$$

is slow with an upper limit, which was determined at room temperature in an ammonia flash photolysis system, of $k(300\ K) \leq 6 \times 10^6$ $cm^3 \cdot mol^{-1} \cdot s^{-1}$ [4]. No difference was found between the reactivity of H_2 and D_2 towards NH(X); i.e., the NH(X) radicals were depleted by H_2 and D_2 with the same rate constant ($k(D_2) \leq 6 \times 10^6$ $cm^3 \cdot mol^{-1} \cdot s^{-1}$) [4]. An activation energy of $E_A = 6.3$ kJ/mol was calculated, and for the formula $k(T) = AT^b \exp(-E_A/RT)$, a value of $b = 0.67$ was computed for the transfer of a hydrogen atom [5]. In this work [5] other NH hydrogen transfer reactions were also calculated, which, however, are not further included in the text.

Nitrogen atoms in their electronic ground state react with NH radicals in a fast reaction

$$NH(X) + N(^4S) \rightarrow N_2 + H$$

[6] with a rate constant $k(300\ K) = 1.5 \times 10^{13}$ $cm^3 \cdot mol^{-1} \cdot s^{-1}$ [7], a value which is a factor of two lower than one from an estimate given in [8].

The reaction of NH(X) with $N_2(X)$

$$NH(X) + N_2 \rightarrow HN_3$$

is very slow. In an NH_3 flash photolysis system the variation of the N_2 pressure up to $p(N_2) = 930$ mbar resulted in an upper limit for the third-order rate constant of the reaction $NH(X) + N_2 + N_2 \rightarrow HN_3 + N_2$ of $k(300\ K) \leq 7.5 \times 10^9$ $cm^6 \cdot mol^{-2} \cdot s^{-1}$ [4]. At the highest pressure applied, $p = 930$ mbar N_2, an upper limit for the second-order room-temperature rate constant of $k \leq 1.8 \times 10^5$ $cm^3 \cdot mol^{-1} \cdot s^{-1}$ was stated [4, 9].

The reaction of the imidogen radical with itself

$$NH(X) + NH(X) \rightarrow \text{products}$$

is of special importance, not only in the N-H system, but also in those which contain additional elements to model the kinetic behavior of NH in these kinetic systems. At room temperature in an NH_3-VUV photolysis system, the NH(X) depletion was followed by the resonance fluorescence technique for different initial NH(X) concentrations. Taking into account all other NH(X) depleting reactions, a value of $k(300\ K) \cong 3 \times 10^{13}\ cm^3 \cdot mol^{-1} \cdot s^{-1}$ was determined [4,10] in agreement with an upper limit determined in a pulse radiolysis kinetic spectroscopy experiment [11]. At high temperatures in the range of 1100 to 2370 K, there are five independent shock tube studies following NH(X) with absorption methods after the pyrolysis of HNCO [3, 12, 13] or HN_3 [14, 50]. A rate constant, which is independent of temperature, of $k(T) = 5 \times 10^{13}\ cm^3 \cdot mol^{-1} \cdot s^{-1}$ can be obtained from these experiments. Together with the room-temperature value of the rate constant, it can be assumed that the rate constant is nearly independent of temperature over the entire temperature range. H atoms were followed as reaction products in this reaction [14] due to $2\ NH(X) \rightarrow N_2 + 2\ H$ in agreement with the interpretation in [8, 12]. For the reaction pathway $NH(X) + NH(X) \rightarrow NH_2 + N$, a rate constant $k(T) = 2 \times 10^{11}\ T^{0.5} \exp(-8.4\ kJ \cdot mol^{-1}/RT)\ cm^3 \cdot mol^{-1} \cdot s^{-1}$ is estimated [6], indicating that this pathway is not significant compared to $2\ NH \rightarrow N_2 + 2\ H$.

The radical-radical reaction

$$NH(X) + NH_2(\tilde{X}) \rightarrow \text{products}$$

was studied at room temperature in an isothermal discharge flow reactor with LMR and LIF detection devices. A rate constant of $k(296\ K) = 8 \times 10^{13}\ cm^3 \cdot mol^{-1} \cdot s^{-1}$ was measured [15, 16]. An earlier estimate, used for a kinetic model of the pulse radiolysis of NH_3-O_2 mixtures, is in agreement with this value [17]. A high-temperature ($2200 < T/K < 2800$) value of the rate constant was obtained from modeling the ammonia pyrolysis behind reflected shock waves; a value $k(T) = 1.5 \times 10^{15}\ T^{-0.5}\ cm^3 \cdot mol^{-1} \cdot s^{-1}$ was obtained with the reaction pathway $NH_2 + NH \rightarrow N_2H_2 + H$ [2]. Extrapolating this value down to room temperatures gives a good agreement with the value determined directly at low temperature.

The reaction

$$NH(X) + NH_3 \rightarrow \text{products}$$

was studied in an indirect way by observing N_2H_4 formation in glow discharge tubes [18]. A direct determination of the rate constant was done in the NH_3 flash photolysis system. An upper limit for the rate constant $k(500\ K) \leq 6 \times 10^7\ cm^3 \cdot mol^{-1} \cdot s^{-1}$ was determined [4].

The reaction of imidogen with hydrazine

$$NH(X) + N_2H_4 \rightarrow \text{products}$$

is also slow at room temperature with $k(300\ K) \leq 6 \times 10^9\ cm^3 \cdot mol^{-1} \cdot s^{-1}$ [4], a value, which is not in contradiction with a rate constant of $k(T) = 1 \times 10^{14} \exp(-41.8\ kJ \cdot mol^{-1}/RT)\ cm^3 \cdot mol^{-1} \cdot s^{-1}$ estimated from hydrazine decomposition at temperatures between 750 and 1000 K in an adiabatic flow reactor [19].

In the N-H-O system, the first reaction to think of is

$$NH(X) + O \rightarrow \text{products}$$

An estimation of the rate constant of this reaction is given in [15] from modeling the measured concentration profiles in the NH-O reaction system with $k(300\ K) = 5 \times 10^{13}\ cm^3 \cdot mol^{-1} \cdot s^{-1}$ [15, 16]. This high value is in agreement with theoretical studies predicting a small activation energy for the reaction [20, 21]. The activation energy was found experimentally to be about $E_A \cong 30$ kJ/mol [22]. In shock tube experiments between 2730 and 3380 K after thermal dissociation of HNCO, a rate constant of $k = 9.2 \times 10^{13}\ cm^3 \cdot mol^{-1} \cdot s^{-1}$ was determined independent of temperature in that range [23]. In a recent photolysis study it was shown that the reaction pathway $NH(X) + O \rightarrow OH + N$ ($k < 1 \times 10^{11}\ cm^3 \cdot mol^{-1} \cdot s^{-1}$) is of minor importance at room temperature [7]. The internal state distribution of the NO product has been determined [24], giving rise to the assumption that the reaction proceeds by the formation and decay of a short-lived HNO complex.

The reaction

$$NH(X) + O_2 \rightarrow \text{products}$$

is of great interest in nitrogen combustion systems and for the oxidation of NH_3 [25]. The room-temperature rate constant was measured after VUV photolysis of NH_3-O_2 mixtures; a value of $k(296\ K) = 5.1 \times 10^9\ cm^3 \cdot mol^{-1} \cdot s^{-1}$ was obtained [4, 26]. In a pulse radiolysis (NH_3-O_2 mixtures) experiment, an upper limit $k(349\ K) \leq 2 \times 10^{10}\ cm^3 \cdot mol^{-1} \cdot s^{-1}$ was determined [17], which is not in contradiction with the earlier value. In a discharge flow reactor the temperature dependence of the rate constant in the temperature range 286 to 543 K was found to be $k(T) = 7.6 \times 10^{10} \exp(-6.4\ kJ \cdot mol^{-1}/RT)\ cm^3 \cdot mol^{-1} \cdot s^{-1}$ [25], independent of pressure in the range 2.3 to 10.5 mbar. OH radicals were detected as reaction products and actually found to be the main product near room temperature [27]. Theoretical studies predict both NO+OH and O+HNO as products at and near room temperature [20, 28], whereas the HNO+O channel dominates at high temperatures and the NO+OH reaction pathway at low temperatures [28], which is in agreement with the experimental observations at low temperatures. The addition of $NH(^3\Sigma^-)$ to $O_2(^3\Sigma_g^-)$ has been investigated by SCF and MRD-CI calculations [29]. The NH-O_2 binding energy is found to be 27 kJ/mol. The complex is predicted to decompose to NO+OH after 1,3 hydrogen migration [29].

The reaction $NH(X) + O_2$ has been studied also at high temperatures [23, 30, 31, 50]. At temperatures between 2200 and 3270 K a significantly higher activation energy $E_A = 71.5$ kJ/mol was observed than near room temperature with a preexponential factor of $A = 3.5 \times 10^{13}\ cm^3 \cdot mol^{-1} \cdot s^{-1}$. Over the entire temperature range the rate constant can be represented by the expression $k(T) = 1.9 \times 10^{13} \exp(-57.1\ kJ \cdot mol^{-1}/RT) + 8.6 \times 10^{10} \exp(-6.6\ kJ \cdot mol^{-1}/RT)\ cm^3 \cdot mol^{-1} \cdot s^{-1}$, which is explained by a difference in the reaction mechanism at low and high temperatures [50].

For the reaction of electronically excited oxygen molecules

$$NH(X) + O_2(a\,^1\Delta_g) \rightarrow \text{products}$$

an upper limit for the room-temperature rate constant $k(295\ K) \leq 6 \times 10^9\ cm^3 \cdot mol^{-1} \cdot s^{-1}$ was published [27]. No products other than in the reaction $NH(X) + O_2(X)$ were found to be formed [27].

For the reactions

$$NH(X) + OH \rightarrow H + HNO \quad (a)$$
$$\rightarrow N + H_2O \quad (b)$$

in the absence of experimental data, estimated rate constants were given in the literature as $k_a(T) = 2 \times 10^{13}\ cm^3 \cdot mol^{-1} \cdot s^{-1}$ independent of temperature in the range 298 to 3000 K and $k_b(T) = 2 \times 10^9\ T^{1.2}\ cm^3 \cdot mol^{-1} \cdot s^{-1}$ for the same temperature range [32]. Also for the rate of the reaction $NH(X) + H_2O \rightarrow OH + NH_2$, no direct measurements are available; the estimate for the temperature range 600 to 3000 K is $k(T) = 1.2 \times 10^8\ T^{1.6} \exp(-117\ kJ \cdot mol^{-1}/RT)\ cm^3 \cdot mol^{-1} \cdot s^{-1}$ [32].

Computer simulation of the NH and NH_2 profiles in a pulse radiolysis system of NH_3-O_2 mixtures provided for the radical-radical reaction

$$NH(X) + HO_2 \rightarrow \text{products}$$

a rate constant of $k(349\ K) = 4.3 \times 10^{13}\ cm^3 \cdot mol^{-1} \cdot s^{-1}$ [17].

The most intensively studied NH(X) reaction is

$$NH(X) + NO \rightarrow products$$

The experimental and theoretical studies are summarized in Table 3. At room temperature NH(X) reacts with a rate constant of $k(300\ K) = 3.5 \times 10^{13}\ cm^3 \cdot mol^{-1} \cdot s^{-1}$ [42], which is nearly independent of temperature [42]. At high temperatures ($T > 1000$ K) the rate of the reaction is significantly slower. The agreement between different experimental studies is not satisfying. A recent determination [23] is, in the absolute value, in reasonable agreement with the extrapolation of the room-temperature data. A rate constant $k(T) = 1.7 \times 10^{14}$ $\exp(-53.2\ kJ \cdot mol^{-1}/RT)\ cm^3 \cdot mol^{-1} \cdot s^{-1}$ was found for the temperature range 2200 to 3350 K [23].

The reaction pathway $NH(X) + NO \rightarrow OH + N_2$ was the only one which was observed at and near room temperature [45]. This product channel is also predicted theoretically [43, 48] at higher temperatures (T = 3500 K) using a potential energy surface (PES) which was improved compared to an earlier calculated PES [47]. A contribution of this product channel of about 30% was observed [44] from the emission of OH(A); it was assumed that OH(A) is also formed directly at high temperatures [49].

Table 3
Experimental and Theoretical Investigations of the $NH(X) + NO \rightarrow \{N_2 + OH,\ H + N_2O\}$ Reaction.

experimental method	experimental conditions p in mbar, T in K	results rate constant in $cm^3 \cdot mol^{-1} \cdot s^{-1}$	Ref.
pulse radiolysis NH_3 [NH(X)](t) absorption $\lambda = 336$ nm	T = 420 $333 \leq p \leq 1330$	$k = 2.3 \times 10^{13}$	[33]
rate constant calculated from $k_1 = K \cdot k_{-1}$	T = 1820 $N_2O + H \rightarrow NH + NO$ (-1)	$k_{-1} = 1.0 \times 10^{12}$	[34]
VUV photolysis NH_3 [NH(X)](t) resonance fluorescence	T = 298 $40 \leq p \leq 931$	$k = 2.8 \times 10^{13}$	[10, 36, 35]
shock tube NH_3/NO/Ar	$1700 \leq T \leq 3000$	$k = 9 \times 10^9\ T^{0.25}$	[37]
flame study premixed $H_2/O_2/CH_3CN$ flame LIF [NH]....	T = 1790 $p = 1.01 \times 10^3$	$k < 4.2 \times 10^{12}$	[1]
shock tube NH_3/H_2/NO/Ar NH* emission $\lambda = 336$ nm	$1760 \leq T \leq 2850$	$k = 8 \times 10^{13}$ $\exp(-123\ kJ \cdot mol^{-1}/RT)$	[38]
	$1330 \leq T \leq 1560$	$k = 1.4 \times 10^{12}$	[39]
	T = 940	$k = 1.05 \times 10^{13}$	[40]

Table 3 (continued)

experimental method	experimental conditions p in mbar, T in K	results rate constant in $cm^3 \cdot mol^{-1} \cdot s^{-1}$	Ref.
quantum chemical calculation (BAC-MP4)		favored products: $H+N_2O$ rather than N_2+OH	[20]
NH(a) from N_2H_4 or HN_3 photolysis quenched to [NH(X)] LIF	T=300 p=6.65 (He)	$k=2.9\times10^{13}$	[41]
photolysis N_2H_4 $\lambda=248$ nm [NH(X)] LIF	$269\leq T\leq377$ p=1.33 He, Ar, N_2, N_2O	$k=3.48\times10^{13}$ independent of temperature independent of carrier gas	[42]
quantum chemical calculation (HF, MP2)		major products: N_2+OH rather than N_2O+H	[43]
shock wave HNCO/Ar NH(A) emission OH(A) emission	T=3500 $NH(X)+NO\rightarrow OH(X)+N_2$ other products	$k_{total}=7.1\times10^{12}$ $k(OH+N_2)/k_{total}=0.32$ $(NH(X)+NO\rightarrow OH(A)+N_2)$	[44]
shock wave HNCO pyrolysis [NH(X)](t) laser absorption $\lambda=336$ nm	$2220\leq T\leq3350$	$k=1.7\times10^{14}$ $\exp(-53.2\ kJ\cdot mol^{-1}/RT)$ contribution of the N_2+OH product channel 19%	[23]
$CHBr_3$/NO photolysis $\lambda=193$ nm [NH] LIF	T=300 p=26.6 (Ar) $[NO]\gg[NH]$	$k=2.8\times10^{13}$ only products: $OH+N_2$ no H atoms produced	[45]
crossed-beam experiment	room temperature	internal state distribution of OH product $[OH(v=1)]/[OH(v=0)]=0.30$ $E_{rot}=25$ kJ/mol	[46]
statistical dynamic method	$NH+NO\rightarrow N_2+OH$ (a) $\rightarrow N_2O+H$ (b) $300\leq T\leq3500$	branching fraction: (b) dominates over the entire temperature range $k_a/(k_a+k_b)=0.19$ (300 K) $=0.30$ (3500 K)	[28]
theoretical characterization of potential energy surface (CASSCF/CI)	NH(X)+NO $^2A'$ surface	N_2+OH predicted products	[48]

The reaction

$$NH(X)+NO_2\rightarrow \text{products}$$

was studied experimentally [42]. The rate constant was determined to be $k(300\ K)=9.7\times10^{12}$ $cm^3\cdot mol^{-1}\cdot s^{-1}$ with a small negative temperature dependence between 269 and 377 K, independent of pressure (i.e., of inert gas He, Ar, N_2, or N_2O) [42]. A theoretical investigation comes to the conclusion that the products might be $HNO+NO$; the formation of N_2O+OH via the intermediate $HNNO_2$ is connected with a barrier which would lead to a positive temperature dependence. Thus from the observed negative temperature dependence, these products can be excluded [43].

In the N-H-C system only a very few direct experimental investigations were published in the literature [4, 11, 15, 41, 51 to 53]. For the reactions of saturated hydrocarbons with NH(X), very small rate constants, considerably less than $k<6\times10^8$ $cm^3\cdot mol^{-1}\cdot s^{-1}$, have been observed in photolysis experiments [41]; this agrees with theoretical studies in which high activation energies are predicted [54]. For CH_4, C_2H_6, n-C_4H_{10}, n-C_5H_{12}, and n-C_6H_{14} reaction with NH(X), some estimates are given relative to the rate of the reaction NH_2+CH_4 [53]. Also methylamine (CH_3NH_2) reacts slowly at room temperature ($k\leq3.6\times10^8$ $cm^3\cdot mol^{-1}\cdot s^{-1}$) [4]. At high temperatures between 1400 and 1770 K the following rate constants were determined recently in a shock tube experiment: $k_{CH_4}(T)=4\times10^{14}$ $\exp(-105\ kJ\cdot mol^{-1}/RT)$ $cm^3\cdot mol^{-1}\cdot s^{-1}$; $k_{C_2H_6}(T)=2.3\times10^{13}$ $\exp(-59\ kJ\cdot mol^{-1}/RT)$ $cm^3\cdot mol^{-1}\cdot s^{-1}$, and $k_{C_3H_8}(T)=1\times10^{13}$ $\exp(-48\ kJ\cdot mol^{-1}/RT)$ $cm^3\cdot mol^{-1}\cdot s^{-1}$ [50, 118]. The extrapolation of these data to room temperature leads to values considerably less than the upper limits given.

For the reaction of unsaturated hydrocarbons (e.g. C_2H_4)

$$NH(X)+C_2H_4\rightarrow products$$

the rate constant has been estimated [51] but also measured directly [4]. For various olefins, rate constants in the order of $k(300\ K)\cong6\times10^8$ $cm^3\cdot mol^{-1}\cdot s^{-1}$ were found [4]. This value contradicts the observation that the rate constant for olefins is considerably less than 6×10^8 $cm^3\cdot mol^{-1}\cdot s^{-1}$ [41]. The reaction of NH(X) with ethylene has been studied theoretically [55, 56]; an activation barrier of about 103 kJ/mol was calculated [56]. In a shock tube experiment a rate constant $k(T)=1.8\times10^{13}$ $\exp(-56\ kJ\cdot mol^{-1}/RT)$ $cm^3\cdot mol^{-1}\cdot s^{-1}$ was determined for the temperature range 1400 to 1770 K [50, 119]. For other olefins ($CH_2{=}CHCH_3$, $CH_2{=}CHC_2H_5$, and $CH{=}CHCH{=}CH_2$) room-temperature rate constants in the order of $k\cong6\times10^8$ $cm^3\cdot mol^{-1}\cdot s^{-1}$ were observed. At high temperatures (in the range given above) the rate constants $k(T)=3.9\times10^{13}$ $\exp(-52\ kJ\cdot mol^{-1}/RT)$ $cm^3\cdot mol^{-1}\cdot s^{-1}$ and $k(T)=4.1\times10^{13}$ $\exp(-40\ kJ\cdot mol^{-1}/RT)$ $cm^3\cdot mol^{-1}\cdot s^{-1}$ were measured for $CH_2{=}CHCH_3$ and $(CH_3)_2C{=}C(CH_3)_2$, respectively [50, 119]. As final products of the reaction of NH(X) with olefins, cyanides are observed [51], which can be understood if the reaction proceeds through an alkyl nitrene intermediate as observed in matrix experiments [57] (see p. 127).

The unsaturated compound C_2H_4NH reacts very slowly with $k(300\ K)=6\times10^8$ $cm^3\cdot mol^{-1}\cdot s^{-1}$ [4], comparable to the other olefins.

In the N-H-C-O system, methanol and acetaldehyde have been studied [4, 50, 118]. CH_3OH reacts with a room-temperature rate constant smaller than $k\leq5.4\times10^8$ $cm^3\cdot mol^{-1}\cdot s^{-1}$ [4]. For CH_3CHO a high-temperature ($1400\leq T/K\leq1770$) rate constant $k(T)=5\times10^{13}$ $\exp(-48\ kJ\cdot mol^{-1}/RT)$ $cm^3\cdot mol^{-1}\cdot s^{-1}$ was determined in shock tube experiments [50, 118].

References on pp. 122/5.

2.1.1.5.1.2 NH(a, b,..., A) Reactions

Imidogen radicals in the electronically excited singlet states a $^1\Delta$, b $^1\Sigma$, and c $^1\Pi$ and in the A $^3\Pi$ triplet state have been studied kinetically. The only one of these states, however, in which depletion has been observed together with the yield of the products is the metastable a $^1\Delta$ state. For the others only the collision-induced depletion, regarded as quenching, has been observed, since there is no clean source available for those states which would easily make the product observation feasable.

For **NH(a $^1\Delta$)** many studies have been published on quenching as well as on elementary chemical reactions. With the noble gases, quenching is the only possible pathway. With the light noble gases He, Ar, and Kr, the quenching processes at and near room temperature are slow with $k_q < 6 \times 10^8$ $cm^3 \cdot mol^{-1} \cdot s^{-1}$ [58 to 62]. This is probably also the case for Ne which has not yet been studied directly.

The collision-induced intersystem crossing

$$NH(a\ ^1\Delta) + Xe \rightarrow NH(X\ ^3\Sigma^-) + Xe$$

is much faster than expected from the nuclear charge of Xe: $k_q = 7.2 \times 10^{12}$ $cm^3 \cdot mol^{-1} \cdot s^{-1}$ [61, 63 to 68], probably due to the formation of an exciplex intermediate. The interaction potential between NH(a) and Ar has, based on ab initio calculations, only a small minimum [69]. The complex NH(a)-Ar has been probed by the LIF technique [70]. The vibration distribution of NH(a, v) is conserved during the quenching process by Xe [65, 68]. The fast quenching of NH(a) by Xe can be used for the calibration of NH(a) relative to NH(X) [65].

Elementary reactions of NH(a) have been studied with two more atoms (N and O) [7]. The reaction

$$NH(a) + N(^4S) \rightarrow \text{products}$$

is not very fast at room temperature: $k = 6 \times 10^{11}$ $cm^3 \cdot mol^{-1} \cdot s^{-1}$ [7]. The product is mainly NH(X) due to a quenching process, $k_q = 6 \times 10^{11}$ $cm^3 \cdot mol^{-1} \cdot s^{-1}$ [7].

The interaction of $O(^3P)$ atoms with NH(a), which proceeds with a rate constant of k(300 K) $= 5.6 \times 10^{12}$ $cm^3 \cdot mol^{-1} \cdot s^{-1}$ is mainly of chemical character

$$\begin{aligned} NH(a) + O(^3P) &\rightarrow NO + H \quad &(a)\\ &\rightarrow OH + N \quad &(b) \end{aligned}$$

with the specific rate constants $k_a = 3.6 \times 10^{12}$ $cm^3 \cdot mol^{-1} \cdot s^{-1}$, $k_b = 7.8 \times 10^{11}$ $cm^3 \cdot mol^{-1} \cdot s^{-1}$, and the quenching rate constant $k_q = 1.2 \times 10^{12}$ $cm^3 \cdot mol^{-1} \cdot s^{-1}$ [7].

The reactions of NH(a) with nine diatomic molecules H_2, D_2, F_2, HF, HCl, N_2, O_2, CO, and NO have been studied. The reaction

$$NH(a) + H_2 \rightarrow \text{products}$$

has attracted experimental [41, 58, 62, 71, 72] as well as theoretical attention [73]. The NH(a) decay rate is reported to proceed with a room-temperature rate constant k(300 K) = 2.5×10^{12} $cm^3 \cdot mol^{-1} \cdot s^{-1}$ [58, 71, 72]. For the reaction NH(a) + D_2, a rate constant k(295 K) $= 1.6 \times 10^{12}$ $cm^3 \cdot mol^{-1} \cdot s^{-1}$ was reported. The branching fraction [H]/[D] was found to be 0.24, indicating that the reaction is dominated by the insertion of NH(a) into the D_2 bond, but vibrational energy of the reaction intermediate $NHD_2^{\ddagger}$ is still localized in newly formed N-D bonds before it passes through the exit barrier into NHD + D or ND_2 + H channels [72].

For the removal of NH(a) by F_2, a rate constant $k = 3.8 \times 10^{11}$ $cm^3 \cdot mol^{-1} \cdot s^{-1}$ was published [58], but no information about the mechanism was given.

The reaction NH(a) + HF is also not very fast at room temperature [58, 74, 75]. For NH(a, v = 0) a depletion rate constant $k(T) = 1.1 \times 10^{12} \exp(-5.1\ kJ \cdot mol^{-1}/RT)$ $cm^3 \cdot mol^{-1} \cdot s^{-1}$ was determined in the temperature range 294 to 479 K [75]. NH(a, v = 1) reacts a factor of three faster than NH(a, v = 0). The deuterated species ND(a) reacts significantly faster with HF than NH, but NH reacts with HF and DF at nearly the same rate. The main product in the reaction NH(a) + HF is $NH_2(\tilde{X})$, whereas the quenching channel contributes less than 5% [75]. The equivalent reaction NH(a) + HCl was found to be about two orders of magnitude faster with $k(300\ K) = 4.7 \times 10^{13}\ cm^3 \cdot mol^{-1} \cdot s^{-1}$ [76].

The intersystem crossing induced by $N_2(X\ ^1\Sigma_g^+)$ molecules

$$NH(a^1\ \Delta) + N_2(X) \rightarrow NH(X\ ^3\Sigma^-) + N_2(X)$$

proceeding via the $HN_3^{\ddagger}$ intermediate, which plays a role in the HN_3-NH(a) source, has been studied intensively [58, 62, 64 to 67, 71, 77 to 79]. The reaction has a small activation energy of about 5 kJ/mol due to a small barrier in the entrance channel [65, 66]. The preexponential factor $A = 4 \times 10^{11}\ cm^3 \cdot mol^{-1} \cdot s^{-1}$ [66] indicates that the intersystem crossing probability is small. The theoretical interpretation of this system was described in [80].

In the isoelectronic system

$$\begin{aligned} NH(a) + CO(X) &\rightarrow NH(X) + CO(X) \qquad (a) \\ &\rightarrow NCO + H \qquad (b) \end{aligned}$$

not only quenching but chemical reaction also was observed [81]. The rate of this reaction has been studied intensively [62, 67, 68, 71, 79, 81]. It has no activation energy and can be represented by $k = 1.2 \times 10^{13}\ cm^3 \cdot mol^{-1} \cdot s^{-1}$ independent of temperature in the range 293 to 459 K [79, 81]. The physical quenching (a) contributes 12%, and the chemical pathway (b) predominates with 88% [81].

The reaction of NH(a) with a triplet diatomic molecule

$$NH(a) + O_2(X\ ^3\Sigma) \rightarrow products$$

has been studied by several authors [61, 64, 71, 78, 79]. The Arrhenius expression obtained in the temperature range 298 to 476 K is $k(T) = 1.1 \times 10^{12}\ \exp(-8.6\ kJ \cdot mol^{-1}/RT)$ $cm^3 \cdot mol^{-1} \cdot s^{-1}$ [79]. The contribution of quenching is $\geq 60\%$ of the overall NH(a) depletion. The formation of OH via $NH(a) + O_2 \rightarrow OH + NO$ contributes less than 4% [61].

The radical-radical reaction

$$NH(a) + NO(X\ ^2\Pi) \rightarrow products$$

is a rapid reaction [64, 77, 82, 83] with a room-temperature rate constant of $k = 1.3 \times 10^{13}$ $cm^3 \cdot mol^{-1} \cdot s^{-1}$ [83]. The physical quenching, which can occur simply by a cis-trans isomerization of the HN-NO intermediate, contributes 40%. The chemical reaction, with OH as one product, remains at 60% [83].

Among the triatomic reactants, H_2O has attracted much attention due to its presence in many natural systems [77, 84, 85]. Also the $NH(a) + D_2O$, $ND(a) + H_2O$, and $ND(a) + D_2O$ reactions have been studied [85]. The reaction is, with $k(298\ K) = 2.9 \times 10^{12}\ cm^3 \cdot mol^{-1} \cdot s^{-1}$, a fast reaction, which proceeds as an insertion reaction with $NH_2(\tilde{X})$ and OH as the main products [77] (the insertion is also observed in the liquid phase). The quenching contributes about 1% of the total NH(a) depletion [77]. In agreement with the insertion dynamics, it was found that $NH(a) + H_2O$ and $ND(a) + H_2O$ react with the same rate, whereas D_2O reacts significantly slower with NH(a) and ND(a) [85].

The insertion mechanism is dominant for all H-atom-containing molecules like HCN [83], H_2S [84], NH_3 [58, 86], and saturated hydrocarbons (see below). HCN reacts with a rate constant of $k(298\ K) = 2.1 \times 10^{13}\ cm^3 \cdot mol^{-1} \cdot s^{-1}$ and leads to the products $NH_2(\tilde{X}) + CN$ [83].

Three additional triatomic molecules have been studied: CO_2 [71, 77, 87], N_2O [71, 77], and NO_2 [77, 87]. CO_2 reacts with NH(a) with a rate constant $k(296\ K) = 1.4 \times 10^{11}$

$cm^3 \cdot mol^{-1} \cdot s^{-1}$ and a quenching ratio $k_q/k_{total} = 0.24$ [77, 87]. N_2O also does not react very fast with NH(a): k(298 K) $= 1.0 \times 10^{12}$ $cm^3 \cdot mol^{-1} \cdot s^{-1}$ with a quenching contribution of 6%. The chemical products are not known. The OH formation relative to the NH(a) depletion is less than 1% [77, 87]. The radical-radical reaction NH(a) + NO_2 is, as expected, very fast with k(298 K) $= 2.2 \times 10^{13}$ $cm^3 \cdot mol^{-1} \cdot s^{-1}$ [77, 87].

In the group of molecules with four atoms are the commonly used NH(a) precursors HN_3 and HNCO and ammonia, the last of which is of special interest in the N-H system. Mainly the reaction

$$NH(a) + HN_3 \rightarrow NH_2(\tilde{A}) + N_3$$

has been intensively studied [41, 58 to 60, 63, 71, 76, 77, 88 to 91], and DN_3 has also been investigated [78, 85]. The formation and emission of $NH_2(\tilde{A})$ has been used to detect NH(a) indirectly [76] (see p. 27). The rate constant for the NH(a) depletion by HN_3 is high; values in the range (5.6 to 11) $\times 10^{14}$ $cm^3 \cdot mol^{-1} \cdot s^{-1}$ were published in the literature [41, 59, 60, 77, 88, 91].

There are significantly less studies [58, 78, 92] on the reaction

$$NH(a) + HNCO \rightarrow products$$

The rate constant k(298 K) $= 9.8 \times 10^{13}$ $cm^3 \cdot mol^{-1} \cdot s^{-1}$ [58] seems to be slightly higher than for the isoelectronic HN_3 molecule; this fact makes HNCO less favored as an NH(a) source.

The reaction

$$NH(a) + NH_3 \rightarrow 2\ NH_2$$

is interesting due to the formation of a highly vibrationally excited electronic ground state hydrazine molecule $N_2H_4^{\ddagger}$. The rate constant of that reaction is high with k(298 K) $= 8.9 \times 10^{13}$ $cm^3 \cdot mol^{-1} \cdot s^{-1}$ [86]. The $N_2H_4^{\ddagger}$ formed with this "chemical activation method" decomposes to form more than 93% NH_2. The quenching pathway is of minor importance with <1% [86].

Hydrazine reacts with NH(a) with a rate constant k(298 K) $= 1 \times 10^{14}$ $cm^3 \cdot mol^{-1} \cdot s^{-1}$; the reaction products were not determined [58].

The polyatomic molecules which have to be discussed are the saturated hydrocarbons CH_4 [41, 71, 73, 76, 93], C_2H_6 [71, 94, 95], C_3H_8 [41, 71, 93, 96], *n*-C_4H_{10} [93]; the unsaturated hydrocarbons C_2H_4 [41, 71, 75, 76, 98], C_3H_8 [41, 75], *trans*-C_4H_8 [75], and acetylene C_2H_2 [75, 99] as well as methylacetylene $CHCCH_3$ [41]. The cyclic compounds *cyclo*-C_6H_{12} [76], benzene C_6H_6 [77, 87], and the substituted methanes CF_4, CF_3H, CF_2H_2, CFH_3 [100] as well as the methane derivatives CH_3OH and CH_3OD [101] were reacted with NH(a).

The reaction

$$NH(a) + CH_4 \rightarrow NH_2 + CH_3$$

has an activation energy of $E_A = 7.8$ kJ/mol in the temperature range 250 to 600 K, and the reaction rate constant is given with a preexponential factor of $A = 5 \times 10^{13}$ $cm^3 \cdot mol^{-1} \cdot s^{-1}$ [41]. The only reaction products are $NH_2(\tilde{X})$ and CH_3 as detected in [93]; the quenching is unimportant with <1% [93]. For the other saturated hydrocarbons mentioned above, rate constants ranging from (1.1 to 3.2) $\times 10^{13}$ $cm^3 \cdot mol^{-1} \cdot s^{-1}$ were observed [41, 71, 75, 93]. The quenching pathway in all cases is unimportant at <1% [93]. The chemical pathway leading to $NH_2(\tilde{X})$ decreases with increasing number of C atoms in the hydrocarbon, which clearly indicates that NH(a) reacts via insertion and not via abstraction [93]. The insertion mechanism is also found for cyclohexane [76]. The reactions of NH(a) with unsaturated hydrocarbons are also very fast reactions with a rate constant for, e.g., C_2H_4 of k(300 K) $=5.1\times10^{13}$ $cm^3\cdot mol^{-1}\cdot s^{-1}$ [75]. The negative temperature dependence of the rate constants [41] indicates that the initial step is an addition to the double bond [98]. The simplest molecule with a triple C-C bond, C_2H_2, reacts with NH(a) with a rate constant k(300 K) $=8.4\times10^{13}$ $cm^3\cdot mol^{-1}\cdot s^{-1}$, independent of temperature in the temperature range 250 to 600 K [41].

Benzene (C_6H_6) depletes NH(a) with a nearly gas kinetic cross section of k(296 K) $=9.9\times10^{13}$ $cm^3\cdot mol^{-1}\cdot s^{-1}$ [87]. The primary products of this reaction are not yet known.

In the reactions of fluoromethanes with NH(a), insertion-elimination reactions are observed which produce HF molecules [100].

For the reaction of NH(a) with methanol a rate constant k(298 K) $=8.2\times10^{13}$ $cm^3\cdot mol^{-1}\cdot s^{-1}$ is reported. The product of this reaction is mainly $NH_2(\tilde{X})$; the physical quenching has been evaluated to be less than 2.4% [101]. In the reaction

$$\begin{aligned} NH(a) + CH_3OD &\rightarrow NH_2(\tilde{X}) + CH_2OD \\ &\rightarrow NHD(\tilde{X}) + CH_3O \end{aligned}$$

which proceeds with nearly the same rate as the reaction $NH(a)+CH_3OH$, a bond selectivity for the NH(a) insertion was detected from the NHD and NH_2 observations. It was found that the O-D bond is about 69 times more reactive towards insertion of NH(a) than a single C-H bond [101].

For the elementary processes of **NH(b $^1\Sigma$)**, only the overall depletion rates (abbreviated as k_q) are published in the literature. There is, in general, no discrimination between chemical reaction and physical quenching. Only for the energy pooling reaction NH(b $^1\Sigma^+$) + O_2 (a $^1\Delta_g$) $\rightarrow$ NH(A $^3\Pi$) + O_2(X $^3\Sigma_g$) were products detected [102]. The experimental results are summarized in tables on pp. 94/5. For the theoretical understanding, different models have been applied. But a potential energy surface, which is necessary for a theoretical understanding, has until now only been calculated for the NH(b $^1\Sigma^-$)-Ar system [69].

The quenching of **NH(c $^1\Pi$)** has been studied experimentally for a large number of collision partners [103 to 112]. Quenching cross sections in the range $(1.4\ \text{to}\ 109)\times10^{-16}$ cm^2 are observed [109]. The results are summarized on pp. 108/11.

For the deactivation/chemical reaction of **NH(A $^3\Pi$)**, similar arguments can be given as for NH(b). The experimental data are summarized in tables on pp. 99/103. The rate constants are about three orders of magnitude larger than for NH(b), since in the case of NH(A) the collision-induced transition is allowed. The quenching cross sections depend on the rotational state, i.e., they decrease with increasing rotational excitation [113]. Thus to compare the various experimentally determined rate constants, the present rotational distribution has to be taken into account. Recently the depletion rates have been measured for different vibronic states [112] and collision-induced changes of other specific quantum states (mixing of Λ doublets [114] and rotational relaxation [115 to 117]).

References:

[1] Morley, C. (Symp. Int. Combust. Proc. **18** [1981] 23/32).
[2] Davidson, D. F.; Kohse-Höinghaus, K.; Chang, A. Y.; Hanson, R. K. (Int. J. Chem. Kinet. **22** [1990] 513/35).
[3] Mertens, J. D.; Chang, A. Y.; Hanson, R. K.; Bowman, C. T. (Int. J. Chem. Kinet. **21** [1989] 1049/67).
[4] Zetzsch, C. (Habilitationsschr. Univ. Bochum 1977, 242 pp.).
[5] Mayer, S. W.; Schieler, L. (J. Phys. Chem. **72** [1968] 236/40).

[6] Hanson, R. K.; Salimian, S. (in: Gardiner, V. C., Jr.; Combustion Chemistry, Chapter 6, Springer, Berlin-Heidelberg-New York 1984).
[7] Hack, W.; Wagner, H. Gg.; Zasypkin, A. (Ber. Bunsen-Ges. Phys. Chem. [1994] to be published).
[8] Miller, J. A.; Bowman, C. T. (Prog. Energy Combust. Sci. **15** [1989] 287/338).
[9] Zetzsch, C.; Stuhl, F. (Ber. Bunsen-Ges. Phys. Chem. **85** [1981] 564/8).
[10] Hansen, I.; Höinghaus, K.; Zetzsch, C.; Stuhl, F. (NBS Spec. Publ. [U.S.] No. 526 [1978] 334/6).

[11] Meaburn, G. M.; Gordon, S. (J. Phys. Chem. **72** [1968] 1592/7).
[12] Kajimoto, O.; Yamamoto, Y.; Fueno, T. (J. Phys. Chem. **83** [1979] 429/35).
[13] Kajimoto, O.; Kondo, O.; Okada, K.; Fujikane, J.; Fueno, T. (Bull. Chem. Soc. Jpn. **58** [1985] 3469/74).
[14] Hori, K.; Oya, M.; Tanaka, H.; Asaba, T. (Symp. Shock Tubes Waves Proc. **15** [1985] 261).
[15] Dransfeld, P.; Hack, W.; Kurzke, H.; Temps, F.; Wagner, H. Gg. (Symp. Int. Combust. Proc. **20** [1985] 655/63).
[16] Temps, F. (Ber. Max-Planck-Inst. Strömungsforsch. **1983** No. 4, 137 pp.).
[17] Pagsberg, P. B.; Erikson, J.; Christensen, H. C. (J. Phys. Chem. **83** [1979] 582/90).
[18] Wannagat, U.; Kohen, H. (Z. Anorg. Allg. Chem. **304** [1960] 276/95).
[19] Eberstein, I. J.; Glassman, I. (Symp. Int. Combust. Proc. **10** [1965] 365/74).
[20] Melius, C. F.; Binkley, J. S. (ACS Symp. Ser. **249** [1984] 103/15).

[21] Walch, P. S. (J. Chem. Phys. **93** [1990] 8036/40).
[22] Dean, A. M.; Chou, M. S.; Stern, D. (ACS Symp. Ser. **249** [1984] 71/84).
[23] Mertens, J. D.; Chang, A. Y.; Hanson, R. K.; Bowman, C. T. (Int. J. Chem. Kinet. **23** [1991] 173/96).
[24] Huang, Y. L.; Dagdigian, P. J. (J. Chem. Phys. **97** [1992] 180/8).
[25] Husain, D.; Norrish, R. G. W. (Proc. R. Soc. [London] A **273** [1963] 145/64).
[26] Zetzsch, C. (J. Photochem. **9** [1978] 151/3).
[27] Hack, W.; Kurzke, H.; Wagner, H. Gg. (J. Chem. Soc. Faraday Trans. II **81** [1985] 949/61).
[28] Miller, J. A.; Melius, C. F. (Symp. Int. Combust. Proc. **24** [1992] 719/26).
[29] Fueno, T.; Yokoyama, K.; Takane, S. (Theor. Chim. Acta **82** [1992] 299/308).
[30] Bian, J.; Vandooren, J.; Van Tiggelen, P. J. (Symp. Int. Combust. Proc. **21** [1986/88] 953/63).

[31] Bian, J.; Vandooren, J.; Van Tiggelen, P. J. (Symp. Int. Combust. Proc. **23** [1990/91] 379/86).
[32] Cohen, N.; Westberg, K. R. (J. Phys. Chem. Ref. Data **20** [1991] 1211/311).
[33] Gordon, S.; Mulac, W.; Nangia, P. (J. Phys. Chem. **75** [1971] 2087/93).
[34] Nip, W. S. (Diss. Univ. Toronto 1974).
[35] Hansen, I.; Höinghaus, K.; Zetzsch, C.; Stuhl, F. (Chem. Phys. Lett. **42** [1976] 370/2).
[36] Höinghaus, K.; Biermann, H. W.; Zetzsch, C.; Stuhl, F. (Z. Naturforsch. **31a** [1976] 239/43).
[37] Roose, T. R.; Hanson, R. K.; Kruger, C. H. (Shock Tube Shock Wave Res. Proc. 11th Int. Symp., Seattle 1977 [1978], pp. 245/53).
[38] Roose, T. R.; Hanson, R. K.; Kruger, C. H. (Symp. Int. Combust. **18** [1981] 853/62).
[39] Peterson, R. C. (Diss. Purdue Univ. 1981).
[40] Kodo, O. (Diss. Osaka Univ. 1982).

[41] Cox, J. W.; Nelson, H. H.; McDonald, J. R. (Chem. Phys. **96** [1985] 175/82).
[42] Harrison, J. A.; Whyte, A. R.; Phillips, L. F. (Chem. Phys. Lett. **129** [1986] 346/52).

[43] Harrison, J. A.; Maclagan, G. A. R. (J. Chem. Soc. Faraday Trans. **86** [1990] 3519/23).
[44] Yokoyama, K.; Sakane, Y.; Fueno, T. (Bull. Chem. Soc. Jpn. **64** [1991] 1738/42).
[45] Yamasaki, K.; Okada, S.; Koshi, M.; Matsui, H. (J. Chem. Phys. **95** [1991] 5087/96).
[46] Patel-Misra, D.; Dagdigian, P. J. (J. Phys. Chem. **96** [1992] 3232/6).
[47] Marshall, P.; Fontijn, A.; Melius, C. F. (J. Chem. Phys. **86** [1987] 5540/9).
[48] Walch, P. S. (J. Chem. Phys. **98** [1993] 1170/7).
[49] Yokoyama, K.; Kitaibe, H.; Fueno, T. (Bull. Chem. Soc. Jpn. **64** [1991] 1731/7).
[50] Röhrig, M. (Diss. Univ. Göttingen 1993, 84 pp.).

[51] Cornell, D. W.; Berry, R. S.; Lwowski, W. (J. Am. Chem. Soc. **88** [1966] 544/50); Cornell, D. W. (Diss. Yale Univ. 1965, pp. 1/77).
[52] Richardson, W. C.; Setser, D. W. (Can. J. Chem. **47** [1969] 2725/7).
[53] Rozenberg, A. S.; Voronko, V. G. (Zh. Fiz. Khim. **46** [1972] 744/6; Russ. J. Phys. Chem. [Engl. Transl.] **46** [1972] 425/6).
[54] Nagase, S.; Fueno, T. (Theor. Chim. Acta **4** [1976] 59/70).
[55] Haines, W. J.; Csizmadia, I. G. (Theor. Chim. Acta **31** [1973] 283/6).
[56] Fueno, T.; Yamaguchi, K.; Kondo, O. (Bull. Chem. Soc. Jpn. **63** [1990] 901/12).
[57] Jacox, M. E.; Milligan, D. E. (J. Am. Chem. Soc. **85** [1963] 278/82).
[58] Bower, R. D.; Jacoby, M. T.; Blauer, J. A. (J. Chem. Phys. **86** [1987] 1954/6).
[59] Piper, L. G.; Krech, R. H.; Taylor, R. L. (J. Chem. Phys. **73** [1980] 791/800).
[60] Rohrer, F.; Stuhl, F. (Chem. Phys. Lett. **111** [1984] 234/7).

[61] Hack, W.; Wilms, A. (J. Phys. Chem. **93** [1989] 3540/6).
[62] Sauder, D. G.; Patel-Misra, D.; Dagdigian, P. J. (J. Chem. Phys. **91** [1989] 5316/23).
[63] Kodama, S. (Bull. Chem. Soc. Jpn. **56** [1983] 2348/54).
[64] Hack, W.; Wilms, A. (Ber. Max-Planck-Inst. Strömungsforsch. **1987** No. 20, 77 pp.).
[65] Hack, W.; Rathmann, K. (J. Phys. Chem. **96** [1992] 47/52).
[66] Nelson, H. H.; McDonald, J. R. (J. Chem. Phys. **93** [1990] 8777/83).
[67] Adams, J. S.; Pasternack, L. (J. Phys. Chem. **95** [1991] 2975/82).
[68] Patel-Misra, D.; Dagdigian, P. J. (J. Chem. Phys. **97** [1992] 4871/80).
[69] Jansen, G.; Hess, B. A. (Chem. Phys. Lett. **192** [1992] 21/8).
[70] Randall, R.; Chuang, C. C.; Lester, I. M. (Chem. Phys. Lett. **200** [1992] 113/20).

[71] Freitag, F.; Rohrer, F.; Stuhl, F. (J. Phys. Chem. **93** [1989] 3170/4).
[72] Tezaki, A.; Okada, S.; Matsui, H. (J. Chem. Phys. **98** [1993] 3876/83).
[73] Fueno, T.; Kajimoto, O.; Bonacic-Koutecky, V. (J. Am. Chem. Soc. **106** [1984] 406/12).
[74] Rathmann, K. (Dipl.-Arbeit Univ. Göttingen 1988).
[75] Rathmann, K. (Diss. Univ. Göttingen 1992).
[76] McDonald, J. R.; Miller, R. G.; Baronavski, A. P. (Chem. Phys. **30** [1978] 133/45).
[77] Hack, W.; Wilms, A. (Z. Phys. Chem. **161** [1989] 107/21).
[78] Bohn, B.; Stuhl, F. (J. Phys. Chem. **97** [1993] 7234/8).
[79] Nelson, H. H.; McDonald, J. R.; Alexander, M. H. (J. Phys. Chem. **94** [1990] 3291/4).
[80] Alexander, M. H.; Werner, H. J.; Dagdigian, P. J. (J. Chem. Phys. **89** [1988] 1388/400).

[81] Hack, W.; Rathmann, K. (J. Phys. Chem. **94** [1990] 3636/9).
[82] Fueno, T.; Fukuda, M.; Yokoyama, K. (Chem. Phys. **124** [1988] 265/72).
[83] Hack, W.; Rathmann, K. (J. Phys. Chem. **94** [1990] 4155/61).
[84] Sudhakar, P. V.; Lammertsma, K. (J. Am. Chem. Soc. **113** [1991] 5219/23).
[85] Orthner, H. (Dipl.-Arbeit Univ. Göttingen 1992).
[86] Hack, W.; Rathmann, K. (Z. Phys. Chem. **176** [1992] 151/60).
[87] Wilms, A. (Diss. Univ. Göttingen 1987).
[88] Paur, R. J.; Bair, E. J. (Int. J. Chem. Kinet. **8** [1976] 139/52).
[89] Paur, R. J.; Bair, E. J. (J. Photochem. **1** [1973] 255/65).

[90] Yamasaki, K.; Watanabe, A.; Tokue, I.; Ito, Y. (Chem. Phys. Lett. **204** [1993] 106/10).

[91] McDonald, J. R.; Miller, R. G.; Baronavski, A. P. (Chem. Phys. Lett. **51** [1977] 57/60).
[92] Drozdoski, W. S.; Baronavski, A. P.; McDonald, J. R. (Chem. Phys. Lett. **64** [1979] 421/5).
[93] Hack, W.; Rathmann, K. (Ber. Bunsen-Ges. Phys. Chem. **94** [1990] 1304/7).
[94] Kajimoto, O.; Fueno, T. (Chem. Phys. Lett. **80** [1981] 484/7).
[95] Kodama, S. (Bull. Chem. Soc. Jpn. **56** [1983] 2355/62).
[96] Kajimoto, O.; Kondo, O.; Fueno, T. (Kokagaku Toronkai Koen Yoshishu **1979** 276/7).
[97] Kitamura, T.; Tsunashima, S.; Sato, S. (Kokagaku Toronkai Koen Yoshishu **1979** 230/1; C.A. **93** [1980] No. 149351).
[98] Kodama, S. (Bull. Chem. Soc. Jpn. **56** [1983] 2363/70).
[99] Kodama, S. (J. Phys. Chem. **92** [1988] 5019/24).
[100] Poole, P. R.; Pimentel, G. C. (J. Chem. Phys. **63** [1975] 1950/8).

[101] Okada, S.; Tezaki, A.; Yamasaki, K.; Matsui, H. (J. Chem. Phys. **98** [1993] 8667/72).
[102] Presser, N.; Zhu, Y.-F.; Gordon, R. J. (J. Phys. Chem. **91** [1987] 4383/8).
[103] Sasaki, S.; Kano, A.; Tsunashima, S.; Sato, S. (Bull. Chem. Soc. Jpn. **59** [1986] 1675/81).
[104] Sasaki, S.; Tsunashima, S.; Sato, S. (Bull. Chem. Soc. Jpn. **59** [1986] 1671/4).
[105] Kawasaki, M.; Hirata, Y.; Tanaka, I. (J. Chem. Phys. **59** [1973] 648/53).
[106] Rohrer, F.; Stuhl, F. (J. Chem. Phys. **86** [1987] 226/33).
[107] Hikida, T.; Maruyama, Y.; Saito, Y.; Mori, Y. (Chem. Phys. **121** [1988] 63/71).
[108] Okabe, H. (J. Chem. Phys. **49** [1968] 2726/33).
[109] Umemoto, H.; Kikuma, J.; Tsunashima, S.; Sato, S. (Chem. Phys. **125** [1988] 397/402).
[110] Umemoto, H.; Kikuma, J.; Tsunashima, S.; Sato, S. (Chem. Phys. **120** [1988] 461/7).

[111] Haak, H. K.; Stuhl, F. (J. Phys. Chem. **88** [1984] 3627/33).
[112] Kenner, R. D.; Kaes, A.; Browarzik, R. K.; Stuhl, F. (J. Chem. Phys. **91** [1989] 1440/5).
[113] Garland, N. L.; Crosley, D. R. (J. Chem. Phys. **90** [1989] 3566/73).
[114] Kaes, A.; Stuhl, F. (J. Chem. Phys. **97** [1992] 7362/70).
[115] Kaes, A.; Stuhl, F. (J. Chem. Phys. **97** [1992] 4661/8).
[116] Alexander, M. H.; Dagdigian, P. J.; Werner, H. J. (Faraday Discuss. Chem. Soc. No. 91 [1991] 319/35).
[117] Chappell, E. L.; Jeffries, J. B.; Crosley, D. R. (J. Chem. Phys. **97** [1992] 2400/5).
[118] Röhrig, M.; Wagner, H. Gg. (Symp. Int. Combust. Proc. **25** in press).
[119] Röhrig, M.; Wagner, H. Gg. (Ber. Bunsen-Ges. Phys. Chem. **98** [1994] in press).

2.1.1.5.2 NH Reactions in Thermal Systems

At the end of the last century the conversion of N_2 from the air to other N-containing compounds in a thermal reaction to produce artificial manures was regarded as a great success of mankind. The thermal reactions of air nitrogen and fuel nitrogen in the combustion in power plants and internal combustion engines are nowadays regarded as causing a severe problem. The aim is to understand and then to avoid the formation of NO_x in combustion systems, which are the most important thermal systems to be discussed in this section.

The active intermediates of nitrogen in combustion are the NH_2 and NH radicals and N atoms due to the general scheme

$$\begin{array}{ccc} \text{fuel-N} \rightarrow \text{HNC} \longrightarrow & NH_i & \xrightarrow{NO} N_2 \\ & \downarrow & \\ & NO & \end{array}$$

with i=0, 1, 2, which describes the fate of nitrogen in flames with respect to the final product NO or N_2. The two radicals NH and NH_2 are closely linked, mainly via the reactions

$$NH_2 + \{H, OH\} \rightleftharpoons NH + \{H_2, H_2O\}$$

Thus many aspects of the nitrogen chemistry in combustion, as described in the paragraph "NH_2 in Flames" (see p. 231), are relevant for the NH radicals and there is no need to repeat the description.

The NH radicals are not only of interest in the processes of fuel-N conversion [1, 2], but also in the $DeNO_x$ [3] and $RAPRENO_x$ (rapid reduction of NO_x) (see p. 233) processes [4, 5]. The reaction NH + NO → products is important in the schemes for the control of NO_x emission from fuel nitrogen [6, 7]. The N_2O formation and removal occur primarily through the reaction $NH + NO \rightarrow N_2O + H$ [8]. The reaction of NH + HCN has to be mentioned, along with $CH_2 + N_2$, as a precursor of rapid NO formation [9]. A detailed reaction mechanism was developed to describe the concentration profiles of NH and other radicals in hydrocarbon-air flames [6,10 to 13].

The NH in flames is mainly in the electronic ground state and can be observed by LIF in, e.g., CH_4-air flames [14] and CH_4-NO_2-O_2 flames [15]. But also electronically excited NH* can be present in flames as was observed by emission [16, 17]. The NH can be formed initially in an electronically excited state in reactions like $CH(X\,^2\Pi) + NO(X) \rightarrow NH(A\,^3\Pi) + CO(X)$ [18]. At high temperatures, around 2000 K, the contribution of electronically excited NH has to be taken into consideration [3]. In special combustion systems, like NF_3-H_2 flames, NH($A\,^3\Pi$-$X\,^3\Sigma^-$) has been observed [19]. In general, however, the modeling of nitrogen chemistry in combustion can be achieved with NH(X) radical reactions [6, 12].

References:

[1] Kimball-Linne, M. A.; Hanson, R. K. (Combust. Flame **64** [1986] 337/51).
[2] Puechberty, D.; Cotterau, M. J. (Combust. Flame **51** [1983] 299/311).
[3] Vandooren, J.; Sarkisov, C. M.; Balakkuin, V. P.; Van Tiggelen, P. J. (Chem. Phys. Lett. **184** [1991] 294/300).
[4] Perry, R. A.; Siebers, D. L. (Nature **324** [1986] 657/8).
[5] Perry, R. A. (Symp. Int. Combust. Proc. **21** [1988] 913/8).
[6] Miller, J. A.; Bowman, C. T. (Prog. Energy Combust. Sci. **15** [1989] 287/338).
[7] Morley, C. (Symp. Int. Combust. Proc. **18** [1981] 23/32).
[8] Roby, R. J.; Bowman, C. T. (Combust. Flame **70** [1987] 119/23).
[9] Sander, W. A.; Lin, C. Y.; Lin, M. C. (Combust. Technol. **51** [1987] 103/8).
[10] Karim, H.; Bozzelli, J.; Dean, A. M. (Chem. Phys. Proces. Combust. **35** [1990] 35-1/35-4; C.A. **114** [1991] No. 188660).
[11] Dean, A. M.; Chou, M. S.; Stern, D. (ACS Symp. Ser. **249** [1984] 71/84).
[12] Miller, J. A.; Kee, R. J. (Annu. Rev. Phys. Chem. **41** [1990] 345/87).
[13] Miller, J. A.; Branch, M. C.; Kee, R. J. (Combust. Flame **43** [1981] 81/98).
[14] Crosley, D. R. (Opt. Eng. **20** [1981] 511/21).
[15] Branch, M. C.; Sadeqi, M. E.; Alfarayedhi, A. A.; Van Tiggelen, P. J. (Combust. Flame **83** [1991] 228/39).
[16] Caralp, F.; Tinel, C.; Loirat, H. (Combust. Flame **33** [1978] 299/303).
[17] Caralp, F.; Loirat, H. (J. Chem. Res. Synop. **8** [1981] 240/1).
[18] Lichtin, D. A.; Berman, M. R.; Lin, M. C. (Chem. Phys. Lett. **108** [1984] 18/24).
[19] Goodfried, P. L.; Woods, H. P. (Combust. Flame **9** [1965] 421/2).

2.1.1.5.3 NH Radicals in Condensed Media (NH(X, A, a, b...))

Reactions of NH radicals in their electronic ground state were studied in the liquid phase. When NH(X) radicals were produced by γ radiation in liquid ammonia a slow reaction, $NH(X) + NH_3 \rightarrow N_2H_4$, was observed [1]. NH(X) reactions in the solid state were assumed when $NH_4ClO_4(s)$ was decomposed by an excimer laser [2]. NH(X) reactions have been observed on a surface [3]. NH reactions with H atoms absorbed on zeolite [4] also have to be mentioned.

In the liquid phase most imidogen reactions were done with NH(a $^1\Delta$) when HN_3 was photolyzed in solutions either at T = 195 K or near room temperature [5 to 14]. In the reactions of NH(a) with the saturated hydrocarbons C_2H_6, C_3H_8, and *iso*-C_4H_{10}, saturated amines were observed as products, due to the insertion of NH(a) into the C-H bond [5 to 8]. For 2-methylbutane $CH_3CH(CH_3)CH_2CH_3$ as reactant, the insertion into the primary, secondary, and tertiary C-H bond was discriminated by the analysis of the final products [13]. The selectivity is higher at lower temperatures. For T = 278 K the ratios of the different rates were: k(primary) = 1, k(secondary) = 1.2, and k(tertiary) = 1.7 and for T = 201 K: k(primary) = 1, k(secondary) = 1.8, and k(tertiary) = 3.0 [13]. Also for the cyclic saturated hydrocarbons cyclobutane [6] and cyclohexane [14], the insertion in the C-H bond was found to be the main reaction pathway. In the interaction of NH(a) with the unsaturated hydrocarbons 1-butene, isobutene, and cis- and trans-2-butene, addition to the double bond and insertion into the C-H bond was observed [12]. For ethylene, propene, and ethylene-ethane, three reaction pathways of NH(a) were discriminated quantitatively: (a) addition to the double bond, (b) insertion into the C-H bond, and (c) quenching to NH(X), with the ratios at T = 195 K of a:b:c = 1:2:3 [9]. Acetic, propionic, and isobutyric acids form alanine due to insertion of NH(a) into the C-H bond [11]; e.g., CH_3CH_2COOH forms α and β alanine with a ratio α/β(298 K) = 1.5 [11]. In the reactions NH(a) + $\{CH_3OH, C_2H_5OH\}$ the products CH_3ONH_2 and $C_2H_5ONH_2$, respectively, are observed due to an insertion into the O-H bond rather than the C-H bond [10]. The selectivity seems to be significantly larger than in the gas phase reaction. In the reaction $NH(a) + H_2O \rightarrow NH_2OH$, the hydroxylamine formation is observed [15], as expected from the gas phase at the high-pressure limit. For acetylene and the acetylene derivative $CH_3C{\equiv}CCH_3$, acetonitrile, propionitrile, and butyronitrile, respectively, are observed as final products, which can be understood if the initial product formed in the reaction of NH(a) with alkynes isomerize to nitriles by the migration of two hydrogen atoms [16].

Also in matrix reactions the imidogen radical has been used in the a $^1\Delta$ state, since there is a direct source for NH(a) by photolyzing HN_3. The quenching of NH(a) as a function of temperature by the matrix atoms Ar, Kr, and Xe has been studied [17]. The chemical reaction NH(a) + NO in an Ar matrix was followed by observing the N_2O product with FTIR spectroscopy. The quantum yield (Φ) of N_2O was found to be $\Phi(N_2O) = 0.7$, indicating that N_2O is the main product under these experimental conditions [18,19].

The reaction of NH(a) with unsaturated hydrocarbons (C_2H_2 and C_2H_4) was one of the first NH(a) reactions studied in matrices [20]. Addition to the C-C double bond was observed as the initial step, with ketenimine as the final product of the $NH(a) + C_2H_2$ reaction [20]. For the reaction of NH(a), again produced in the photolysis of HN_3 at T = 12 K, with $CH_3C{\equiv}CCH_3$, an interesting difference was observed between the Ar and Xe matrices. In Ar 3,3-dimethylketenimine is formed on the singlet surface, whereas in the Xe matrix *N*-methyl-1-amino-1-propyne is found as the major product, which is formed on the triplet surface after intersystem crossing of NH(a) to NH(X) induced by Xe [21].

References:

[1] Delcourt, M. O.; Belloni, J. (Radiochem. Radioanal. Lett. **13** [1973] 329/38).
[2] Chen, J.; Quinones, E.; Dagdigian, P. J. (J. Chem. Phys. **93** [1990] 4033/42).
[3] Fisher, E. R.; Ho, P.; Breiland, W. G.; Buss, R. J. (J. Phys. Chem. **96** [1992] 9855/61).
[4] Uyama, H.; Uchibura, T.; Niijima, H.; Matsumoto, O. (Chem. Lett. **1987** No. 4, pp. 555/8).
[5] Tsunashima, S.; Hotta, M.; Hamada, J.; Sato, S. (Proc. Yamada 3rd Conf. Free Radicals, Sandra, Jpn., 1979, pp. 269/70).
[6] Hamada, J.; Tsunashima, S.; Sato, S. (Bull. Chem. Soc. Jpn. **55** [1982] 1739/42).
[7] Tsunashima, S.; Hotta, M.; Sato, S. (Chem. Phys. Lett. **64** [1979] 435/9).
[8] Tsunashima, S.; Hamada, J.; Hotta, M.; Sato, S. (Bull. Chem. Soc. Jpn. **53** [1980] 2443/7).
[9] Kitamura, T.; Tsunashima, S.; Sato, S. (Bull. Chem. Soc. Jpn. **54** [1981] 55/9).
[10] Kawai, J.; Tsunashima, S.; Sato, S. (Bull. Chem. Soc. Jpn. **55** [1982] 3312/6).

[11] Tsunashima, S.; Kitamura, T.; Sato, S. (Bull. Chem. Soc. Jpn. **54** [1981] 2869/71).
[12] Hamada, J.; Tsunashima, S.; Sato, S. (Bull. Chem. Soc. Jpn. **56** [1983] 662/6).
[13] Kawai, J.; Tsunashima, S.; Sato, S. (Nippon Kagaku Kaishi No. 1 [1984] 32/6).
[14] McDonald, J. R.; Miller, R. G.; Baronavski, A. P. (Chem. Phys. **30** [1978] 133/45).
[15] Kawai, J.; Tsunashima, S.; Sato, S. (Chem. Lett. **1983** No. 6, pp. 823/6).
[16] Kawai, J.; Tsunashima, S.; Sato, S. (Chem. Phys. Lett. **110** [1984] 655/8).
[17] Ramsthaler-Sommer, A.; Eberhardt, K. E.; Schurath, U. (J. Chem. Phys. **85** [1986] 3760/9).
[18] Yokoyama, K.; Kitaibe, H.; Fueno, T. (Bull. Chem. Soc. Jpn. **64** [1991] 1731/7).
[19] Yokoyama, K.; Sakane, Y.; Fueno, T. (Bull. Chem. Soc. Jpn. **64** [1991] 1738/42).
[20] Jacox, M. E.; Milligan, D. E. (J. Am. Chem. Soc. **85** [1963] 278/82).

[21] Collins, S. T.; Pimentel, G. C. (J. Phys. Chem. **88** [1984] 4258/64).

2.1.2 The Imidogen Cation, NH^+

CAS Registry Numbers: NH^+ *[19067-62-0]*, ND^+ *[18987-16-1]*, $^{15}NH^+$ *[19121-55-2]*

2.1.2.1 Formation in the Gas Phase

2.1.2.1.1 By Ion-Molecule Reactions

Reaction $N^+ + H_2 \rightarrow NH^+ + H$. The nearly thermoneutral hydrogen-atom abstraction reaction

$$N^+ + H_2 \rightarrow NH^+ + H \quad (1)$$

and its isotopic variants

$$N^+ + D_2 \rightarrow ND^+ + D \quad (2)$$
$$N^+ + HD \rightarrow NH^+ + D \quad (3a)$$
$$\rightarrow ND^+ + H \quad (3b)$$

have been studied by various techniques in the temperature range 300 to 8 K. Reaction (1), on one hand, has served as a model system in the study of the dynamics and kinetics of ion-molecule reactions (for a somewhat recent review see [1]) and is, on the other hand, of astrochemical relevance, since it has been assumed to be the first step of ammonia formation in interstellar clouds (see e.g. [2]).

The dynamics of reactions (1) to (3) at room temperature have been studied in several ion-beam collision experiments by mass-spectroscopic detection of the NH^+ or ND^+ products and investigation of their angular and velocity distributions for initial relative energies ranging from 0.79 to 33 eV [3 to 8]. It was found that for initial relative energies above 2 eV the NH^+ is generated predominantly through direct reaction [3, 5, 7, 8], while a long-lived NH_2^+ complex is apparently involved in the reaction at energies below 1 eV [5]. The predominant reactive species was assumed to be the ground-state ion $N^+(^3P)$, although there is evidence for the metastable species $N^+(^1D, ^1S)$ to participate at very high (>6 eV) [7] and low (<1 eV) [6] initial relative energies. NH^+ was mainly formed in the a $^4\Sigma^-$ state at <4 eV and in the X $^2\Pi_r$ state at >6.5 eV [3]. The formation of NH^+ and ND^+ in their excited states B $^2\Delta$ from the reactions of $N^+(^1D)$ with H_2 and D_2 was demonstrated by observing the B $^2\Delta \rightarrow$ X $^2\Pi_r$ chemiluminescent transition in the visible and near UV (cf. p. 147) [9]. Theoretical studies of reaction (1), i.e., ab initio SCF MO [10] and CI [11 to 16] calculations of the potential energy surface (involving various electronic states of the NH_2^+ intermediate; cf. p. 240) and a classical trajectory calculation [17] (using the SCF potential function [10]) are available.

The rate constants of reaction (1) were measured at 300 K using different techniques, and values (in 10^{-10} $cm^3 \cdot molecule^{-1} \cdot s^{-1}$) of $k_1 = 5.6$ (flowing afterglow reaction tube) [18], 4.8 ± 0.2 (ion cyclotron resonance) [19, 20], and 4.8 (SIFT = selected-ion flow tube) [21] were obtained (the values $k_1 = 6.4$ [22] and 6.2 [23] presumably include other reaction channels; see also [1]).

The rate constants and the reaction cross sections of reactions (1) to (3) have been measured at 300 K as a function of the ion-molecule center-of-mass kinetic energy $KE_{c.m.}$ from near thermal energies up to ~0.5 eV applying the SIFDT (selected-ion flow drift tube) method [24 to 26] and up to 30 eV using a guided ion beam apparatus [27]; plots of k (or reaction cross section) versus $KE_{c.m.}$ gave the reaction thermicities ΔH, showing channels (1), (2), and (3a) to be very slightly endoergic by a few tens of meV and channel (3b) either endoergic [27] or exoergic [24] (both with large uncertainties). To simulate interstellar conditions, experimental techniques that allow measurements below 80 K have been applied: the ion trap method at 11 to 20 K [28, 29] and the CRESU method (Cinétique de Réaction en Écoulement Supersonique Uniforme) at 8 to 163 K [30, 31]; Arrhenius-type plots of k versus T^{-1} give ΔH. In order to derive small endothermicities, the exact knowledge of the energetical states of the reactants is important, as there are the fine-structure levels of ground state $N^+(^3P_{0,1,2})$, the metastable ionic states $N^+(^1D, ^1S, ^5S)$, and the rotational levels of the H_2 molecule [24, 27, 32]. The energy spacings of the latter are of the same order of magnitude as the endothermicity itself; therefore, the 3:1 populations of *ortho*- and *para*-H_2 had to be accounted for in analyzing low-temperature data [28 to 30], or else experiments with pure *para*-H_2 (J = 0), such as

$$N^+ + para\text{-}H_2 \rightarrow NH^+ + H \qquad (4),$$

have been carried out [31].

The results obtained from SIFDT [24], guided ion beam [27], ion trap [28, 29], and CRESU [31] experiments are compiled in Table 4, p. 130. **Fig. 3**, p. 130, shows Arrhenius-type plots of ln k versus $KE_{c.m.}^{-1}$ obtained from the SIFDT experiments at 300 K [24].

Reaction of N^+ with H_2S. A minor product channel for the reaction of N^+ ions with H_2S is the formation of $NH^+ + SH$ with a branching ratio of 3% compared to the formation of $H_2S^+ + N$, $SH^+ + NH$, and $S^+ + NH_2$ with branching ratios of 56, 29, and 12%, respectively; measurements with a selected-ion flow tube (SIFT) at 300 K gave a total rate constant of 1.9×10^{-9} $cm^3 \cdot molecule^{-1} \cdot s^{-1}$ [1, 21].

Reaction $N + H_2^+ \rightarrow NH^+ + H$. The dynamics of the reaction $N + D_2^+ \rightarrow ND^+ + D$ has been investigated in a merged-beam apparatus over the range of relative kinetic energies from ~0.005 to 10 eV. The reaction was found to proceed by a direct mechanism at relative energies as low as 0.031 eV with the ground-state $N(^4S)$ atom as the reactive species. The experimental findings were interpreted in terms of a qualitative electronic state correla-

Table 4
Reaction $N^+ + H_2 \rightarrow NH^+ + H$ (1) and Variants (2), (3a), (3b), and (4).
Rate constants k_i in 10^{-10} $cm^3 \cdot molecule^{-1} \cdot s^{-1}$, reaction thermicities ΔH_i in meV.

	SIFDT[a)] [24]	guided ion beam[b)] [27]	ion trap[c)] [28, 29]	CRESU[d)] [31]
k_1	3.7	3.9±0.8	4.86 exp[−(85±10)/T]	4.16 exp(−41.9/T)
k_2	1.3	1.7±0.4	−	2.37 exp(−197.9/T)
k_{3a}	0.5	} 2.7±0.6	−	−
k_{3b}	3	} (34:66)	−	3.17 exp(−16.3/T)
k_4	−	−	−	8.35 exp(−168.5/T)
ΔH_1	11±3	33±24	7.4±0.8	18±2
ΔH_2	33±4	61±30	−	−
ΔH_{3a}	43±6	69±30	−	−
ΔH_{3b}	−10±4	19±27	−	−

[a)] Rate constants k_i at T = 300 K (thermalized reactants); the ΔH_i values measured at 300 K were corrected for the rotational energies of H_2, HD, and D_2. − [b)] Thermal rate constants at T = 300 K; ΔH_0° values. − [c)] Measurements at 11 to 20 K [28, 29] combined with an average value for k_1 at 300 K from measurements of [18, 21, 24]. The original result, $k_1 = 6.48 \times 10^{-10}$ exp[−(85±10)/T], was corrected in a private communication to [1]. − [d)] Results for k_1 at T = 8 to 163 K, k_2 at T = 45 to 163 K, k_{3b} at 20 K, k_4 at 20 to 163 K [31] combined with the SIFDT data for k_1, k_2, and k_{3b} at 300 K from [24]; ΔH_1 is for the reaction $N^+(^3P_0) + H_2(X\,^1\Sigma_g^+, J=0) \rightarrow NH^+(X\,^2\Pi_r, J=0) + H(^2S)$.

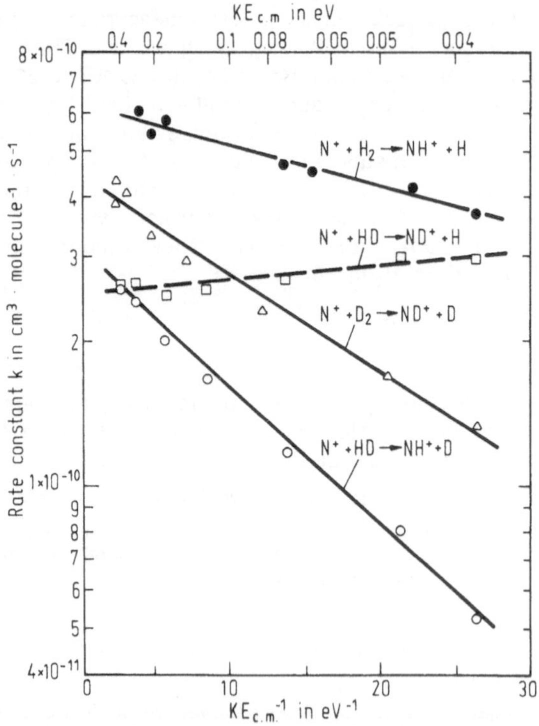

Fig. 3. Arrhenius-type plots of the rate constant versus the center-of-mass kinetic energy $KE_{c.m.}$ for the reaction $N^+ + H_2 \rightarrow NH^+ + H$ and its isotopic variants obtained from SIFDT experiments at 300 K [24].

tion diagram for $N + H_2^+ \rightarrow NH_2^+ \rightarrow NH^+ + H$, assuming collinear collisions as well as a C_{2v} approach [33]. Theoretical studies of the reaction, i.e., ab initio SCF MO [34] and CI [15, 16] calculations of the potential energy surface (involving various electronic states of the NH_2^+ intermediate; cf. p. 241) are available.

References:

[1] Ikezoe, Y.; Matsuoka, S.; Takebe, M.; Viggiano, A. (Gas Phase Ion-Molecule Reaction Rate Constants Through 1986, Maruzen, Tokyo 1987, pp. 1/224, 120).
[2] Herbst, E.; DeFrees, D. J.; McLean, A. D. (Astrophys. J. **321** [1987] 898/906).
[3] Gislason, E. A.; Mahan, B. H.; Tsao, C. W.; Werner, A. S. (J. Chem. Phys. **54** [1971] 3897/901).
[4] Eisele, G.; Henglein, A.; Botschwina, P.; Meyer, W. (Ber. Bunsenges. Phys. Chem. **78** [1974] 1090/7).
[5] Fair, J. A.; Mahan, B. H. (J. Chem. Phys. **62** [1975] 515/9).
[6] Farrar, J. M.; Hansen, S. G.; Mahan, B. H. (J. Chem. Phys. **65** [1976] 2908/10).
[7] Mahan, B. H.; Ruska, W. E. W. (J. Chem. Phys. **65** [1976] 5044/51).
[8] Hansen, S. G.; Farrar, J. M.; Mahan, B. H. (J. Chem. Phys. **73** [1980] 3750/62).
[9] Kusunoki, I.; Ottinger, C. (J. Chem. Phys. **80** [1984] 1872/81).
[10] Gittins, M. A.; Hirst, D. M. (Chem. Phys. Letters **35** [1975] 534/6).

[11] Bender, C. F.; Meadows, J. H.; Schaefer, H. F., III (Faraday Discuss. Chem. Soc. No. 62 [1977] 59/66).
[12] Bender, C. F.; Meadows, J. H.; Schaefer, H. F., III (LBL-5114 [1976] 23 pp. from ERDA Energy Res. Abstr. **2** No. 2 [1977] No. 4047; C.A. **86** [1977] No. 146291).
[13] Gittins, M. A.; Hirst, D. M.; Guest, M. F. (Faraday Discuss. Chem. Soc. No. 62 [1977] 67/76, 138, 140, 142/3).
[14] Hirst, D. M. (DL-SCI-R [1977/78] 9/14; C.A. **79** [1979] No. 44805).
[15] Hirst, D. M. (Mol. Phys. **35** [1978] 1559/68).
[16] Hirst, D. M. (NATO Adv. Study Inst. Ser. B **40** [1979] 24/9).
[17] Gittins, M. A.; Hirst, D. M. (Chem. Phys. Lett. **65** [1979] 507/10).
[18] Fehsenfeld, F. C.; Schmeltekopf, A. L.; Ferguson, E. E. (J. Chem. Phys. **46** [1967] 2802/8).
[19] Kim, J. K.; Theard, L. P.; Huntress, W. T., Jr. (J. Chem. Phys. **62** [1975] 45/52).
[20] Huntress, W. T., Jr. (Astrophys. J. Suppl. Ser. **33** [1977] 495/514).

[21] Smith, D.; Adams, N. G.; Miller, T. M. (J. Chem. Phys. **69** [1978] 308/18).
[22] Adams, N. G.; Smith, D. (Int. J. Mass Spectrom. Ion Phys. **21** [1976] 349/59).
[23] Tichý, M.; Rakshit, A. B.; Lister, D. G.; Twiddy, N. D.; Adams, N. G.; Smith, D. (Int. J. Mass Spectrom. Ion Phys. **29** [1979] 231/47).
[24] Adams, N. G.; Smith, D. (Chem. Phys. Lett. **117** [1985] 67/70).
[25] Adams, N. G.; Smith, D.; Millar, T. J. (Mon. Not. R. Astron. Soc. **211** [1984] 857/65).
[26] Smith, D.; Adams, N. G. (in: Millar, T. J.; Williams, D. A.; Rate Coefficients in Astrochemistry, Kluwer, Dordrecht-Boston-London 1988, pp. 153/71).
[27] Ervin, K. M.; Armentrout, P. B. (J. Chem. Phys. **86** [1987] 2659/73).
[28] Luine, J. A.; Dunn, G. H. (Astrophys. J. **299** [1985] L 67/L 70).
[29] Barlow, S. E.; Luine, J. A.; Dunn, G. H. (Int. J. Mass Spectrom. Ion Processes **74** [1986] 97/128).
[30] Marquette, J. B.; Rowe, B. R.; Dupeyrat, G.; Roueff, E. (Astron. Astrophys. **147** [1985] 115/20).

[31] Marquette, J. B.; Rebrion, C.; Rowe, B. R. (J. Chem. Phys. **89** [1988] 2041/7).
[32] Rowe, B. R. (in: Millar, T. J.; Williams, D. A.; Rate Coefficients in Astrochemistry, Kluwer, Dordrecht-Boston-London 1988, pp. 135/52).

[33] McClure, D. J.; Douglass, C. H.; Gentry, R. W. (J. Chem. Phys. **66** [1977] 2079/93).
[34] Hirst, D. M. (Chem. Phys. Lett. **53** [1978] 125/7).

2.1.2.1.2 By Decomposition of NH_3

Electron Impact. Mass-spectrometric observations of NH^+ upon electron impact on NH_3 suggested that the ion is formed by the two processes $NH_3+e^- \rightarrow NH^+ + H_2 + 2e^-$ and $NH_3+e^- \rightarrow NH^+ + 2H + 2e^-$. The energy difference of the two appearance potentials, $AP(NH^+)=22.9\pm0.5$ and 27.2 ± 0.5 eV, derived by measuring the partial ionization cross section for the production of NH^+ as a function of the electron energy E_{el} ($\sim$10 to 183 eV), amounts to about the dissociation energy of H_2 [1]. NH^+ formation above 20 eV was also observed by [2] ($E_{el}=12.5$ to 80 eV) and [3] ($E_{el}=5$ to 60 eV). Two other studies reported $AP=17.1\pm0.1$ and 21.6 ± 0.1 eV (extrapolated voltage difference technique) [4] and AP = 17.2 (quoted in [5]) and $\sim$22.7 eV (first differential ionization curve at electron energies of 16 to 37 eV) [6]. The latter results were reinterpreted by [1], and APs of $\sim$23 and 27 eV were read from the ionization curve presented by [6]. Most recently however, thermochemical thresholds of 17.5 and 22.0 eV were derived for the formation of the ion in the ground state $X\,^2\Pi_r$ (and values for NH^+ in the excited states A, B, and C) [7] using the ionization potential of NH as measured by [8] and thermochemical data from [9] (see also the paragraph on NH^+ formation by photodissociation and photoionization of NH_3).

Relative abundances between 2.4 and 4.7 for the primary NH^+ ions at electron energies of 100 and 50 eV (compared to relative abundances of 51 to 54 for NH_3^+ and 42 for NH_2^+) were reported [10, 11]. A rate constant of 1.6×10^{-9} $cm^3\cdot molecule^{-1}\cdot s^{-1}$ for the production of NH^+ from NH_3 at $E_{el}=80$ eV was measured [2].

Electron-impact dissociation and ionization of NH_3 (E_{el} between 40 and 150 eV) was also applied for spectroscopic studies on NH^+. The UV-visible emission systems A $^2\Sigma^-$, B $^2\Delta$, and C $^2\Sigma^+ \rightarrow X\,^2\Pi_r$ [7, 12 to 14], the A $^2\Sigma^-$ and C $^2\Sigma^+ \leftarrow X\,^2\Pi_r$ transitions in the high-resolution translational energy spectrum (TES) [15], and the hyperfine structure in the purely rotational spectrum [16] were observed (cf. pp. 146/8).

Alpha-Particle Impact. The relative abundance of primary NH^+ ions in the mass spectrum of NH_3 bombarded with 1- to 3.5-MeV α particles at pressures of 0.01 to 0.04 Torr is 3.0 ± 0.9 compared to 53 for NH_3^+ and 44 for NH_2^+ [11].

Photodissociation and Photoionization. The partial cross sections for NH^+ production by dissociative photoionization of NH_3 using synchrotron radiation of 8 to 112 nm showed thresholds at 68.0 and 54.5 nm, $AP=18.23\pm0.05$ and 22.75 ± 0.1 eV, for the processes $NH_3 \rightarrow NH^+ + H_2$ and $NH_3 \rightarrow NH^+ + 2H$ [17].

UV laser photodissociation at 6.42 eV of ND_3^+ ions (produced by electron impact on ND_3) showed a weak ND^+ signal, even though thermochemical data predict the reaction $NH_3^+ \rightarrow NH^+ + H_2$ to be endothermic by $\sim$7 eV [18].

Hollow Cathode Discharges. Discharges through flowing NH_3 or NH_3-He mixtures produce NH^+ ions in the electronic ground and various excited states. This technique has been applied for spectroscopic studies of NH^+, as there are observations of the A $^2\Sigma^-$, B $^2\Delta$, and C $^2\Sigma^+ \rightarrow X\,^2\Pi_r$ emission band systems in the visible and UV [19 to 23] (cf. pp. 146/7), the $v=1\leftarrow0$ vibration-rotation transitions in the X $^2\Pi_r$ and a $^4\Sigma^-$ states, and transitions between the X and a states [24, 25] (cf. p. 146).

References:

[1] Märk, T. D.; Egger, F.; Cheret, M. (J. Chem. Phys. **67** [1977] 3795/802).
[2] Derwish, G. A. W.; Galli, A.; Giardini-Guidoni, A.; Volpi, G. G. (J. Chem. Phys. **39** [1963] 1599/605).
[3] Wight, G. R.; Van der Wiel, M. J.; Brion, C. E. (J. Phys. B At. Mol. Phys. **10** [1977] 1863/73).
[4] Reed, R. I.; Snedden, W. (J. Chem. Soc. **1959** 4132/3).
[5] Levin, R. D.; Lias, S. G. (NSRDS-NBS-71 [1982] 1/628, 123).
[6] Morrison, J. D.; Traeger, J. C. (Int. J. Mass Spectrom. Ion Phys. **11** [1973] 277/88).
[7] Müller, U.; Schulz, G. (J. Chem. Phys. **96** [1992] 5924/37).
[8] Foner, S. N.; Hudson, R. L. (J. Chem. Phys. **74** [1981] 5017/21).
[9] Franklin, J. L.; Dillard, J. G.; Rosenstock, H. M.; Herron, J. T.; Draxl, K.; Field, F. H. (NSRDS-NBS-26 [1969] 1/285; C.A. **71** [1969] No. 33523).
[10] Melton, C. E. (J. Chem. Phys. **45** [1966] 4414/24).

[11] Fluegge, R. A.; Landman, D. A. (J. Chem. Phys. **54** [1971] 1576/91).
[12] Müller, U.; Schulz, G. (Chem. Phys. Lett. **170** [1990] 401/5).
[13] Brzozowski, J.; Elander, N.; Erman, P.; Lyyra, M. (Phys. Scr. **10** [1974] 241/3).
[14] Smith, W. H. (J. Chem. Phys. **51** [1969] 520/4).
[15] Hamdan, M.; Mazumdar, S.; Marathe, V. R.; Badrinathan, C.; Brenton, A. G.; Mathur, D. (J. Phys. B At. Mol. Opt. Phys. **21** [1988] 2571/84).
[16] Edwards, C. P.; Maclean, C. S.; Sarre, P. J. (J. Chem. Phys. **76** [1982] 3829/31).
[17] Samson, J. A. R.; Haddad, G. N.; Kilcoyne, L. D. (J. Chem. Phys. **87** [1987] 6416/22).
[18] Kutina, R. E.; Edwards, A. K.; Pandolfi, R. S.; Berkowitz, J. (J. Chem. Phys. **80** [1984] 4112/9).
[19] Colin, R.; Douglas, A. E. (Can. J. Phys. **46** [1968] 61/73).
[20] Balasubramanian, T. K.; Bhale, G. L.; Gopal, S.; Krishnamurthy, G.; Lakshminarayana, G.; Saksena, M. D.; Saraswathy, P.; Shetty, B. J.; Singh, M. (Symp. Int. Astron. Union No. 120 **1985** [1987] 91/2).

[21] Krishnamurthy, G.; Saraswathy, M. (Pramana **6** [1976] 235/43).
[22] Feast, M. W. (Astrophys. J. **114** [1951] 344/55).
[23] Lunt, W.; Pearse, R. W. B.; Smith, E. C. W. (Nature **136** [1935] 32).
[24] Kawaguchi, K.; Amano, T. (J. Chem. Phys. **88** [1988] 4584/91).
[25] Amano, T. (Philos. Trans. R. Soc. London A **324** [1988] 163/78).

2.1.2.1.3 From NH_2, N_2H_4, NH_2F, NHF_2, HN_3, and H_2-N_2 Mixtures

The threshold for the dissociative photoionization process $NH_2 + h\nu \rightarrow NH^+ + H + e^-$ was observed at 71.25 ± 0.02 nm or 17.401 ± 0.005 eV; corrected for the internal energy of NH_2, the 0 K threshold is at 17.440 ± 0.005 eV. The extremely weak process $N_2H_4 + h\nu \rightarrow NH^+ + NH_3 + e^-$ has its onset at ~78 to 79 nm [1]. The onset of the process $NH_2F + h\nu \rightarrow NH^+ + HF + e^-$ is at 14.5 eV; the abundancies of NH^+ from NH_2F and NHF_2 are 2% and 1% at $h\nu = 20.65$ eV [2]. The formation of NH^+ ions via $HN_3 \rightarrow NH^+ + N_2 + e^-$ with an abundance of 16.8% relative to HN_3^+ was observed at $AP = 14.4 \pm 0.2$ eV upon electron impact on HN_3 [3].

Hollow cathode discharges through N_2H_4-He mixtures allowed the observation of the UV-visible A, B, and C → X emission systems of NH^+ [4]. Hollow cathode discharges through H_2-N_2 mixtures were used for spectroscopic studies in the IR ($v = 1 \leftarrow 0$ transitions in the X $^2\Pi_r$ and a $^4\Sigma^-$ states of NH^+ and transitions between the X and a states) [5, 6] and in the submillimeter range (purely rotational $J=3/2 \leftarrow 1/2$ transition of $NH^+(X\ ^2\Pi_r,\ v=0)$) [7].

References:

[1] Gibson, S. T.; Greene, J. P.; Berkowitz, J. (J. Chem. Phys. **83** [1985] 4319/28).
[2] Baumgärtel, H.; Jochims, H. W.; Ruehl, E.; Bock, H.; Dammel, R.; Minkwitz, J.; Nass, R. (Inorg. Chem. **28** [1989] 943/9).
[3] Franklin, J. L.; Dibeler, V. H.; Reese, R. M.; Krauss, M. (J. Am. Chem. Soc. **80** [1958] 298/302).
[4] Colin, R.; Douglas, A. E. (Can. J. Phys. **46** [1968] 61/73).
[5] Kawaguchi, K.; Amano, T. (J. Chem. Phys. **88** [1988] 4584/91).
[6] Amano, T. (Philos. Trans. R. Soc. London A **324** [1988] 163/78).
[7] Verhoeve, P.; Ter Meulen, J. J.; Meerts, W. L.; Dymanus, A. (Chem. Phys. Lett. **132** [1986] 213/7).

2.1.2.1.4 By Ionization of NH

A weak peak at 13.49 eV in the He I photoelectron spectrum (PES) of NH_2 (produced from $F + NH_3$) is attributed to the ionization process $NH^+(X\ ^2\Pi_r) \leftarrow NH(X\ ^3\Sigma^-)$ with NH arising from the secondary reaction $F + NH_2 \rightarrow NH + HF$ [1].

Cross sections for photoionization of the 1π orbital of NH leading to $NH^+(X\ ^2\Pi_r)$ and of the 3σ orbital of NH leading to $NH^+(a\ ^4\Sigma^-)$ and $NH^+(A\ ^2\Sigma^-)$ at photon energies between 10 and 60 eV were calculated applying quantum-theoretical methods [2].

Resonance-enhanced multiphoton ionization (REMPI) of NH and ND and detection of the NH^+ or ND^+ ions with a time-of-flight (TOF) mass spectrometer [3 to 8] or TOF detection of the photoelectrons (REMPI-PES) [5, 9, 10] enabled highly excited states of the NH (ND) radical (cf. pp. 83/7) and vibrational and rotational populations of the resulting NH^+ or ND^+ ions to be characterized.

References:

[1] Dunlavey, S. J.; Dyke, J. M.; Jonathan, N.; Morris, A. (Mol. Phys. **39** [1980] 1121/35).
[2] Wang, K.; Stephens, J. A.; McKoy, V. (J. Chem. Phys. **93** [1990] 7874/82).
[3] Johnson, R. D., III; Hudgens, J. W. (J. Chem. Phys. **92** [1990] 6420/5).
[4] Ashfold, M. N. R.; Clement, S. G.; Howe, J. D.; Western, C. M. (J. Chem. Soc. Faraday Trans. **87** [1991] 2515/23).
[5] Clement, S. G.; Ashfold, M. N. R.; Western, C. M.; De Beer, E.; De Lange, C. A.; Westwood, N. P. C. (J. Chem. Phys. **96** [1992] 4963/73).
[6] Clement, S. G.; Ashfold, M. N. R.; Western, C. M.; Johnson, R. D., III; Hudgens, J. W. (J. Chem. Phys. **96** [1992] 5538/40).
[7] Clement, S. G.; Ashfold, M. N. R.; Western, C. M. (J. Chem. Soc. Faraday Trans. **88** [1992] 3121/8).
[8] Clement, S. G.; Ashfold, M. N. R.; Western, C. M.; Johnson, R. D., III; Hudgens, J. W. (J. Chem. Phys. **97** [1992] 7064/72).
[9] De Beer, E.; Born, M.; De Lange, C. A.; Westwood, N. P. C. (Chem. Phys. Lett. **186** [1991] 40/6).
[10] Wang, K.; Stephens, J. A.; McKoy, V.; De Beer, E.; De Lange, C. A.; Westwood, N. P. C. (J. Chem. Phys. **97** [1992] 211/21).

2.1.2.2 Formation in a Solid

A paramagnetic species detected by ESR in γ-irradiated $(NH_4)_2HPO_4$ single crystals and characterized by hyperfine lines arising from one N and one H nucleus ($g_{xx}=2.0089$, $g_{yy}=2.0026$, $g_{zz}=2.0048$, $g_{av}=2.0054$ and (in MHz) $A_{xx}(^{14}N)=22.6$, $A_{yy}(^{14}N)=101.0$, $A_{zz}(^{14}N)=10.4$, $A_{iso}(^{14}N)=45\pm5$; $A_{xx}(^{1}H)=-31.6$, $A_{yy}(^{1}H)=-63.4$, $A_{zz}(^{1}H)=-101.1$, $A_{iso}(^{1}H)=-65\pm5$) [1] has been conjectured to be NH^+ [1] or NH^- [2]; an ab initio UHF calculation supported the assignment to NH^+ [3].

References:

[1] Morton, J. R. (J. Phys. Chem. Solids **24** [1963] 209/12).

[2] Atkins, P. W.; Symons, M. C. R. (The Structure of Inorganic Radicals, Elsevier, London 1967, pp. 107/8).

[3] Claxton, T. A. (Trans. Faraday Soc. **66** [1970] 1540/3).

2.1.2.3 Heat of Formation. Thermodynamic Functions

A recent critical review of the **heat of formation** of NH (cf. pp. 56/8) and NH_2 (cf. p. 193) [1] recommended the excellently agreeing values for the heats of formation of NH^+,

$$\Delta_fH^\circ_0(NH^+)=1660\pm3 \text{ and } 1658\pm1 \text{ kJ/mol.}$$

The first value is based on kinetics measurements (guided ion beam tandem mass spectroscopy) for the reaction $N^+ + H_2 \rightarrow NH^+ + H$ (cf. pp. 129/30) by combining the endothermicity of the reaction with $\Delta_fH^\circ_0(H)$ and $\Delta_fH^\circ_0(N^+)$ [2]. It confirms the earlier result, $\Delta_fH^\circ_0(NH^+)=1661\pm4$ kJ/mol, based on a CRESU study of the same reaction [3]. The second value was obtained from photoionization mass-spectrometric studies of the dissociative photoionization process $NH_2 + h\nu \rightarrow NH^+ + H + e^-$ (cf. p. 133) by combining the threshold for NH^+ formation with $\Delta_fH^\circ_0(NH_2)$ [4]. In another critical review of gas-phase thermochemical data, the photoionization results of [4] were used to derive $\Delta_fH^\circ_0(NH^+)=1678.1$ kJ/mol [5]. A slightly lower heat of formation, $\Delta_fH^\circ_0(NH^+)=1648\pm1$ kJ/mol, was obtained from mass-spectrometric studies of the process $H_2NF + h\nu \rightarrow NH^+ + HF + e^-$ [6]. The previously recommended value, $\Delta_fH^\circ_{298}(NH^+)=1598$ kJ/mol [7], based on the appearance potentials of NH^+ from NH_3 under electron impact [8], and another value, $\Delta_fH^\circ(NH^+)=1690$ kJ/mol, based on the appearance potential of NH^+ from HN_3 under electron impact [9], can thus be disregarded.

Theoretical values from ab initio calculations are in fair agreement with the experimental results: $\Delta_fH^\circ_0(NH^+)=1664$ kJ/mol was obtained from Møller-Plesset perturbation calculations [10, 11], $\Delta_fH^\circ_0(NH^+)=1643$ kJ/mol from a CI calculation [12].

The **heat capacity** C°_p, the **thermodynamic functions**, entropy S°, Gibbs free energy $-(G^\circ-H^\circ_0)/T$, enthalpy $H^\circ-H^\circ_0$ (based on $\Delta_fH^\circ_0=1602.852$ kJ/mol), and the equilibrium constant K° (for $NH^+ + e^- \rightleftharpoons N + H$) are listed for the temperature range 298.15 to 20000 K [13]. Selected values are as follows:

T in K	C°_p	S°	$-(G^\circ-H^\circ_0)/T$	$H^\circ-H^\circ_0$ in kJ/mol	log K°
	in $J\cdot mol^{-1}\cdot K^{-1}$				
298.15	32.705	187.616	155.762	9.497	164.1535
400	31.748	197.085	165.144	12.777	123.1211
600	31.083	209.778	178.056	19.033	83.0149

Table (continued)

T in K	C_p^o	S^o	$-(G^o-H_0^o)/T$	$H^o-H_0^o$ in kJ/mol	log K^o
	in $J\cdot mol^{-1}\cdot K^{-1}$				
800	31.617	218.775	187.162	25.290	62.9156
1000	32.585	225.931	194.223	31.708	50.8275
2000	36.329	249.836	216.629	66.413	26.4866
3000	38.185	264.944	230.365	103.738	18.2350
4000	39.798	276.147	240.458	142.716	14.0335
5000	41.547	285.215	248.538	183.388	11.4653
6000	42.973	292.925	255.309	225.697	9.7215
10000	40.675	314.803	275.142	396.609	6.1114
20000	28.637	338.600	301.954	732.928	3.2482

The partition function Q and the equilibrium constant K ($N^+ + H \rightleftharpoons NH^+$) have been calculated, and polynomial expansions as functions of T fitted to the tabulated values for T = 1000 to 10000 K by [14] and for T = 1000 to 9000 K by [15].

References:

[1] Anderson, W. R. (J. Phys. Chem. **93** [1989] 530/6).

[2] Ervin, K. M.; Armentrout, P. B. (J. Chem. Phys. **86** [1987] 2659/73).

[3] Marquette, J. B.; Rowe, B. R.; Dupeyrat, G.; Roueff, E. (Astron. Astrophys. **147** [1985] 115/20).

[4] Gibson, S. T.; Greene, J. P.; Berkowitz, J. (J. Chem. Phys. **83** [1985] 4319/28).

[5] Lias, S. G.; Bartmess, J. E.; Liebman, J. F.; Holmes, J. L.; Levin, R. D.; Mallard, W. G. (J. Phys. Chem. Ref. Data **17** Suppl. 1 [1988] 1/861, 617).

[6] Baumgärtel, H.; Jochims, H. W.; Ruehl, E.; Bock, H.; Dammel, R.; Minkwitz, J.; Nass, R. (Inorg. Chem. **28** [1989] 943/9).

[7] Franklin, J. L.; Dillard, J. G.; Rosenstock, H. M.; Herron, J. T.; Draxl, K.; Field, F. H. (NSRDS-NBS-26 [1969] 1/285; C.A. **71** [1969] No. 33523).

[8] Reed, R. I.; Snedden, W. (J. Chem. Soc. **1959** 4132/3).

[9] Franklin, J. L.; Dibeler, V. H.; Reese, R. M.; Krauss, M. (J. Am. Chem. Soc. **80** [1958] 298/302).

[10] Pople, J. A.; Curtiss, L. A. (J. Phys. Chem. **91** [1987] 3637/9).

[11] Pople, J. A.; Curtiss, L. A. (J. Phys. Chem. **91** [1987] 155/62).

[12] Power, D.; Brint, P.; Spalding, T. R. (J. Mol. Struct. **110** [1984] 155/66 [THEOCHEM **19**]).

[13] Glushko, V. P.; Gurvich, L. V.; Bergman, G. A.; Veits, I. V.; Medvedev, V. A.; Khachkuruzov, G. A.; Yungman, V. S. (Thermodynamic Properties of Individual Substances, Vol. 1, Book 2, Nauka, Moscow 1978, pp. 225/6).

[14] Tarafdar, S. P. (J. Quant. Spectrosc. Radiat. Transfer **17** [1977] 537/42).

[15] Sauval, A. J.; Tatum, J. B. (Astrophys. J. Suppl. Ser. **56** [1984] 193/209).

2.1.2.4 Electronic Structure. Ionization Potential. Dipole Moment

Electron Configuration. Electronic States. The NH^+ ion is isoelectronic with the CH radical. Its ground state X $^2\Pi_r$ results from the ionization of the 1π MO in the NH(X $^3\Sigma^-$) radical (cf. p. 31), and the lowest excited valence states arise from $1\pi \leftarrow 3\sigma$, $1\pi \leftarrow 2\sigma$, and

$4\sigma \leftarrow 3\sigma$ excitations. Experimentally observed are the ground state and the lowest four excited states with the following main electron configurations and dissociation limits [1]:

$(1\sigma)^2\ (2\sigma)^2\ (3\sigma)^2\ (1\pi)^1$	$X\ ^2\Pi_r$	$N^+(^3P)+H(^2S)$
$(1\sigma)^2\ (2\sigma)^2\ (3\sigma)^1\ (1\pi)^2$	$a\ ^4\Sigma^-$	$N(^4S^o)+H^+$
$(1\sigma)^2\ (2\sigma)^2\ (3\sigma)^1\ (1\pi)^2$	$A\ ^2\Sigma^-$	$N^+(^3P)+H(^2S)$
$(1\sigma)^2\ (2\sigma)^2\ (3\sigma)^1\ (1\pi)^2$	$B\ ^2\Delta$	$N(^2D^o)+H^+$
$(1\sigma)^2\ (2\sigma)^2\ (3\sigma)^1\ (1\pi)^2$	$C\ ^2\Sigma^+$	$N^+(^1D^o)+H(^2S)$

The ground state X has been identified with the lower, the states A, B, and C with the upper states of the three emission systems A→X, B→X, and C→X in the visible and UV region (cf. pp. 146/8). Perturbations of ground-state rotational-vibrational levels observed in the spectra were found to arise from strong interactions between the quasidegenerate $X\ ^2\Pi_r$ and $a\ ^4\Sigma^-$ states. Rotational-vibrational transitions in the X and a states as well as between X and a have been observed in the IR (cf. p. 146). Term values of the excited states are listed together with the spectroscopic constants in Tables 5 and 6, pp. 140/2.

A number of ab initio MO calculations referred to in the various bibliographies of quantum-chemical calculations (cf. p. 31) deal with the ground state and some excited states of NH^+. To be mentioned among the various ground-state calculations are the early SCF MO calculation at the near-Hartree-Fock limit [2, 3] and PNO-CI and CEPA studies [4]. Excited states are included in calculations by the SCF MO (including a semiempirical estimate of the correlation energy) method [5], by CI methods [6 to 8], by perturbation theory [9 to 11], and by an algebraic approach for solving the Schrödinger equation [12]. Theory predicts an additional bound valence state, which lies 13.8 eV above the ground state,

$(1\sigma)^2\ (2\sigma)^1\ (3\sigma)^1\ (1\pi)^3$ $^4\Pi$ $N(^4P)+H^+$ [5]

and a number of repulsive states [5 to 7, 11].

The vertical excitation energies for the $a\ ^4\Sigma^-$, $A\ ^2\Sigma^-$, $B\ ^2\Delta$, $C\ ^2\Sigma^+ \leftarrow X\ ^2\Pi_r$ transitions at $r_e = 1.9614\ a_0$ (the experimental value for the equilibrium internuclear distance in the NH molecule) were calculated by the many-body perturbation theory of second and third order (H^v study) [9].

Ionization Potential. Charge stripping reactions of NH^+ ions (obtained by 70-eV electron impact on NH_3) with N atoms result in unstable NH^{2+} ions. The minimum energy for the ionization of the NH^+ ion in a vertical transition, $E_i = 25 \pm 1$ eV, was estimated from the position of the N^+ and H^+ peaks in the mass spectrum [13]. Theoretical results are $E_i(ad) = 26.9$ eV from an MRSD CI calculation [14, 15] and $E_i(ad) = 26.4$ eV and $E_i(vert) = 26.9$ eV from a Møller-Plesset perturbation calculation of fourth order (MP4) [16].

Dipole Moment. The dipole moments of $NH^+(X\ ^2\Pi_r)$ and $NH^+(a\ ^4\Sigma^-)$ at internuclear distances r = 0.7 to 20.0 Å were derived by ab initio MO calculations using a single configuration (cf. above) for the a state, but three configurations ($4\sigma \leftarrow 3\sigma$ and $(4\sigma)^2 \leftarrow (3\sigma)^2$ excitations in addition to the dominant ground-state configuration) for the X state. At r = 1.1 Å ($\approx r_e$ of both states), the dipole moments (with the origin at the center of mass) are $\mu = 6.8461 \times 10^{-30}$ C·m for the ground state X and $\mu = 5.5832 \times 10^{-30}$ C·m for the $a\ ^4\Sigma^-$ state (= 2.05 and 1.67 D, respectively) [10].

References:

[1] Herzberg, G. (Molecular Spectra and Molecular Structure, Vol. 1, Spectra of Diatomic Molecules, Van Nostrand, Princeton, N. J., 1950, p. 341).

[2] Cade, P. E.; Huo, W. M. (J. Chem. Phys. **47** [1967] 614/48).
[3] Cade, P. E.; Huo, W. (At. Data Nucl. Data Tables **12** [1973] 415/66, 433).
[4] Rosmus, P.; Meyer, W. (J. Chem. Phys. **66** [1977] 13/9).
[5] Liu, H. P. D.; Verhaegen, G. (J. Chem. Phys. **53** [1970] 735/45).
[6] Guest, M. F.; Hirst, D. M. (Mol. Phys. **34** [1977] 1611/21).
[7] Kusunoki, I.; Yamashita, K.; Morokuma, K. (Chem. Phys. Lett. **123** [1986] 533/6).
[8] Yamashita, K.; Yabushita, S.; Morokuma, K.; Kusunoki, I. (Chem. Phys. Lett. **137** [1987] 193/4).
[9] Sun, H.; Sheppard, M. G.; Freed, K. F. (J. Chem. Phys. **74** [1981] 6842/8).
[10] Farnell, L.; Ogilvie, J. F. (J. Mol. Spectrosc. **101** [1983] 104/32).

[11] Park, J. K.; Sun, H. (Bull. Korean Chem. Soc. **11** [1990] 34/41).
[12] Frank, A.; Lemus, R.; Iachello, F. (J. Chem. Phys. **91** [1989] 29/41).
[13] Proctor, C. J.; Porter, C. J.; Ast, T.; Bolton, P. D.; Beynon, J. H. (Org. Mass Spectrom. **16** [1981] 454/8).
[14] Pope, S. A.; Hillier, I. H.; Guest, M. F.; Kendric, J. (Chem. Phys. Lett. **95** [1983] 247/9).
[15] Pope, S. A.; Hillier, I. H.; Guest, M. F. (Faraday Symp. Chem. Soc. No. 19 [1984] 109/23).
[16] Koch, W.; Schwarz, H. (Int. J. Mass Spectrom. Ion Processes **68** [1986] 49/56).

2.1.2.5 Spectroscopic Constants

Hyperfine Constants. The far-IR spectrum of the purely rotational transition $J=3/2 \leftarrow 1/2$ of NH^+ in the X $^2\Pi_r$, v=0 ground state (cf. p. 146) [1] has been analyzed using rotational, fine-structure, and Λ-doubling parameters [2] obtained from a fit of all available spectroscopic data for NH^+ and using hyperfine and quadrupole coupling Hamiltonians that contain the Frosch-Foley parameters a, b, c, and d for both nuclei [3] and the coupling constants eq_0Q and eq_2Q for the N nucleus, respectively. Since the parameters a, b, and c could not be evaluated independently, the linear combinations $2[a-0.5(b+c)]$ and $0.65\,b+0.25[a+0.5(b+c)]$ have been introduced (for the definition and the exact evaluations of the multiplicators ~0.65 and ~0.25 in the second expression, see equations (4) to (7) of [1]). The experimental hyperfine parameters [1] are given below together with the hyperfine parameters obtained from a many-body perturbation calculation of 3rd order [4] (all values in MHz):

hf parameter	experimental results [1]		theoretical results [4]	
	^{14}N nucleus	^1H nucleus	^{14}N nucleus	^1H nucleus
$2[a-0.5(b+c)]$	301.09(4.5)	142.8(6.4)	308.1$^{a)}$	131.0$^{b)}$
$0.65\,b+0.25[a+0.5(b+c)]$	61.9(1.6)	−51.4(2.0)	63.3$^{a)}$	−51.3$^{b)}$
d	165.0(2.4)	41.7(3.3)	171.2	39.8
eq_0Q	−9.6(6.5)	–	–	–
eq_2Q	−37.2(9.8)	–	–	–

a) Individual values: $a=138.50$, $b_F=b+c/3=22.62$, $c=-80.57$ MHz. – b) Individual values: $a=64.08$, $b_F=-69.89$, $c=100.00$ MHz.

Using highly correlated wave functions from ab initio coupled-cluster (UCCD(ST)) [5] and quadratic configuration interaction (QCISD(T)) [6] calculations, the isotropic hf constants A_{iso} $(=b_F=b+c/3)=18.9$ and 19.9 MHz for the ^{14}N and −72.3 and −75.3 MHz for the ^1H nucleus were calculated.

Fine-Structure Constants. For the electronic ground state X $^2\Pi_r$ and the very low lying excited state a $^4\Sigma^-$, fine-structure splittings were analyzed in the high-resolution IR absorption spectrum of the v=1←0 vibration-rotation transitions in these states and between them. The fine-structure arises from spin-orbit coupling (constant A), spin-rotation coupling (constant γ), and Λ-type doubling (constants p and q) in the X $^2\Pi_r$ state, from spin-rotation (γ) and spin-spin coupling (λ) in the a $^4\Sigma^-$ state, and from spin-orbit coupling, $^2\Pi_{3/2}$-$^4\Sigma^-_{3/2}$ and $^2\Pi_{1/2}$-$^4\Sigma^-_{1/2}$, between the two states (coupling parameters $\xi_{3/2}$ and $\xi_{1/2}$). The high-resolution IR spectrum even enabled the inclusion of centrifugal distortion terms for the spin-rotation and the Λ-doubling constants (γ_D and p_D, q_D) and for the spin-orbit coupling parameters (ξ_D). The following constants were determined for the $^{14}NH^+$ ion by a merged least-squares fit (with suitable weight factors) [7] of the IR data [7], of term values from UV-visible emission (A, B, C→X) spectra [8], and of the two far-IR, purely rotational transitions [1] (all constants in cm^{-1}):

state	constant	v=0 state	v=1 state
X $^2\Pi_r$	A	81.6568(21)	81.7670(15)
	p	0.23330(20)	0.22748(36)
	q	0.055039(39)	0.05561(11)
	$p_D \cdot 10^3$	−0.0316(44)	0.0142(54)
	$q_D \cdot 10^3$	−0.00866(33)	−0.01542(93)
	γ	−0.08312(44)	−0.07896(46)
	$\gamma_D \cdot 10^3$ *)	0.0148	0.0042
a $^4\Sigma^-$	γ	0.00243(36)	0.00187(31)
	λ	0.12332(67)	0.10214(61)
	$\lambda_D \cdot 10^3$	0.057(41)	0.408(33)
X $^2\Pi_r$-a $^4\Sigma^-$	$\xi_{3/2}$	81.328(22)	77.1609(24)
	$\xi_{1/2}$	87.162(34)	83.0025(39)
	ξ_D	−	−0.004533(55)

*) From $(D+\gamma_D/2)\cdot 10^3 = 1.6387(25)$ and 1.5913(26), $D \cdot 10^3 = 1.6313(15)$ and 1.5892(17) for v=0 and 1, respectively, (D=centrifugal distortion constant; cf. pp. 140/1) given in [7].

On the basis of UV-visible emission data (A, B, C→X) for $^{14}NH^+$, $^{15}NH^+$, and $^{14}ND^+$ [8] and IR data for $^{14}NH^+$ [7], the perturbations occurring between the X $^2\Pi_r$ and a $^4\Sigma^-$ states have been studied and a set of rotational (cf. pp. 140/1) and fine-structure constants for the three isotopic species in their X $^2\Pi_r$, v=0 and 1 and a $^4\Sigma^-$, v=0 and 1 states derived. They are, of course, less accurate than those listed above for $^{14}NH^+$. The constants γ_D for X $^2\Pi_r$, λ, λ_D, and γ for a $^4\Sigma^-$, and ξ_D have been set equal to zero, and p_D and q_D could not be derived for the two heavier isotopes [9].

For the higher excited states, the optical emission spectra gave the following spin-orbit and spin-rotation coupling constants (in cm^{-1}) [8]:

constant	state		$^{14}NH^+$	$^{14}ND^+$	$^{15}NH^+$
A	B $^2\Delta$	v=0	−3.6	−3.5	−3.6
γ	A $^2\Sigma^-$	v=0	−0.097	−0.052	−0.101
		v=1	−0.100	−0.053	−
	C $^2\Sigma^+$	v=0	+0.119*)	−	−

*) γ= +0.105, +0.108, and +0.111 for v=0, 1, and 2, respectively [10].

Ab initio calculations of spin-orbit splittings in NH^+ were carried out for the X $^2\Pi_r$ state by using a relativistic (Dirac-Fock) SCF wave function [11] and for the X $^2\Pi_r$ [12] and B $^2\Delta$ states [13] by using nonrelativistic (Hartree-Fock) SCF wave functions. MCSCF calculations of the spin-orbit splittings in $^2\Pi$ and $^3\Pi$ states of the first-row hydrides XH^+, XH, XH^- (including X=N) are available [14] which use effective nuclear charges Z_{eff} for the X atoms based, however, on the experimental A values quoted by Huber and Herzberg [15] (in the case of NH^+ from [8]).

Rotational and Vibrational Constants. Internuclear Distance. Rotational and vibrational constants were obtained from the high-resolution IR absorption spectrum of $^{14}NH^+$ in the region of the $v=1 \leftarrow 0$ transitions in the ground state X $^2\Pi_r$, in the lowest excited state a $^4\Sigma^-$, and between the two states [7], from the UV-visible emission spectra A $^2\Sigma^-$, B $^2\Delta$, C $^2\Sigma^+ \rightarrow$ X $^2\Pi_r$ of $^{14}NH^+$, $^{14}ND^+$, and $^{15}NH^+$ [8, 10, 16, 17], and by including the two lowest rotational transitions (J=3/2, d)$\leftarrow$(J=1/2, d) and (J=3/2, c)$\leftarrow$(J=1/2, c) observed in the far IR [1]. The strong perturbation of the electronic ground state X $^2\Pi_r$ by the very low-lying a $^4\Sigma^-$ state was early discovered and had to be accounted for in the rotational-vibrational analyses of the spectra. Detailed studies of the perturbations using matrix diagonalization techniques showed spin-orbit interaction between the two states (see above) to be the major cause; the effective Hamiltonian matrix elements used for fitting the rotational levels of the interacting X $^2\Pi_r$ and a $^4\Sigma^-$ states are listed in [7, 9]. The rotational constants B and the centrifugal distortion constants D, H (with subscripts e or 0 denoting the molecular equilibrium or the lowest vibrational level, respectively), the corresponding rotation-vibration interaction constants α_e, γ_e, β_e, the vibrational constants ω_e (or $\Delta G_{1/2}=\nu_0$), $\omega_e x_e$, $\omega_e y_e$, and the internuclear distance r_e or r_0 obtained by converting the rotational constant B_e or B_0 are compiled in Table 5, below, for the X $^2\Pi_r$ and a $^4\Sigma^-$ states and in Table 6, pp. 141/2, for the A $^2\Sigma^-$, B $^2\Delta$, and C $^2\Sigma^+$ states. The compilation of Huber and Herzberg [15] is based on the earlier UV-visible emission studies [8, 10, 16].

The resonance-enhanced multiphoton ionization-photoelectron spectra (REMPI-PES) of NH(a $^1\Delta$) with the resonant intermediate Rydberg states $\{NH^+(X\,^2\Pi_r)\}$ (3pσ/3dσ) i $^1\Pi$ and $\{NH^+(B\,^2\Delta)\}$ (3sσ)/$\{NH^+(X\,^2\Pi_r)\}$ (3dπ) j $^1\Delta$ (cf. pp. 33/4) give $\omega_e=3093(3)$ cm^{-1} and $\omega_e x_e=86.3(10)$ cm^{-1} for the X $^2\Pi_r$ state, $\omega_e=2350(7)$ cm^{-1} and $\omega_e x_e=69.4(18)$ cm^{-1} for the B $^2\Delta$ state [18].

A few ab initio calculations of spectroscopic constants (and potential curves; see pp. 143/4) are available, for $NH^+(X\,^2\Pi_r)$ using PNO-CI and CEPA wave functions [19], for the X, a, A, B, and C states using MBPT [20], CI [21], and SCF (and semiempirical estimate of the correlation energy) [22] wave functions, and for the A, B, and C states using an MRSD CI wave function [23]. All these calculations, except obviously that of [20], omit the strong interaction between the X and the a state. A number of further ab initio calculations (SCF, CI, MP) for the ground state [24 to 30] and the excited a state [29] of NH^+ include geometry optimizations and give r_e values.

Table 5
NH^+(X $^2\Pi_r$, a $^4\Sigma^-$). Spectroscopic Constants (cm^{-1}) and Internuclear Distance r_e (Å).

constant	$^{14}NH^+$	$^{14}NH^+$	$^{15}NH^+$	$^{14}ND^+$	$^{14}NH^+$	$^{14}ND^+$
X $^2\Pi_r$						
B_e	15.6891	15.6911	15.6253	8.3781	15.67	8.35
B_0	15.331128(46)	—	—	—	—	—
α_e	—	0.6075	0.6068	0.2332	0.64	0.25
$D_0 \cdot 10^3$	1.6313(15)	1.63	1.65	0.464	—	—

Table 5 (continued)

constant	$^{14}NH^+$	$^{14}NH^+$	$^{15}NH^+$	$^{14}ND^+$	$^{14}NH^+$	$^{14}ND^+$
$H_0 \cdot 10^7$	–	1.35	2.12	–	–	–
ω_e ($\Delta G_{1/2}$)	(2903.1718(21))	3047.58	3040.77	2226.93	3038	2218
$\omega_e x_e$	–	72.19	71.87	38.55	58	31
r_e	1.0692 ± 0.0002	1.0692	1.0691	1.0692	–	–
a $^4\Sigma^-$						
T_e	–	509.41	509.78	537.85	–	–
T_0	323.8976(45)	323.90	324.57	401.86	–	–
B_e	15.0523	15.013	14.982	7.926	–	–
B_0	14.66001(14)	–	–	–	–	–
α_e	–	0.712	0.770	0.200	–	–
$D_0 \cdot 10^3$	1.8270(33)	1.78	1.80	~0.51	–	–
ω_e ($\Delta G_{1/2}$)	(2544.3004(40))	2672.57	2666.60	1952.90	–	–
$\omega_e x_e$	–	64.23	63.94	34.30	–	–
r_e	1.0924 ± 0.0001	1.093	1.092	1.099	–	–
remark	a)	b)	b)	b)	c)	c)
Ref.	[7]	[9]	[9]	[9]	[17]	[17]

a) Merged least-squares fit (with suitable weight factors) of the own high-resolution IR data (v=1←0 transitions in the X $^2\Pi_r$ and a $^4\Sigma^-$ states and between them) [7], term values from the UV-visible emission spectra (A $^2\Sigma^-$, B $^2\Delta$, C $^2\Sigma^+ \rightarrow$ X $^2\Pi_r$) [8], and the two purely rotational far-IR transitions [1] for the $^{14}NH^+$ species alone.

b) Extending the treatment of [7] to the other two isotopic species $^{15}NH^+$ and $^{14}ND^+$, which results in less accurate constants, since no rotational-vibrational or purely rotational data exist for these isotopes [9].

c) Rotational-vibrational analysis of the B $^2\Delta \rightarrow$ X $^2\Pi_r$ emission resulting from the chemiluminescent reaction of $N^+(^1D)$ ions with H_2 and D_2 [17].

Table 6
NH^+(A $^2\Sigma^-$, B $^2\Delta$, C $^2\Sigma^+$). Spectroscopic Constants (cm^{-1}) and Internuclear Distance r (Å).

constant	$^{14}NH^+$	$^{15}NH^+$	$^{14}ND^+$	$^{14}NH^+$	$^{14}ND^+$	$^{14}NH^+$
A $^2\Sigma^-$						
T_e	~22200	–	~22200	–	–	–
T_0	21567.67	21569.13	21750.59	–	–	–
B_e	11.4553	–	6.1206	–	–	–
B_0	11.1105	11.0635	5.9830	–	–	–
α_e	0.6897	–	0.2752	–	–	–
$D_0 \cdot 10^3$	2.02	2.01	0.58	–	–	–
$H_0 \cdot 10^8$	18.5	–	–	–	–	–
ω_e	~1706.9	–	–	–	–	–
$\Delta G_{1/2}$	1585.49	–	1182.40	–	–	–
$\omega_e x_e$	~60.7	–	–	–	–	–
r_e	1.2511	–	1.2507	–	–	–
r_0	1.2704	1.2731	1.265	–	–	–

Table 6 (continued)

constant	$^{14}NH^+$	$^{15}NH^+$	$^{14}ND^+$	$^{14}NH^+$	$^{14}ND^+$	$^{14}NH^+$
B $^2\Delta$						
T_e	~23300	–	~23300	(22885)	–	–
T_0	22960.46	22961.40	23063.83	–	–	–
B_e (B_0)	(13.516)	(13.456)	(7.2715)	13.8	7.4	–
α_e	–	–	–	0.63	0.25	–
$D_0 \cdot 10^3$	1.9	1.9	0.55	–	–	–
$H_0 \cdot 10^8$	–	–	1.5	–	–	–
ω_e	~2280	–	~1672	2371	1732	–
$\omega_e x_e$, $\omega_e y_e$	–	–	–	74, 1.0	39, 0.3	–
r_0	1.1519	1.1544	1.1475	–	–	–
C $^2\Sigma^+$						
T_e	~35000	–	–	–	–	–
T_0	34561.38	–	–	–	–	34561.27
B_e (B_0)	(12.8766)	–	–	–	–	13.2652
α_e, γ_e	–	–	–	–	–	0.7891, 0.0089
D_e (D_0) $\cdot 10^3$	(2.01)	–	–	–	–	2.00
$\beta_e \cdot 10^4$	–	–	–	–	–	−0.8
$H_e \cdot 10^8$	–	–	–	–	–	~11
ω_e ($\Delta G_{1/2}$)	(2004.0)	–	–	–	–	2150.56
$\omega_e x_e$	–	–	–	–	–	73.07
r_e (r_0)	(1.1801)	–	–	–	–	1.1626
remark	a)	a)	a)	b)	b)	c)
Ref.	[8]	[8]	[8]	[17]	[17]	[10]

a) The constants for the v=0 levels result from rotational-vibrational analyses of the UV-visible emission systems A $^2\Sigma^-$(v=0, 1), B $^2\Delta$(v=0) → X $^2\Pi_r$(v=0) of $^{14}NH^+$ and $^{14}ND^+$, A $^2\Sigma^-$(v=0), B $^2\Delta$ (v=0) → X $^2\Pi_r$ (v=0) of $^{15}NH^+$, and C $^2\Sigma^+$(v=0) → X $^2\Pi_r$(v=0) of $^{14}NH^+$; ω_e and $\omega_e x_e$ for $^{14}NH^+$(A $^2\Sigma^-$) have been calculated from the $\Delta G_{1/2}$ values of $^{14}NH^+$ and $^{14}ND^+$ [8]; $\Delta G_{1/2}$ for $^{14}NH^+$(C $^2\Sigma^+$) is taken from [16]. The other equilibrium values have been evaluated by Huber and Herzberg [15].

b) Rotational-vibrational analysis of the B $^2\Delta$ → X $^2\Pi_r$ emission resulting from the chemiluminescent reaction of $N^+(^1D)$ ions with H_2 and D_2 [17].

c) Rotational-vibrational analysis of the C $^2\Sigma^+$(v=0, 1, 2) → X $^2\Pi_r$(v=0, 1) emission spectrum in the UV [10]. The constants D_e, β_e, and H_e have been evaluated by Huber and Herzberg [15].

References:

[1] Verhoeve, P.; Ter Meulen, J. J.; Meerts, W. L.; Dymanus, A. (Chem. Phys. Lett. **132** [1986] 213/7).
[2] Brown, J. M. (unpublished results; private communication to [1]).
[3] Frosch, R. A.; Foley, H. M. (Phys. Rev. [2] **88** [1952] 1337/49).
[4] Kristiansen, P.; Veseth, L. (J. Chem. Phys. **84** [1986] 6336/44).
[5] Carmichael, I. (J. Phys. Chem. **94** [1990] 5734/40).

[6] Carmichael, I. (J. Phys. Chem. **95** [1991] 108/11).
[7] Kawaguchi, K.; Amano, T. (J. Chem. Phys. **88** [1988] 4584/91).
[8] Colin, R.; Douglas, A. E. (Can. J. Phys. **46** [1968] 61/73).
[9] Colin, R. (J. Mol. Spectrosc. **136** [1989] 387/401).
[10] Krishnamurthy, G.; Saraswathy, M. (Pramana **6** [1976] 235/43).

[11] Baeck, K. K.; Lee, Y. S. (J. Chem. Phys. **93** [1990] 5775/82).
[12] Abegg, P. W. (Mol. Phys. **30** [1975] 579/96).
[13] Lefebvre-Brion, H.; Bessis, N. (Can. J. Phys. **47** [1969] 2727/30).
[14] Koseki, S.; Schmidt, M. W.; Gordon, M. S. (J. Phys. Chem. **96** [1992] 10768/72).
[15] Huber, K. P.; Herzberg, G. (Molecular Spectra and Molecular Structure, Vol. 4, Constants of Diatomic Molecules, Van Nostrand Reinhold, New York 1979, pp. 460/1).
[16] Feast, M. W. (Astrophys. J. **114** [1951] 344/55).
[17] Kusunoki, I.; Ottinger, C. (J. Chem. Phys. **80** [1984] 1872/81).
[18] Clement, S. G.; Ashfold, M. N. R.; Western, C. M.; De Beer, E.; De Lange, C. A.; Westwood, N. P. C. (J. Chem. Phys. **96** [1992] 4963/73).
[19] Rosmus, P.; Meyer, W. (J. Chem. Phys. **66** [1977] 13/9).
[20] Park, J. K.; Sun, H. (Bull. Korean Chem. Soc. **11** [1990] 34/41).

[21] Guest, M. F.; Hirst, D. M. (Mol. Phys. **34** [1977] 1611/21).
[22] Liu, H. P. D.; Verhaegen, G. (J. Chem. Phys. **53** [1970] 735/45).
[23] Kusunoki, I.; Yamashita, K.; Morokuma, K. (Chem. Phys. Lett. **123** [1986] 533/6).
[24] Lathan, W. A.; Hehre, W. J.; Curtiss, L. A.; Pople, J. A. (J. Am. Chem. Soc. **93** [1971] 6377/87).
[25] Lathan, W. A.; Curtiss, L. A.; Hehre, W. J.; Lisle, J. B.; Pople, J. A. (Progr. Phys. Org. Chem. **11** [1974] 175/261, 206).
[26] Pople, J. A. (Int. J. Mass Spectrom. Ion Phys. **19** [1976] 89/106).
[27] Pople, J. A.; Curtiss, L. A. (J. Phys. Chem. **91** [1987] 155/62).
[28] Pope, S. A.; Hillier, I. H.; Guest, M. F. (Faraday Symp. Chem. Soc. No. 19 [1984] 109/23).
[29] Power, D.; Brint, P.; Spalding, T. R. (J. Mol. Struct. **110** [1984] 155/66 [THEOCHEM **19**]).
[30] Farnell, L.; Pople, J. A.; Radom, L. (J. Chem. Phys. **87** [1983] 79/82).

2.1.2.6 Potential Energy Functions

Rydberg-Klein-Rees (RKR) potential energy functions for the two lowest states, X $^2\Pi_r$ and a $^4\Sigma^-$, were computed using the most accurate term values and equilibrium rotational and vibrational constants for $^{14}NH^+$ and $^{14}ND^+$ which are based on high-resolution IR absorption and UV-visible emission spectra (cf. Table 5, pp. 140/1); for the region above the v=1 level of the $^{14}NH^+$ species (no spectroscopic measurements for v>1), Morse potentials were assumed for the X $^2\Pi_r$ state which dissociates to $N^+(^3P)+H(^2S)$ and for the a $^4\Sigma^-$ state which dissociates to $N(^4S)+H^+$. Numerical values for the energy and a graphical representation for r=0.900 to 1.350 Å showing the observed vibrational levels v=0 and 1 of NH^+ and ND^+ are given [1]. For the X $^2\Pi_r$ and B $^2\Delta$ states, RKR potential curves at r=0.8 to 3.0 Å (graphical presentation with v''=0 to 5 and v'=0 to 6 levels) were evaluated based on the term values and equilibrium rotational and vibrational constants for NH^+ and ND^+, derived from the B $^2\Delta$(v'=0 to 3)→X $^2\Pi_r$(v''=0 to 2) emission spectrum [2]. **Fig. 4**, p. 144, is reproduced from [1] and [2].

Potential energy functions for the bound states of NH^+ have also been calculated by ab initio MO methods, for the ground state X by SCF, PNO-CI, and CEPA [3], for the X, A, B, and C states by MRSD CI [4], and for the X, a, A, B, and C states by MBPT [5],

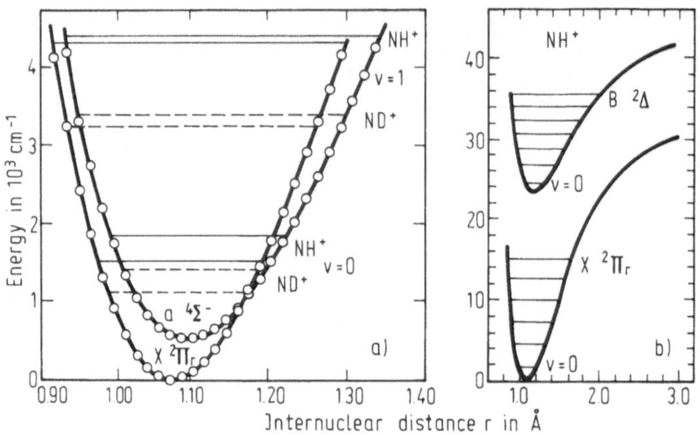

Fig. 4. Rydberg-Klein-Rees potential energy curves of NH^+ and ND^+; a) X $^2\Pi_r$ and a $^4\Sigma^-$ states from [1], b) X $^2\Pi_r$ and B $^2\Delta$ states from [2].

CI [6], and SCF (including a semiempirical estimate of the correlation energy) [7]. Whereas almost all of these calculations omit the strong interaction between the quasidegenerate X and a states, which predicts the a-state potential curve to lie below that of the X state [6, 7], an effective valence Hamiltonian study (H^v, based on MBPT), for the first time, resulted in the correct energetical ordering of the two states in the region around the potential minimum [5]. Including the spin-orbit coupling between the X and a states in the MRSD CI calculation [8] and in an MP3 calculation [9], however, also led to the wrong ordering of the X and a states.

Theoretical potential curves for a number of unobserved states are also available; these are a high-lying bound $^4\Pi$ state correlating with $N(^4P)+H^+$ [7], repulsive $^4\Pi$ and $^2\Sigma^-$ states correlating with $N^+(^3P)+H(^2S)$ [5 to 7], repulsive $^2\Pi_i$ and $^2\Sigma^-$ states correlating with $N(^2D)+H^+$ [4 to 7], and repulsive $^2\Pi$ and $^2\Delta$ states correlating with $N^+(^1D)+H(^2S)$ [4, 5] (for tabulated energy values or graphical representations, see the quoted publications).

References:

[1] Colin, R. (J. Mol. Spectrosc. **136** [1989] 387/401).
[2] Kusunoki, I.; Ottinger, C. (J. Chem. Phys. **80** [1984] 1872/81).
[3] Rosmus, P.; Meyer, W. (J. Chem. Phys. **66** [1977] 13/9).
[4] Kusunoki, I.; Yamashita, K.; Morokuma, K. (Chem. Phys. Lett. **123** [1986] 533/6).
[5] Park, J. K.; Sun, H. (Bull. Korean Chem. Soc. **11** [1990] 34/41).
[6] Guest, M. F.; Hirst, D. M. (Mol. Phys. **34** [1977] 1611/21).
[7] Liu, H. P. D.; Verhaegen, G. (J. Chem. Phys. **53** [1970] 735/45).
[8] Yamashita, K.; Yabushita, S.; Morokuma, K.; Kusunoki, I. (Chem. Phys. Lett. **137** [1987] 193/4).
[9] Farnell, L.; Ogilvie, J. F. (J. Mol. Spectrosc. **101** [1983] 104/32).

2.1.2.7 Dissociation Energy

The endothermicity of the reaction $N^+ + H_2 \rightarrow NH^+ + H$ obtained by various experimental methods (cf. pp. 128/30) combined with literature values for the ionization potentials of N and H and the dissociation energy of H_2 result in the proton affinity of atomic N or the dissociation energy of NH^+: $D_0^\circ(N\text{-}H^+) = 3.531 \pm 0.003$ eV [1] (SIFDT apparatus), 3.524 ± 0.003 eV [2] (preliminary value 3.5095 ± 0.0007 eV [3]; CRESU studies), and 3.520 eV [4] (ion-trap technique). $\Delta_f H_0^\circ(NH^+)$ (cf. p. 135) combined with the heats of formation of N and H^+ gives $D_0^\circ(N\text{-}H^+) = 3.51 \pm 0.03$ eV [5]. A reliable value for the first ionization potential of NH, $E_i(NH) = 13.47 \pm 0.05$ eV (see p. 137), combined with $E_i(H)$ and $D_0^\circ(NH)$ ($= 3.46 \pm 0.05$ eV; cf. pp. 56/8) results in $D_0^\circ(N\text{-}H^+) = 3.59 \pm 0.07$ eV [6].

Ab initio SCF MO and CI calculations indicate dissociation of ground-state $NH^+(X\ ^2\Pi_r)$ into $N^+(^3P) + H(^2S)$; this limit lies 0.95 eV above that of $N(^4S) + H^+$ which correlates only with the a $^4\Sigma^-$ state. Energies of dissociation into the lowest limit, $D_0^\circ(N\text{-}H^+) = 3.11$ eV (MRSD CI) [7], 3.48 eV (CEPA), 3.50 eV (SCF) [8], 3.39 eV (SCF and semiempirical estimate of the correlation energy) [9], and $D_e(NH^+) = 3.57$ eV (CI) [10], were calculated.

References:

[1] Adams, N. G.; Smith, D. (Chem. Phys. Lett. **117** [1985] 67/70).
[2] Marquette, J. B.; Rebrion, C.; Rowe, B. R. (J. Chem. Phys. **89** [1988] 2041/7).
[3] Marquette, J. B.; Rowe, B. R.; Dupeyrat, G.; Roueff, E. (Astron. Astrophys. **147** [1985] 115/20).
[4] Luine, J. A.; Dunn, G. H. (Astrophys. J. **299** [1985] L 67/L 70).
[5] Ervin, K. M.; Armentrout, P. B. (J. Chem. Phys. **86** [1987] 2659/73).
[6] Foner, S. N.; Hudson, R. L. (J. Chem. Phys. **74** [1981] 5017/21).
[7] Kusunoki, I.; Yamashita, K.; Morokuma, K. (Chem. Phys. Lett. **123** [1986] 533/6).
[8] Rosmus, P.; Meyer, W. (J. Chem. Phys. **66** [1977] 13/9).
[9] Liu, H. P. D.; Verhaegen, G. (J. Chem. Phys. **53** [1970] 735/45).
[10] Guest, M. F.; Hirst, D. M. (Mol. Phys. **34** [1977] 1611/21).

2.1.2.8 Spectra

Microwave Spectrum. Due to the possible astrophysical relevance of the NH^+ ion, there is interest in its microwave spectrum in the wavelength range of available radiotelescopes. The ion, however, has not yet been observed in any extraterrestrical object, and laboratory measurements in the MW region have not yet been performed. Using the molecular constants of NH^+ in the X $^2\Pi_r$ and a $^4\Sigma^-$ states based on spectroscopic measurements in the far IR [1], IR [2, 3], and UV-visible regions [4] (cf. pp. 139/41), the frequencies and relative intensities of the Λ-doubling transitions between various hyperfine levels of the four lowest rotational levels of $^{14}NH^+(X\ ^2\Pi_{1/2}, v=0)$ have been calculated. The following wavelength (λ) and frequency (ν) regions, numbers of transitions (no.) and strongest transitions (ν_{max}) have been predicted (for future radioastronomical research) [2]:

λ in cm	transition	no.	ν in MHz	ν_{max} in MHz
2.2	J=1/2, f−J=1/2, e	13	13375.1 to 13751.5	13734.4
1.5	J=3/2, f−J=3/2, e	25	20073.3 to 20403.8	20403.8
1.25	J=5/2, f−J=5/2, e	26	23765.0 to 24072.3	24061.9
1.12	J=7/2, f−J=7/2, e	26	26586.1 to 26884.6	26865.9

Prior to the laser-spectroscopic studies of NH^+, the Λ-doubling frequencies of $^{14}NH^+$, $^{15}NH^+$, and $^{14}ND^+$ in the J=1/2 to 7/2 levels of the ground state X $^2\Pi_r$ have been predicted from ab initio calculations; the significance of the X $^2\Pi_r$-a $^4\Sigma^-$ interaction was already recognized [5, 6].

The Λ-doublet population inversion in the J = 1/2 and 3/2 levels of $NH^+(X\ ^2\Pi_r)$ and excitation of MW emission by collisions with H atoms (expected to occur in interstellar clouds) was treated by a scattering theoretical approach [7, 8].

Far-IR Spectrum. The purely rotational transition J = 3/2 ← 1/2 of NH^+ in the X $^2\Pi_r$, v = 0 ground state was detected at 34 cm^{-1} (~0.3 mm) for the first time using a submillimeter tunable laser sideband spectrometer. The NH^+ ions were produced by hollow-cathode discharges in a fast-flowing mixture of He with ~2% admixture of equal amounts of H_2 and N_2. The measurements were performed in the 1019- and 1012-GHz regions, and seven hyperfine components of the (J = 3/2, d) ← (J = 1/2, d) transition at 1019.2 GHz and five hyperfine components of the (J = 3/2, c) ← (J = 1/2, c) transition at 1012.5 GHz could be recorded (using the coupling scheme $J + I_N = F_1$, $F_1 + I_H = F$, where the nuclear spins are $I_N = 1$ and $I_H = 1/2$, and considering the selection rule $\Delta F = 0, \pm 1$, seventeen transitions are expected in each parity component ($c = \Pi^+$, $d = \Pi^-$) of the J = 3/2 ← 1/2 transition). The spectrum has been analyzed to give the Frosch-Foley hyperfine constants a, b, c, and d for both nuclei, the quadrupole coupling constants eq_0Q and eq_2Q for the N nucleus (cf. p. 138), and the following hyperfine-free purely rotational frequencies [1]: $\nu_{0d}(3/2 \leftarrow 1/2)$ = 1019210.8(6) MHz and $\nu_{0c}(3/2 \leftarrow 1/2)$ = 1012539.5(8) MHz.

IR Spectrum. The v = 1 ← 0 vibration-rotation transitions in the X $^2\Pi_r$ and a $^4\Sigma^-$ states of NH^+ as well as v = 1 ← 0 transitions between these quasidegenerate states were observed at high resolution in the range 3.2 to 4.1 μm (3100 to 2450 cm^{-1}). The NH^+ ions were generated by hollow-cathode discharges either in a mixture of N_2 (~10 mTorr), H_2 (~10 mTorr), and He (~1.4 Torr) or in a mixture of NH_3 (~10 mTorr) and He (~1.4 Torr). The absorption lines were observed with a tunable difference-frequency laser as a radiation source; the accuracy of the wavenumber measurements was estimated to be about 0.002 cm^{-1} which enabled fine-structure splittings to be resolved. In the range 3100 to 2700 cm^{-1}, 39 absorption lines, almost all of them being e ← e, f ← f doublets in the P and R branches and one being an e ← f, f ← e doublet in the Q branch (Λ-type doubling), were assigned to the v = 1 ← 0 transitions (J″ = 1/2 to 11/2) in the X $^2\Pi_{1/2}$ (22 lines) and X $^2\Pi_{3/2}$ (17 lines) substates (ν_0 = 2903.1718(21) cm^{-1}). These are overlapped in the low-energy part by the v = 1 (J′, N′) ← 0 (J″) (J″ = 3/2 to 11/2; e ← e and f ← f) P- and R-branch transitions of the a $^4\Sigma^-$ ← X $^2\Pi_{1/2}$ and a $^4\Sigma^-$ ← X $^2\Pi_{3/2}$ systems (eight and four lines, respectively). The region between 2700 and 2450 cm^{-1} shows 18 P- and R-branch (e ← e and f ← f) lines of the v = 1 ← 0 transition (J″ = 1/2 to 13/2) in the a $^4\Sigma^-$ state (ν_0 = 2544.3004(40) cm^{-1}) and, at the high-energy end, the two X $^2\Pi_{1/2}$, X $^2\Pi_{3/2}$ ← a $^4\Sigma^-$, v = 1 ← 0 (J″ = 3/2 and 5/2, e ← e and f ← f) R-branch transitions [2]. Analysis of the IR spectrum, including data from UV-visible [4] and far-IR [1] spectra, yielded the most accurate rotational, vibrational, and fine-structure constants so far for the X $^2\Pi_r$ and a $^4\Sigma^-$ states of $^{14}NH^+$ [2] (cf. pp. 140/1 and 139).

Prior to the laser-spectroscopic studies, the transitions between the v = 0 levels of the X $^2\Pi_r$ and a $^4\Sigma^-$ states and the v = 1 ↔ 0 transitions within the X $^2\Pi_r$ and a $^4\Sigma^-$ states and between them, were predicted by an ab initio MO calculation (Møller-Plesset perturbation theory of third order) that included the spin-orbit interaction between the two states [9].

UV-Visible Spectra. Three emission band systems of NH^+ have been detected in the visible and UV region between 550 and 260 nm and identified as the A $^2\Sigma^-$ → X $^2\Pi_r$ (550 to 430 nm), B $^2\Delta$ → X $^2\Pi_r$ (500 to 360 nm), and C $^2\Sigma^+$ → X $^2\Pi_r$ (300 to 260 nm) transitions.

Hollow-cathode discharges through streaming NH_3 first revealed the UV emission bands which were tentatively attributed to NH^+ [10] and later shown (by rotational-vibrational analysis of spectrograms of moderate dispersion) to arise from a $^2\Sigma^+ \rightarrow {}^2\Pi$ transition analogous to the C $^2\Sigma^+ \rightarrow$ X $^2\Pi$ transition of the isoelectronic CH radical [11].

High-resolution spectra of the A→X, B→X, and C→X emission systems of $^{14}NH^+$, $^{15}NH^+$, and $^{14}ND^+$ were photographed from hollow-cathode discharges through rapidly flowing He with small amounts of $^{14}NH_3$, $^{15}NH_3$, $^{14}N_2H_4$, or $^{14}N_2D_4$. The wavenumbers (absolute error ± 0.05 cm^{-1}) of some 500 lines for the $^{14}NH^+$ species are listed in [4], those for the two other isotopic species are available from the Depository of Unpublished Data, National Science Library, National Research Council of Canada in Ottawa. The spectra clearly show (as already pointed out earlier [11]) that the very low lying a $^4\Sigma^-$ state causes strong perturbations in the v=0 and 1 levels of the ground state X $^2\Pi_r$.

In the A→X system, three red-degraded bands peaking at 534.94, 462.84, and 431.27 nm have been identified with the v=(0, 1), (0, 0), and (1, 0) transitions. Their rotational and fine structures indicate a $^2\Sigma \rightarrow {}^2\Pi$ transition with the upper state belonging to and the lower state approaching Hund's case (b), i.e., the six main branches P_{11d}, Q_{11c}, R_{11d} ($^2\Sigma^- \rightarrow {}^2\Pi_{1/2}$) and P_{22d}, Q_{22c}, and R_{22d} ($^2\Sigma^- \rightarrow {}^2\Pi_{3/2}$), where the indices c and d are for the Π^+ and Π^- components of the Λ-doublets, are accompanied by the satellite branches P_{12c}, Q_{12d}, and R_{21c} in the (0, 0) band, by the P_{12c} and R_{21c} branches in the (0, 1) and (1, 0) bands. Rotational lines with J up to 31/2, 19/2, and 23/2 in the (0, 0), (1, 0), and (0, 1) bands, respectively, were observed. Four extra branches of R_1, P_1, R_2, and P_2 type with J values below 15/2 in the (0, 1) band obviously arose from the perturbations by the a $^4\Sigma^-$ state. An additional 14 lines in that region remained unassigned [4], ten of which could later be identified by means of an ab initio MO calculation for the X $^2\Pi_r$ and a $^4\Sigma^-$ states and the interaction between them [9].

In the red-degraded B $^2\Delta \rightarrow$ X $^2\Pi_r$, v=(0, 0) band with its head at 434.85 nm, more than 200 rotational lines were observed in the 450- to 430-nm region and attributed to the twelve main branches P_{iic}, P_{iid}, Q_{iic}, Q_{iid}, R_{iic}, R_{iid} (i=1 and 2; J up to 29/2) and the corresponding twelve satellite branches P_{ijc}, P_{ijd}, Q_{ijc}, Q_{ijd}, R_{ijc}, R_{ijd} (ij=12 and 21; J up to 15/2) as expected for a doublet-doublet $^2\Delta \rightarrow {}^2\Pi$ transition with both states approaching Hund's case (b) and allowing for Λ-type splitting in the $^2\Pi$ state [4]. Chemiluminescence in the 360- to 500-nm region, resulting from $N^+ + H_2$ or $N^+ + D_2$ collisions at center-of-mass kinetic energies of 1 to 9 eV, was identified as the B $^2\Delta$, v′=0 to 3→X $^2\Pi_r$, v″=0 to 2 spectrum of NH^+ (ND^+) with the following band heads λ_H [12]:

v′, v″	λ_H in nm	v′, v″	λ_H in nm	v′, v″	λ_H in nm
0, 0	434.8	1, 0	397.0	2, 0	367.0
1, 1	448.5	2, 1	411.0	3, 1	381.1
2, 2	464.0	3, 2	426.5		

The C $^2\Sigma^+ \rightarrow$ X $^2\Pi_r$ emission system showed up weakly in the hollow-cathode discharge source; in the red-degraded (0, 0) band at ~290 nm, rotational lines in the main branches P_{11c}, Q_{11d}, R_{11c}, and P_{22c}, Q_{22d}, R_{22c} with J up to 15/2 and in the satellite branch Q_{21c} with J up to 9/2 could be assigned [4]. In a condensed discharge through flowing NH_3, five red-degraded bands appeared in the 300- to 260-nm region. Confirming and extending the earlier results of [4, 10, 11], these have been identified as the (0, 0), (1, 0), (1, 1), (2, 1), and (2, 0) bands of the C→X system, exhibiting rotational lines in the main branches with J up to 43/2, 43/2, 29/2, 31/2, and 41/2, respectively. The band origins are at 289.3, 273.5, 297.2, 281.6, and 260.25 nm [13].

Hyperfine structure in the electronic spectra of NH^+ (and other XH^+ ions) was observed by high-resolution laser-photofragment spectroscopy; results for NH^+ have not yet been published [14].

The A, B, and C→X emission systems of NH^+ were also observed following electron impact on NH_3. **Emission cross sections** at 100 eV incident energy were determined for the A→X, v=(0, 0), (1, 0), B→X, v=(0, 0), (1, 0), and C→X, v=(0, 0), (1, 0), (1, 1), (0, 1) bands [15, 16]. Using electron bombardment at energies of 7 keV and the high-frequency deflection (HFD) technique, the following **radiative lifetimes** τ were measured for the excited states [17]:

transition	v', v''	λ in nm ("focussed")	τ in ns
$A\ ^2\Sigma^- \rightarrow X\ ^2\Pi_r$	0, 0	473	1090 ± 100
	1, 0	431	1080 ± 100
$B\ ^2\Delta \rightarrow X\ ^2\Pi_r$	0, 0	445	980 ± 100
$C\ ^2\Sigma^+ \rightarrow X\ ^2\Pi_r$	0, 0	288.5	400 ± 40
	1, 0	275	410 ± 50
	2, 1	283.5	390 ± 50

Based on the spectroscopic studies, **Franck-Condon factors, r-centroids,** and **oscillator strengths** have been calculated for eighteen $B\ ^2\Delta$, v', J'=2→$X\ ^2\Pi_r$, v'', J''=1 transitions with v'=0 to 6 and v''=0 to 3 [12]. The rotational dependence of the Franck-Condon factors was calculated for $A\ ^2\Sigma^-$, v', J'→$X\ ^2\Pi_r$, v'', J'' and $C\ ^2\Sigma^+$, v', J'→$X\ ^2\Pi_r$, v'', J'' transitions with v'=0 to 6, v''=0 and 1, and J''=1, 6, 11, 16, 21, 26, 31 [18].

Transition moments as a function of the internuclear distance r have been derived by ab initio MRSD CI calculations for the band systems $A\ ^2\Sigma^- \leftrightarrow X\ ^2\Pi_r$ and $B\ ^2\Delta \leftrightarrow X\ ^2\Pi_r$ [19] and, in response to a comment [20] which criticized to neglect the strong spin-orbit interaction between the $X\ ^2\Pi_r$ and a $^4\Sigma^-$ states, for the $A\ ^2\Sigma^-_{1/2} \leftrightarrow X\ ^2\Pi_{1/2}$, $A\ ^2\Sigma^-_{1/2} \leftrightarrow a\ ^4\Sigma^-_{1/2}$ and $B\ ^2\Delta_{3/2} \leftrightarrow X\ ^2\Pi_{3/2}$, $B\ ^2\Delta_{3/2} \leftrightarrow a\ ^4\Sigma^-_{3/2}$ transitions [21]. The A↔X and B↔X transition moments are not drastically affected by the perturbation in contrast to the spin-forbidden A↔a and B↔a transitions, which are notably perturbed and have significant transition moments [21].

Translational Energy Spectrum (TES). In the high-resolution TES recorded from grazing collisions of a fast NH^+ ion beam with He as the neutral target gas, two inelastic peaks at energy losses of 2.7 and 4.1 eV were observed and assigned to the $A\ ^2\Sigma^- \leftarrow X\ ^2\Pi_r$ and $C\ ^2\Sigma^+ \leftarrow X\ ^2\Pi_r$ transitions of NH^+; the peak at 4.1 eV showed partially resolved vibrational structure ($\Delta E \approx 0.35$ eV ≈ 2800 cm^{-1}) of the ground state [22].

References:

[1] Verhoeve, P.; Ter Meulen, J. J.; Meerts, W. L.; Dymanus, A. (Chem. Phys. Lett. **132** [1986] 213/7).
[2] Kawaguchi, K.; Amano, T. (J. Chem. Phys. **88** [1988] 4584/91).
[3] Amano, T. (Philos. Trans. R. Soc. London A **324** [1988] 163/78).
[4] Colin, R.; Douglas, A. E. (Can. J. Phys. **46** [1968] 61/73).
[5] Wilson, I. D. L. (Mol. Phys. **36** [1978] 597/610).
[6] Wilson, I. D. L.; Richards, W. G. (Nature [London] **271** [1978] 137).
[7] Dixon, R. N.; Field, D. (Mon. Not. R. Astron. Soc. **189** [1979] 583/91).
[8] Dixon, R. N.; Field, D. (Symp.-Int. Astron. Union No. 87 [1980] 583/7).

[9] Farnell, L.; Ogilvie, J. F. (J. Mol. Spectrosc. **101** [1983] 104/32).
[10] Lunt, W.; Pearse, R. W. B.; Smith, E. C. W. (Nature **136** [1935] 32).
[11] Feast, M. W. (Astrophys. J. **114** [1951] 344/55).
[12] Kusunoki, I.; Ottinger, C. (J. Chem. Phys. **80** [1984] 1872/81).
[13] Krishnamurthy, G.; Saraswathy, M. (Pramana **6** [1976] 235/43).
[14] Edwards, C. P.; Maclean, C. S.; Sarre, P. J. (J. Chem. Phys. **76** [1982] 3829/31).
[15] Müller, U.; Schulz, G. (Chem. Phys. Lett. **170** [1990] 401/5).
[16] Müller, U.; Schulz, G. (J. Chem. Phys. **96** [1992] 5924/37).
[17] Brzozowski, J.; Elander, N.; Erman, P.; Lyyra, M. (Phys. Scr. **10** [1974] 241/3).
[18] Singh, P. D.; De Almeida, A. A. (J. Quant. Spectrosc. Radiat. Transfer **27** [1982] 471/9).
[19] Kusunoki, I.; Yamashita, K.; Morokuma, K. (Chem. Phys. Lett. **123** [1986] 533/6).
[20] Farnell, L.; Ogilvie, J. F. (Chem. Phys. Lett. **137** [1987] 191/2).
[21] Yamashita, K.; Yabushita, S.; Morokuma, K.; Kusunoki, I. (Chem. Phys. Lett. **137** [1987] 193/4).
[22] Hamdan, M.; Mazumdar, S.; Marathe, V. R.; Badrinathan, C.; Brenton, A. G.; Mathur, D. (J. Phys. B At. Mol. Opt. Phys. **21** [1988] 2571/84).

2.1.2.9 Reactions

The rate constants and product ion distributions of the binary reactions of NH^+ with 14 (diatomic to seven-atomic) molecules [1] and of NH^+ and ND^+ with C_2H_4 and C_2D_4 [2] have been measured at 300 K using an SIFT (selected-ion flow tube) apparatus; the NH^+ ions were generated in a low-pressure electron impact ion source (electron energy ~70 eV) containing NH_3 at a pressure of ~1 mTorr. The results, quoted also in the review [3], are given in Table 7 together with the rate constants for the reactions $NH^+ + H_2 \rightarrow NH_2^+ + H$ and $NH^+ + N_2 \rightarrow N_2H^+ + N$ from a flowing-afterglow study (FA) [4], for the reactions $NH^+ + H_2 \rightarrow NH_2^+ + H$ and $NH^+ + NH_3 \rightarrow$ products obtained by ion cyclotron resonance (ICR) [5, 6], and for reactions of NH^+ with its parent molecule NH_3 observed mass-spectrometrically (MS) upon electron impact [7, 8] and α-particle irradiation [9] of NH_3 (reactants are arranged in order of their ionization potentials).

The dynamics of the reactions of NH^+ with H_2 and D_2 was studied by measuring the angular and energy distributions of the product ions NH_2^+, NHD^+, and ND_2^+ at collision energies (center-of-mass kinetic energies) of $KE_{c.m.} = 0.4$ to 1 eV [10].

A theoretical study of the charge exchange reaction $NH + NH^+ \rightarrow NH^+ + NH$ (asymptotic theory of resonance charge exchange; near-Hartree-Fock wave functions for NH and NH^+ from [11]) gave the charge transfer cross sections at collisional energies of 0.03 to 10000 eV [12].

Table 7
Reactions of NH^+ and ND^+ (*)) with Various Molecules.
Rate constants in 10^{-9} $cm^3 \cdot molecule^{-1} \cdot s^{-1}$, product ion percentages in parentheses.

reactant	product ions and percentages	rate constant	Ref.
CH_3NH_2	H_4CN^+ (45), $CH_3NH_2^+$ (20), $CH_3NH_3^+$ (20), H_2CN^+ (10), H_3CN^+ (5)$^{a)}$	2.1	[1]
NO	NO^+ (80), N_2H^+ (20)	0.89	[1]
NH_3	NH_3^+ (75), NH_4^+ (25)	2.4	[1]
NH_3	unknown product distribution	2.1 ± 0.22	[5, 6]

Table 7 (continued)

reactant	product ions and percentages	rate constant	Ref.
NH_3	$N_2H_3^+$, $N_2H_2^+$, N_2H^+, possibly NH_4^+	d)	[7]
NH_3	$N_2H_2^+$ (other channels not considered)	0.003	[8]
NH_3	product distribution not given	2.1 ± 0.5	[9]
H_2S	H_2S^+ (55), H_2NS^+ (15), SH^+ (15), HNS^+ (15)	1.7	[1]
C_2H_4	$C_2H_3^+$ (25), $C_2H_4^+$ (25), H_2CN^+ (20), $C_2H_2^+$ (10), H_3CN^+ (10), $H_3C_2N^+$ (10)	1.5	[2]
C_2D_4	$C_2D_3^+$ (25), $C_2D_4^+$ (25), $HDCN^+$ (20), $C_2D_2^+$ (10), D_3CN^+ (10), $D_3C_2N^+$ (5), $HD_2C_2N^+$ (5)	1.5	[2]
C_2H_4*)	$C_2H_3^+$ (25), $C_2H_4^+$ (25), $HDCN^+$ (20), $C_2H_2^+$ (10), H_3CN^+ (10), $H_3C_2N^+$ (5), $H_2DC_2N^+$ (5)	1.5	[2]
C_2D_4*)	$C_2D_3^+$ (25), $C_2D_4^+$ (25), D_2CN^+ (20), $C_2D_2^+$ (10), D_3CN^+ (10), $D_3C_2N^+$ (10)	1.5	[2]
CH_3OH	H_3CO^+ (70), HCO^+ (15), $CH_3OH_2^+$ (10), H_2CO^+ (5)[b]	3.0	[1]
H_2CO	HCO^+ (55), H_2CO^+ (30), H_3CO^+ (15)	3.3	[1]
COS	COS^+ (85), NS^+ (5), SH^+ (5), $HCOS^+$ (5)	1.8	[1]
O_2	O_2^+ (55), NO^+ (25), HO_2^+ (20)	0.82	[1]
H_2O	H_3O^+ (30), H_2O^+ (30), NH_2^+ (25), HNO^+ (10), NH_3^+ (5)[c]	3.5	[1]
CH_4	H_2CN^+ (70), NH_2^+ (20), CH_5^+ (10)	0.96	[1]
CO_2	HCO_2^+ (35), HNO^+ (35), NO^+ (30)	1.1	[1]
CO	NCO^+ (55), HCO^+ (45)	0.98	[1]
H_2	NH_2^+ (85), H_3^+ (15)	1.5	[1]
H_2	NH_2^+ (100)	0.95 ± 0.10	[5, 6]
H_2	NH_2^+ (100)	~1	[4]
N_2	N_2H^+ (100)	0.65	[1]
N_2	N_2H^+ (100)	~1	[4]

a), b), c) Obvious misprints, H_2CN^+ (20)[a], H_2CO^+ (15)[b], and NH_2^+ (5)[c], in [1, Table 1] and copied by [3]. — d) Cross sections for the reactions forming $N_2H_3^+ + H$, $N_2H_2^+ + H_2$, and $N_2H^+ + H + H_2$ are (in 10^{-18} cm^2/molecule) 35, 78, and 89, respectively [7].

References:

[1] Adams, N. G.; Smith, D.; Paulson, J. F. (J. Chem. Phys. **72** [1980] 288/97).
[2] Smith, D.; Adams, N. G. (Chem. Phys. Lett. **76** [1980] 418/23).
[3] Ikezoe, Y.; Matsuoka, S.; Takebe, M.; Viggiano, A. (Gas Phase Ion-Molecule Reaction Rate Constants Through 1986, Maruzen, Tokyo 1987, pp. 1/224, 91/2).
[4] Fehsenfeld, F. C.; Schmeltekopf, A. L.; Ferguson, E. E. (J. Chem. Phys. **46** [1967] 2802/8).
[5] Kim, J. K.; Theard, L. P.; Huntress, W. T., Jr. (J. Chem. Phys. **62** [1975] 45/52).
[6] Huntress, W. T., Jr. (Astrophys. J. Suppl. Ser. **33** [1977] 495/514).
[7] Derwish, G. A. W.; Galli, A.; Giardini-Guidoni, A.; Volpi, G. G. (J. Chem. Phys. **39** [1963] 1599/605).
[8] Melton, C. E. (J. Chem. Phys. **45** [1966] 4414/24).
[9] Fluegge, R. A.; Landman, D. A. (J. Chem. Phys. **54** [1971] 1576/91).
[10] Eisele, G.; Henglein, A.; Botschwina, P.; Meyer, W. (Ber. Bunsenges. Phys. Chem. **78** [1974] 1090/7).

[11] Cade, P. E.; Huo, W. (At. Data Nucl. Data Tables **12** [1973] 415/66).
[12] Yevseyev, A. V.; Radtsig, A. A.; Smirnov, B. M. (J. Phys. B **15** [1982] 4437/52).

2.1.3 Adducts of NH^+ with NH_3

The adducts $[NH(NH_3)_n]^+$ with n = 1 to 4 were identified in high-pressure mass spectrometry of NH_3 at $p \geq 0.04$ Torr. The ion intensity was low relative to the main products $[NH_4(NH_3)_m]^+$. The peaks assigned to the $[NH(NH_3)_n]^+$ adducts disappeared at pressures above 1 Torr.

Reference:

Long, J. W.; Franklin, J. L. (Int. J. Mass Spectrom. Ion Phys. **12** [1973] 403/10).

2.1.4 NH^{n+} Ions (n = 2 to 6)

CAS Registry Numbers: NH^{2+} *[75648-91-8]*, NH^{3+} *[74341-68-7]*

NH^{2+}. Charge stripping reactions of NH^+ ions (generated by 70-eV electron impact on NH_3) with N atoms resulted in unstable NH^{2+} ions; a minimum energy of 25 ± 1 eV needed to ionize NH^+ in a vertical transition has been estimated by analyzing the mass spectra of the fragments $N^{2+} + H$ or $N^+ + H^+$ [1]. Long-lived NH^{2+} ions could be observed in the mass spectrum following dissociative ionization of NH_3 by 100-eV electron impact. An attempt was made to study electronically excited states by means of high-resolution translational energy spectrometry (TES) after collisions with He atoms; in contrast to some other XH^{2+} ions, no structure indicating electronic transitions to excited states could be observed for NH^{2+} for energy changes up to ± 20 eV; vertically accessible states either lie at such low energies above the ground state that the instrumental resolution is insufficient to enable their detection, or the cross sections for excitation processes are extremely small (an ab initio SCF MO calculation predicted the first excited state $^1\Pi$ $(1\sigma)^2$ $(2\sigma)^2$ $(3\sigma)^1$ $(1\pi)^1$ to lie 8 eV above the ground state; cf. below) [2].

A few quantum-chemical ab initio calculations predicted structure, stability, and some other molecular properties for the ion. SCF, CASSCF, and MRSD CI calculations with various basis sets up to triple-zeta plus polarization quality gave repulsive potential curves for the ground state X $^1\Sigma^+$ $(1\sigma)^2$ $(2\sigma)^2$ $(3\sigma)^2$, energies of −848.9 (SCF) or −872.8 (CI) kJ/mol, and a zero activation energy for the deprotonation (or exothermic charge-separation) reaction $NH^{2+} \rightarrow N^+ + H^+$ [3, 4]. On the other hand, ground-state potential curves with a very shallow minimum at rather long internuclear distances of $r_e = 1.28$ and 1.249 Å were obtained, respectively, by an SCF calculation [2] and a Møller-Plesset perturbation calculation of second order (MP2) [5]. The latter calculation also rendered the transition structure $(N \cdots H)^{2+}$ with $r_e = 1.692$ Å for the charge separation reaction with a barrier of only 9.2 kJ/mol which, however, vanishes after including the zero-point energy (harmonic vibrational frequency $\omega_e = 1426$ cm^{-1}). Thus, spontaneous dissociation into N^+ and H^+, in this case with $\Delta H = -570.3$ kJ/mol, is very likely; the heat of formation, $\Delta_f H^\circ = 4217$ kJ/mol, was also obtained [5].

NH^{3+}. The charge transfer reaction between N^{3+} ions and H atoms at thermal energies was studied theoretically by an ab initio CI calculation of the possible potential energy curves; the reaction was found to take place mainly via NH^{3+} $(1s\sigma)^2$ $(2s\sigma)^2$ $(3s\sigma)^1$, $^2\Sigma^+$ with three probable dissociation limits to H^+ and, respectively, $N^{2+}(2s^2\ 3s,\ ^2S)$, $N^{2+}(2s\ 2p^2,\ ^2S)$, and $N^{2+}(2s\ 2p^2,\ ^2D)$ [6].

NH^{4+}. The charge transfer reaction between N^{4+} ions and H atoms at kinetic temperatures of 30 to 10^5 K was studied theoretically; for the "low energy" regime (appropriate for astrophysics), Hartree-Fock and IVO (Improved Virtual Orbital) calculations result in NH^{4+} $(1s\sigma)^2$ $(2s\sigma)^1$ $(3\sigma)^1$, $^1\Sigma$ and $^3\Sigma$ states that dissociate into $N^{4+}(1s^2\,2s,\,^2S)+H$, $N^{3+}(1s^2\,2s\,3s,\,^{1,3}S)+H^+$, $N^{3+}(1s^2\,2s\,2p,\,^{1,3}P^o)+H^+$, and $N^{3+}(1s^2\,2s\,3d,\,^{1,3}D)+H^+$[7].

NH^{5+}. Electron capture in collisions of N^{5+} ions with H atoms has been theoretically investigated by employing quantum-chemical (valence-bond Cl) and semiclassical MO expansion methods; for the transient quasi-molecule NH^{5+}, thirteen repulsive Σ and Π states have been found that correlate with N^{4+}(3s, 3p, 3d, 4s, 4p, 4d, 4f)$+H^+$ and, in one case, with $N^{5+}+H$ [8].

NH^{6+}. In a theoretical study with the aim of interpreting Hund's rule for molecular systems, SCF calculations were carried out for the low-lying singlet and triplet states $^{1,3}\Sigma^+$ of two-electron, first-row hydride cations including NH^{6+} [9].

References:

[1] Proctor, C. J.; Porter, C. J.; Ast, T.; Bolton, P. D.; Beynon, J. H. (Org. Mass Spectrom. **16** [1981] 454/8).
[2] Hamdan, M.; Mazumdar, S.; Marathe, V. R.; Badrinathan, C.; Brenton, A. G.; Mathur, D. (J. Phys. B At. Mol. Opt. Phys. **21** [1988] 2571/84).
[3] Pope, S. A.; Hillier, I. H.; Guest, M. F. (Faraday Symp. Chem. Soc. No. 19 [1984] 109/23).
[4] Pope, S. A.; Hillier, I. H.; Guest, M. F.; Kendric, J. (Chem. Phys. Lett. **95** [1983] 247/9).
[5] Koch, W.; Schwarz, H. (Int. J. Mass Spectrom. Ion Processes **68** [1986] 49/56).
[6] Heil, T. G.; Butler, S. E.; Dalgarno, A. (Phys. Rev. [3] A **27** [1983] 2365/83).
[7] Feickert, C. A.; Blint, R. J.; Surratt, G. T.; Watson, W. D. (Astrophys. J. **286** [1984] 371/6).
[8] Shimakura, N.; Kimura, M. (Phys. Rev. [3] A **44** [1991] 1659/67).
[9] Liu, S.-B.; Liu, X.-Y.; Yang, Q.-S.; Yu, X.-Y. (J. Mol. Struct. **251** [1991] 271/81 [THEOCHEM **83**]).

2.1.5 The Imide Ion, NH^-

CAS Registry Numbers: NH^- *[23841-33-0]*, $^{15}NH^-$ *[107145-37-9]*

Formation

Gas Phase. For laser spectroscopic studies (for photodetachment at 488 nm and high-resolution IR absorption and autodetachment spectra of NH^-, see below), beams of NH^- ions were generated in hot-cathode discharge sources containing gaseous NH_3 [1] or HN_3 [2 to 10]. Electron-impact excitation of NH_3 and ND_3 results in NH^- (ND^-) formation by dissociative electron attachment; the maxima of the ion currents measured as a function of the electron energy (up to 14 eV) were observed at ~10.2 and ~10.6 eV for NH^- and ND^- [11] and at ~10.5 eV for NH^- [12]. NH^- ions are also formed from HN_3 by resonance capture of electrons with an appearance potential of 0.8 ± 0.3 eV [13]. NH^- ions could be observed in a sputter source for negative ions, in which a K^+ ion beam was accelerated towards a titanium hollow cone containing NH_3 gas; attempts to generate a beam of NH^- ions were unsuccessful, however [14]. The mass spectrum of NH_3 irradiated with 100-eV electrons at a pressure of 2×10^{-7} Torr shows 0.4% negative ions (H^-, NH^-, and NH_2^-) as primary products; the relative abundance of NH^- is 0.13 compared to 100 for H^-. At a pressure of 1 Torr, where secondary ion-molecule reactions take place, the NH^- ion was observed with an abundance of 0.24% compared to ~52% for NH_3^-, ~31% for H^-, and ~16% for NH_2^- [15].

Solution. NH^- ions are formed by γ radiolysis of "alkaline" NH_3 solutions, i.e., liquid NH_3 containing $NaNH_2$ or KNH_2. The NH_2^- ion can cause ionization of the NH_2 radical (from the primary radiolytic process) according to $NH_2 + NH_2^- \rightarrow NH^- + NH_3$ [16, 17].

Solid. Adsorption of H atoms on the surface of solid NaN_3 at 77 K and subsequent reaction of the H atoms with the solid after slightly heating to ~125 K was studied by electron spin resonance; the paramagnetic species observed at ~125 K was identified as the NH^- radical anion, and the formation process $H + NaN_3 \rightarrow NaN_3 \cdot H_{ads} \rightarrow NaNH + N_2$ was postulated [18]. A paramagnetic species, detected by ESR in γ-irradiated $(NH_4)_2HPO_4$ single crystals and characterized by hyperfine lines arising from one N and one H nucleus (see p. 135) [19], has been conjectured to be NH^+ [19] or NH^- [20]; an ab initio UHF calculation supported the assignment to NH^+ [21].

Heat of Formation

$\Delta_f H_{298}(NH^-) = 340.21$ kJ/mol is given in a collection of critically evaluated data on the heats of formation of positive and negative ions in the gas phase; the value was derived from the thermochemical relation $\Delta_f H(NH^-) = \Delta_{acid}H(NH_2) - \Delta_f H(NH_2) + \Delta_f H(H^+)$, where the acidity of NH_2 is given by $\Delta_{acid}H(NH_2) = D(HN\text{-}H) + E_i(H) - EA(NH)$; the electron affinity value $EA(NH) = 0.381 \pm 0.014$ eV (from laser spectroscopic studies of NH^- [2]) and literature data for the remaining $\Delta_f H$'s, for the bond dissociation energy D of NH_2, and for the ionization potential E_i of H were used [22]. A larger, but less reliable value, $\Delta_f H \approx 380$ kJ/mol, is based on the appearance potential of NH^- from HN_3 under electron impact (see above) and on the assumption of dissociative electron attachment, $HN_3 + e^- \rightarrow NH^- + N_2$ [13].

Electronic Structure

The NH^- ion is isoelectronic with the extensively studied OH radical. An inverted ground state X $^2\Pi_i$ correlating with $N^-(^3P) + H(^2S)$ and represented by the electron configuration $(1\sigma)^2\ (2\sigma)^2\ (3\sigma)^2\ (1\pi)^3$ (i.e., attachment of an electron into the half-filled highest MO of the NH(X $^3\Sigma^-$) radical, cf. p. 31) and the excited valence states $(1\sigma)^2\ (2\sigma)^2\ (3\sigma)^1\ (1\pi)^4$ A $^2\Sigma^+$, $(1\sigma)^2\ (2\sigma)^2\ (3\sigma)^2\ (1\pi)^2\ (4\sigma)^1$ $^4\Sigma^-$, $^2\Sigma^-$, $^2\Delta$ are predicted by MO theory. The ground state was observed by IR autodetachment spectroscopy in the region of the fundamental vibration-rotation band, and a number of molecular constants were derived (see below) [3, 5 to 10]. The ground state has also been the subject of a few ab initio SCF MO [21, 23 to 27], PNO-CI and CEPA [28], MP4 [29, 30], combined MP4 and CI (G1, G2) [31, 32], and a series of MP2 to MP4 and multireference CI (CIPSI, MR-CISD) [33] calculations. MCSCF-CI and SCEP-CEPA calculations for the ground state and MCSCF-CI calculations for the lowest excited state A $^2\Sigma^+$ deal with the potential curves of both states (see below) and predict a term value $T_e = 29153\ cm^{-1}$ for the bound, but still unobserved A state [34]. MBPT (many-body perturbation theory) calculations on the five lowest valence states of NH^- at $r = 1.9614\ a_0$, the equilibrium internuclear distance of NH(X $^3\Sigma^-$), give the vertical excitation energies ΔE(A $^2\Sigma^+$, $^4\Sigma^-$, $^2\Sigma^-$, $^2\Delta \leftarrow$ X $^2\Pi_i$) = 4.53, 8.21, 9.65,10.40 eV, respectively [35].

Ionization Potential

Photodetachment of NH^- at 488 nm in a crossed ion beam-laser beam apparatus [1, 2] and autodetachment of NH^- upon excitation of the fundamental vibration-rotation band around 3000 cm^{-1} in a coaxial ion beam-laser beam spectrometer [3, 6, 7, 10] yielded the energies for ionizing $^{14}NH^-$ and $^{15}NH^-$ to $^{14,15}NH(X\ ^3\Sigma^-)$ and $^{14}NH^-$ to $^{14}NH(a\ ^1\Delta)$. The values are compiled in the Section "Electron Affinity" of NH, p. 38.

Proton Affinity

An "experimental" value, $PA_0 = 1659 \pm 19$ kJ/mol, was calculated from the relation $PA_0(NH^-) = \Delta_fH_0(H^+) + \Delta_fH_0(NH^-) - \Delta_fH_0(NH_2)$ using data from recent thermochemical compilations; the difference of the total molecular energies of NH^- and NH_2 (Møller-Plesset perturbation calculation MP4) gives $PA_e = 1698$ kJ/mol and, after correcting for the zero-point energies, $PA_0 = 1668$ kJ/mol [30].

Dipole Moment

The dipole moment functions $\mu(r)$ for the ground state X $^2\Pi_i$ and the excited state A $^2\Sigma^+$ at r = 1.5 to 2.9 a_0 were derived by an MCSCF-CI calculation; at r = 2.0 a_0, the approximate equilibrium internuclear distance of both states, values of $\mu(X\ ^2\Pi_i) \approx 0.24\ ea_0$, $\mu(A\ ^2\Sigma^+) \approx 0.20\ ea_0$ (calculated with the N atom at the center of the coordinate system) and $\mu(X\ ^2\Pi_i) \approx 0.36\ ea_0$ (relative to the center-of-mass of the NH^- ion) were obtained; the positive values demonstrate charge flux from the N towards the H atom (μ becomes negative at $r > 2.5\ a_0$) [34].

Spectroscopic Constants

Hyperfine Coupling Constants. Hyperfine structure (hfs) due to dipole-dipole interaction between the nuclear magnetic moments of ^{14}N (I = 1), ^{15}N (I = 1/2), and ^1H (I = 1/2) and the electron spin (S = 1/2) magnetic moment was observed in the Q branch and in the lowest J transitions of the P and R branches of the fundamental vibration-rotation spectrum of the X $^2\Pi_i$ electronic ground state. An analysis of the $Q_2(1/2, e \rightarrow f)$ transition, which shows the best-resolved hfs, resulted in the following Frosch-Foley [36] hyperfine coupling parameters for ^{14}N, ^{15}N, and ^1H in $^{14}NH^-$ and $^{15}NH^-$ [8, 9], [10, pp. 92/132, 125] (parameters in MHz):

nucleus	parameter	in $^{14}NH^-$	in $^{15}NH^-$
N	a − (b + c)/2	79.89	−111.72
	d	107.61	−150.525
H	a − (b + c)/2	50.79	50.87
	d	44.47	44.86

Fine-Structure Constants. Analysis of 115 transitions in the P, Q, and R branches of the fundamental band of $^{14}NH^-$ [7], [10, p. 85] and of 52 Q- and R-branch transitions in the fundamental band of $^{15}NH^-$ [6], [10, p. 85] gave the **spin-orbit coupling constants** A_v for the lowest vibrational levels v = 0 and 1 of the inverted electronic ground state X $^2\Pi_i$. Slightly higher are the values derived in the first study of the fundamental vibration-rotation spectrum of $^{14}NH^-$ (48 R-branch transitions) by autodetachment spectroscopy [3]. The following results were obtained:

ion	A_0 in cm^{-1}	A_1 in cm^{-1}	Ref.
$^{14}NH^-$	−48.4264(370)	−48.8737(386)	[7], [10, p. 85]
	−48.83(11)	−49.34(14)	[3]
$^{15}NH^-$	−48.4074(210)	−48.8418(225)	[6], [10, p. 85]

Thus, the value $A(X\ ^2\Pi_i) = -63\ cm^{-1}$ estimated by the isoelectronic extrapolation $A(NH^-, X\ ^2\Pi_i)/A(NH, A\ ^3\Pi_i) \approx A(OH, X\ ^2\Pi_i)/A(OH^+, A\ ^3\Pi_i)$ [2] and quoted by Huber and Herzberg [37] is obsolete.

Theoretical results for $^{14}NH^-(A\ ^2\Pi_i)$ in the equilibrium position, $A_e = -57.8\ cm^{-1}$ [26] and $-53.3\ cm^{-1}$ [27], were calculated using the ab initio near-Hartree-Fock wave function of [23, 24], and $|A_e| = 58.20\ cm^{-1}$ [38] results from MCSCF calculations for $^2\Pi$ and $^3\Pi$ states of first-row hydrides XH^+, XH, XH^- using effective nuclear charges Z_{eff} for the X atoms (which, however, are based on the experimental A values quoted in [37]).

The lower J transitions of the fundamental vibration-rotation spectra of $^{14}NH^-$ and $^{15}NH^-$, both in the electronic ground state $X\ ^2\Pi_i$, show well-resolved **Λ-type doubling**; the following constants p_v and q_v (in cm^{-1}) for the vibrational levels v = 0 and 1 were derived:

ion	$10^2 \cdot p_0$	$10^2 \cdot q_0$	$10^2 \cdot p_1$	$10^2 \cdot q_1$	Ref.
$^{14}NH^-$	9.219(66)	−2.1214(109)	8.413(63)	−1.8167(88)	[7], [10, p. 85]
	8.9(4)	−2.13(5)	8.1(4)	−1.82(4)	[3]
$^{15}NH^-$	9.338(81)	−2.129(21)	8.619(76)	−1.806(25)	[6], [10, p. 85]

Rotational and Vibrational Constants. Internuclear Distance. Analysis of the IR spectra of $^{14}NH^-$ and $^{15}NH^-$ in the region of the fundamental vibration-rotation band around 3000 cm^{-1}, observed by high-resolution autodetachment spectroscopy, gave for each isotopic species the rotational constant B_e, the centrifugal stretching constants D_e and H_e, and the corresponding rotation-vibration interaction constants α_e, β_e, and η_e. The vibrational constants ω_e and $\omega_e x_e$, derived from the band origins T_1 $(= G(1) - G(0) = \omega_e - 2\omega_e x_e)$ for $^{14}NH^-$ and $^{15}NH^-$ and using isotopic relations, and the equilibrium internuclear distances r_e, obtained by converting the rotational constants B_e, depend on the charge localization in the anion (because of the different reduced masses of N^-H and NH^- resulting from the placement of the extra electron mass). The results from more recent high-resolution IR studies of Miller et al. [6, 7], [10, pp. 87/90] and the less precise results from the pioneering work of Neumark et al. [3] are given in Table 8 together with some theoretical data obtained from highly correlated ab initio wave functions [28, 34]. Another experimental result for the internuclear distance was derived from a Franck-Condon factor analysis of the photodetachment transitions $NH^-(X\ ^2\Pi_i) \rightarrow NH(X\ ^3\Sigma^-)$, $NH(a\ ^1\Delta)$ with the known r_e of $NH(X\ ^3\Sigma^-)$ to be $r_e(^{14}NH^-, X\ ^2\Pi_i) = 1.047 \pm 0.002$ Å [2]. However, this value seems to be misprinted according to [28] and was proposed to be corrected to 1.037 ± 0.01 Å.

For the lowest excited (unobserved) state $A\ ^2\Sigma^+$, predicted to be bound, the spectroscopic constants $B_e = 16.57\ cm^{-1}$, $\alpha_e = 0.739\ cm^{-1}$, $\omega_e = 3157\ cm^{-1}$, $\omega_e x_e = 85\ cm^{-1}$, and $r_e = 1.040$ Å were derived by an ab initio MCSCF-CI calculation [34].

Table 8
$NH^-(X\ ^2\Pi_i)$. Rotational and Vibrational Constants in cm^{-1}, Internuclear Distance r_e in Å.

constant	experimental $^{14}NH^-$ [7, 10]	$^{15}NH^-$ [6, 10]	$^{14}NH^-$ [3]	theoretical $^{14}NH^-$ [34][a)]	[34][b)]	[28][c)]
B_e	16.57971(46)	16.50566(25)	16.607	16.41	16.48	16.61
α_e	0.70618(14)	0.70191(30)	0.712	0.876	0.731	0.691
$D_e \cdot 10^3$	1.7879(41)	1.7623(48)	−	−	−	−
$\beta_e \cdot 10^5$	−1.67(11)	−1.546(54)	−	−	−	−

Table 8 (continued)

constant	experimental $^{14}NH^-$ [7, 10]	$^{15}NH^-$ [6, 10]	$^{14}NH^-$ [3]	theoretical $^{14}NH^-$ [34]^a)	[34]^b)	[28]^c)
$H_e \cdot 10^7$	1.83(7)	{1.805}^d)	—	—	—	—
$\eta_e \cdot 10^8$	−1.56(4)	{−1.54}^d)	—	—	—	—
ω_e	3191.52^e) 3193.15^f)	3184.40^e) 3186.01^f)	g)	3155	3173	3226
$\omega_e x_e$	85.56^e) 86.38^f)	85.18^e) 85.99^f)	g)	88	89	85
r_e	1.039940(14)^e) 1.039678(14)^f)	1.039942(8)^e) 1.039678(8)^f)	1.039	1.043	1.045	1.039

a) MCSCF-CI. — b) SCEP-CEPA. — c) CEPA. — d) Calculated from the constants of $^{14}NH^-$. — e) With charge localized on N. — f) With charge localized on H. — g) Term value of first vibrational level: $T_1 = \omega_e - 2\omega_e x_e = 3020.36(1)\ cm^{-1}$.

Potential Energy Functions

The potential energy curves for the ground state X $^2\Pi_i$ and the lowest excited state A $^2\Sigma^+$ obtained by an ab initio MCSCF-CI calculation [34] are depicted in **Fig. 5** together with the potential energy curve for the neutral NH molecule in its ground state X $^3\Sigma^-$ from a CEPA calculation [39].

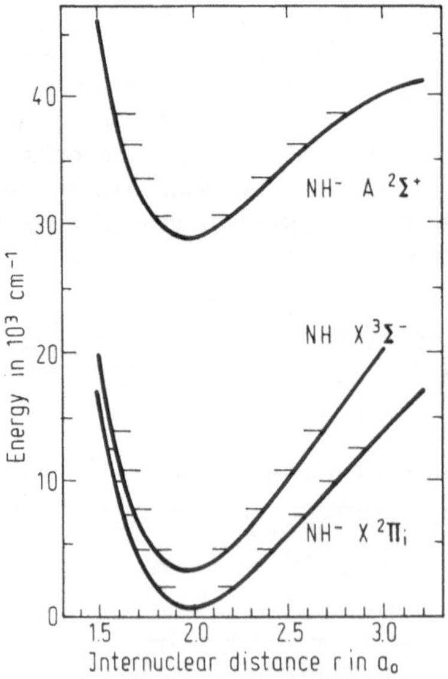

Fig. 5. MCSCF-CI potential energy functions for the X $^2\Pi_i$ and A $^2\Sigma^+$ states of NH^- [34]. The potential energy function of the X $^3\Sigma^-$ state of NH has been taken from [39] and shifted to match the experimental electron affinity of NH [3].

Dissociation Energy

An "experimental" result, $D_0(N^--H) = 3.85 \pm 0.15$ eV, was derived [28] via a cycle from the electron affinities of N (−0.07 eV) [40] and NH (0.38 eV) [2] and a recommended D_0(N-H) (3.40 eV) [39]; the dissociation energy $D_0(N-H^-)$ is about 0.8 eV lower (EA(H) = 0.75 eV [40]). Theoretical $D_0(N^--H)$ values are 3.92 eV (MP4) [29], 3.51 eV (PNO-CI), and 3.60 eV (CEPA) [28].

Spectra

IR Spectrum. Laser autodetachment spectroscopy has been used to measure the vibration-rotation spectra of $^{14}NH^-$ and $^{15}NH^-$ in the region of the fundamental vibration ($v = 1 \leftarrow 0$) around 3000 cm^{-1}. This technique, which is very appropriate for spectroscopic studies of negative ions [4, 41], was applied to extract an NH^- ion beam from a hot-cathode discharge source containing HN_3 and coaxially merge it with a color-center laser beam. The photons excite the fundamental N-H vibration and populate vibrational-rotational levels that are energetically higher than the ground state of the neutral NH molecule (EA(NH) = 0.374 eV ≙ 3017 cm^{-1}, see p. 38; see also Fig. 5) which leads to autodetachment. By monitoring the remaining fast neutral molecules or electrons as a function of laser energy, the vibration-rotation transitions of NH^- appear as sharp features on a slowly varying background arising from direct photodetachment.

The observed rotational fine structure of the fundamental band is intermediate between Hund's case (a) and Hund's case (b) coupling of angular momenta. Spin-orbit splitting into the F_1 ($\Omega = 3/2$) and F_2 ($\Omega = 1/2$) manifolds of the inverted ground state X $^2\Pi_i$ was observed for transitions at low J (case (a)) followed by a rapid convergence of the F_1/F_2 ($J = N \pm 1/2$) doublets with increasing J due to the uncoupling of the electronic spin from the internuclear axis (case(b)). Furthermore, Λ-type doubling of the rotational levels into two sublevels, labeled e (for parity $(-1)^{J-1/2}$) and f (for parity $-(-1)^{J-1/2}$), and transitions according to the selection rules $e \rightarrow f$ or $f \rightarrow e$ for $\Delta J = 0$ and $e \rightarrow e$ or $f \rightarrow f$ for $\Delta J = \pm 1$ were observed.

The first IR autodetachment study of NH^- and of any molecular anion was carried out by Neumark et al. [3] using an F-center laser in the spectral range 3019 to 3326 cm^{-1} (3.312 to 3.020 μm). With a resolution of better than 20 MHz (0.0007 cm^{-1}) allowing the resolution of spin-orbit splitting and Λ-type doubling, they measured 48 R-branch transitions, R_1(J, $f \rightarrow f$ and $e \rightarrow e$) with J = 3/2 to 27/2 (the J = 27/2, $e \rightarrow e$ line was later discarded by [7]) and R_2(J, $f \rightarrow f$ and $e \rightarrow e$) with J = 1/2 to 25/2 (the J = 25/2, $f \rightarrow f$ line was later assigned to the F_1 manifold by [7]). At low resolution, four Q-branch transitions (J = 1/2 to 5/2) were also observed, whereas the P-branch transitions lie beyond the low end of the laser frequency range. Analysis of the linewidths gave an insight into the dynamics of the autodetachment process and indicated a propensity for autodetachment out of the upper Λ-doublet levels of NH^-(v = 1) [3]. This observation was accounted for by a first-order perturbation theory model [42].

Subsequent extensive studies by Miller et al. [5 to 10] in the frequency range 2941 to 3326 cm^{-1} (≙ 3.40 to 3.01 μm; tunable Li:RbCl color-center laser) and at a higher resolution of 7 MHz enabled the measurement (accuracy of 0.01 cm^{-1}) and assignment of 115 R-, Q-, and P-branch lines (J up to 71/2, 21/2, and 5/2, respectively) of $^{14}NH^-$ [5, 7], [10, pp. 48/91] and of 52 R-, Q-, and P-branch lines (J up to 37/2, 13/2, and 3/2, respectively) of $^{15}NH^-$ [6], [10, pp. 48/91] with resolved spin-orbit splitting and, at low J values, Λ-type doubling for both isotopic species. The linewidths of the higher-energy Λ-doublet transitions, terminating on the upper Λ doublet of the v = 1 levels, increase rapidly with increasing

J, while those of the transitions terminating on the lower Λ doublet of the v=1 levels are narrow and independent of J. It may be noted that the rotational levels of the v=0 state with J>27/2 have energies above the autodetachment threshold and that these rotationally excited metastable levels were observed up to J=71/2, which is 1.8 eV into the autodetachment continuum. The hyperfine structure (hfs) due to dipole-dipole interaction between the nuclear magnetic moments of ^{14}N (I=1), ^{15}N (I=1/2), and ^{1}H (I=1/2) and that of the electron spin (S=1/2) could be resolved in the Q branch and in the lowest J transitions of the P and R branches of the $^{14}NH^-$ and $^{15}NH^-$ spectra; the Q_2(1/2, e→f) transitions of $^{14}NH^-$ and $^{15}NH^-$, which show the best-resolved hfs, have been analyzed [8, 9], [10, pp. 92/132].

The rotational, vibrational, fine-structure, and hyperfine-structure constants for the NH^- ion, derived from the high-resolution IR autodetachment spectra, are given above; the electron affinity of the NH molecule is given on p. 38.

A theoretical study (ab initio MCSCF-CI calculation) of the ground state X $^2\Pi_i$ and the first excited valence state A $^2\Sigma^+$ of NH^- gives the rotationless dipole matrix elements for transitions between low-lying vibrational levels of these states, R_v^v, R_v^{v+1}, R_v^{v+2} with v up to 3, and Einstein coefficients of spontaneous IR emission in the X and A states, A_v^{v+1}, A_v^{v+2} with v up to 2, all for NH^- and ND^- [34].

UV Spectrum. The as yet unobserved A $^2\Sigma^+$↔X $^2\Pi_i$ transition has been predicted to occur at ν_{00}=29155 cm^{-1} (≙343 nm; for T_e and the vibrational constants of X and A, see above). Its electronic transition moment function at r=1.5 to 2.9 a_0 was obtained by an ab initio MCSCF-CI calculation. From the latter and the potential energy functions for the X and A states (see Fig. 5, p. 156), the radiative transition probabilities, i.e., the radiative lifetimes τ for the two lowest vibrational levels v'=0 and 1 and the absorption oscillator strengths $f_{v',0}$, have been derived for NH^- and ND^- [34]:

	NH^-		ND^-	
	τ in ns	$f_{v',0}$	τ in ns	$f_{v',0}$
v'=0	226	0.377×10^{-2}	221	0.388×10^{-2}
v'=1	263	0.147×10^{-3}	245	0.947×10^{-4}

References:

[1] Celotta, R. J.; Bennett, R. A.; Hall, J. L. (J. Chem. Phys. **60** [1974] 1740/5).
[2] Engelking, P. C.; Lineberger, W. C. (J. Chem. Phys. **65** [1976] 4323/4).
[3] Neumark, D. M.; Lykke, K. R.; Andersen, T.; Lineberger, W. C. (J. Chem. Phys. **83** [1985] 4364/73).
[4] Lykke, K. R.; Murray, K. K.; Neumark, D. M.; Lineberger, W. C. (Philos. Trans. R. Soc. London A **324** [1988] 179/96, 184/8).
[5] Al-Za'al, M.; Miller, H. C.; Farley, J. W. (Chem. Phys. Lett. **131** [1986] 56/9).
[6] Miller, H. C.; Farley, J. W. (J. Chem. Phys. **86** [1987] 1167/71).
[7] Al-Za'al, M.; Miller, H. C.; Farley, J. W. (Phys. Rev. A Gen. Phys. **35** [1987] 1099/112).
[8] Miller, H. C.; Al-Za'al, M.; Farley, J. W. (Phys. Rev. Lett. **58** [1987] 2031/4).
[9] Miller, H. C.; Al-Za'al, M.; Farley, J. W. (AIP Conf. Proc. No. 160 [1987] 354/6).
[10] Miller, H. C. (Diss. Univ. Oregon 1988, pp. 1/243, 48/132; Diss. Abstr. Intern. B **49** [1989] 5366/7).

[11] Compton, R. N.; Stockdale, J. A.; Reinhardt, P. W. (Phys. Rev. [2] **180** [1969] 111/20).
[12] Marx, R.; Mauclaire, G. (Int. J. Mass Spectrom. Ion Phys. **10** [1972/73] 213/26).

[13] Franklin, J. L.; Dibeler, V. H.; Reese, R. M.; Krauss, M. (J. Am. Chem. Soc. **80** [1957] 298/302).
[14] Middleton, R.; Adams, C. T. (Nucl. Instrum. Methods **118** [1974] 329/36).
[15] Melton, C. E. (J. Chem. Phys. **45** [1966] 4414/24).
[16] Dainton, F. S.; Skwarski, T.; Smithies, D.; Wezranowski, E. (Trans. Faraday Soc. **60** [1964] 1068/86).
[17] Belloni, J.; Saito, E. (Electrons Fluids Nat. Met.-Ammonia Solutions 3rd Colloq. Weyl, Kibbutz-Hanita, Israel, 1972 [1973], pp. 461/71; C.A. **82** [1975] No. 118139).
[18] Vasil'ev, A. A.; Lisetskii, V. N.; Kulikov, N. F.; Savel'ev, G. G. (Khim. Vys. Energ. **21** [1987] 189/90; High Energy Chem. [USSR] **21** [1987] 154/5).
[19] Morton, J. R. (J. Phys. Chem. Solids **24** [1963] 209/12).
[20] Atkins, P. W.; Symons, M. C. R. (The Structure of Inorganic Radicals, Elsevier, London 1967, pp. 107/8).

[21] Claxton, T. A. (Trans. Faraday Soc. **66** [1970] 1540/3).
[22] Lias, S. G.; Bartmess, J. E.; Liebman, J. F.; Holmes, J. L.; Levin, R. D.; Mallard, W. G. (J. Phys. Chem. Ref. Data **17** Suppl. 1 [1988] 1/861, 781).
[23] Cade, P. E.; Huo, W. (At. Data Nucl. Data Tables **12** [1973] 415/66, 434).
[24] Cade, P. E. (Proc. Phys. Soc. [London] **91** [1967] 842/54).
[25] Deakyne, C. A. (AD-A133656 [1983] 39 pp. from Gov. Rep. Announce. Index [U. S.] **84** No. 3 [1984] 63; C.A. **100** [1984] No. 162045).
[26] Walker, T. E. H.; Richards, W. G. (Symp. Faraday Soc. **1968** 64/8).
[27] Trivedi, H. P.; Richards, W. G. (J. Chem. Phys. **72** [1980] 3438/9).
[28] Rosmus, P.; Meyer, W. (J. Chem. Phys. **69** [1978] 2745/51).
[29] Frenking, G.; Koch, W. (J. Chem. Phys. **84** [1986] 3224/9).
[30] Pople, J. A.; Schleyer, P. von R.; Kaneti, J.; Spitznagel, G. W. (Chem. Phys. Lett. **145** [1988] 359/64).

[31] Pople, J. A.; Head-Gordon, M.; Fox, D. J.; Raghavachari, K.; Curtiss, L. A. (J. Chem. Phys. **90** [1989] 5622/9).
[32] Curtiss, L. A.; Raghavachari, K.; Trucks, G. W.; Pople, J. A. (J. Chem. Phys. **94** [1991] 7221/30).
[33] Novoa, J. J.; Mota, F.; Arnau, F. (J. Phys. Chem. **95** [1991] 3096/105).
[34] Mänz, U.; Zilch, A.; Rosmus, P.; Werner, H.-J. (J. Chem. Phys. **84** [1986] 5037/44).
[35] Sun, H.; Freed, K. F. (J. Chem. Phys. **76** [1982] 5051/9).
[36] Frosch, R. A.; Foley, H. M. (Phys. Rev. [2] **88** [1937] 1337/49).
[37] Huber, K. P.; Herzberg, G. (Molecular Spectra and Molecular Structure, Vol. 4, Constants of Diatomic Molecules, Van Nostrand Reinhold, New York 1979, pp. 460/1).
[38] Koseki, S.; Schmidt, M. W.; Gordon, M. S. (J. Phys. Chem. **96** [1992] 10768/72).
[39] Meyer, W.; Rosmus, P. (J. Chem. Phys. **63** [1975] 2356/75).
[40] Hotop, H.; Lineberger, W. C. (J. Phys. Chem. Ref. Data **4** [1975] 539/76, 542, 547).

[41] Andersen, T. (Phys. Scr. T **34** [1991] 23/35).
[42] Chalasinski, G.; Kendall, R. A.; Taylor, H.; Simons, J. (J. Phys. Chem. **92** [1988] 3986/91).

2.1.6 The NH^{2-} Ion

CAS Registry Number: *[32323-01-6]*

There is no experimental proof yet available for the existence of a free NH^{2-} ion. A few quantum-chemical calculations were carried out on the ground state X $^1\Sigma^+$ $(1\sigma)^2$ $(2\sigma)^2$ $(3\sigma)^2$ $(1\pi)^4$, mostly however with the intent to test computational methods (cf. bibliography of quantum-chemical calculations, p. 31). A heat of formation of $\Delta_fH^\circ_{298} = 1396$ kJ/mol was obtained from an ab initio SCF MO calculation for the closed-shell reaction 1/2 N_2 + 3/2 H_2 → NH^{2-} + 2 H^+ [1]. Energy differences between NH^{2-} and NH_2^- from SCF MO [2, 3] and simpler calculations [4, 5] gave values for the proton affinity.

References:

[1] Hopkinson, A. C.; Yates, K.; Csizmadia, I. G. (Theor. Chim. Acta **23** [1971/72] 369/77).
[2] Kozmutza, C.; Kapuy, E.; Robb, M. A.; Daudel, R.; Csizmadia, I. G. (J. Comput. Chem. **3** [1982] 14/22).
[3] Hopkinson, A. C.; Holbrook, N. K.; Yates, K.; Csizmadia, I. G. (J. Chem. Phys. **49** [1968] 3596/601).
[4] Tamassy-Lentei, I.; Szaniszlo, J. (Acta Phys. **35** [1974] 201/11).
[5] Kruglyak, Yu. A.; Sapiro, I. L. (Zh. Strukt. Khim. **7** [1966] 262/6; J. Struct. Chem. [USSR] **7** [1966] 254/8).

2.1.7 The Amidogen Radical, NH_2

Other names: Nitrogen dihydride, amino, aminyl
CAS Registry Numbers: NH_2 *[13770-40-6]*, $^{15}NH_2$ *[15021-32-6]*, ND_2 *[15117-84-7]*, $^{15}ND_2$ *[28486-56-8]*, $^{13}NH_2$ *[106083-60-7]*

2.1.7.1 Production and Detection of NH_2 Radicals

Walter Hack
Max-Planck-Institut für Strömungsforschung
Göttingen

The experimental methods that are applied to study radical reactions in general are also used for NH_2 radicals. These include photolysis systems, shock tube experiments, and isothermal flow reactors and are described in great detail in several publications. Here only NH_2-specific problems will be discussed. Elementary reactions of NH_2 radicals can be part of complex reaction systems, like nitrogen-containing fuel flames, or the NH_2 radicals are present due to specific laboratory-based production reactions in photolysis or flow reactors. In the latter case, NH_2 is produced in elementary reactions. In complex and in isolated systems, it is useful to detect NH_2 directly.

2.1.7.1.1 NH_2 Radical Sources in the Gas Phase

NH_2 radicals play some role in the environment. They are generated by photochemical processes in the atmosphere and by combustion processes. Moreover, $NH_2(\tilde{X}\ ^2B_1)$ can be produced by many chemical reactions and via energy disposal in appropriate precursor molecules.

Among the chemical reactions,

$$OH + NH_3 \rightarrow NH_2 + H_2O$$

is widespread (see pp. 231, 233). This reaction has been the subject of many experimental studies [1 to 16]. Recently, an ab initio study for the reaction $OH + NH_3 \rightleftharpoons NH_2 + H_2O$ was also published [17]. The experimental methods applied include discharge flow reactors, photolysis, shock tube experiments, and flame simulations; thus, a wide temperature range between 298 and 2360 K is experimentally covered. Assuming Arrhenius behavior, an activation energy of $E_A = 8$ kJ/mol and a preexponential factor of $k = 3 \times 10^{12}$ $cm^3 \cdot mol^{-1} \cdot s^{-1}$ is a good approximation between room temperature and 1000 K. Over the entire temperature range, a three-parameter fit, $k(T) = 9.6 \times 10^6 \ T^{1.8} \exp(-2.1 \ kJ \cdot mol^{-1}/RT)$ $cm^3 \cdot mol^{-1} \cdot s^{-1}$ [15], has been recommended.

The reaction $OH + NH_3 \rightarrow NH_2 + H_2O$ can be used as an NH_2 source in kinetic studies; it produces besides the desired NH_2 radical only H_2O. The reaction with a room-temperature rate constant of about 1×10^{11} $cm^3 \cdot mol^{-1} \cdot s^{-1}$ is fast enough to suppress the consecutive reaction $NH_2 + OH \rightarrow$ products ($k = 1.1 \times 10^{13}$ $cm^3 \cdot mol^{-1} \cdot s^{-1}$, see below) by a sufficiently high excess of NH_3 over OH. This reaction has the disadvantage that another chemical reaction, $F + H_2O \rightarrow OH + HF$ or $H + NO_2 \rightarrow OH + NO$, or a photolysis process, $H_2O_2 \xrightarrow{h\nu} 2\,OH$, $NHO_3 \xrightarrow{h\nu} OH + NO_2$, is needed to produce OH radicals. The reaction $OH + NH_3$ plays an important role in flames (see p. 231).

The reaction

$$H + NH_3 \rightarrow NH_2 + H_2$$

is an important NH_2 radical source only in high-temperature systems, in particular in fuel-rich, ammonia-seeded flames. The room-temperature rate constant $k(300\ K) \cong 7 \times 10^3$ $cm^3 \cdot mol^{-1} \cdot s^{-1}$ indicates that this reaction can only contribute at high temperatures. The reaction rate for the reactions $H + NH_3$ and $D + ND_3$ was determined with three different techniques: flash photolysis shock tube [18 to 21], high-temperature flow reactor [22], and high-temperature photochemistry [23 to 25]. In the narrow temperature range 500 to 1000 K, the reaction rate can be described by an Arrhenius expression with an activation energy of $E_A = 60$ kJ/mol and a preexponential factor of 10^{14} $cm^3 \cdot mol^{-1} \cdot s^{-1}$. Over the temperature range 500 to 1800 K, the deviation from linear Arrhenius behavior becomes significant and a three-parameter fit, $k(T) = 5.4 \times 10^5 \ T^{2.4} \exp(-41.5 \ kJ \cdot mol^{-1}/RT)$ $cm^3 \cdot mol^{-1} \cdot s^{-1}$, should be used. The activation energy for the deuterium reaction $D + ND_3$, measured in the same temperature range and with the same method as $H + NH_3$ [23, 24], is only insignificantly higher (<1%); in these studies the ratio of the preexponential factors was found to be $k(D)/k(H) = 0.57$. The agreement between the theoretically calculated rate constants for $H + NH_3$ and $D + ND_3$ is good, i.e. the computed barrier height is overestimated by about 4 kJ/mol [26].

The reaction

$$O(^3P) + NH_3 \rightarrow NH_2 + OH,$$

which may be important in ammonia flames under lean conditions, has been studied with various methods. A recent critical review is given in [27] which recommends a rate constant of $k(T) = 1.1 \times 10^6 \ T^{2.1} \exp(-21.8 \ kJ \cdot mol^{-1}/RT)$ $cm^3 \cdot mol^{-1} \cdot s^{-1}$ at temperatures above 400 K. In a more recent flash photolysis shock tube investigation, a high-temperature value was obtained [28] in good agreement with the recommended value.

The hydrogen abstraction from ammonia by F atoms via

$$F + NH_3 \rightarrow NH_2 + HF$$

is an ideal NH_2 source, since the reaction is fast with $k(T) = 1.5 \times 10^{14} \exp(-5.4 \ kJ \cdot mol^{-1}/RT)$ $cm^3 \cdot mol^{-1} \cdot s^{-1}$ [29]. Even at room temperature ($k(300\ K) = 2 \times 10^{13}$ $cm^3 \cdot mol^{-1} \cdot s^{-1}$) nearly every collision leads to reaction. The reaction is fast enough to generate a supersonic beam of free NH_2 radicals [30]. The consecutive reaction $F + NH_2 \rightarrow NH + HF$ (see p. 225) is also a fast reaction, $k(300\ K) = 2.3 \times 10^{13}$ $cm^3 \cdot mol^{-1} \cdot s^{-1}$ [29], but it can easily be suppressed by a sufficient excess of $[NH_3]$ over [F]. A stationary NH_2 concentration $[NH_2]_{st} = (k(F + NH_3)/k(F + NH_2))\ [NH_3]$ can be obtained, if only these two reactions are taken into account; thus, at room temperature, $[NH_2]_{st} \cong [NH_3]$. This indicates that the consecutive reaction is not the limiting process for the available NH_2 concentration; other NH_2-consuming reactions and the F atom concentration are essential to maximize the available NH_2 concentration. The F atoms are produced in a microwave discharge (F_2/He) in a flow reactor [31, 32] or by pulse radiolysis of SF_6 [33]. The reaction $F+NH_3$ provides a clean NH_2 source, since besides the radical NH_2 wanted only the inert HF molecule is formed. Another hydrogen abstraction reaction, $CF_3+NH_3 \rightarrow CF_3H+NH_2$, also produces NH_2 radicals [34], however without being regarded to be an NH_2 source.

The chemical reactions described up to now all use NH_3 as the precursor molecule, but N_2H_4 [35, 36] can also be used in a two-step mechanism

$$H+N_2H_4 \rightarrow H_2+N_2H_3$$

$$H+N_2H_3 \rightarrow 2\ NH_2$$

to produce NH_2 radicals. The rate constant for the initial step is $k=1.3\times10^{13}$ $\exp(-10.5\ kJ\cdot mol^{-1}/RT)\ cm^3\cdot mol^{-1}\cdot s^{-1}$ and for the second step $k(300\ K)=1.6\times10^{12}$ $cm^3\cdot mol^{-1}\cdot s^{-1}$ [36], i.e. both reactions are fast at room temperature.

The reaction

$$H+HNCO \rightarrow NH_2+CO,$$

which is important in flames (see p. 233), is a fast NH_2-producing process at high temperatures with $k(T)=2.1\times10^{14}\exp(-70\ kJ\cdot mol^{-1}/RT)\ cm^3\cdot mol^{-1}\cdot s^{-1}$ [37], e.g. in shock tube experiments.

The reaction of H atoms with HN_3, which is isoelectronic with HNCO, produces electronically excited, highly vibrational-excited $NH_2(\tilde{A}\ ^2A,\ v_2 \leq 15)$ radicals in a direct elementary process [38].

An interesting way to produce NH_2 radicals is the electronic chemical activation method, starting with the electronically excited species $O(^1D)$ or $NH(a\ ^1\Delta)$ and react them with NH_3. The reaction $O(^1D)+NH_3 \rightarrow NH_2OH^{\ddagger} \rightarrow NH_2+OH$ has been studied by several authors [39 to 44]. $O(^1D)$ atoms were produced by O_3 photolysis either with a conventional flash lamp [40], by exciplex laser photolysis ($\lambda=248$ nm) [42], or by Nd:YAG laser pulses ($\lambda=266$ nm) [41]. Every $O(^1D)$-NH_3 collision leads to reaction, even at room temperature ($k(300\ K)=2\times10^{14}\ cm^3\cdot mol^{-1}\cdot s^{-1}$) [43]. The OH radicals, which are partly vibrationally excited, react in these systems with NH_3 and form additional NH_2 radicals. The reaction pathway $O(^1D)+NH_3 \rightarrow NH(a)+H_2O$, which is also allowed on the electronic singlet surface, contribute about 7% relative to $OH(X\ ^2\Pi,\ v=0)$ yield [41]. (The isovalent reaction $S(^1D)+NH_3$, with $S(^1D)$ obtained by CS_2 photolysis, has also been studied [45].) The $NH(a\ ^1\Delta)$, which is isoelectronic with $O(^1D)$, reacts with NH_3 on the singlet surface $NH(a)+NH_3 \rightarrow N_2H_4^{\ddagger} \rightarrow 2\ NH_2$ and produces $NH_2(\tilde{X})$ radicals [46]. The reaction is fast with $k(300\ K)=8.9\times10^{13}\ cm^3\cdot mol^{-1}\cdot s^{-1}$ [46]. The yield of NH(X) is below 1%. This electronic chemical activation method is an elegant way to produce NH_2 in the absence of other radicals. NH(a) is produced by photolysis of HN_3.

NH_2 radicals are produced via energy disposal in precursor molecules like NH_3 or N_2H_4 and less often amines. The energy is transmitted to the precursor molecule most often as photons (single photon UV, multiphoton IR), but dissociation can also be caused by radiolysis, by electron impact, or energy transfer from excited atoms. In pyrolysis systems, NH_2 radicals are formed by thermal dissociation of the precursor molecules.

The $NH_3(\tilde{X})$ photolysis with a single UV photon is a well-known technique to produce $NH_2(\tilde{X})$ radicals [47]. The results were described in several review publications [48 to 50].

The photolysis of NH_3 has been studied at different levels of sophistication [19, 51 to 67]. An ab initio SCF CI treatment of NH_3 photodissociation is given in [68]. The absorption spectrum of the NH_3 transition $\tilde{A}\ ^1A_2'' - \tilde{X}\ ^1A_1$ and the photodissociation were studied theoretically in [69, 70].

The first electronic state of $NH_3(\tilde{A}\ ^1A_2'')$ produced during the photolysis correlates adiabatically with the dissociation products $NH_2(\tilde{A}\ ^2A_1) + H(^2S)$. Thus, the fragmentation channel leading to the electronically excited $NH_2(\tilde{A})$ product starts to contribute as soon as the energetic threshold for this channel is exceeded ($\lambda \leq 206$ nm). $NH_3(\tilde{A})$ molecules that pass through a region of conical intersection between the $\tilde{A}$ and the $\tilde{X}$ states of NH_3 emerge on the lower $\tilde{X}$ surface and yield $NH_2(\tilde{X})$ fragments [65 to 67, 69, 70]. The energy distribution of the NH_2 fragment has been examined by several authors [56, 62, 71]. The H atoms are formed with excess translational energy [55]. The photodissociation dynamics of the $\tilde{A}$ state of NH_3 and ND_3 was investigated using the technique of H (D) atom photofragment translational spectroscopy [65].

Other channels leading to NH(X, a, b, A...) + H_2 (2 H) were observed below 160 nm. Above 160 nm, no NH was detected [72]. NH can, however, appear even at 193 nm, if a significantly high fluence is applied leading to more than one photon absorption [73]. The quantum yield for NH_2 was found to be close to unity in several independent studies [74 to 76]. A combined flash photolysis shock tube technique was applied to study NH_2 reactions at high temperatures in the range 900 to 1850 K [19].

Hydrazine was also used as a precursor molecule in UV photolysis [77 to 83]. In the excimer laser photolysis of N_2H_4 and N_2D_4, the primary photoproducts were found to be $N_2H_3 + H$. The NH_2 radical is then formed via the reaction $N_2H_3 + H \rightarrow 2\ NH_2$ [79]. The electronically excited radicals $NH_2(\tilde{A})$ and $ND_2(\tilde{A})$ were obtained mainly due to two-photon processes [81]. At short wavelengths (105 nm), electronically excited $NH_2(\tilde{A}\ ^2A_1)$ radicals are generated by direct photodissociation [82].

The infrared multiple photon dissociation (IRMPD) of NH_3 [84 to 89], N_2H_4 [88, 90], and amines (CH_3NH_2 [85, 86, 88, 90 to 94], $C_6H_5CH_2NH_2$ [95]) leads to $NH_2(\tilde{X})$ radicals [96 to 103]. Most of the experiments were performed with pulsed CO_2 lasers; only in [89] a cw-CO_2 laser was applied. In the IRMPD of NH_3 the $NH_2(\tilde{X})$ yield was about a factor of 20 lower than in the experiment in which CH_3NH_2 was irradiated with the same fluence [85]. For small precursor molecules above threshold, the dissociaton rates are comparable with the laser up-pumping rate. Thus in the cases of NH_3 and N_2H_4, no internal excitation of the NH_2 fragment was observed (<1%) [90]. Molecules with a larger number of degrees of freedom have a longer lifetime and therefore sufficient time to absorb further photons; this explains why for CH_3NH_2 substantially more vibrationally excited $NH_2(0 \leq v_1 \leq 1, 0 \leq v_2 \leq 3, 0)$ was observed than for N_2H_4 [90, 91]. The amount of internal excitation depends, for a given molecule, on the laser fluence. IRMPD can be used to produce NH_2 radicals in vibrationally excited states $v_2 = 0,1$, in particular if CH_3NH_2 is used as a precursor [91].

Pulse radiolysis is another method to form NH_2 radicals as a fragment of NH_3. High concentrations of NH_2 can be produced by short pulses of 2-MeV electrons [104, 105]. The simultaneous production of other active species like NH and H can lead to complications when this method is used for kinetic studies. Dissociation of NH_3 by electron impact at low energies (0 to 100 eV) leads to NH_2 radicals (E > 6.2 eV) [106] but also to electronically excited $NH_2(\tilde{A})$ and NH(c,A...) fragments [107]. Radiolysis of gaseous NH_3 with ^{60}Co γ-radiation was applied to produce NH_2 radicals in the gas phase [108].

Photolysis of NH_3 in shock tubes (flash photolysis shock tube technique (FPST)) can be applied to study $NH_2(\tilde{X})$ radical reactions at high temperatures [18, 19]. NH_2 radicals are produced in a shock tube by thermal decomposition of precursor molecules like N_2H_4 [109 to 113], NH_3 [114 to 116], and CH_3NH_2 [117, 118]. Thermal dissociation can be attained in a plasma [119].

Energy disposal can also be achieved by energy transfer from Ar* [120 to 122], $Ar^* + NH_3 \rightarrow NH_2(\tilde{X}) + H + Ar$. In this energy transfer process, electronically excited NH_2 radicals are obtained [122]. $NH_2(\tilde{X})$ radicals were observed upon energy transfer from $N_2(A)$ to NH_3 [123].

The heterogeneous depletion of NH_2 is a critical parameter in flow reactor experiments. The activity of the wall is described by a coefficient γ which is defined as the number of NH_2 wall collisions leading to NH_2 depletion relative to all NH_2 wall collisions; γ was determined for different wall materials and wall coatings (quartz [124], pyrex, halocarbon wax, teflon [125], and aluminum oxide [8]). A value $\gamma(300\ K) = 3.4 \times 10^{-4}$ was obtained for teflon [126]. Halocarbon wax is nearly as good as teflon at room temperature, $\gamma(300\ K) = 4 \times 10^{-4}$, but slightly better at higher temperatures, $\gamma(358\ K) = 5.9 \times 10^{-4}$ [126]. Above 600 K quartz gives values of γ in the range 1×10^{-3} to 3×10^{-3} [125]. Above 1000 K aluminum oxide has to be used with $\gamma(300\ K) = 1.2 \times 10^{-2}$ to $\gamma(1050\ K) = 1.3 \times 10^{-2}$ [8]. The γ values are not optimal but nearly independent of temperature.

Which NH_2 source to use depends on the kinetic problem and the specific facilities available. In flow reactors, heterogeneous depletion of NH_2 and other intermediate radicals can not be avoided, thus giving erroneous kinetic parameters, if the γ_i values (as defined above) depend on the reactant concentrations. The photolytic NH_2 production is disadvantageous, if the reactant also absorbs the photolysis light; in the case of 1,3-butadiene [127], for example, the reaction of excited species or of photofragments may interfere. NH_2 production in flames is part of a complex system; here it is difficult to identify an elementary process, for which the kinetic parameters are to be determined.

References:

[1] Kurylo, M. J. (Chem. Phys. Lett. **23** [1973] 467/71).
[2] Stuhl, F. (J. Chem. Phys. **59** [1973] 635/7).
[3] Zellner, R.; Smith, I. W. M. (Chem. Phys. Lett. **26** [1974] 72/4).
[4] Hack, W.; Hoyermann, K.; Wagner, H. Gg. (Ber. Bunsen-Ges. Phys. Chem. **78** [1974] 386/91).
[5] Cox, R. A.; Darwent, R. G.; Holt, P. M. (Chemosphere **4** [1975] 201/5).
[6] Smith, I. W. M.; Zellner, R. (Int. J. Chem. Kinet. **7** [1975] 341/51).
[7] Perry, R. A.; Atkinson, R.; Pitts, J. N., Jr. (J. Chem. Phys. **64** [1976] 3237/9).
[8] Silver, J. A.; Kolb, C. E. (Chem. Phys. Lett. **75** [1980] 191/5).
[9] Fujii, N.; Miyama, H.; Koshi, M.; Asaba, T. (Symp. Int. Combust. Proc. **18** [1981] 873/83).
[10] Niemitz, K. J.; Wagner, H. Gg.; Zellner, R. (Z. Phys. Chem. [Munich] **124** [1981] 155/70).

[11] Salimian, S.; Hanson, R. K.; Kruger, C. H. (Int. J. Chem. Kinet. **16** [1984] 725/39).
[12] Zabielski, M. F.; Serry, D. J. (Int. J. Chem. Kinet. **17** [1985] 1191/9).
[13] Fujii, N.; Chiba, K.; Uchida, S.; Miyama, H. (Chem. Phys. Lett. **127** [1986] 141/4).
[14] Dransfeld, P.; Hack, W.; Jost, W.; Rouveirolles, P.; Wagner, H. Gg. (Khim. Fiz. **6** [1987] 1668/76).
[15] Jeffries, J. P.; Smith, G. P. (J. Phys. Chem. **90** [1986] 487/91).
[16] Diau, E. W. G.; Tso, T. L.; Lee, Y. P. (J. Phys. Chem. **94** [1990] 5261/5).
[17] Gimenez, X.; Moreno, M.; Lluch, J. M. (Chem. Phys. **165** [1992] 41/6).
[18] Michael, J. V.; Klemm, R. B.; Brobst, W. D.; Bosco, S. R.; Nava, D. F. (J. Phys. Chem. **89** [1985] 3335/7).

[19] Michael, J. V.; Sutherland, J. W.; Klemm, R. B. (Int. J. Chem. Kinet. **17** [1985] 315/26).
[20] Michael, J. V.; Sutherland, J. W.; Klemm, R. B. (J. Phys. Chem. **90** [1986] 497/500).

[21] Sutherland, J. W.; Michael, J. V. (J. Chem. Phys. **88** [1988] 830/4).
[22] Hack, W.; Rouveirolles, P.; Wagner, H. Gg. (J. Phys. Chem. **90** [1986] 2505/11).
[23] Marshall, P.; Fontijn, A. (J. Chem. Phys. **85** [1986] 2637/43).
[24] Marshall, P.; Fontijn, A. (J. Phys. Chem. **91** [1987] 6297/9).
[25] Ko, T.; Marshall, P.; Fontijn, A. (J. Phys. Chem. **94** [1990] 1401/4).
[26] Garrett, B. C.; Koszykowski, M. L.; Melius, C. F.; Page, M. (J. Phys. Chem. **94** [1990] 7096/106).
[27] Cohen, N. (Int. J. Chem. Kinet. **19** [1987] 319/62).
[28] Sutherland, J. W.; Patterson, P. M.; Klemm, R. B. (J. Phys. Chem. **94** [1990] 2471/5).
[29] Walther, C. D.; Wagner, H. Gg. (Ber. Bunsen-Ges. Phys. Chem. **87** [1983] 403/9).
[30] Farthing, J. W.; Flechter, I. W.; Whitehead, J. C. (J. Phys. Chem. **87** [1983] 1663/5).

[31] Hack, W.; Schacke, H.; Schröter, M.; Wagner, H. Gg. (Symp. Int. Combust. Proc. **17** [1978] 505/13).
[32] Hack, W.; Schröter, M. R.; Wagner, H. Gg. (Ber. Max-Planck-Inst. Strömungsforsch. **1979** No. 13).
[33] Pagsberg, P. B.; Sztuba, B.; Ratajczak, E.; Sillesen, A. (Acta Chem. Scand. **45** [1991] 329/34).
[34] Pasteris, L.; Staricco, E. H. (Int. J. Chem. Kinet. **17** [1985] 1221/30).
[35] Ghosh, P. K.; Bair, E. J. (J. Chem. Phys. **45** [1966] 4738/41).
[36] Gehring, M.; Hoyermann, K.; Wagner, H. Gg.; Wolfrum, J. (Ber. Bunsen-Ges. Phys. Chem. **75** [1971] 1287/94).
[37] Mertens, J. D.; Kohse-Höinghaus, K.; Hanson, R. K.; Bowman, C. T. (Int. J. Chem. Kinet. **23** [1991] 655/68).
[38] Kajimoto, O.; Kawajri, T.; Iumo, T. (Chem. Phys. Lett. **76** [1980] 315/8).
[39] Davidson, J. A.; Schiff, H. I; Streit, G. E.; McAfee, J. R.; Schmeltekopf, A. L.; Howard, C. J. (J. Chem. Phys. **67** [1977] 5021/5).
[40] Bulatov, V. P.; Buloyan, A. A.; Cheskis, S. G.; Kozliner, M. Z.; Sarkisov, O. M.; Trostin, A. L. (Chem. Phys. Lett. **74** [1980] 288/92).

[41] Sanders, N. D.; Butler, J. E.; McDonald, J. R. (J. Chem. Phys. **73** [1980] 5381/3).
[42] Patrick, P.; Golden, D. M. (J. Phys. Chem. **88** [1984] 491/5).
[43] Cheskis, S. G.; Iogansen, A. A.; Sarkisov, O. M.; Titov, A. A. (Chem. Phys. Lett. **120** [1985] 45/9).
[44] Cheskis, S. G.; Iogansen, A. A.; Kulakov, P. V.; Sarisov, O. M.; Titov, A. A. (Chem. Phys. Lett. **143** [1988] 348/52).
[45] Krishnamachari, S. L. N. G.; Venkitachalam, T. V. (Chem. Phys. Lett. **41** [1976] 183/4).
[46] Hack, W.; Rathmann, K. (Z. Phys. Chem. [Munich] **176** [1992] 151/60).
[47] Herzberg, G.; Ramsay, D. A. (J. Chem. Phys. **20** [1952] 347).
[48] McNesby, J. R.; Tanaka, I.; Okabe, H. (J. Chem. Phys. **36** [1962] 605/7).
[49] Calvert, J. G.; Pitts, J. N., Jr. (Photochemistry, Wiley, New York 1966, 899 pp.).
[50] Okabe, H. (Photochemistry of Small Molecules, Wiley, New York 1978, p. 269).

[51] Stuhl, F.; Welge, K. H. (Z. Naturforsch. **18a** [1963] 900/6).
[52] Beyer, K. D.; Welge, K. H. (Z. Naturforsch. **22a** [1967] 1161/70).
[53] Okabe, H.; Lenzi, M. (J. Chem. Phys. **47** [1967] 5241/6).
[54] Di Stefano, G.; Lenzi, M.; Margani, A.; Chieu, N. X. (J. Chem. Phys. **67** [1977] 3832/3).
[55] Back, R. A.; Koda, S. (Can. J. Chem. **55** [1977] 1387/95).
[56] Donnelly, V. M.; Baronavski, A. P.; McDonald, J. R. (Chem. Phys. **43** [1979] 283/93).
[57] Xuan, N.; Di Stefano, G.; Lenzi, M.; Margani, A. (J. Chem. Phys. **74** [1981] 6219/23).

[58] Vinogradov, I. P.; Firsov, V. V. (Opt. Spectrosc. **53** [1982] 46/9).
[59] Suto, M.; Lee, L. C. (J. Chem. Phys. **78** [1983] 4515/22).
[60] Koplitz, B.; Xu, Z.; Wittig, C. (Chem. Phys. Lett. **137** [1987] 505/9).

[61] Biesner, J.; Schneider, L.; Schmeer, J.; Ahlers, G.; Xie, X.; Welge, K. H.; Ashfold, M. N. R. (J. Chem. Phys. **88** [1988] 3607/16).
[62] Fuke, K.; Yamada, H.; Yoshida, Y.; Kaya, K. (J. Chem. Phys. **88** [1988] 5238/40).
[63] Kenner, R. D.; Browarzik, R. K.; Stuhl, F. (Chem. Phys. **121** [1988] 457/71).
[64] Xie, X.; Biesner, J.; Welge, K. H. (Huaxue Wuli Xuebao **1** [1988] 167/76).
[65] Biesner, J.; Schneider, L.; Ahlers, G.; Xie, X.; Welge, K. H.; Ashfold, M. N. R.; Dixon, R. N. (J. Chem. Phys. **91** [1989] 2901/11).
[66] Ashfold, M. N. R.; Dixon, R. N.; Irving, S. J.; Koeppe, H. M.; Meier, W.; Nightingale, J. R.; Schneider, L.; Welge, K. H. (Philos. Trans. R. Soc. London A **332** [1990] 375/86).
[67] Woodbridge, E. L.; Ashfold, M. R.; Leone, S. R. (J. Chem. Phys. **94** [1991] 4195/204).
[68] Runau, R.; Peyerimhoff, S. D.; Buenker, R. J. (J. Mol. Spectrosc. **68** [1977] 253/68).
[69] Rosmus, P.; Botschwina, P.; Werner, H. J.; Vaida, V.; Engelking, P. C.; McCarthy, M. I. (J. Chem. Phys. **86** [1987] 6677/92).
[70] McCarthy, M. I.; Rosmus, P.; Werner, H. J.; Botschwina, P.; Vaida, V. (J. Chem. Phys. **86** [1987] 6693/700).

[71] Dressler, K.; Ramsay, D. A. (Philos. Trans. R. Soc. London A **251** [1959] 553/602).
[72] Zetzsch, C.; Stuhl, F. (Ber. Bunsen-Ges. Phys. Chem. **85** [1981] 564/8).
[73] Donnelly, V. M.; Baronavski, A. P.; McDonald, J. R. (Chem. Phys. **43** [1979] 271/81).
[74] Groth, W. E.; Schurath, U.; Schindler, R. N. (J. Phys. Chem. **72** [1968] 3914/20).
[75] Schurath, U.; Tiedemann, P.; Schindler, R. N. (J. Phys. Chem. **73** [1969] 456/9).
[76] Lesclaux, R.; Khe, P. V. (J. Chim. Phys. Phys.-Chim. Biol. **70** [1973] 119/25).
[77] Arvis, M.; Devillers, C.; Gillois, M.; Curtat, M. (J. Phys. Chem. **78** [1974] 1356/60).
[78] Curtat, M.; Souil, F.; Devillers, C. (Propr. Hydrazine Ses. Appl. Source Energ. Colloq. Int., Poitiers 1974 [1975], pp. 41/5).
[79] Hawkins, W. G.; Houston, P. L. (J. Phys. Chem. **86** [1982] 704/9).
[80] Husain, D.; Norrish, R. G. W. (Proc. R. Soc. London A **273** [1963] 145/64).

[81] Lindberg, P.; Raybone, D.; Salthouse, J. A.; Watkinson, T. M.; Whitehead, J. C. (Mol. Phys. **62** [1987] 1297/306).
[82] Biehl, H.; Stuhl, F. (J. Photochem. Photobiol. A **59** [1991] 135/42).
[83] Hopkirk, A.; Salthouse, J. A.; White, R. W. P.; Whitehead, J. C.; Winterbottom, F. (Chem. Phys. Lett. **188** [1992] 399/404).
[84] Campbell, J. D.; Hancock, G.; Halpern, J. B.; Welge, K. H. (Opt. Commun. **17** [1976] 38/42).
[85] Hancock, G.; Hennessy, R. J.; Villis, T. (J. Photochem. **9** [1978] 197/9).
[86] Hancock, G.; Hennessy, R. J.; Villis, T. (J. Photochem. **10** [1979] 305/14).
[87] Hanazaki, I.; Kasatani, K.; Kuwata, K. (Chem. Phys. Lett. **75** [1980] 123/7).
[88] Xiang, T. X.; Torres, L. M.; Guillory, W. A. (J. Chem. Phys. **83** [1985] 1623/9).
[89] Masanet, J.; Deson, J.; Lalo, C.; Lempereur, T.; Tardieu de Maleissye, J. (J. Photochem. **36** [1987] 1/10).
[90] Filseth, S. V.; Danon, J.; Feldmann, D.; Campbell, J. D.; Welge, K. H. (Chem. Phys. Lett. **63** [1979] 615/20).

[91] Messing, L.; Sadowski, C. M.; Filseth, S. V. (Chem. Phys. Lett. **66** [1979] 95/9).
[92] Hancock, G.; Villis, T. (Springer Ser. Chem. Phys. **6** [1979] 190/2).
[93] Zacharias, H.; Schmiedl, R.; Boettner, R.; Geilhaupt, M.; Meier, U.; Welge, K. H. (Springer Ser. Opt. Sci. **21** [1979] 329/37).
[94] Xiang, T. X.; Guillory, W. A. (J. Chem. Phys. **85** [1986] 2019/28).

[95] Reisler, H.; Pessine, F. B. T.; Wittig, C. (Chem. Phys. Lett. **99** [1983] 388/93).
[96] Letokhov, V. S.; Ryabov, E. A.; Tumanov, O. A. (JETP **36** [1963] 1069/73).
[97] Ambartzumian, R. V.; Letokhov, V. S. (Acc. Chem. Res. **10** [1977] 61/7).
[98] Campbell, J. D.; Hancock, G.; Halpern, J. B.; Welge, K. H. (Chem. Phys. Lett. **44** [1976] 404/10).
[99] Campbell, J.; Hancock, H.; Halpern, J.; Welge, K. H. (Opt. Commun. **18** [1976] 34/5).
[100] Avouris, Ph.; Loy, M. M. T.; Chan, J. Y. (Chem. Phys. Lett. **63** [1979] 624/9).

[101] Ashfold, M. N. R.; Hancock, G.; Ketley, G. (Faraday Discuss. Chem. Soc. **67** [1979] 204/11).
[102] Duncanson, J. A.; Gericke, K. H.; Torres, L. M.; Guillory, W. A. (Int. Conf. Photochem. **11** [1983]).
[103] Gericke, K. H.; Torres, L. M.; Guillory, W. A. (J. Chem. Phys. **80** [1984] 6134/40).
[104] Gordon, S.; Mulac, W.; Nangia, P. (J. Phys. Chem. **75** [1971] 2087/93).
[105] Pagsberg, P. B.; Eriksen, J.; Christensen, H. C. (J. Phys. Chem. **83** [1979] 582/90).
[106] Marx, R.; Mauclaire, G.; Wallart, M.; Deraulede, A. (Adv. Mass Spectrom. **6** [1974] 735/42).
[107] Müller, U.; Schulz, G. (J. Chem. Phys. **96** [1992] 5924/37).
[108] Dzantiev, B. G.; Ermakov, A. N.; Lomako, L. T.; Molokanova, T. A. (Vestsi Akad Navuk BSSR Ser. Fiz. Energ. Navuk **1974** No. 3, pp. 36/41).
[109] Gilbert, M. (Combust. Flame **2** [1958] 149/56).
[110] Diesen, R. W. (J. Chem. Phys. **39** [1963] 2121/8).

[111] Michel, K. W. (Symp. Int. Combust. Proc. **10** [1965] 351).
[112] Meyer, E.; Oleschewski, H. A.; Troe, J.; Wagner, H. Gg. (Symp. Int. Combust. Proc. **12** [1969] 345/55).
[113] Meyer, E.; Wagner, H. Gg. (Z. Phys. Chem. [Munich] **89** [1974] 329/31).
[114] Yumura, M.; Asaba, T. (Symp. Int. Combust. Proc. **18** [1981] 863/72).
[115] Holzrichter, K. (Diss. Göttingen 1980).
[116] Holzrichter, K.; Wagner, H. Gg. (Symp. Int. Combust. Proc. **18** [1981] 769/75).
[117] Dorko, E. A.; Pohelkin, N. R.; Wert, J. C., III; Mueller, G. W. (J. Phys. Chem. **83** [1979] 297/302).
[118] Higashihara, T.; Gardiner, W. C., Jr.; Hwang, S. M. (J. Phys. Chem. **91** [1987] 1900/5).
[119] Levitskii, A. A.; Ovsyannikov, A. A.; Polak, L. S. (Commun. 3rd Symp. Int. Chim. Plasmas, Limoges, Fr., 1977, Vol. 2, p. 9).
[120] Stedman, D. H. (J. Chem. Phys. **52** [1970] 3966/70).

[121] Sekiya, H.; Nishiyama, N.; Tsuji, M.; Nishimura, Y. (J. Chem. Phys. **86** [1987] 163/9).
[122] Ohashi, K.; Kasai, T.; Che, D. C.; Kuwata, K. (J. Phys. Chem. **93** [1989] 5484/7).
[123] Hack, W.; Kurzke, H.; Ottinger, C.; Wagner, H. Gg. (Chem. Phys. **126** [1988] 111/24).
[124] Bulatov, V. P.; Buloyan, A. A.; Iogansen, A. A.; Sarkisov, O. M.; Cheskis, S. (Khim. Fiz. **4** [1982] 513/5).
[125] Hack, W.; Kurzke, H.; Rouveirolles, P.; Wagner, H. Gg. (Ber. Bunsen-Ges. Phys. Chem. **90** [1986] 1210/9).
[126] Hack, W. (unpublished results).
[127] Lesclaux, R.; Khe, P. V.; Soulignac, J. C.; Joussot-Dubien, J. (Int. Conf. Photochem., Edmonton, Can., 1975, Vol. 7).

2.1.7.1.2 NH_2 Radical Sources in Condensed Phases

The NH_2 production in condensed phases will be discussed in Section 2.1.7.5.3.

2.1.7.1.3 Detection Methods

Detection methods for free radicals are summarized in [1]. Many of the detection methods for NH_2 are based on the $\tilde{A}(^2A_1)-\tilde{X}(^2B_1)$ transition, using specific rotational lines of the (0,9,0) − (0,0,0) band. This spectrum in the visible 400- to 900-nm range was first observed during flash photolysis of ammonia [2] and has since been analyzed in detail [3, 4]. $\tilde{A} \leftarrow \tilde{X}$ absorption with a white light source and a monochromator, i.e. conventional resonant absorption spectrometry, has been applied to detect NH_2 in flames [5] and in pulse radiolysis [6], discharge [7], and flash photolysis systems [8 to 11]. The sensitivity is limited, since the spectral width of the analyzing light is broader than the absorption line. Narrow-band absorption lasers were used to measure the concentration quantitatively [12], i.e. to determine the oscillator strength of the transition used (see p. 204). The sensitivity of the absorption technique can be increased by a multiple reflection cell outside the cavity of a laser or by intracavity absorption. The latter device has been used to detect NH_2 in a dye laser cavity [13 to 17] and applied to study NH_2 reactions [18]. Both experimental arrangements yield nearly the same sensitivity [16].

The visible emission $\tilde{A} \rightarrow \tilde{X}$ [19, 20], which was discovered more than a century ago [21], can also be used to detect NH_2, if NH_2 is present in the $\tilde{A}$ state, e.g. in shock tubes at high temperatures ($T \approx 2000$ K) [22 to 25], during photolytic NH_2 generation (e.g. NH_3 photolysis at $\lambda = 193$ nm) [26, 27], or if $NH_2(\tilde{X})$ is excited to the $\tilde{A}$ state [28 to 30].

The laser-induced fluorescence method (LIF), based on the transition $\tilde{A} \leftarrow \tilde{X}$, uses a tunable cw dye laser as excitation source [29] or a pulsed dye laser source [30, 31]. The most intense emission was obtained by exciting the rotational lines within the (0,9,0) − (0,0,0) vibronic band at 597.72 nm. At low pressures the LIF method is the most sensitive NH_2 detection method ($[NH_2] \geq 10^{-15}$ mol/cm^3) and has been applied for kinetic measurements in flow reactors [32]. At higher pressures ($p > 100$ mbar) the quenching of the $\tilde{A}$ state [26, 27] significantly reduces the fluorescence quantum yield and thus the sensitivity. If the concentration of the quencher changes in the observed area it becomes difficult to obtain signals proportional to the NH_2 concentrations via LIF measurements. Nevertheless, the laser-induced fluorescence method has been used to detect NH_2 in atmospheric-pressure flames with an Nd:YAG-pumped dye laser [33] or a fixed frequency Kr^+ laser [34]. Using this Kr^+ laser two rotational lines of the (0,11,0) − (0,2,0) vibrational hot band at 647.1 nm were excited. In order to convert the LIF signals into absolute or even relative NH_2 concentrations, the quenching rate for $NH_2(\tilde{A})$ was assumed to be independent of position in the flame, although the concentration of the quenching molecules varies substantially within the flame. Determining even relative concentrations in an environment, in which the assumption of constant quenching is not valid, requires either laser-induced saturated fluorescence (i.e. high laser fluence, thus the stimulated emission is much faster than all the quenching processes) or laser-induced predissociation fluorescence (i.e. the lifetime of the emitting state is short due to predissociation and thus a collision-free situation is obtained up to high pressures). In both cases the fluorescence quantum yield and thus the sensitivity are lower. Even though these methods are in principal possible for NH_2, they have not yet been used for this radical.

Rotational transitions in the wavelength range around 316 μm have been used to detect NH_2 in a supersonic jet ($\cong 5$ K) with a very high sensitivity ($[NH_2] \geq 2 \times 10^{-16}$ mol/cm^3) using a tunable, far-infrared laser device [35].

The Zeeman effect was applied to detect NH_2 by using transitions between the Zeeman components of a single rotational state (electron spin resonance (ESR)) [36 to 39] and transitions between Zeeman components of two rotational states (laser magnetic resonance

(LMR)) [40 to 46]. The ESR spectra are always measured in the condensed phase, i.e. NH_2 is produced in the gas phase and then trapped at 77 K in ammonia matrices. The LMR in the far-infrared region (FIR LMR) (λ=78, 108, and 118.6 μm H_2O, D_2O dc-discharge laser) [40, 41] and in the 9 to 10 μm region (CO laser and CO_2 laser) [42 to 45] has been used to detect NH_2 in the gas phase directly. The deuterated radicals NHD [46] and ND_2 [42, 43] were also detected in the gas phase by LMR. Since an LMR spectrometer is based on an intracavity technique and phase-sensitive detection (due to a fast modulation of the magnetic field), it is a very sensitive method for detecting NH_2 below 3×10^{-16} mol/cm^3 [47].

NH_2 detection with coherent anti-Stokes-Raman scattering (CARS) was reported in [48]. The author found spectroscopic and kinetic evidence for an NH_2 CARS spectrum in the wavelength range 3195 to 3225 cm^{-1} during laser photolysis of NH_3 [48].

NH_2 was mass-spectrometrically detected by electron impact ionization [49 to 52] and photoionization [53]. In photoelectron spectroscopy experiments in the vacuum UV (He Iα radiation), the $NH_2^+(\tilde{X}\ ^3B_1, ^1A_1, ^1B_1) \leftarrow NH_2(\tilde{X}\ ^2B_1)$ transitions were observed [54] (see p. 197), on which the photoionization mass spectrometry in the wavelength range 75 to 112.5 nm [53] is based.

References:

[1] Hack, W. (Int. Rev. Phys. Chem. **4** [1985] 165/200).
[2] Herzberg, G.; Ramsay, D. A. (J. Chem. Phys. **20** [1952] 347).
[3] Dressler, K.; Ramsay, D. A. (Philos. Trans. R. Soc. London A **251** [1959] 553/602).
[4] Johns, J. W. C.; Ramsay, D. A.; Ross, S. C. (Can. J. Chem. **54** [1976] 1804/14).
[5] Nadler, M. P.; Wang, V. K.; Kaskan, W. E. (J. Phys. Chem. **74** [1970] 917/22).
[6] Gordon, S.; Mulac, W.; Nangia, P. (J. Phys. Chem. **75** [1971] 2087/93).
[7] Hanes, M. H.; Bair, E. J. (J. Chem. Phys. **38** [1963] 672/6).
[8] Arvis, M.; Devillers, C.; Gillois, M.; Curtat, M. (J. Phys. Chem. **78** [1974] 1356/60).
[9] Mantei, K. A.; Bair, E. J. (J. Chem. Phys. **49** [1968] 3248/56).
[10] Lesclaux, R.; Soulignac, J. C.; Khe, P. V. (Chem. Phys. Lett. **43** [1976] 520/3).

[11] Lesclaux, R.; Khe, P. V. (Informal Conf. Photochem. **12** [1976] P 3, 1/3).
[12] Kohse-Höinghaus, K.; Davidson, D. F.; Hanson, R. K. (J. Quant. Spectrosc. Radiat. Transfer **42** [1989] 1/17).
[13] Atkinson, G. H.; Laufer, A. H.; Kurylo, M. J. (J. Chem. Phys. **59** [1973] 350/4).
[14] Sviridenkov, E. A.; Frolov, M. P. (Kratk. Soobshch. Fiz. **1976** No. 3, p. 8).
[15] Cheskis, S. G.; Sarkisov, O. M. (Chem. Phys. Lett. **62** [1979] 72/6).
[16] Mulenko, S. A. (Rev. Roum. Phys. **32** [1987] 173/8).
[17] Mulenko, S. A.; Smirnov, V. N. (Kratk. Soobshch. Fiz. **1980** No. 12, pp. 24/31).
[18] Nadtochenko, V. L.; Sarkisov, O. M.; Cheskis, S. G. (Fiz.-Khim. Protsessy Gazov. Kondens. Fazakh **1979** 21; C.A. **93** [1980] No. 70522).
[19] Pearse, R. W. B.; Gaydon, A. G. (The Identification of Molecular Spectra, Chapman and Hall, London 1963).
[20] Okabe, H.; Lenzi, M. (J. Chem. Phys. **47** [1967] 5241/6).

[21] Dibbits, H. C. (Ann. Phys. [Leipzig] [2] **122** [1864] 497/545).
[22] Roose, T. R.; Hanson, R. K.; Kruger, C. H. (Symp. Int. Combust. Proc. **18** [1981] 853/62).
[23] Roose, T. R. (Diss. Stanford Univ. 1981).
[24] Salimian, S.; Hanson, R. K.; Kruger, C. H. (Combust. Flame **56** [1984] 83/95).
[25] Salimian, S.; Hanson, R. K.; Kruger, C. H. (Int. J. Chem. Kinet. **16** [1984] 725/39).
[26] Donnelly, V. M.; Baronavski, A. P.; McDonald, J. R. (Chem. Phys. **43** [1979] 283/93).

[27] Donnelly, V. M.; Baronavski, A. P.; McDonald, J. R. (Chem. Phys. **43** [1979] 271/81).
[28] Kroll, M. (J. Chem. Phys. **63** [1975] 319/25, 1803/9).
[29] Hancock, G.; Lange, W.; Lenzi, M.; Welge, K. H. (Chem. Phys. Lett. **33** [1975] 168/72).
[30] Halpern, J. B.; Hancock, G.; Lenzi, M.; Welge, K. H. (J. Chem. Phys. **63** [1975] 4808/16).
[31] Alwahabi, Z. T.; Harkin, C. G; Mc Caffery, A. J.; Whithaker, B. J. (J. Chem. Soc. Faraday Trans. II **85** [1989] 1003/15).
[32] Hack, W.; Schröter, M. R.; Wagner, H. Gg. (Ber. Max-Planck-Inst. Strömungsforsch. **1979** No. 13).
[33] Copeland, R. A.; Crosley, D. R.; Smith, G. P. (Symp. Int. Combust. Proc. **20** [1984] 1195/203).
[34] Wong, K. N.; Anderson, W. R.; Vanderhoff, J. A.; Kotlar, A. J. (J. Chem. Phys. **86** [1987] 93/101).
[35] Cohen, R. C.; Busarow, K. L.; Schmuttenmaer, C. A.; Lee, Y. T.; Saykally, R. J. (Chem. Phys. Lett. **164** [1989] 321/4).
[36] Marx, R.; Mauclaire, G. (Adv. Chem. Ser. **82** [1968] 212/21).
[37] Froben, F. W. (J. Phys. Chem. **78** [1974] 2047/50).
[38] Michaut, J. P.; Roncin, J.; Marx, R. (Chem. Phys. Lett. **36** [1975] 599/605).
[39] Mishra, S. P.; Symons, M. C. R. (J. Chem. Res. Synop. **1977** No. 2, p. 48).
[40] Davies, P. B.; Russel, D. K.; Thrush, B. A.; Wayne, F. D. (J. Chem. Phys. **62** [1975] 3739/42).
[41] Davies, P. B.; Russell, D. K.; Thrush, B. A.; Radford, H. E. (Chem. Phys. Lett. **42** [1976] 35/8).
[42] Hills, G. W.; McKellar, A. R. W. (J. Mol. Spectrosc. **74** [1979] 224/7).
[43] Hills, G. W.; McKellar, A. R. W. (J. Chem. Phys. **71** [1979] 3330/7).
[44] Brown, J. M.; Buttenshaw, J.; Carrington, A.; Parent, C. R. (Mol. Phys. **33** [1977] 589/92).
[45] Kawaguchi, K.; Yamada, C.; Hirota, E.; Brown, J. M.; Buttenshaw, J.; Parent, C. R.; Sears, T. J. (J. Mol. Spectrosc. **81** [1980] 60/72).
[46] Carrington, A.; Geiger, J. S.; Smith, D. R.; Bonnett, J. D.; Brown, C. (Chem. Phys. Lett. **90** [1982] 6/8).
[47] Temps, F. (Ber. Max-Planck-Inst. Strömungsforsch. **1983** No. 4, 140 pp.).
[48] Dreier, Th. (Ber. Max-Planck-Inst. Strömungsforsch. **1983** No. 17, 135 pp.).
[49] Foner, S. N.; Hudson, R. L. (J. Chem. Phys. **29** [1958] 442/3).
[50] Foner, S. N. (Adv. At. Mol. Phys. **2** [1966] 385/461).
[51] Wagner, H. Gg.; Wolfrum, J. (Angew. Chem. **83** [1971] 561/77; Angew. Chem. Int. Ed. Engl. **10** [1971] 604/19).
[52] Donovan, R. J.; Husain, D.; Kirsch, L. J. (Annu. Rep. Progr. Chem. A **69** 1972 [1973] 19/74).
[53] Gibson, S. T.; Greene, J. P.; Berkowitz, J. (J. Chem. Phys. **83** [1985] 4319/28).
[54] Dunlavey, S. J.; Dyke, J. M.; Jonathan, N.; Morris, A. (Mol. Phys. **39** [1980] 1121/35).

2.1.7.2 Molecular Properties

2.1.7.2.1 Electronic Structure. Potential Energy Functions

Valence States

The Ground State $\tilde{X}\,^2B_1$ and the Excited State $\tilde{A}\,^2A_1$. The $\tilde{X}\,^2B_1$ state of the nine-electron radical NH_2 (ND_2) belongs to the point group C_{2v}. The ground state of NHD (point group C_s) is $1\,^1A''$. According to the orbital diagram given by Walsh [1], the electron configuration is $(1\,a_1)^2\,(2\,a_1)^2\,(1\,b_2)^2\,(3\,a_1)^2\,(1\,b_1)$. For the ordering of the orbitals, see p. 175.

The electron configuration of the first excited state $\tilde{A}\,^2A_1$ (NH_2, ND_2) or $1\,^1A'$ (NHD) is $(1a_1)^2\,(2a_1)^2\,(1b_2)^2\,(3a_1)\,(1b_1)^2$; see for example [2, 3]. Experimental evidence of this state was obtained from the electronic absorption and emission spectra ($\tilde{A}\,^2A_1 \leftrightarrow \tilde{X}\,^2B_1$); see pp. 203/8. The term value $T_0 = 11122.6\ cm^{-1}$ [4, 5] was determined from the $\tilde{A} \leftarrow \tilde{X}$ absorption band systems [3]. $T_0 = 11220.137\ cm^{-1}$ was obtained [6] by adjusting the $T(0,0,0)_{K_a=1}(J=1)$ value (taken from [7]) to $J=0$ using rotational constants derived in [5] (for notation, see p. 195).

NH_2 is one of the first and one of the best examples of the **Renner-Teller effect**. The very detailed analysis of the NH_2 absorption spectrum by Dressler and Ramsay [3] showed that the spectrum is based on a transition between the two electronic states $\tilde{X}\,^2B_1$ and $\tilde{A}\,^2A_1$. These widely separated electronic states rise from a hypothetical $^2\Pi$ state of a linear molecule whose degeneracy is removed by strong coupling between the electronic and vibrational-bending angular momentum as predicted by Herzberg and Teller [8] and discussed in detail by Renner [9]. This vibronic interaction was quantitatively interpreted by Pople and Longuet-Higgins [10]. A schematic diagram of the vibronic energy levels of the bending vibration of a linear triatomic molecule [3, 11] is given in **Fig. 6**, p. 172. For the Π state ($\Lambda = 1$) the electronic angular momentum $h/2\pi$ about the internuclear axis couples with the bending-vibrational angular momentum $l_2h/2\pi$ to form a resultant $Kh/2\pi$, where $K = |\pm l_2 \pm 1|$. Possible values of K are v_2+1, v_2-1, $v_2-3, \ldots$, 1 or 0, depending on whether v_2+1 is odd or even; see for example [3, 12]. The bending vibration of NH_2 in the ground state $\tilde{X}\,^2B_1(0,v_2'',0)$ was studied from $v_2''=4$ to 10 by laser-induced fluorescence. The change of the K-type structure at $v_2''=9$ indicates the transition from bent to linear geometry and marks the onset of strong Renner-Teller interaction [7]. For theoretical treatments of the Renner-Teller effect large-amplitude bending vibrations must be considered, as shown by vibronic (configuration interaction) calculations [13].

The Excited State $\tilde{B}\,^2B_2$. No experimental results were reported for the NH_2 (ND_2) $\tilde{B}\,^2B_2$ state (C_{2v} symmetry) or the NHD $2\,^2A'$ state (C_s symmetry) with the electron configuration $(1a_1)^2\,(2a_1)^2\,(1b_2)\,(3a_1)^2\,(1b_1)^2$. Term energies of the $\tilde{B}$ state $T_e = 37585\ cm^{-1}$ [14] and $38634\ cm^{-1}$ [15] were calculated by ab initio CI methods. An earlier ab initio study inferred the $\tilde{B}\,^2B_2$ state to be unstable [16]. Ab initio multireference double-CI calculations were carried out for potential surfaces and the vibronic coupling of the $\tilde{B}\,^2B_2$ and $\tilde{A}\,^2A_1$ states in the region of conical intersection [17, 18]. Since the electric-dipole forbidden transition $\tilde{B}\,^2B_2 \leftarrow \tilde{X}\,^2B_1$ should become allowed due to the vibronic coupling between $\tilde{A}\,^2A_1$ and $\tilde{B}\,^2B_2$, information on $\tilde{B}\,^2B_2$ might be obtained from the expected absorption system in the NH_2 spectrum with the origin near 259 nm [15].

Potential Energy Functions. The bending potential energy curves of the ground and first excited electronic states fitted by an empirical method [5] are given in **Fig. 7**, p. 173. Using the spectroscopic data of Dressler and Ramsey [3], the barrier to linearity in the $\tilde{X}\,^2B_1$ state was determined to be 12024(15) cm^{-1} [5]. Dixon [19] found the interesting result that NH_2 is not linear at equilibrium in the $\tilde{A}\,^2A_1$ state but has a small barrier to linearity of 730(14) cm^{-1} [5]. Bending potential energy curves were calculated using ab initio SCF [20 to 22] and CI [14, 20, 23] methods. The bending potential curves for NH_2 in the $\tilde{X}\,^2B_1$ and $\tilde{A}\,^2A_1$ states including a-axis rotation were obtained from laser-induced fluorescence studies for $K_a = 0$, 10, and 20. The barrier to linearity increases rapidly with increasing K_a [24].

Potential curves for the symmetric and antisymmetric stretching were calculated by ab initio SCF and CI methods [20]. Potential curves for NH_2 as a function of one NH bond length (second bond length and bond angle are fixed) were obtained by ab initio SCF calculations up to the dissociation limit [21, 25]. The shape of the potential energy curve

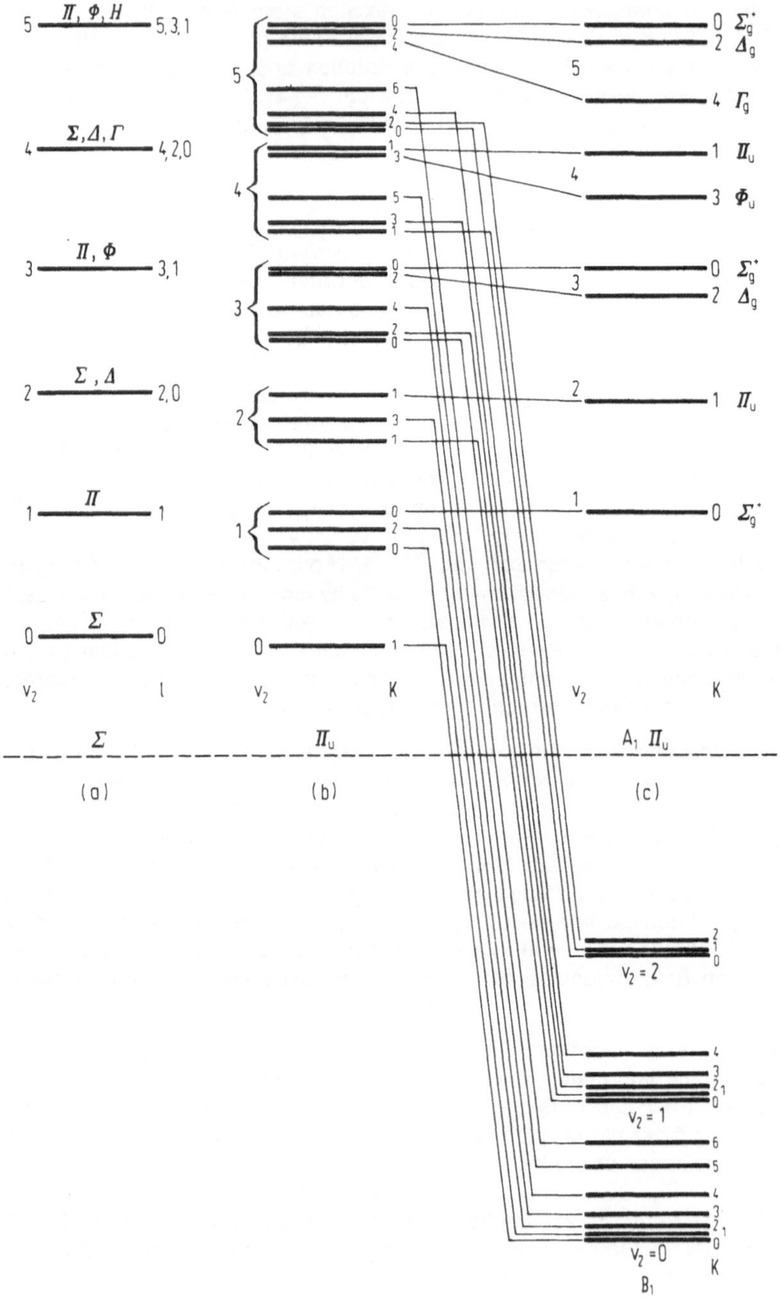

Fig. 6. Energy levels for the bending vibration of a linear molecule for (a) an electronic Σ state and (b) an electronic Π_u state. In (c) the energy levels are given for two electronic states A_1 (linear structure) and B_1 (bent structure) resulting from vibronic splitting of the Π_u state (from [3]).

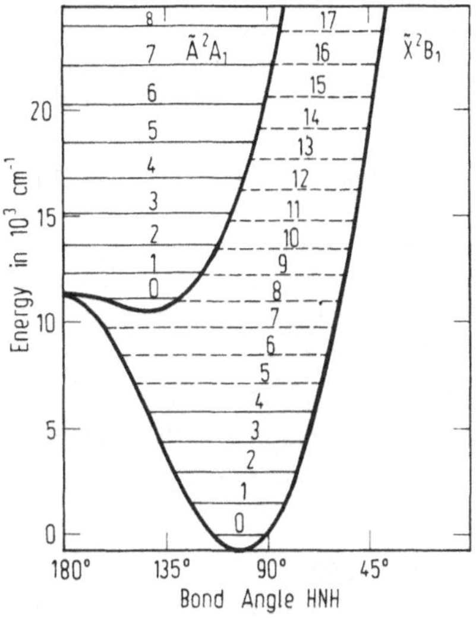

Fig. 7. Bending potential energy functions for the $\tilde{X}\,^2B_1$ and $\tilde{A}\,^2A_1$ states [5]. Experimentally determined (————) and calculated (-----) vibrational energy levels $(0,v_2,0)_{K_a=0}$ are indicated using the bent-molecule notation for both states.

with increasing bond length for the three lowest electronic states according to

$$HNH(\tilde{X}\,^2B_1) \rightarrow HN-H(1\,^2A'') \rightarrow NH(X\,^3\Sigma^-) + H(^2S),$$
$$HNH(\tilde{A}\,^2A_1) \rightarrow HN-H(1\,^2A') \rightarrow NH(a\,^1\Delta) + H(^2S),$$
$$HNH(\tilde{B}\,^2B_2) \rightarrow HN-H(2\,^2A') \rightarrow NH(b\,^1\Sigma^+) + H(^2S),$$

is depicted in [25].

Potential energy contour plots of NH_2 with C_{2v} symmetry in the $\tilde{X}\,^2B_1$, $\tilde{A}\,^2A_1$, and $\tilde{B}\,^2B_2$ states are presented for varying r(H···H) and R(N···H_2) distances (R is the distance between the center of mass of H_2 and N) [26]. Parameters from an analytic expression for the three-dimensional potential energy surfaces of NH_2 (related to NH stretching and bending vibrations) in the electronic ground state $\tilde{X}\,^2B_1$ were determined by fitting the potential expansion to potential energy points calculated by an ab initio CI method [27].

Rydberg States

Excitation from the ground state configuration to the first member of the Rydberg s series gives rise to one quartet state and four doublet states. These states and the corresponding configurations are shown in the following table together with the term energies T_e and bond angles α_e calculated by ab initio CI methods for the constant bond length r_e(N-H) = 1.024 Å [14]. The Renner-Teller states are combined by a bracket. The correlation with linear NH_2 is given in the last column.

state	configuration	T_e in cm^{-1}	α_e	state
$3\,^2A_1$	$(1a_1)^2\,(2a_1)^2\,(1b_2)^2\,(1b_1)^2$ 3s	76719	180°	$^2\Sigma_g^+$
$3\,^2B_1$ *)	$(1a_1)^2\,(2a_1)^2\,(1b_2)^2\,(3a_1)\,(1b_1)$ 3s	66718	180° }	$^2\Delta_g$
$2\,^2A_1$ *)	$(1a_1)^2\,(2a_1)^2\,(1b_2)^2\,(3a_1)^2$ 3s	61943	≈100° }	
$2\,^2B_1$	$(1a_1)^2\,(2a_1)^2\,(1b_2)^2\,(3a_1)\,(1b_1)$ 3s	54555	154°	$^2\Sigma_g^-$
$1\,^4B_1$	$(1a_1)^2\,(2a_1)^2\,(1b_2)^2\,(3a_1)\,(1b_1)$ 3s	49264	180°	$^4\Sigma_g^-$

*) States $2\,^2A_1$ and $3\,^2B_1$ were exchanged in the original.

Total energies of twelve core-hole states were calculated by using ab initio CI methods. These Rydberg states belong to the vertical excitations from a $1a_1$ electron in the ground state $\tilde{X}\,^2B_1$ into four quartet (two $^4B_1$, one $^4A_1$, and one $^4A_2$) and eight doublet (four $^2B_1$, two $^2A_1$, and two $^2A_2$) states [28].

Potential Energy Functions. The bending potential curves obtained by ab initio CI calculations indicated linear (or almost linear) geometry for four Rydberg 3s members. Only NH_2 in the $2\,^2A_1$ state (correlated with the linear $^2\Delta_g$ state) was found to be decidedly bent ($\alpha \approx 100°$) [14]. Potential curves for NH_2 as a function of one NH bond length (the other bond length and the bond angle are fixed) were obtained up to the dissociation limit by ab initio MCSCF calculations for the $2\,^2B_1$ and $2\,^2A_1$ states [25] according to

$$HNH(2\,^2B_1) \rightarrow HN-H(2\,^2A'') \rightarrow NH(a\,^1\Delta) + H(^2S)$$
$$HNH(2\,^2A_1) \rightarrow HN-H(3\,^2A') \rightarrow NH(A\,^3\Pi) + H(^2S).$$

References:

[1] Walsh, A. D. (J. Chem. Soc. **1953** 2260/6).
[2] Herzberg, G. (The Spectra and Structures of Simple Free Radicals: An Introduction to Molecular Spectroscopy, Cornell University Press, Ithaca 1971, pp. 1/226, 98, 139).
[3] Dressler, K.; Ramsay, D. A. (Philos. Trans. R. Soc. London A **251** [1959] 553/604).
[4] Jacox, M. E. (J. Phys. Chem. Ref. Data **17** [1988] 269/511, 282).
[5] Jungen, C.; Hallin, K. E. J.; Merer, A. J. (Mol. Phys. **40** [1980] 25/63).
[6] Martin, J. M. L.; François, J. P.; Gijbels, R. (J. Chem. Phys. **97** [1992] 3530/6).
[7] Vervloet, M. (Mol. Phys. **63** [1988] 433/49).
[8] Herzberg, G.; Teller, E. (Z. Phys. Chem. B **21** [1933] 410/46).
[9] Renner, R. (Z. Phys. **92** [1934] 172/93).
[10] Pople, J. A.; Longuet-Higgins, H. C. (Mol. Phys. **1** [1958] 372/83).

[11] Dressler, K.; Ramsay, D. A. (J. Chem. Phys. **27** [1957] 971/2).
[12] Herzberg, G. (Molecular Spectra and Molecular Structure III: Electronic Spectra and Electronic Structure of Polyatomic Molecules, Van Nostrand, New York 1966, pp. 1/745, 26).
[13] Perić, M.; Peyerimhoff, S. D.; Buenker, R. J. (Mol. Phys. **49** [1983] 379/400).
[14] Peyerimhoff, S. D.; Buenker, R. J. (Can. J. Chem. **57** [1979] 3182/9).
[15] Bell, S.; Schaefer, H. F., III (J. Chem. Phys. **67** [1977] 5173/7).
[16] Thomson, C.; Brotchie, D. (Mol. Phys. **28** [1974] 301/3).
[17] Petrongolo, C.; Buenker, R. J.; Peyerimhoff, S. D. (Chem. Phys. Lett. **115** [1985] 249/52).
[18] Petrongolo, C.; Hirsch, G.; Buenker, R. J. (Mol. Phys. **70** [1990] 825/34).
[19] Dixon, R. N. (Mol. Phys. **9** [1965] 357/66).
[20] Buenker, R. J.; Perić, M.; Peyerimhoff, S. D.; Marian, R. (Mol. Phys. **43** [1981] 987/1014).

[21] Brown, R. D.; Williams, G. R. (Mol. Phys. **25** [1973] 673/94).
[22] Bender, C. F.; Schaefer, H. F., III (J. Chem. Phys. **55** [1971] 4798/803).
[23] Nakatsuji, H.; Izawa, M. (J. Chem. Phys. **97** [1992] 435/9).
[24] Dixon, R. N.; Irving, S. J.; Nightingale, J. R.; Vervloet, M. (J. Chem. Soc. Faraday Trans. **87** [1991] 2121/33).
[25] Saxon, R.P.; Lengsfield, B. H., III; Liu, B. (J. Chem. Phys. **78** [1983] 312/20).
[26] Polak, R.; Vojtik, J. (Chem. Phys. **114** [1987] 43/54).
[27] Jensen, P.; Buenker, R. J.; Hirsch, G.; Rai, S. N. (Mol. Phys. **70** [1990] 443/54).
[28] Cambi, R.; Ciullo, G.; Sgamellotti, A.; Tarantelli, F.; Guest, M. F. (Chem. Phys. Lett. **91** [1982] 178/84).

2.1.7.2.2 Ionization Potential E_i in eV

Ionization potentials of NH_2 were measured by photoelectron spectroscopy (PES), by photoionization (PI) and electron impact (EI) mass spectrometry, and were calculated by ab initio methods (MCSTEP, MP4, MCSCF). The first two adiabatic ionization potentials of NH_2 are listed in the following table.

ionized	experimental method				ab initio calculations		
orbital	PI	PI	PES	EI	MCSTEP	MP4	MCSCF
3 a_1	11.14(1)	11.17(5)	11.46*)	11.4(1)	11.00	11.20	10.9
1 b_1	12.445(2)	–	12.45	–	12.16	12.48	–
Ref.	[1]	[2]	[3]	[4]	[5]	[6]	[7]

*) This value was questioned [1, 5] because the spectra of all species present in the sampling region were superimposed in the photoelectron spectrum.

Vertical ionization potentials E_i were determined by photoelectron spectroscopy [3] and calculated by ab initio procedures [3, 5].

ionized orbital	3 a_1	1 b_1	3 a_1	1 b_2	1 b_2	3 a_1
ionic state	$\tilde{X}\,^3B_1$	$\tilde{a}\,^1A_1$	$\tilde{b}\,^1B_1$	$^3A_2$	$^1A_2$	$^3B_2$
E_i(PES) [3]*)	12.00	12.45	14.27	–	–	–
E_i(ab initio MCSTEP) [5]	11.77	12.22	14.03	16.86	18.26	19.39
E_i(ab initio SCF CI) [3]	11.37	12.10	13.71	16.35	17.66	–

*) The probable error in determining the adiabatic ionization potential from the PE spectrum [3] indicates that the vertical E_i may also be too high [5].

Additional values of ionization potentials were obtained from ab initio [8 to 10] and semiempirical [11 to 16] calculations.

References:

[1] Gibson, S. T.; Greene, J. P.; Berkowitz, J. (J. Chem. Phys. **83** [1985] 4319/28).
[2] McCulloh, K. E. (Int. J. Mass Spectrom. Ion Phys. **21** [1976] 333/42).
[3] Dunlavey, S. J.; Dyke, J. M.; Jonathan, N.; Morris, A. (Mol. Phys. **39** [1980] 1121/35).
[4] Foner, S. N. (J. Chem. Phys. **29** [1958] 442/3).
[5] Graham, R. L.; Golab, J. T.; Yeager, D. L. (J. Chem. Phys. **88** [1988] 2572/81).
[6] Pople, J. A.; Curtiss, L. A. (J. Phys. Chem. **91** [1987] 155/62).
[7] Pope, S. A.; Hillier, I. H.; Guest, M. F. (Faraday Symp. Chem. Soc. **19** [1984] 109/23).
[8] Power, D.; Brint, P.; Spalding, T. R. (J. Mol. Struct. **110** [1984] 155/66 [THEOCHEM **19**]).
[9] Heaton, M. M.; Cowdry, R. (J. Chem. Phys. **62** [1975] 3002/9).
[10] Lathan, W. A.; Hehre, W. J.; Curtiss, L. A.; Pople, J. A. (J. Am. Chem. Soc. **93** [1971] 6377/87).

[11] Degtyarev, L. S.; Stetsenko, A. A. (Teor. Eksp. Khim. **19** [1983] 555/65; Theor. Exp. Chem. [Engl. Transl.] **19** [1983] 513/22).
[12] Boehm, M. C.; Sen, K. D.; Schmidt, P. C. (Chem. Phys. Lett. **78** [1981] 357/60).
[13] Nakajima, Y.; Ogata, M.; Ichikawa, H. (Shitsuryo Bunseki **28** [1980] 243/7; C.A. **94** [1981] No. 191423).
[14] Bews, J. R.; Glidewell, C. (Inorg. Chim. Acta **39** [1980] 217/25).

[15] Bischof, P. (J. Am. Chem. Soc. **98** [1976] 6844/9).
[16] Zahradník, R.; Čársky, P. (Theor. Chim. Acta **27** [1972] 121/34).

2.1.7.2.3 Electron Affinity A in eV. Proton Affinity A_p in eV

The **electron affinity** $A = 0.771 \pm 0.005$ was determined from a high-resolution (5 to 25 meV) photoelectron spectrum of NH_2^- [1]. $A = 0.779 \pm 0.037$ was obtained from a photoelectron spectrum at lower resolution [2]. Photodetachment-threshold measurements gave $A = 0.744 \pm 0.022$ [3] and 0.76 ± 0.04 [4]. $A = 0.76 \pm 0.14$ was determined from pulsed ICR studies of ion-molecule reactions in ammonia [5]. A considerably higher value, A = 1.2, resulted from applying the magnetron method [6].

Adiabatic and vertical electron affinities A(ad) = 0.69 and A(vert) = 0.68 according to $NH_2(\tilde{X}\,^2B_1) + e^- \rightarrow NH_2^-(\tilde{X}\,^1A_1)$ were calculated by ab initio (MCSTEP) procedures [7]. Additional ab initio [8 to 12] and semiempirical [13 to 17] calculations of the electron affinity were reported.

The **proton affinity** $A_p = 8.09 \pm 0.09$ was determined from a pulsed ion cyclotron resonance spectrum [5]. Semiempirically calculated A_p values were also published [13, 17].

References:

[1] Wickham-Jones, C. T.; Ervin, K. M.; Ellison, G. B.; Lineberger, W. C. (J. Chem. Phys. **91** [1989] 2762/3).
[2] Celotta, R. J.; Bennett, R. A.; Hall, J. L. (J. Chem. Phys. **60** [1974] 1740/5).
[3] Smyth, K. C.; Brauman, J. I. (J. Chem. Phys. **56** [1972] 4620/5).
[4] Feldmann, D. (Z. Naturforsch. A **26** [1971] 1100/1).
[5] DeFrees, D. J.; Hehre, W. J.; McIver, R. T. J.; McDaniel, D. H. (J. Phys. Chem. **83** [1979] 232/7).
[6] Page, F. M. (Trans. Faraday Soc. **57** [1961] 1254/8).
[7] Yeager, D. Y.; Nichols, J. A.; Golab, J. T. (J. Chem. Phys. **97** [1992] 8441/8).
[8] Merchan, M.; Roos, B. O. (Chem. Phys. Lett. **184** [1991] 346/52).
[9] Ortiz, J. V. (J. Chem. Phys. **86** [1987] 308/12).
[10] Ortiz, B. J. V.; Ohrn, Y. (Chem. Phys. Lett. **77** [1981] 548/54).
[11] Andersen, E.; Simons, J. (J. Chem. Phys. **65** [1976] 5393/7).
[12] Heaton, M. M.; Cowdry, R. (J. Chem. Phys. **62** [1975] 3002/9).
[13] Degtyarev, L. S.; Stetsenko, A. A. (Teor. Eksp. Khim. **19** [1983] 555/65; Theor. Exp. Chem. [Engl. Transl.] **19** [1983] 513/22).
[14] Boehm, M. C.; Sen, K. D.; Schmidt, P. C. (Chem. Phys. Lett. **78** [1981] 357/60).
[15] Dewar, M. J. S.; Rzepa, H. S. (J. Am. Chem. Soc. **100** [1978] 784/90).
[16] Zahradník, R.; Čársky, P. (Theor. Chim. Acta **27** [1972] 121/34).
[17] Lathan, W. A.; Hehre, W. J.; Curtiss, L. A.; Pople, J. A. (J. Am. Chem. Soc. **93** [1971] 6377/87).

2.1.7.2.4 Dipole Moment μ in D. Quadrupole Moment Θ in 10^{-26} esu·cm^2

The **dipole moment** of NH_2 in the $\tilde{X}\,^2B_1$ state was determined as $\mu = 1.82 \pm 0.05$ by optical Stark spectroscopy of NHD [1]. Ab initio calculations were performed on NH_2 in the $\tilde{X}\,^2B_1$, $\tilde{A}\,^2A_1$, and $\tilde{B}\,^2B_2$ states. The calculated values for the ground state agree well with the experimental value (UCC, unitary coupled cluster).

$\mu(\tilde{X}\ ^2B_1)$	2.013	2.095	1.894	1.83	1.77
$\mu(\tilde{A}\ ^2A_1)$	–	0.821	–	0.82	0.84
$\mu(\tilde{B}\ ^2B_2)$	–	3.002	–	0.90	–
method	UCC	SCF MO	SCF MO	SCF MO	SCF MO
Ref.	[2]	[3]	[4]	[5]	[6]

Semiempirical calculations [7 to 12] were also reported.

Only theoretical values of the **quadrupole moment** Θ are available. Ab initio calculations gave the tensor elements $\Theta_{xx}=0.9701$, $\Theta_{yy}=-0.3670$, and $\Theta_{zz}=-0.6030$ (y ⊥ molecular plane, $z=C_2$ axis with the positive direction from H atoms to N) [4].

References:

[1] Brown, J. M.; Chalkley, S. W.; Wayne, F. D. (Mol. Phys. **38** [1979] 1521/37).
[2] Watts, J. D.; Trucks, G. W.; Bartlett, R. J. (Chem. Phys. Lett. **164** [1989] 502/8).
[3] Bell, S.; Schaefer, H. F., III (J. Chem. Phys. **67** [1977] 5173/7).
[4] Brown, R. D.; Williams, G. R. (Chem. Phys. **3** [1974] 19/34).
[5] Brown, R. D.; Williams, G. R. (Mol. Phys. **25** [1973] 673/94).
[6] Del Bene, J. (J. Chem. Phys. **54** [1971] 3487/90).
[7] Dixit, A. N. (Indian J. Pure Appl. Phys. **29** [1991] 41/2).
[8] Bews, J. R.; Glidewell, C. (Inorg. Chim. Acta **39** [1980] 217/25).
[9] Zahradník, R.; Čársky, P. (Theor. Chim. Acta **27** [1972] 121/34).
[10] Phillips, A. J. (Diss. Howard Univ. 1969, 105 pp. from Diss. Abstr. Int. B **30** [1970] 3814/5).
[11] Pople, J. A.; Beveridge, D. L.; Dobosh, P. A. (J. Chem. Phys. **47** [1967] 2026/33).
[12] Trinajstic, N. (Croat. Chem. Acta **38** [1966] 287/91).

2.1.7.2.5 Hyperfine Interaction Constants

2.1.7.2.5.1 Nuclear Quadrupole Coupling Constants eqQ in MHz

Two of the three constants eqQ_{aa}, eqQ_{bb}, and eqQ_{cc} were determined from microwave optical double resonance (MODR) spectra of NH_2 [1 to 4], microwave absorption spectra of ND_2 [5], and by ab initio (UHF) calculations [6]. The third constant follows from $eqQ_{aa}+eqQ_{bb}+eqQ_{cc}=0$ [3]. The nuclear quadrupole coupling constants listed in the following table are given in terms of the principle axis system (b ∥ C_2, c ⊥ molecular plane).

state	sample	eqQ_{aa}	eqQ_{bb}	eqQ_{cc}	Ref.
$\tilde{X}\ ^2B_1(0,0,0)$	$ND_2(^{14}N)$	0.21 ± 0.08	-3.75 ± 0.11	3.54 ± 0.20	[5]
	$NH_2(^{14}N)$	0.29 ± 0.80	-3.99 ± 1.51	3.70 ± 0.51	[1]
	$NH_2(^{14}N)$	0.299	−2.711	2.412	[6]
	$NH_2(^{1}H)$	−0.1445	−0.1516	0.2961	[6]
$\tilde{A}\ ^2A_1(0,10,0)_{K_a=1}$	$NH_2(^{14}N)$	1.4 ± 1.0	6.0 ± 1.0	-7.4 ± 1.0	[4]

References:

[1] Hills, G. W.; Cook, J. M. (J. Mol. Spectrosc. **94** [1982] 456/60).
[2] Cook, J. M.; Hills, G. W.; Curl, R. F., Jr. (J. Chem. Phys. **67** [1977] 1450/61).
[3] Hills, G. W.; Cook, J. M.; Curl, R. F., Jr.; Tittel, F. K. (J. Chem. Phys. **65** [1976] 823/8).

[4] Hills, G. W.; Brazier, C. R.; Brown, J. M.; Cook, J. M.; Curl, R. F., Jr. (J. Chem. Phys. **76** [1982] 240/52).
[5] Kanada, M.; Yamamoto, S.; Saito, S. (J. Chem. Phys. **94** [1991] 3423/8).
[6] Brown, R. D.; Williams, G. R. (Chem. Phys. **3** [1974] 19/34).

2.1.7.2.5.2 Nuclear Spin-Rotation Interaction Constants C in MHz

The analysis of microwave absorption bands of ND_2 gave the ^{14}N nuclear spin-rotation interaction constants $C_{aa}=0.269\pm0.027$, $C_{bb}=0.045\pm0.021$, and $C_{cc}=0.019\pm0.022$.

Reference:

Kanada, M.; Yamamoto, S.; Saito, S. (J. Chem. Phys. **94** [1991] 3423/8).

2.1.7.2.5.3 Magnetic Hyperfine Coupling Constants. Electronic g Factor

Gas Phase. The isotropic or Fermi contact hyperfine (hf) coupling constant A_{iso} and the anisotropic or dipole hf tensor elements A_{aa}, A_{bb}, and A_{cc} of NH_2 in the $\tilde{X}\,^2B_1(0,0,0)$ and $\tilde{A}\,^2A_1(0,v_2,0)$ states are compiled in Table 9. The experimentally determined constants were derived from microwave optical double resonance (MODR), microwave absorption (MWA), laser magnetic resonance (LMR), infrared optical double resonance (IODR), optical optical double resonance (OODR), saturation, and magnetic level-crossing spectra.

A significant difference between the ^{14}N Fermi contact terms in the (0,9,0) and (0,10,0) levels of the $\tilde{A}\,^2A_1$ state, obtained from saturation spectra [1] and magnetic level-crossing spectra [2, 3], could not be confirmed by the MODR study [4].

The g factors $g_x=2.0023$, $g_y=2.0039$, $g_z=2.0083$ (x, y, z$\triangleq$c, a, b), and $g_{av}=2.0048$ were determined [5] from rotation and spin-rotation constants (B_i and B_i^s) using the relation $g_i=2.0023-B_i^s/2B_i$. The average value was confirmed by a LMR study which gave $g_{av}=2.0050$ [6]. Values of $g'=2.0023-3^{-1/2}(g_{aa}+g_{bb}+g_{cc})$ for NH_2 in several $(0,v_2,0)$ levels of the $\tilde{A}\,^2A_1$ state were obtained from IODR studies (for notation, see p. 195).

level	$(0,9,0)2_{20}$	$(0,10,0)2_{11}$	$(0,11,0)_{K_a=0}$	$(0,12,0)_{K_a=1}$
g′	1.9963(19)	1.99470(62)	2.106(63)	1.891(74)
Ref.	[7]	[7]	[8]	[8]

Ab initio-calculated hf coupling constants for the states $\tilde{X}\,^2B_1$ and $\tilde{A}\,^2A_1$ of NH_2 [9, 10] and isotopomers [10] are in good agreement with experimental data (Table 9). Additional ab initio calculations [11 to 27] were performed on NH_2 (or isotopomers [14]) in the ground state $\tilde{X}\,^2B_1$ and in the excited states $\tilde{A}\,^2A_1$ [11, 20] and $\tilde{B}\,^2B_2$ [20]. For semiempirical calculations, see [21, 28 to 37].

Table 9
NH_2 and NHD: Fermi Contact hf Coupling Constant A_{iso} and Dipole hf Coupling Constants A_{aa}, A_{bb}, and A_{cc} (in MHz) from Microwave Spectra.
Anisotropic constants are given in terms of the principle axis system (b ∥ C_2, c ⊥ molecular plane).

radical	A_{iso}	A_{aa}	A_{bb}	A_{cc}	remarks	Ref.
ground state $\tilde{X}$						
NH_2 (^{14}N)	27.88(20)	−43.38(30)	−44.75	88.13(18)	a) b) c) d)	[38]
(H)	−67.59(28)	17.54(63)	−12.75	−4.79(47)	a) c)	[38]

Table 9 (continued)

radical	A_{iso}	A_{aa}	A_{bb}	A_{cc}	remarks	Ref.
NHD (^{14}N)	28.0	−43.0	−44.4	87.5	a)	[39]
(H)	−67.2	−25.5	30.5	−5.2	a)	[39]
(D)	−10.3	8.2	−7.4	−0.8	a)	[39]
ND_2 (^{14}N)	28.055(33)	−43.136(48)	−44.272(63)	87.408	e) f)	[40]
(D)	−10.241(28)	2.874(45)	−2.108(69)	−0.766	e) f)	[40]
Ã $^2A_1$(0,9,0)						
NH_2 (^{14}N)	154.6(26)	−48.7(49)	–	–	a) g) h) i)	[4]
(H)	49.4(55)	48.5(109)	–	–	a)	[4]
Ã $^2A_1$(0,10,0)						
NH_2 (^{14}N)	153.0(13)	−39.5(33)	76.9(14)	−37.4(19)	a) d) g)	[4]
(H)	52.2(14)	59.5(34)	−9.5(16)	−50.0(20)	a)	[4]
Ã $^2A_1$(0,11,0)						
NH_2 (^{14}N)	151(10)	−26(9)	79(6)	–	k)	[8]
Ã $^2A_1$(0,12,0)						
NH_2 (^{14}N)	155(8)	−60(7)	–	–	k)	[8]

a) From MODR spectra. – b) Additional values from MODR spectra [41, 42]. – c) Additional values from LMR spectra [6, 43]. – d) Additional values from saturation spectra [1]. – e) From MWA spectra. – f) Additional values from MODR spectra [44]. – g) Additional values from IODR spectra [7]. – h) Additional values from OODR spectra [45]. – i) Additional values from magnetic level-crossing spectra [2, 3]. – k) From IODR spectra.

Solid Phase. There exists in the literature a wide range of isotropic and anisotropic hf coupling constants which were determined from electron spin resonance (ESR) spectra of NH_2 in the solid state. Isotropic g factors and hf coupling constants (in MHz) are compiled in the following table:

solid matrix	Ar	NH_3	NH_3	$K^{15}NH_2SO_3$ [a)]
T in K	4.2	4.2	77	77
g_{iso}	2.0048(1)	2.0038	2.0046(3)	–
$A_{iso}(^{14}N)$	28.9 ± 0.2	31.7	42.6	52.1 ± 5 (^{15}N)
A_{iso}(H)	−67.03 ± 0.2 [b)]	−67.0	−71.3 [b)]	−76.5 ± 5 [b)]
Ref.	[46]	[47]	[48]	[49]

[a)] Single crystal. – [b)] The negative sign of A_{iso}(H) was established in [47].

Calculations showed the increase of the ^{14}N hf coupling constant A_{iso} from the gaseous to the solid state is caused by distortion of the geometry of NH_2. The nearly equal A_{iso} values of NH_2 in the gas phase and Ar matrix (at 4 K) indicate free rotation of NH_2 in the matrix [46, 50].

Additional values of hf coupling constants and/or the g factor determined from ESR spectra were reported for NH_2 and/or ND_2 in solid Ar [50, 51], Kr [51, 52], Xe [51], NH_3 [48, 53, 54], ND_3 [48, 53, 55], frozen ammonia-water system (77 K) [54, 56, 57], aqueous glasses containing NaN_3 [58], or single crystals of NH_3OHCl and $(NH_3OH)_2SO_4$ [59]. Several hf coupling constants were derived from ESR spectra of NH_2 and/or isotopomers in zeolites [60 to 62].

References:

[1] Hills, G. W.; Philen, D. L.; Curl, R. F., Jr.; Tittel, F. K. (Chem. Phys. **12** [1976] 107/11).
[2] Dixon, R. N.; Field, D. (Mol. Phys. **34** [1977] 1563/76).
[3] Dixon, R. N.; Field, D. (Lasers Chem. Proc. Conf., London 1977, pp. 145/9; C.A. **88** [1978] No. 81234).
[4] Hills, G. W.; Brazier, C. R.; Brown, J. M.; Cook, J. M.; Curl, R. F., Jr. (J. Chem. Phys. **76** [1982] 240/52).
[5] Dixon, R. N. (Mol. Phys. **10** [1965] 1/6).
[6] Davies, P. B.; Russell, D. K.; Thrush, B. A.; Radford, H. E. (Proc. R. Soc. London A **353** [1977] 299/318).
[7] Amano, T.; Kawaguchi, K.; Kakimoto, M.; Saito, S.; Hirota, E. (J. Chem. Phys. **77** [1982] 159/67).
[8] Kawaguchi, K.; Suzuki, T.; Saito, S.; Hirota, E. (J. Opt. Soc. Am. B Opt. Phys. **4** [1987] 1203/11).
[9] Engels, B.; Perić, M.; Reuter, W.; Peyerimhoff, S. D.; Grein, F. (J. Chem. Phys. **96** [1992] 4526/35).
[10] Perić, M.; Engels, B. (J. Chem. Phys. **97** [1992] 4996/5006).

[11] Nakatsuji, H.; Izawa, M. (J. Chem. Phys. **97** [1992] 435/9).
[12] Nakano, T.; Morihashi, K.; Kikuchi, O. (Bull. Chem. Soc. Jpn. **65** [1992] 603/5).
[13] Nakano, T.; Morihahsi, K.; Kikuchi, O. (Chem. Phys. Lett. **186** [1991] 572/6).
[14] Funken, K.; Engels, B.; Peyerimhoff, S. D.; Grein, F. (Chem. Phys. Lett. **172** [1990] 180/6).
[15] Sekino, H.; Bartlett, R. J. (J. Chem. Phys. **82** [1985] 4225/9).
[16] Bird, S.; Claxton, T. A. (J. Chem. Soc. Faraday Trans. II **80** [1984] 851/60).
[17] Thuomas, K. A.; Eriksson, A.; Lund, A. (J. Magn. Reson. **37** [1980] 223/9).
[18] Hinchliffe, A.; Bounds, D. G. (J. Mol. Struct. **54** [1979] 231/8).
[19] Brown, R. D.; Williams, G. R. (Chem. Phys. **3** [1974] 19/34).
[20] Brown, R. D.; Williams, G. R. (Mol. Phys. **25** [1973] 673/94).

[21] Zhidomirov, G. M.; Chuvylkin, N. D. (Theor. Chim. Acta **30** [1973] 197/204).
[22] Claxton, T. A.; McWilliams, D.; Smith, N. A. (Chem. Phys. Lett. **4** [1970] 505/6).
[23] Claxton, T. A. (Trans. Faraday Soc. **66** [1970] 1537/9).
[24] Meyer, W. (J. Chem. Phys. **51** [1969] 5149/62).
[25] Chung, A. L. H. (J. Chem. Phys. **46** [1967] 3144/58).
[26] Sutcliffe, B. T. (J. Chem. Phys. **39** [1963] 3322/6).
[27] Higuchi, J. (J. Chem. Phys. **35** [1961] 2270/1).
[28] Vetter, R.; Kuehnel, W.; Gey, E. (J. Mol. Struct. **90** [1982] 71/80 [THEOCHEM **7**]).
[29] Mehrotra, P. K.; Chandrasekhar, J.; Manoharan, P. T.; Subramanian, S. (Chem. Phys. Lett. **68** [1979] 219/21).
[30] Pandey, P. K. K.; Chandra, P. (Can. J. Chem. **57** [1979] 3126/34).

[31] Tino, J.; Klimo, V. (Chem. Phys. Lett. **25** [1974] 427/30).
[32] Chuvylkin, N. D.; Zhidomirov, G. M. (J. Magn. Reson. **11** [1973] 367/72).
[33] Zhidomirov, G. M.; Chuvylkin, N. D. (Chem. Phys. Lett. **14** [1972] 52/6).
[34] Zhidomirov, G. M.; Chuvylkin, N. D.; Brotikovskii, O. I. (Chem. Phys. Lett. **10** [1971] 341/4).
[35] Chiu, M. F.; Gilbert, B. C.; Sutcliffe, B. T. (J. Phys. Chem. **76** [1972] 553/64).
[36] Cook, D. B.; Hinchliffe, A.; Palmieri, P. (Chem. Phys. Lett. **3** [1969] 223/5).
[37] Chuvylkin, N. D.; Zhidomirov, G. M. (Mol. Phys. **25** [1973] 1233/5).
[38] Hills, G. W.; Cook, J. M. (J. Mol. Spectrosc. **94** [1982] 456/60).
[39] Steimle, T. C.; Brown, J. M.; Curl, R. F., Jr. (J. Chem. Phys. **73** [1980] 2552/8).
[40] Kanada, M.; Yamamoto, S.; Saito, S. (J. Chem. Phys. **94** [1991] 3423/8).

[41] Cook, J. M.; Hills, G. W.; Curl, R. F., Jr. (J. Chem. Phys. **67** [1977] 1450/61).
[42] Hills, G. W.; Cook, J. M.; Curl, R. F., Jr.; Tittel, F. K. (J. Chem. Phys. **65** [1976] 823/8).
[43] Davies, P. B.; Russell, D. K.; Thrush, B. A.; Radford, H. E. (Chem. Phys. Lett. **42** [1976] 35/8).
[44] Cook, J. M.; Hills, G. W. (J. Chem. Phys. **78** [1983] 2144/53).
[45] Dixon, R. N.; Field, D.; Noble, M. (Proc. 3rd Yamada Conf. Free Radicals, Sanda, Jpn., 1979, pp. 239/42; C.A. **93** [1980] No. 57382).
[46] Foner, S. N.; Cochran, E. L.; Bowers, V. A.; Jen, C. K. (Phys. Rev. Lett. **1** [1958] 91/2).
[47] Michaut, J. P.; Roncin, J.; Marx, R. (Chem. Phys. Lett. **36** [1975] 599/605).
[48] Smith, D. R.; Seddon, W. A. (Can. J. Chem. **48** [1970] 1938/42).
[49] Morton, J. R.; Smith, D. R. (Can. J. Chem. **44** [1966] 1951/5).
[50] Cochran, E. L.; Adrian, F. J.; Bowers, V. A. (J. Chem. Phys. **51** [1969] 2759/61).

[51] Fischer, P. H. H.; Charles, S. W.; McDowell, C. A. (J. Chem. Phys. **46** [1967] 2162/6).
[52] Coope, J. A. R.; Farmer, J. B.; Gardner, C. L.; McDowell, C. A. (J. Chem. Phys. **42** [1965] 2628/9).
[53] Marx, R.; Maruani, J. (J. Chim. Phys. **61** [1964] 1604/9).
[54] Symons, M. C. R.; Rao, K. V. S. (J. Chem. Soc. A **1971** 2163/6).
[55] Vertii, A. A.; Ivanchenko, I. V.; Popenko, N. A.; Tarapov, S. I.; Shestopalov, V. P.; Belyaev, A. A.; Get'man, V. A.; Dzyubak, A. P.; Karnaukhov, I. M.; Lukhanin, A. A.; Sorokin, P. V.; Sporov, E. A.; Tolmachev, I. A. (Dokl. Akad. Nauk SSSR **314** [1990] 1389/91; Dokl. Phys. [Engl. Transl.] **35** [1990] 899/901).
[56] Al-Naimy, B. S.; Moorthy, P. N.; Weiss, J. J. (J. Phys. Chem. **70** [1966] 3654/60).
[57] Tupikov, V. I.; Pshezhetskii, S. Ya. (Zh. Fiz. Khim. **38** [1964] 2511/3; Russ. J. Phys. Chem. **38** [1964] 1364/5).
[58] Ginns, I. S.; Symons, M. C. R. (J. Chem. Soc. Faraday Trans. II **68** [1972] 631/7).
[59] Koksal, F.; Cakir, O.; Gumrukcu, I.; Birey, M. (Z. Naturforsch. **40a** [1985] 903/5).
[60] Suzuno, Y.; Shiotani, M.; Sohma, J. (Chem. Phys. Lett. **44** [1976] 177/9).

[61] Vansant, E. F.; Lunsford, J. H. (J. Phys. Chem. **76** [1972] 2716/8).
[62] Sorokin, Yu. A.; Kotov, A. G.; Pshezhetskii, S. Ya. (Dokl. Akad. Nauk SSSR **159** [1964] 1385/8; Dokl. Phys. Chem. [Engl. Transl.] **154/159** [1964] 1163/5).

2.1.7.2.6 Rotational Constants. Centrifugal Distortion Constants. Spin-Rotation Coupling Constants

Electronic Ground State

NH_2 Radical. The NH_2 radical is an asymmetric top with the asymmetry parameter $\kappa = (2B-A-C)/(A-C) = -0.38$ (axes $b \parallel C_2$, $c \perp$ molecular plane). An increase of the rotational quantum number N leads to a change from prolate- to oblate-top behavior. The rotational constants A, B, and C, the centrifugal distortion constants Δ_K, Δ_{NK}, Δ_N, δ_K, and δ_N, and the spin-rotational coupling constants A_s, B_s, and C_s, for the vibrational ground and excited states are listed in Table 10, p. 182. The rotational Hamiltonian used for fitting the spectroscopic data is a combination of the A-reduced asymmetric rotor $\hat{H}_{ROT}$ [1] and the spin-rotation Hamiltonian $\hat{H}_{SR}$ [2]:

$$\hat{H}_{ROT} = A\hat{N}_a^2 + B\hat{N}_b^2 + C\hat{N}_c^2 - \Delta_N\hat{N}^4 - \Delta_{NK}\hat{N}^2\hat{N}_a^2 - \Delta_K\hat{N}_a^4 - 1/2[\delta_N\hat{N}^2 + \delta_K\hat{N}_a^2, \hat{N}_+^2 + \hat{N}_-^2]_+ + \ldots,$$
$$\hat{H}_{SR} = A_s\hat{N}_a\hat{S}_a + B_s\hat{N}_b\hat{S}_b + C_s\hat{N}_c\hat{S}_c - \ldots.$$

Higher order terms for rotation (sextic centrifugal distortion constants H_K, H_{KN}, H_{NK}, H_N, h_K, h_{NK}, h_N plus the eigth-power term L_K) and spin-rotation coupling (quartic centrifugal distortion constants Δ^s_K, Δ^s_{KN}, Δ^s_{NK}, Δ^s_N, δ^s_K, δ^s_N including the sextic term H^s_K) were reported, for example, in [3 to 7].

Table 10
$NH_2(\tilde{X}\,^2B_1)$. Rotational Constants A, B, and C, Spin-Rotation Coupling Constants A_s, B_s, and C_s, and Centrifugal Distortion Constants Δ_K, Δ_{NK}, Δ_N, δ_K, and δ_N for the Vibrational Ground State and Excited States (in cm^{-1}).

constant	vibrational state (v_1,v_2,v_3)				
	(0,0,0)	(0,1,0)	(0,2,0)	(1,0,0)	(0,0,1)
A	23.69299(14)	25.96802(21)	28.87719(42)	23.04758(24)	22.64606(43)
B	12.952008(88)	13.11085(10)	13.26849(39)	12.76028(28)	12.87697(25)
C	8.172737(55)	8.033726(56)	7.89878(38)	8.00949(28)	8.05816(15)
$\Delta_K \times 10^2$	2.1953(14)	3.5676(32)	5.8714(42)	2.0566(27)	2.0010(79)
$\Delta_{NK} \times 10^2$	−0.41603(77)	−0.5427(11)	−0.6938(24)	−0.3986(21)	−0.4248(22)
$\Delta_N \times 10^2$	0.10582(19)	0.11790(20)	0.13317(71)	0.10607(47)	0.10880(43)
$\delta_K \times 10^2$	0.09943(97)	0.2478(14)	0.5116(45)	0.0971(33)	0.0879(27)
$\delta_N \times 10^3$	0.4252(11)	0.4860(11)	0.5625(45)	0.4281(30)	0.4351(22)
A_s	−0.30890(12)	−0.40339(42)	−0.5452(17)	−0.29452(42)	−0.28212(89)
B_s	−0.045214(55)	−0.04819(28)	−0.5015(69)	−0.04446(20)	−0.04545(47)
$C_s \times 10^3$	0.404(34)	0.72(17)	0.78(58)	0.16(13)	−0.12(41)
remarks	a)	a), b)	c), d)	c), e)	c), f)
Ref.	[4]	[4]	[3]	[3]	[3]

a) From the fit of infrared frequencies of the v_2 band [4] and of MODR data [2]. – b) Additional values in [7 to 11]. – c) From an analysis of the coupled (1,0,0), (0,0,1), and (0,2,0) vibrational states. Frequencies from the IR absorption spectrum [3, 12] and the $\tilde{A}\rightarrow\tilde{X}(0,2,0)$, $\tilde{X}(1,0,0)$ visible and near-infrared band systems [3]. – d) Additional values in [7, 11]. – e) Additional values in [6, 11, 12]. – f) Additional values in [12].

Additional concurring constants of $NH_2(0,0,0)$ were determined from MODR [13 to 16], LMR [8, 17, 18], IR [4] as well as optical [7, 10, 19] spectra, and combinations of some of these investigations [20].

The (0,0,0) parameters [4] in Table 10 were suggested as the best available for describing the rotational levels with $N<10$ and $K<6$ [3]. For a wider range of rotational levels with higher N and K values, the (0,0,0)-state parameters [7], obtained from a larger set of spectroscopic data, may be more reliable [4].

The values of A and A_s constants increase rapidly with increasing v_2 as the molecule approaches linearity:

constant	vibrational state $(0,v_2,0)$			
	(0,3,0)	(0,4,0)	(0,5,0)	(0,6,0)
A in cm^{-1}	32.4915(40)	37.564(23)	44.266(51)	50.57(11)
A_s in cm^{-1}	−0.764(24)	−1.16(5)	−2.00(6)	−4.00(28)
remarks	a)	b), c)	b)	b)

a) From a rotational analysis of the $\tilde{A}\,^2A_1(0,v_2',0)\rightarrow\tilde{X}\,^2B_1(0,3,0)$ band systems [5]. – b) From laser-induced fluorescence (LIF) spectra [21]. – c) Additional data in [11].

Rotational constants determined from LIF spectra were reported for NH_2 in the (1,1,0) state [11]. Rotational, centrifugal distortion, and spin-rotation coupling constants were determined from LIF experiments for NH_2 in the (1,3,0) and (2,1,0) states [21] and were derived for $^{14}NH_2$ and $^{15}NH_2$ in the (0,0,0) state from an ab initio-calculated general quartic valence force field [22].

ND_2 and NHD Radicals. Rotational, centrifugal distortion, and spin-rotation coupling constants of ND_2 are given (in cm^{-1}) in the following table. Additional constants were determined from LMR [20] and MODR spectra [23].

constant	vibrational state		constant	vibrational state	
	(0,0,0) [24] a)	(0,1,0) [25] b)		(0,0,0) [24] a)	(0,1,0) [25] b)
A	13.342093(3)	14.2344(5)	$\Delta_K \times 10^3$	6.5636(11)	9.23(4)
B	6.487443(5)	6.5513(5)	$\Delta_{NK} \times 10^3$	−1.1278(8)	−1.31(5)
C	4.290101(2)	4.2379(5)	$\Delta_N \times 10^3$	0.2443(3)	0.280(12)
A_s	−0.171045(4)	−0.2054(27)	$\delta_K \times 10^3$	0.2101(13)	0.49(12)
B_s	−0.022299(3)	0.0233(11)	$\delta_N \times 10^3$	0.101(1)	0.115(7)
$C_s \times 10^3$	0.114(2)	0.108 c)			

a) From MW spectroscopic data. — b) From a fit of MODR, LMR, IODR, and optical spectroscopic data [25]. — c) Fixed at the (0,0,0) value [25].

The analysis of MODR spectra of NHD gave the vibrational ground state constants (in cm^{-1}): A=20.1150(17), B=8.1114(6), C=5.6674(3), A_s= −0.2348(2), B_s= −0.0384(1), C_s= $2.6(13) \times 10^{-4}$, $\Delta_K = 8.81(19) \times 10^{-3}$, $\Delta_{NK} = 9.7(5) \times 10^{-4}$, $\Delta_N = 3.2(1) \times 10^{-4}$, $\delta_K = 1.52(4) \times 10^{-3}$, and $\delta_N = 1.02(2) \times 10^{-4}$ [26]. These values are in good agreement with constants obtained from an analysis of the $\tilde{A}\,^2A_1(0,v_2',0) \rightarrow \tilde{X}\,^2B_1(0,0,0)$ emission band systems [27].

Spin-rotation coupling constants $^{14}ND_2$ and ^{14}NHD were calculated [28] using a reduced spin-rotation Hamiltonian [2]. Rotational, centrifugal distortion, and spin-rotation coupling constants of $^{14}ND_2$ and ^{14}NHD in the (0,0,0) state which were derived from an ab initio-calculated general valence force field of 4th order [22] agree well with experimentally derived constants.

Electronically Excited State $\tilde{A}\,^2A_1$

The rotational constants B=8.78 for NH_2 and B=4.41 cm^{-1} for ND_2 were obtained from the analysis of the $\tilde{A}\,^2A_1(0,v_2',0) \leftarrow \tilde{X}\,^2B_1(0,0,0)$ absorption band systems for a wide range of bending vibrational excitations [19].

References:

[1] Watson, J. K. G. (in: Durig, J. R.; Vibrational Spectra and Structure, Vol. 6, Dekker, New York 1977, pp. 1/89).

[2] Brown, J. M.; Sears, T. J. (J. Mol. Spectrosc. **75** [1979] 111/33).

[3] McKellar, A. R. W.; Vervloet, M.; Burkholder, J. B.; Howard, C. J. (J. Mol. Spectrosc. **142** [1990] 319/35).

[4] Burkholder, J. B.; Howard, C. J.; McKellar, A. R. W. (J. Mol. Spectrosc. **127** [1988] 415/24).

[5] Merienne-Lafore, M. F.; Vervloet, M. (J. Mol. Spectrosc. **108** [1984] 160/2).

[6] Vervloet, M.; Merienne-Lafore, M. F. (Can. J. Phys. **60** [1982] 49/55).
[7] Birss, F. W.; Merienne-Lafore, M. F.; Ramsay, D. A.; Vervloet, M. (J. Mol. Spectrosc. **85** [1981] 493/5).
[8] Kawaguchi, K.; Yamada, C.; Hirota, E.; Brown, J. M.; Buttenshaw, J.; Parent, C. R.; Sears, T. J. (J. Mol. Spectrosc. **81** [1980] 60/72).
[9] Vervloet, M.; Merienne-Lafore, M. F.; Ramsay, D. A. (Chem. Phys. Lett. **57** [1978] 5/7).
[10] Johns, J. W. C.; Ramsay, D. A.; Ross, S. C. (Can. J. Phys. **54** [1976] 1804/14).

[11] Kroll, M. (J. Chem. Phys. **63** [1975] 319/25).
[12] Amano, T.; Bernath, P. F.; McKellar, A. R. W. (J. Mol. Spectrosc. **94** [1982] 100/13).
[13] Hills, G. W.; Cook, J. M.; Curl, R. F., Jr.; Tittel, F. K. (J. Chem. Phys. **65** [1976] 823/8).
[14] Cook, J. M.; Hills, G. W.; Curl, R. F., Jr. (J. Chem. Phys. **67** [1977] 1450/61).
[15] Hills, G. W.; Lowe, R. S.; Cook, J. M.; Curl, R. F., Jr. (J. Chem. Phys. **68** [1978] 4073/6).
[16] Hills, G. W.; Cook, J. M. (J. Mol. Spectrosc. **94** [1982] 456/60).
[17] Davies, P. B.; Russell, D. K.; Thrush, B. A.; Radford, H. E. (Chem. Phys. Lett. **42** [1976] 35/8).
[18] Davies, P. B.; Russell, D. K.; Thrush, B. A.; Radford, H. E. (Proc. R. Soc. London A **353** [1977] 299/318).
[19] Dressler, K.; Ramsay, D. A. (Philos. Trans. R. Soc. London A **251** [1959] 553/604).
[20] Birss, F. W.; Ramsay, D. A.; Ross, S. C.; Zauli, C. (J. Mol. Spectrosc. **78** [1979] 344/6).

[21] Vervloet, M. (Mol. Phys. **63** [1988] 433/49).
[22] Martin, J. M. L.; François, J. P.; Gijbels, R. (J. Chem. Phys. **97** [1992] 3530/6).
[23] Cook, J. M.; Hills, G. W. (J. Chem. Phys. **78** [1983] 2144/53).
[24] Kanada, M.; Yamamoto, S.; Saito, S. (J. Chem. Phys. **94** [1991] 3423/8).
[25] Muenchausen, R. E.; Hills, G. W.; Merienne-Lafore, M. F.; Ramsay, D. A.; Vervloet, M.; Birss, F. W. (J. Mol. Spectrosc. **112** [1985] 203/10).
[26] Steimle, T. C.; Brown, J. M.; Curl, R. F., Jr. (J. Chem. Phys. **73** [1980] 2552/8).
[27] Ramsay, D. A.; Wayne, F. D. (Can. J. Phys. **57** [1979] 761/6).
[28] Brown, J. M.; Sears, T. J.; Watson, J. K. G. (Mol. Phys. **41** [1980] 173/82).

2.1.7.2.7 Moments of Inertia. Bond Length. Bond Angle

Electronic Ground State $\tilde{X}\,^2B_1$. The following moments of inertia I_A, I_B, I_C and inertial defect Δ ($=I_C-I_A-I_B$), all in amu·Å^2, were calculated for NH_2 and ND_2 from the rotational constants of the (0,0,0) state.

radical	I_A	I_B	I_C	Δ	Ref.
NH_2	0.7117(5)	1.3032(6)	2.0611(8)	0.0462	[1]
ND_2	1.2646	2.597	3.920	0.058	[2]

The bond lengths r_0(NH) = 1.02454, 1.02458 [3], 1.0244 Å [4] and the bond angle α_0(HNH) = 103°20′ [3, 4] were derived from rotational constants obtained from LMR [3, 4] and $\tilde{A} \leftarrow \tilde{X}$ absorption [3, 5] spectra. The angle α decreases by 7′ on complete deuteration of NH_2 [2]. LIF experiments showed a decrease of α to 92.2° on excitation of the rotation around the a axis up to $K_a = 21$ [6].

The NH_2 geometric structure was calculated by ab initio [7 to 17] and semiempirical [18 to 22] methods.

Electronically Excited State $\tilde{A}\,^2A_1$. Theoretical studies on the effects of orbital angular momentum in nonlinear molecules [23, 24] showed NH_2 to be nonlinear at equilibrium ($\alpha_e = 144° \pm 5°$), with a small barrier to linearity (see p. 171) [24]. An empirical fit of the bending potential function confirmed the slightly bent nuclear arrangement yielding $\alpha_e = 144.2°$ and $r_e = 1.007$ Å [25]. Ab initio treatments established this structure [10, 16]. These results corrected the earlier conclusion of a linear structure indicated by an analysis of the $\tilde{X} \leftarrow \tilde{A}$ absorption spectrum [2].

Additional Electronically Excited States. Ab initio calculations on the valence state $\tilde{B}\,^2B_2$ gave the geometry $r_e = 1.162$ Å and $\alpha_e = 47.5°$ [14]. Extended theoretical studies verified the $\tilde{B}\,^2B_2$ state structure to be in the range 1.130 Å $\leq r_e \leq$ 1.227 Å and $46.5° \leq \alpha_e \leq 50.0°$ [13, 22]. A considerably smaller α_e (26.5°) and higher r_e (1.62 Å) value resulted from an earlier ab initio study [15].

For the geometry of NH_2 in the Rydberg states $^4\Sigma_g^-(^4B_1)$, $^2\Sigma_g^-(2\,^2B_1)$, $^2\Delta_g(2\,^2A_1, 3\,^2B_1)$ and $^2\Sigma_g^+(^2A_1)$ [13], see p. 173. Geometry optimization was carried out on core-Rydberg excited states of NH_2 by ab initio CI calculations [9, 11].

References:

[1] Duxbury, G. (J. Mol. Spectrosc. **25** [1968] 1/11).
[2] Dressler, K.; Ramsay, D. A. (Philos. Trans. R. Soc. London A **251** [1959] 553/604).
[3] Davies, P. B.; Russell, D. K.; Thrush, B. A.; Radford, H. E. (Proc. R. Soc. London A **353** [1977] 299/318).
[4] Davies, P. B.; Russell, D. K.; Thrush, B. A.; Radford, H. E. (Chem. Phys. Lett. **42** [1976] 35/8).
[5] Johns, J. W. C.; Ramsay, D. A.; Ross, S. C. (Can. J. Phys. **54** [1976] 1804/14).
[6] Dixon, R. N.; Irving, S. J.; Nightingale, J. R.; Vervloet, M. (J. Chem. Soc. Faraday Trans. **87** [1991] 2121/33).
[7] Jensen, P.; Buenker, R. J.; Hirsch, G.; Rai, S. N. (Mol. Phys. **70** [1990] 443/54).
[8] Watts, J. D.; Trucks, G. W.; Bartlett, R. J. (Chem. Phys. Lett. **164** [1989] 502/8).
[9] Cambi, R.; Ciullo, G.; Sgamellotti, A.; Tarantelli, F.; Guest, M. F. (Chem. Phys. Lett. **91** [1982] 178/84).
[10] Buenker, R. J.; Perić, M.; Peyerimhoff, S. D.; Marian, R. (Mol. Phys. **43** [1981] 987/1014).

[11] Clark, D. T.; Sgamellotti, A.; Tarantelli, F. (Chem. Phys. **52** [1980] 1/9).
[12] Hinchliffe, A.; Bounds, D. G. (J. Mol. Struct. **54** [1979] 231/8).
[13] Peyerimhoff, S. D.; Buenker, R. J. (Can. J. Chem. **57** [1979] 3182/9).
[14] Bell, S.; Schaefer, H. F., III (J. Chem. Phys. **67** [1977] 5173/7).
[15] Brown, R. D.; Williams, G. R. (Mol. Phys. **25** [1973] 673/94).
[16] Bender, C. F.; Schaefer, H. F., III (J. Chem. Phys. **55** [1971] 4798/803).
[17] Del Bene, J. (J. Chem. Phys. **54** [1971] 3487/90).
[18] Bews, J. R.; Glidewell, C. (Inorg. Chim. Acta **39** [1980] 217/25).
[19] Canuto, S. R. A.; Vyanna, J. D. M. (J. Phys. B **8** [1975] 2987/94).
[20] Zahradník, R.; Čársky, P. (Theor. Chim. Acta **27** [1972] 121/34).

[21] Gordon, M. S.; Pople, J. A. (J. Chem. Phys. **49** [1968] 4643/50).
[22] Polak, R.; Vojtik, J. (Chem. Phys. **114** [1987] 43/54).
[23] Pople, J. A.; Longuet-Higgins, H. C. (Mol. Phys. **1** [1958] 372/83).
[24] Dixon, R. N. (Mol. Phys. **9** [1965] 357/66).
[25] Jungen, C.; Hallin, K. E. J.; Merer, A. J. (Mol. Phys. **40** [1980] 25/63).

2.1.7.2.8 Molecular Vibrations

2.1.7.2.8.1 Fundamental, Overtone, and Combination Frequencies

Electronic Ground State

NH_2 Radicals. The following table lists the frequencies of NH_2 in the gas phase of the symmetric stretching $\nu_1(A_1)$, the symmetric bending $\nu_2(A_1)$, the antisymmetric stretching $\nu_3(B_1)$, and the $2\nu_2$ overtone. The fundamentals were determined from high-resolution IR spectra [1, 2], and the overtone was derived from the emission spectrum in the visible and near-IR region [1]. The following band origins correspond to the $0_{0,0}$ rotational level:

	ν_1	ν_2	$2\nu_2$	ν_3
ν in cm^{-1}	3219.3723(4)	1497.3184(3)	2961.2369(26)	3301.1098(7)
remarks	a), b), c)	e)	a), d)	b)
Ref.	[1]	[2]	[1]	[1]

a) By Fermi interaction between the ν_1 and $2\nu_2$ vibrations perturbed band origins. The hypothetical unperturbed band origins, $2\nu_2=2964.5026$ and $\nu_1=3216.1066$ cm^{-1}, correspond to the Fermi interaction parameter $W=28.85(8)$ cm^{-1} [1]. — b) By Coriolis coupling between the ν_1 and ν_3 vibrations perturbed band origins. A further IR study gave $\nu_1=3219.371$ and $\nu_3=3301.110$ cm^{-1} [3]. — c) $\nu_1=3219.36$ cm^{-1} from the analysis of $\tilde{A}\,^2A_1\rightarrow\tilde{X}\,^2B_1(1,0,0)$ emission band systems [4]. — d) $2\nu_2=2961.210(4)$ cm^{-1} from visible emission study [5]. — e) Earlier ν_2 measurements from laser magnetic resonance (LMR) [6], visible absorption [7, 8] and emission [5, 9] studies.

Frequencies of the bending fundamental and overtones $(0,\nu_2,0)$ up to the region of Renner-Teller interactions (see p. 171) and of combination vibrations are compiled in Table 11. Levels are perturbed by Fermi resonance between (ν_1,ν_2,ν_3) and $(\nu_1\pm1,\nu_2\mp2,\nu_3)$ states and by Darling-Denison resonance between (ν_1,ν_2,ν_3) and $(\nu_1\pm2,\nu_2,\nu_3\mp2)$ states [10]. Evidence for Fermi perturbation of (0,3,0) by (1,1,0) was found near $K''_a=8$ [11]. The relatively large difference (28 cm^{-1}) between the observed and calculated (0,8,0) frequency at $K_a=1$ indicated the resonance of the (0,8,0) with the (0,3,2) level [10].

Table 11
$NH_2(\tilde{X}\,^2B_1)$: Frequencies of Fundamental, Overtone, and Combination Vibrations (ν_1,ν_2,ν_3) (in cm^{-1}) for the Rotational Ground State $0_{0,0}$.

ν_2	$(0,\nu_2,0)$	$(1,\nu_2,0)$	$(2,\nu_2,0)$	$(0,\nu_2,2)$
10	13448.64 a)	–	–	–
9	12181 b)	–	–	–
8	10948.47 c)	–	–	–
7	9716.90 c)	–	–	–
6	8451.42 c)	–	–	–
5	7140.35 d)	10281.48 a)	–	–
4	5785.57 c)	8942.59 a)	–	–
3	4391.35(1) e)	7564.63 a)	–	–
2	2961.2369(26) f)	6150.27 a)	9227.14 a)	9421.48 a)
1	1497.3184(3) g)	4705.8 ± 1 h)	7804.69 a)	8000.36 a)
0	–	3219.3723(4) f)	6332.62 a)	6539 a)

a) From laser-induced fluorescence spectra [10]. — b) Value extrapolated because of the strong perturbation due to resonance between the states (0,9,0) and (0,3,2) for $K_a=0$ [10]. — c) From electronic emission spectra [11]. — d) Extrapolated [10]. — e) From a rotational analysis of the $\tilde{A}\,^2A_1(0,v_2',0) \rightarrow \tilde{X}\,^2B_1(0,3,0)$ emission band system [12]. — f) From IR spectrum [1]. — g) From IR spectrum [2]. — h) From LIF spectrum [13].

Ab initio-calculated vibrational frequencies are published for $v_1 \leq 2$, $v_2 \leq 6$, and $v_3 \leq 2$ [14]. Bending overtones $(0,v_2,0)$ (up to $v_2=22$) were calculated by using empirical potential curves [15, 16] or ab initio procedures [17 to 19].

The fundamental vibrations of NH_2 in a matrix [20, 21] are given on p. 201.

ND_2 and NHD Radicals. Experimental frequencies of the bending vibration were reported for gaseous ND_2 and for NHD in a matrix. Stretching and bending vibrational frequencies of $^{14}ND_2$ and ^{14}NHD were derived from an ab initio-computed general valence force field of 4th order [22] which reproduces the experimental NH_2 fundamentals to within a few cm^{-1}. The frequencies of ν_1, ν_2, $2\nu_2$, and ν_3 vibrations (in cm^{-1}) are compiled in the following table:

radical	ν_1	ν_2		$2\nu_2$	ν_3
$^{14}ND_2$	2359.2	1108.7450(16)	1112.1	2195.75	2458.0
^{14}NHD	2404.1	1321	1322.7	—	3258.2
method	ab initio a)	exp b)	ab initio a)	exp c)	ab initio a)

a) From an all-electron (quadratic configuration interaction with all single and double excitations) potential [22]. — b) For gaseous ND_2 derived [23] from LMR [24] and IODR frequencies [25]. For NHD in a nitrogen matrix from an IR spectrum [20]. — c) For gaseous ND_2 from an LMR spectrum [24].

Overtone and combination frequencies of ND_2 and NHD were ab initio-calculated for $v_1 \leq 2$, $v_2 \leq 6$, $v_3 \leq 3$ [14] and for $v_1=v_3=0$, $v_2 \leq 20$ [17].

Vibrational frequencies of ND_2 in a matrix [20, 21] are given on p. 201.

Electronically Excited State $\tilde{A}\,^2A_1$

Fundamental vibrations of NH_2 and ND_2 were derived from $\tilde{A} \leftarrow \tilde{X}$ absorption band systems [26]: $\nu_1(A_1)=3325$ and $\nu_2(A_1)=633$ cm^{-1} for NH_2 [27]; $\nu_1(A_1)=2520$ and $\nu_2(A_1)=430$ cm^{-1} for ND_2 [28].

The fundamental vibration ν_2 and the overtones $n\nu_2$ may be evaluated from the vibronic term values $T(0,v_2,0)=T_e+G(0,v_2,0)$. $T(0,v_2,0)$ values of NH_2 were determined from the $\tilde{A}$-$\tilde{X}$ absorption [8, 26] and emission [11] bands (see p. 206) and calculated using empirical bending vibration potential curves [15, 16] or ab initio procedures [17 to 19, 29]. $T(0,v_2,0)$ values were determined from the $\tilde{A}$-$\tilde{X}$ absorption bands of ND_2 [26] and NHD [30] (see p. 205) and were calculated for ND_2 and NHD by ab initio methods [17].

References:

[1] McKellar, A. R. W.; Vervloet, M.; Burkholder, J. B.; Howard, C. J. (J. Mol. Spectrosc. **142** [1990] 319/35).

[2] Burkholder, J. B.; Howard, C. J.; McKellar, A. R. W. (J. Mol. Spectrosc. **127** [1988] 415/24).

[3] Amano, T.; Bernath, P. F.; McKellar, A. R. W. (J. Mol. Spectrosc. **94** [1982] 100/13).
[4] Vervloet, M.; Merienne-Lafore, M. F. (Can. J. Phys. **60** [1982] 49/55).
[5] Birss, F. W.; Merienne-Lafore, M. F.; Ramsay, D. A.; Vervloet, M. (J. Mol. Spectrosc. **85** [1981] 493/5).
[6] Kawaguchi, K.; Yamada, C.; Hirota, E.; Brown, J. M.; Buttenshaw, J.; Parent, C. R.; Sears, T. J. (J. Mol. Spectrosc. **81** [1980] 60/72).
[7] Ramsay, D. A. (Proc. 10th Colloq. Spectros. Int., College Park, Md., 1962 [1963] pp. 583/96; C.A. **61** [1964] 10184).
[8] Johns, J. W. C.; Ramsay, D. A.; Ross, S. C. (Can. J. Phys. **54** [1976] 1804/14).
[9] Vervloet, M.; Merienne-Lafore, M. F.; Ramsay, D. A. (Chem. Phys. Lett. **57** [1978] 5/7).
[10] Vervloet, M. (Mol. Phys. **63** [1988] 433/49).

[11] Dixon, R. N.; Irving, S. J.; Nightingale, J. R.; Vervloet, M. (J. Chem. Soc. Faraday Trans. **87** [1991] 2121/33).
[12] Merienne-Lafore, M. F.; Vervloet, M. (J. Mol. Spectrosc. **108** [1984] 160/2).
[13] Kroll, M. (J. Chem. Phys. **63** [1975] 319/25).
[14] Jensen, P.; Buenker, R. J.; Hirsch, G.; Rai, S. N. (Mol. Phys. **70** [1990] 443/54).
[15] Duxbury, G.; Dixon, R. N. (Mol. Phys. **43** [1981] 255/74).
[16] Jungen, C.; Hallin, K. E. J.; Merer, A. J. (Mol. Phys. **40** [1980] 25/63).
[17] Perić, M.; Engels, B. (J. Chem. Phys. **97** [1992] 4996/5006).
[18] Perić, M.; Buenker, R. J.; Peyerimhoff, S. D. (Mol. Phys. **59** [1986] 1283/303).
[19] Perić, M.; Peyerimhoff, S. D.; Buenker, R. J. (Mol. Phys. **49** [1983] 379/400).
[20] Milligan, D. E.; Jacox, M. E. (J. Chem. Phys. **43** [1965] 4487/93).

[21] Jacox, M. E. (J. Phys. Chem. Ref. Data **13** [1984] 945/1068).
[22] Martin, J. M. L.; François, J. P.; Gijbels, R. (J. Chem. Phys. **97** [1992] 3530/6).
[23] Muenchausen, R. E.; Hills, G. W.; Merienne-Lafore, M. F.; Ramsay, D. A.; Vervloet, M.; Birss, F. W. (J. Mol. Spectrosc. **112** [1985] 203/10).
[24] Hills, G. W.; McKellar, A. R. W. (J. Chem. Phys. **71** [1979] 3330/7).
[25] Muenchausen, R. E.; Hills, G. W. (Chem. Phys. Lett. **99** [1983] 335/41).
[26] Dressler, K.; Ramsay, D. A. (Philos. Trans. R. Soc. London A **251** [1959] 553/604).
[27] Jacox, M. E. (J. Phys. Chem. Ref. Data **19** [1990] 1387/546, 1396).
[28] Jacox, M. E. (J. Phys. Chem. Ref. Data **17** [1988] 269/511, 283).
[29] Nakatsuji, H.; Izawa, M. (J. Chem. Phys. **97** [1992] 435/9).
[30] Ramsay, D. A.; Wayne, F. D. (Can. J. Phys. **57** [1979] 761/6).

2.1.7.2.8.2 Harmonic Frequencies and Anharmonicity Constants

Electronic Ground State. Harmonic frequencies of NH_2, $\omega_1^0 = 3274.65(4.25)$, $\omega_3^0 = 3332.15(5.70)\ cm^{-1}$, and anharmonicity constants, $x_{11}^0 = -53.82(2.06)$, $x_{12}^0 = -16.02(0.66)$, $x_{33}^0 = -31.04(3.06)$, and $x_{23}^0 = -19.68(0.98)\ cm^{-1}$, were calculated by least-squares fitting of frequencies from laser-induced fluorescence (LIF) spectra [1]. $\omega_2^0 = 1514.9 \pm 1.5$ and $x_{22}^0 = 17.1 \pm 1.5\ cm^{-1}$ were determined from an earlier LIF study [2]. The constants ω_2 and x_{22} of ND_2 in $G(v_2) = \omega_2(v_2 + 1/2) + x_{22}(v_2 + 1/2)^2$ were determined from fundamental $(0,1,0) \leftarrow (0,0,0)$ and hot band $(0,2,0) \leftarrow (0,1,0)$ frequencies from laser magnetic resonance (LMR) spectra [3]. A reanalysis of the bands gave the revised values for ND_2 of $\omega_2 = 1127.75$ and $x_{22} = -9.5\ cm^{-1}$ [4]. $\omega_3 = 2515\ cm^{-1}$ was calculated from ω_2 and the inertial defect in the (0,0,0) and (0,1,0) states [3].

A complete set of harmonic frequencies and anharmonicity constants of $^{14}NH_2$, $^{15}NH_2$, $^{14}ND_2$, and ^{14}NHD (in cm^{-1}) were derived from an ab initio-calculated general quartic valence force field [5] which reproduces the experimental fundamentals to within a few cm^{-1}:

constant	ω_1	ω_2	ω_3	$-x_{11}$	$-x_{22}$	$-x_{33}$	$-x_{12}$	$-x_{13}$	$-x_{23}$
$^{14}NH_2$	3375.4	1545.6	3470.0	38.88	14.75	43.96	9.25	148.26	18.87
$^{14}ND_2$	2439.5	1135.4	2551.9	20.15	8.57	25.12	2.45	77.57	9.88
^{14}NHD	2494.3	1356.7	3424.3	39.83	9.75	75.51	10.08	11.04	19.04

The harmonic frequencies ω_1, ω_2, and ω_3 were calculated by ab initio methods for NH_2 [6 to 8] and for ND_2 and NHD [6].

Electronically Excited State $\tilde{A}\ ^2A_1$. Vibrational constants of NH_2, $\omega_1^0 + x_{11}^0 = 3325$, $\omega_2^0 = 622$, $x_{22}^0 = 11.4$, $x_{12}^0 = 5\ cm^{-1}$, and of ND_2, $\omega_1^0 + x_{11}^0 \approx 2520$, $\omega_2^0 = 422$, $x_{22}^0 = 8.1\ cm^{-1}$, were determined from $\tilde{A}\ ^2A_1 \leftarrow \tilde{X}\ ^2B_1$ absorption band systems using the relations $\nu(1,v_2',0) - \nu(0,v_2',0) = \omega_1^0 + x_{11}^0 + x_{12}^0 v_2'$ and fitting the band origins to the formula $\nu(0,v_2',0) = T(0,0,0) + \omega_2^0 v_2' + x_{22}^0 v_2'^2$ [9]. Anharmonicity constants (in cm^{-1}) were estimated on the basis of a normal-coordinate analysis [10]:

constants	x_{11}	x_{22}	x_{33}	x_{12}	x_{13}	x_{23}
$^{14}NH_2$	−37	11	−45	5	−153	−46
$^{14}ND_2$	−19	6	−27	4	−80	−30

The harmonic frequencies ω_1, ω_2, and ω_3 were ab initio-calculated for NH_2 [6, 8] and for ND_2 and NHD [6].

Additional Excited States. Vibrational frequencies of NH_2 in the $\tilde{B}\ ^2B_2$ state were calculated by ab initio procedures [8]. Stretching and bending vibrational frequencies of NH_2 were ab initio-calculated for the core-hole states $^3B_1$, $^1B_1$, $^3\Pi_u$, and $^1\Pi_u$ [11].

References:

[1] Vervloet, M. (Mol. Phys. **63** [1988] 433/49).
[2] Kroll, M. (J. Chem. Phys. **63** [1975] 319/25).
[3] Hills, G. W.; McKellar, A. R. W. (J. Chem. Phys. **71** [1979] 3330/7).
[4] Muenchausen, R. E.; Hills, G. W.; Merienne-Lafore, M. F.; Ramsay, D. A.; Vervloet, M.; Birss, F. W. (J. Mol. Spectrosc. **112** [1985] 203/10).
[5] Martin, J. M. L.; François, J. P.; Gijbels, R. (J. Chem. Phys. **97** [1992] 3530/6).
[6] Buenker, R. J.; Perić, M.; Peyerimhoff, S. D.; Marian, R. (Mol. Phys. **43** [1981] 987/1014).
[7] Watts, J. D.; Trucks, G. W.; Bartlett, R. J. (Chem. Phys. Lett. **164** [1989] 502/8).
[8] Brown, R. D.; Williams, G. R. (Mol. Phys. **25** [1973] 673/94).
[9] Dressler, K.; Ramsay, D. A. (Philos. Trans. R. Soc. London A **251** [1959] 553/604).
[10] Kuchitsu, K.; Morino, Y. (Bull. Chem. Soc. Jpn. **38** [1965] 805/13).
[11] Cambi, R.; Ciullo, G.; Sgamellotti, A.; Tarantelli, F.; Guest, M. F. (Chem. Phys. Lett. **91** [1982] 178/84).

2.1.7.2.8.3 Force Constants

The force constants $f_{rr} = 6.503$ mdyn/Å, $f_{\alpha\alpha} = 0.706$ mdyn·Å, $f_{rr'} = -0.068$ mdyn/Å, and $f_{r\alpha} = 0.706$ mdyn were taken from an ab initio-computed general valence force field of 4th order in internal coordinates of NH_2 in the $\tilde{X}\ ^2B_1$ state [1]. Force constants were also calculated by ab initio [2 to 5] and semiempirical [6, 7] procedures for the $\tilde{X}\ ^2B_1$ and $\tilde{A}\ ^2A_1$ states.

Cubic and quartic force constants of linear radicals were estimated on the basis of a normal-coordinate analysis for NH_2 [8] and ND_2 [8, 9] in the (quasi-linear) $\tilde{A}\,^2A_1$ state.

References:

[1] Martin, J. M. L.; François, J. P.; Gijbels, R. (J. Chem. Phys. **97** [1992] 3530/6).
[2] Clark, D. T.; Sgamellotti, A.; Tarantelli, F. (Chem. Phys. **52** [1980] 1/9).
[3] Thuomas, K. A.; Eriksson, A.; Lund, A. (J. Magn. Reson. **37** [1980] 223/9).
[4] Brown, R. D.; Williams, G. R. (Mol. Phys. **25** [1973] 673/94).
[5] Bender, C. F.; Schaefer, H. F., III (J. Chem. Phys. **55** [1971] 4798/803).
[6] Canuto, S. R. A.; Vyanna, J. D. M. (J. Phys. B **8** [1975] 2987/94).
[7] Lathan, W. A.; Hehre, W. J.; Curtiss, L. A.; Pople, J. A. (J. Am. Chem. Soc. **93** [1971] 6377/87).
[8] Kuchitsu, K.; Morino, Y. (Bull. Chem. Soc. Jpn. **38** [1965] 805/13).
[9] Makushkin, Y. S.; Ulenikov, O. N.; Cheglokov, A. E. (Izv. Vyssh. Uchebn. Zaved. Fiz. **20** [1977] 54/8; Sov. Phys. J. [Engl. Transl.] **20** [1977] 463/7).

2.1.7.2.9 Bond Dissociation Energy. Atomization Enthalpy

Bond Dissociation Energy D(HN-H) in kJ/mol. The bond dissociation energy $D^\circ_0 = 380.9 \pm 2.1$ was calculated from $\Delta_fH^\circ_0(NH) = 356.4$ kJ/mol, $\Delta_fH^\circ_0(H) = 216.1$ kJ/mol, and $\Delta_fH^\circ_0(NH_2) = 191.6$ kJ/mol [1]. Three decades earlier, $D_0 = 368 \pm 17$ had been evaluated from thermochemical data [2]. D_e(HN-H) = 412.1 was determined from enthalpies of formation and spectroscopic data [3]. Ab initio calculations gave D_0(HN-H) = 385 ± 8 [1] and D_e(HN-H) = 378 [3]. Additional D values were obtained by ab initio [4, 5] and semiempirical [6, 7] calculations.

Atomization Enthalpy $\Delta_{at}H$ in kJ/mol. The atomization enthalpies $\Delta_{at}H = 756.9 \pm 6.3$ (without zero point energy) [8] and $\Delta_{at}H^\circ_{298} = 716.3$ [9] were determined from experimental values of dissociation energies and enthalpies of formation. $\Delta_{at}H^\circ_0 = 709.6$ was derived [10] from JANAF enthalpies of formation [11]. The sum of ab initio-calculated bond dissociation energies D_e gave the atomization enthalpy $\Delta_{at}H = 754.0$ [8] and 759.6 [12].

References:

[1] Gibson, S. T.; Greene, J. P.; Berkowitz, J. (J. Chem. Phys. **83** [1985] 4319/28).
[2] Altshuller, A. P. (J. Chem. Phys. **22** [1954] 1947/8).
[3] Pople, J. A.; Binkley, J. S.; Seeger, R. (Int. J. Quantum Chem. Quantum Chem. Symp. No. 10 [1976] 1/19).
[4] Klimo, V.; Tino, J. (Collect. Czech. Chem. Commun. **49** [1984] 1731/5).
[5] Saxon, R. P.; Lengsfield, Byron H., III; Liu, B. (J. Chem. Phys. **78** [1983] 312/20).
[6] Polak, R.; Vojtik, J. (Chem. Phys. **114** [1987] 43/54).
[7] Pamuk, H. O.; Trindle, C. (Int. J. Quantum Chem. Quantum Chem. Symp. No. 12 [1978] 271/82).
[8] Pople, J. A.; Frisch, M. J.; Luke, B. T.; Binkley, J. S. (Int. J. Quantum Chem. Quantum Chem. Symp. No. 17 [1983] 307/20).
[9] Leroy, G.; Peeters, D.; Wilante, C. (J. Mol. Struct. **88** [1982] 217/33 [THEOCHEM **5**]).
[10] Irwin, A. W. (Astron. Astrophys. Suppl. Ser. **74** [1988] 145/60).
[11] Chase, M. W., Jr.; Davies, C. A.; Downey, J. R., Jr.; Frurip, D. J.; McDonald, R. A.; Syveryd, A. N. (JANAF Thermochemical Tables, 3rd Ed., J. Phys. Chem. Ref. Data **14** Suppl. No. 1 [1985] 1/1856, 1270).
[12] Martin, J. M. L.; Francois, J. P.; Gijbels, R. (Chem. Phys. Lett. **163** [1989] 387/91).

2.1.7.2.10 Relaxation Processes. Radiative Lifetimes and Quenching Rates

The notation used for NH_2 in the excited state $\tilde{A}\,^2A_1$ is related to a linear molecule (v_2(linear) = $2v_2$(bent) + K_a + 1), see p. 195.

Relaxation of NH_2 in the State $\tilde{A}\,^2A_1(0,v_2,0)$. Radiative lifetime and quenching of the $\tilde{A}\,^2A_1$ state of NH_2 were investigated by time-resolved laser-induced fluorescence (LIF) [1 to 5] or fluorescence from $NH_2(\tilde{A}\,^2A_1)$ formed by UV photolysis of NH_3 [6 to 8]. The zero-pressure lifetime τ_0 and the quenching rate constant k_q (in $cm^3 \cdot molecule^{-1} \cdot s^{-1}$) were determined for the $3_{0,3}$, $4_{0,4}$, and $5_{0,5}$ rotational levels of the $\tilde{A}\,^2A_1(0,9,0)$ state for a variety of quenching gases [5]:

gas	Ar	He	N_2	CH_4	CO	H_2
τ_0 in μs	11.9(23)	10.0(20)	10.8(20)	10.2(19)	11.3(15)	10.8(20)
10^{10} k_q	1.52(27)	1.45(29)	4.01(25)	3.06(70)	4.66(70)	4.57(87)

For He, substantially lower rate constants $k_q = 2.5 \times 10^{-11}$ [3] and 2.6×10^{-11} $cm^3 \cdot molecule^{-1} \cdot s^{-1}$ [1] were obtained for quenching the $(0,9,0)1_{0,1}$ level. For quenching by nitrogen, the k_q value given above is twice the average value determined for the $2_{0,2}$ and $7_{0,7}$ levels of the (0,9,0) state. The reason for the discrepancy is not clear [1]. Rates of quenching by added gases relative to spontaneous emission were measured using the Stern-Volmer method [8]. Lifetimes and quenching rate constants for various bending-vibration excitations are given in Table 12.

The lifetime $\tau \approx 10$ μs of the $\tilde{A}\,^2A_1(0,9,0)$ state is rather long compared to the mean time interval between successive collisions with a neighboring molecule for pressures of about 0.1 Torr. In this case, the decay of $\tilde{A}\,^2A_1(0,9,0)$ rovibronic states is dominated by collisional transfer and quenching mechanisms [9]. The effects were explained in detail by the combination of (1) radiative decay to the $\tilde{X}\,^2B_1$ ground state, (2) irreversible quenching to the $\tilde{X}\,^2B_1$ ground state, (3) reversible collisional transfer to other rovibronic levels within the $\tilde{A}\,^2A_1(0,9,0)N_{K_a}$ states with $K_a = 0$ or 2, and (4) reversible collisional transfer to high-lying levels of the $\tilde{X}\,^2B_1$ state vibronically coupled to the $\tilde{A}\,^2A_1(0,9,0)N_0$ state [9]. With H atoms as collision partners, weak collisions predominate in the rotational energy transfer induced in electronically excited NH_2 [4, 10].

The quenching rate constant of $\tilde{A}\,^2A_1(0,9,0)_0$ by helium was found to increase monotonically from 2×10^{-11} to 7×10^{-11} $cm^3 \cdot molecule^{-1} \cdot s^{-1}$ with increasing rotational quantum number from N = 0 to 7, respectively. Quenching of the vibronic levels $(0,7,0)_0$, $(0,8,0)_1$, and $(0,10,0)_1$ showed no significant rotational level dependence [1].

Table 12
$NH_2(\tilde{A}\,^2A_1)$: Zero-Pressure Lifetimes τ_0 and Quenching Rate Constants k_q (in $cm^3 \cdot molecule^{-1} \cdot s^{-1}$), Determined by Fluorescence from $(0,v_2,0)N_{K_a,K_c}$ Levels.

v_2	N_{K_a,K_c}	λ in nm	gas	τ_0 in μs	10^{10} k_q	Ref.
5	$3_{0,3}$	741.6 a)	NH_3	45.6 ± 5.7	5.03 ± 0.20	[6]
6	$1_{1,1}$, $2_{1,2}$, $3_{1,3}$, $4_{1,4}$	696.9 a)	NH_3	31.6 ± 3.6	6.21 ± 0.23	[6]
7	$4_{4,0}$, $5_{4,1}$, $6_{4,3}$	710.0 a)	NH_3	30.2 ± 1.8	5.83 ± 0.20	[6]
7	$3_{0,3}$	661.93 b)	NH_3	$23 \pm^{41}_{9}$	11.8 ± 1.5	[5]
7	$3_{0,3}$	–	He	–	0.59	[1]
8	$3_{1,3}$, $4_{1,4}$, $5_{1,5}$	630.08 b)	NH_3	13 ± 2.5	9.4 ± 0.9	[5]
8	$3_{1,3}$	–	He	–	0.39	[1]

Table 12 (continued)

v_2	N_{K_a,K_c}	λ in nm	gas	τ_0 in μs	10^{10} k_q	Ref.
9	$3_{0,3}$, $4_{0,4}$, $5_{0,5}$	597.72 [b)]	NH_3	10.0 ± 1.7	10.0 ± 1.0	[5]
9	$3_{0,3}$	–	H	–	2.7	[4]
9	$3_{0,3}$	–	He	–	0.41	[1]
10	$4_{1,4}$	570.70 [b)]	NH_3	18 ± 4	13.6 ± 1.5	[5]
10	$4_{1,4}$	–	He	–	0.51	[1]
11	$2_{0,2}$, $3_{0,3}$, $4_{0,4}$	542.9 [b)]	NH_3	8.1 ± 2	9.5 ± 2.1	[5]
12	$4_{1,4}$	516.61 [b)]	NH_3	6.6 ± 1.1	9.9 ± 1.4	[5]

[a)] Fluorescence line from $NH_2(\tilde{A}\,^2A_1(0,v_2,0)N_{K_a,K_c})$ formed through ArF excimer laser photolysis of NH_3 at 193 nm. – [b)] Wavelength of pulsed tunable dye laser used for excitation into the $\tilde{A}\,^2A_1(0,v_2,0)N_{K_a,K_c}$ state.

The following radiative lifetimes τ of single $\tilde{A}\,^2A_1(0,v_2,0)1_{1,0}(J=3/2)$ levels of NH_2 in the supersonic free jet were determined from the decay curve of laser-induced fluorescence (λ is the excitation line) [2]:

v_2	8	10	12	14
τ in μs	$10.3\pm^{2.9}_{1.9}$	$7.2\pm^{0.2}_{0.3}$	5.0 ± 0.2	4.4 ± 0.2
λ in nm	628.8	569.5	515.6	471.3

The decrease of τ with decreasing λ was explained by the correlation of the transition energy ΔE with the Einstein coefficient $A = 1/\tau_0 \sim (\Delta E)^3 \sim \lambda^{-3}$ [2, 5]. A long-lived fluorescence component with $\tau \geq 100$ μs was interpreted by a weak coupling between some levels of the $\tilde{A}\,^2A_1$ and $\tilde{X}\,^2B_1$ states [6]. A very short radiative lifetime of the $(0,10,0)_1$ level, τ = 0.35 μs, measured by polarized fluorescence spectroscopy [11], is in disagreement with τ = 18 ± 4 μs [5] (see Table 12) and τ ≈ 15 μs from the phase lag of emission of $NH_2(\tilde{A})$ generated by UV photolysis of NH_3 [7]. Radiative lifetimes for several vibronic levels were calculated [12] from absolute transition moments obtained by ab initio calculations [13]. The calculated lifetimes are about 10 to 20 per cent shorter than the observed lifetimes [12].

Relaxation of ND_2 in the State $\tilde{X}\,^2B_1(0,1,0)$. The vibrational relaxation rate constants were measured for $ND_2(0,1,0)$ by time-resolved laser magnetic resonance technique. The rate constants for ND_2 quenched by ND_3, CF_4, and Ar are $(27\pm10)\times10^{-12}$, $(19\pm4)\times10^{-12}$, and $(2.4\pm0.7)\times10^{-10}$ $cm^3\cdot molecule^{-1}\cdot s^{-1}$, respectively [14].

References:

[1] Wysong, I. J.; Jeffries, J. B.; Crosley, D. R. (J. Chem. Phys. **93** [1990] 237/41).
[2] Mayama, S.; Hiraoka, S.; Obi, K. (J. Chem. Phys. **80** [1984] 7/12).
[3] Dearden, S. J.; Dixon, R. N.; Field, D. (J. Chem. Soc. Faraday Trans. II **78** [1982] 1423/32).
[4] Dixon, R. N.; Field, D. (Proc. R. Soc. London A **366** [1979] 247/76).
[5] Halpern, J. B.; Hancock, G.; Lenzi, M.; Welge, K. H. (J. Chem. Phys. **63** [1975] 4808/16).
[6] Donnelly, V. M.; Baronavski, A. P.; McDonald, J. R. (Chem. Phys. **43** [1979] 283/93).
[7] Koda, S. (Bull. Chem. Soc. Jpn. **50** [1977] 1683/6).
[8] Lenzi, M.; McNesby, J. R.; Mele, A.; Nguyen Xuan, C. (J. Chem. Phys. **57** [1972] 319/23).
[9] Fuyuki, T.; Allain, B.; Perrin, J. (J. Appl. Phys. **68** [1990] 3322/37).
[10] Alwahabi, Z. T.; Harkin, C. G.; McCaffery, A. J.; Whitaker, B. J. (J. Chem. Soc. Faraday Trans. II **85** [1989] 1003/15).

[11] Kroll, M. (J. Chem. Phys. **63** [1975] 1803/9).
[12] Jungen, C.; Hallin, K. E. J.; Merer, A. J. (Mol. Phys. **40** [1980] 25/63).
[13] Peyerimhoff, S. D.; Buenker, R. J. (Can. J. Chem. **57** [1979] 3182/9).
[14] Chichinin, A. I.; Krasnoperov, L. N. (Chem. Phys. Lett. **115** [1985] 343/8).

2.1.7.3 Enthalpy of Formation. Thermodynamic Functions

Enthalpy of Formation Δ_fH in kJ/mol

NH_2(g). The most recent experimental values of the enthalpy of formation are compiled in the following table:

$\Delta_fH^\circ_{298}$	$\Delta_fH^\circ_0$	determination from	Ref.
189.5	192.5	the third-law analysis of equilibrium constants of the reaction $H+NH_3 \leftrightarrow NH_2+H_2$(T range 900 to 1620 K)	[1]
192	–	rate constants of reaction $NH_2+H_2O \leftrightarrow NH_3+OH$ (T range 639 to 1140 K)	[2]
192 ± 7	–	rate constants of reaction $H+NH_3 \leftrightarrow NH_2+H_2$ (T range 673 to 1003 K)	[3]
–	191.6 ± 1.3	analysis of a photoionization mass spectrum of NH_2	[4]

In a review of published enthalpies of formation, $\Delta_fH^\circ_0 = 192 \pm 1$ is recommended [5]. The error limit of ± 1, which is considerably lower than the error ± 6.3 in JANAF tables [6], was arrived at from the good agreement of the recent results [5].

The first values of Δ_fH, determined from the kinetics of the decomposition of hydrazine [7], methylhydrazine [8], phenylhydrazine [9], and benzylamine [10] were too low ($\Delta_fH^\circ \leq 172$). More recent experimental data of Δ_fH range from 179 to 201 [11 to 18].

Several Δ_fH values were calculated on the basis of ab initio methods [19 to 24]. Semiempirically calculated Δ_fH values are too low [25 to 28].

ND_2(g). $\Delta_fH^\circ_{298} = 185.4 \pm 8.4$ and $\Delta_fH^\circ_0 = 188.3 \pm 8.4$ are given in the JANAF thermochemical tables [6, 29].

Heat Capacity C_p. Thermodynamic Functions

NH_2(g). The heat capacity and thermodynamic functions for NH_2 as an ideal gas were calculated for a standard state pressure of 0.1 MPa. Values of the molar heat capacity C°_p, entropy S°, Gibbs energy function $-(G^\circ - H^\circ_{298})/T$ (all in $J \cdot mol^{-1} \cdot K^{-1}$), and enthalpy $H^\circ - H^\circ_{298}$ (in kJ/mol) were selected from the JANAF tables [6] and are given below.

T in K	0	100	200	298.15	400	600	800
C°_p	0	33.259	33.280	33.572	34.395	36.838	39.715
S°	0	158.327	181.381	194.707	204.680	219.066	230.054
$-(G^\circ - H^\circ_{298})/T$	∞	224.351	197.762	194.707	196.034	201.448	207.274
$H^\circ - H^\circ_{298}$	−9.929	−6.602	−3.276	0	3.458	10.571	18.224

T in K	1000	1500	2000	3000	4000	5000	6000
C°_p	42.597	48.392	52.346	58.162	62.011	63.785	64.137
S°	239.229	257.675	272.168	294.555	311.866	325.928	337.604
$-(G^\circ - H^\circ_{298})/T$	212.771	224.807	234.899	251.239	264.312	275.275	284.718
$H^\circ - H^\circ_{298}$	26.458	49.303	74.539	129.947	190.215	253.265	317.312

Additional values of thermodynamic functions were listed for T up to 6000 K [30] and 3000 K [31]. The effect of the $\tilde{A}\,^2A_1$ state becomes significant around 2000 K [32]. The JANAF data were used to form polynomial expansions of the partition function for the range 1000 to 6000 K (of astrophysical interest) [33]. The dissociation function computed from thermodynamic data was presented graphically for the range 1000 to 6000 K [34].

ND_2(g). The heat capacity and thermodynamic functions for ND_2 as an ideal gas were calculated by standard methods (pressure 0.1 MPa) for $T \leq 6000$ K. For $T = 298.15$ K, $C^\circ_p = 34.422$ and $S^\circ = -(G^\circ - H^\circ_{298})/T = 204.291\ J \cdot mol^{-1} \cdot K^{-1}$ [6].

References:

[1] Sutherland, J. W.; Michael, J. V. (J. Chem. Phys. **88** [1988] 830/4).
[2] Baetz, P.; Ehbrecht, J.; Hack, W.; Rouveirolles, P.; Wagner, H. G. (Symp. Combust. Proc. **22** 1988 [1989] 1107/56).
[3] Hack, W.; Rouveirolles, P.; Wagner, H. G. (J. Phys. Chem. **90** [1986] 2505/11).
[4] Gibson, S. T.; Greene, J. P.; Berkowitz, J. (J. Chem. Phys. **83** [1985] 4319/28).
[5] Anderson, W. R. (J. Phys. Chem. **93** [1989] 530/6).
[6] Chase, M. W., Jr.; Davies, C. A.; Downey, J. R., Jr.; Frurip, D. J.; McDonald, R. A.; Syveryd, A. N. (JANAF Thermochemical Tables, 3rd Ed., J. Phys. Chem. Ref. Data **14** Suppl. No. 1 [1985] 1/1856, 1005, 1270).
[7] Szwarc, M. (Proc. R. Soc. London A **198** [1949] 267/84).
[8] Kerr, J. A.; Sekhar, R. C.; Trotman-Dickenson, A. F. (J. Chem. Soc. **1963** 3217/25).
[9] Kerr, J. A.; Trotman-Dickenson, A. F.; Wolter, M. (J. Chem. Soc. **1964** 3584/8).
[10] Szwarc, M. (Proc. R. Soc. London A **198** [1949] 285/92).

[11] Niemitz, K. J.; Wagner, H. G.; Zellner, R. (Z. Phys. Chem. [Munich] **124** [1981] 155/70).
[12] Holzrichter, K.; Wagner, H. G. (Symp. Combust. Proc. **18** 1980 [1981] 769/75).
[13] Cardy, H.; Liotard, D.; Dargelos, A.; Poquet, E. (Nouv. J. Chim. **4** [1980] 751/6).
[14] DeFrees, D. J.; Hehre, W. J.; McIver, Robert T. J.; McDaniel, D. H. (J. Phys. Chem. **83** [1979] 232/7).
[15] Tsang, W. (Int. J. Chem. Kinet. **10** [1978] 41/66).
[16] Carson, A. S.; Laye, P. G.; Yurekli, M. (J. Chem. Thermodyn. **9** [1977] 827/9).
[17] Bohme, D. K.; Hemsworth, R. S.; Rundle, H. W. (J. Chem. Phys. **59** [1973] 77/81).
[18] Golden, D. M.; Solly, R. K.; Gac, N. A.; Benson, S. W. (J. Am. Chem. Soc. **94** [1972] 363/9).
[19] Melius, C. F.; Ho, P. (J. Phys. Chem. **95** [1991] 1410/9).
[20] Melius, C. F.; Binkley, J. S. (Symp. Combust. Proc. **21** 1987 [1988]1953/63).

[21] Martin, J. M. L.; Francois, J. P.; Gijbels, R. (Chem. Phys. Lett. **163** [1989] 387/91).
[22] Pople, J. A.; Luke, B. T.; Frisch, M. J.; Binkley, J. S. (J. Phys. Chem. **89** [1985] 2198/203).
[23] Klimo, V.; Tino, J. (Collect. Czech. Chem. Commun. **49** [1984] 1731/5).
[24] Power, D.; Brint, P.; Spalding, T. R. (J. Mol. Struct. **110** [1984] 155/66 [THEOCHEM **19**]).
[25] Nakajima, Y.; Ogata, M.; Ichikawa, H. (Shitsuryo Bunseki 28 [1980] 243/7; C.A. **94** [1981] No. 191423).
[26] Bews, J. R.; Glidewell, C. (Inorg. Chim. Acta **39** [1980] 217/25).
[27] Bischof, P. (J. Am. Chem. Soc. **98** [1976] 6844/9).
[28] Dewar, M. J. S.; Rzepa, H. S. (J. Am. Chem. Soc. **100** [1978] 784/90).
[29] Chase, M. W., Jr.; Curnutt, J. L.; Downey, J. R., Jr.; McDonald, R. A.; Syverud, A. N.; Valenzuela, E. A. (J. Phys. Chem. Ref. Data 11 [1982] 695/940, 790).

[30] Yungman, V. S.; Gurvich, L. V.; Rtishcheva, N. P. (Tr. Gos. Inst. Prikl. Khim. No. 49 [1962] 20/37; C.A. **60** [1964] 1179).

[31] Milligan, D. E.; Jacox, M. E. (J. Chem. Phys. **43** [1965] 4487/93).

[32] Martin, J. M. L.; François, J. P.; Gijbels, R. (J. Chem. Phys. **97** [1992] 3530/6).

[33] Irwin, A. W. (Astron. Astrophys. Suppl. Ser. **74** [1988] 145/60).

[34] Sharp, C. M. (NATO ASI Ser. C **157** [1985] 661/72).

2.1.7.4 Spectra

2.1.7.4.1 Notation

The rotational-vibrational levels of the asymmetric top molecule in a given electronic state (e.g. $\tilde{X}\,^2B_1$, $\tilde{A}\,^2A_1$) are characterized (apart from spin-splitting, see below) by $(v_1,v_2,v_3)N_{K_a,K_c}(J)$. (v_1,v_2,v_3) indicates the vibrational state, N is the quantum number of the rotational angular momentum apart from spin (S), and K_a and K_c are the quantum numbers of the projections of N on the symmetry axis of the limiting prolate and oblate symmetric top, respectively; J is the quantum number of the total (spin and rotation) angular momentum. Rotational lines are characterized by, for example, ^PQ_K and ^RQ_K for lines of the Q branch ($\Delta N=0$) with $\Delta K=-1$ and $\Delta K=+1$, respectively. The splitting of each rotational level into a doublet F_1 and F_2 by electron spin-rotation interaction (fine structure) is characterized by $^1N_{K_a,K_c}$ with $F_1=N-1/2$ and $^2N_{K_a,K_c}$ with $F_2=N+1/2$ [1, 2].

For the quantum numbers of the bending vibration, the correlation between linear- and bent-molecule notation is $v_2(\text{linear})=2v_2(\text{bent})+K_a+1$; see for example [3, 4]. Most authors prefer the linear molecule notation which is used in this chapter. Energy levels due to vibronic splitting (Renner-Teller effect, see p. 171) [5] are shown in Fig. 6, p. 172.

References:

[1] Herzberg, G. (Molecular Spectra and Molecular Structure III: Electronic Spectra and Electronic Structure of Polyatomic Molecules, Van Nostrand, New York 1966, pp. 1/745).

[2] Herzberg, G. (The Spectra and Structures of Simple Free Radicals: An Introduction to Molecular Spectroscopy, Cornell University Press, Ithaca 1971, pp. 1/226).

[3] Vervloet, M. (Mol. Phys. **63** [1988] 433/49).

[4] Ross, S. C.; Birss, F. W.; Vervloet, M.; Ramsay, D. A. (J. Mol. Spectrosc. **129** [1988] 436/70).

[5] Dressler, K.; Ramsay, D. A. (Philos. Trans. R. Soc. London A **251** [1959] 553/604).

2.1.7.4.2 Electron Spin Resonance Spectrum

The NH_2 radical trapped in various matrices was extensively studied by electron spin resonance (ESR) spectroscopy; see Table 13, p. 196. The ESR spectrum of NH_2 in an Ar matrix at 4.2 K consists of the expected nine lines [1, 2] assigned to the threefold hyperfine splittings of 67 and 29 MHz due to the two protons and ^{14}N, respectively. The peak intensities within each triplet are roughly 1:1:1 instead of the expected 1:2:1. An interpretation is given in [3]. Spectral lines of deuterated NH_2 in Ar formerly assigned to ND_2 [1] were reassigned to NHD [2].

Table 13
ESR Spectra of NH_2 and Isotopomers Trapped in a Matrix.
The method of generation of the radicals is given in the footnotes.

radical	matrix (T in K)
$^{14}NH_2$, $^{14}ND_2$, ^{14}NHD [a)]	ammonia in Ar [b)] (4.2) [1, 2]
$^{14}NH_2$	$H^{14}N_3$ in Ar and Xe [c)] [4], and in Kr (4.2) [c)] [4, 5]
$^{14}NH_2$	pure ammonia, single or polycrystal [d)] (4.2, 77) [6]
$^{14}NH_2$	pure ammonia [e)] (98, 103, 118) [7] and (77) [d)] [8], [f)] [9]
$^{14,15}NH_2$, $^{14,15}ND_2$	pure ammonia [g)] (77, 118) [10]
$^{14}ND_2$	pure ammonia (98, 118) [e)], (88, 138) [d)] [7], and (0.25 to 4.2) [h)] [11]
$^{14}NH_2$, $^{14}ND_2$ [i)]	frozen ammonia-water system [d)] (77) [6, 8, 12, 13]
$^{14}NH_2$, $^{14}ND_2$, ^{14}NHD [j)]	$Na^{14}N_3$ in aqueous glasses [d)] (77) [14, 15]
$^{14}NH_2$	frozen H_2O_2 solution in $^{14}NH_3$ [d), k)] (77) [16]
$^{15}NH_2$	$K^{15}NH_2SO_3$ single crystal [d)] (77) [17]
$^{14}NH_2$	NH_3OHCl and $(NH_3OH)_2SO_4$ single crystal [d)] (298) [18]
$^{14}NH_2$	ammonia in zeolites [d)] (77) [19, 20], (153) [20]
$^{14}NH_2$, $^{15}NH_2$, $^{14}ND_2$	ammonia in zeolites [d)] (77) [21]
$^{14}NH_2$, $^{14}ND_2$	ammonia-silica gel system [d)] (77) [22]

[a)] In [2] only. — [b)] Photolyzed with light of 184.9 nm. — [c)] Photolyzed with Hg lamp. — [d)] γ-irradiated (Co source). — [e)] Electron bombarded with 100 eV. — [f)] Electron bombarded between 8 and 30 eV. — [g)] 1 MeV He^+ ion bombarded. — [h)] Electron bombarded. — [i)] In [13] only. — [j)] In [14] only. — [k)] Photolyzed with light of wavelengths $\geq$290 nm.

References:

[1] Foner, S. N.; Cochran, E. L.; Bowers, V. A.; Jen, C. K. (Phys. Rev. Lett. **1** [1958] 91/2).
[2] Cochran, E. L.; Adrian, F. J.; Bowers, V. A. (J. Chem. Phys. **51** [1969] 2759/61).
[3] McConnell, H. M. (J. Chem. Phys. **29** [1958] 1422).
[4] Fischer, P. H. H.; Charles, S. W.; McDowell, C. A. (J. Chem. Phys. **46** [1967] 2162/6).
[5] Coope, J. A. R.; Farmer, J. B.; Gardner, C. L.; McDowell, C. A. (J. Chem. Phys. **42** [1965] 2628/9).
[6] Michaut, J. P.; Roncin, J.; Marx, R. (Chem. Phys. Lett. **36** [1975] 599/605).
[7] Marx, R.; Maruani, J. (J. Chim. Phys. **61** [1964] 1604/9).
[8] Symons, M. C. R.; Rao, K. V. S. (J. Chem. Soc. A **1971** 2163/6).
[9] Marx, R.; Mauclaire, G. (Adv. Chem. Ser. **82** [1968] 212/21).
[10] Smith, D. R.; Seddon, W. A. (Can. J. Chem. **48** [1970] 1938/42).

[11] Vertii, A. A.; Ivanchenko, I. V.; Popenko, N. A.; Tarapov, S. I.; Shestopalov, V. P.; Belyaev, A. A.; Get'man, V. A.; Dzyubak, A. P.; Karnaukhov, I. M.; Lukhanin, A. A.; Sorokin, P. V.; Sporov, E. A.; Tolmachev I. A. (Dokl. Akad. Nauk SSSR **314** [1990] 1389/91; Dokl. Phys. [Engl. Transl.] **35** [1990] 899/901).
[12] Tupikov, V.I.; Pshezhetskii, S.Ya. (Zh. Fiz. Khim. **38** [1964] 2511/3; Russ. J. Phys. Chem. **38** [1964] 1364/5).
[13] Al-Naimy, B. S.; Moorthy, P. N.; Weiss, J. J. (J. Phys. Chem. **70** [1966] 3654/60).
[14] Ginns, I. S.; Symons, M. C. R. (J. Chem. Soc. Faraday Trans. II **68** [1972] 631/7).
[15] Moorthy, P. N.; Rao, K. N.; Shankar, J. (Nature [London] Phys. Sci. **234** [1971] 17/8).

[16] Roginskii, V. A.; Kotov, A. G. (Zh. Fiz. Khim. **40** [1966] 175/7; Russ. J. Phys. Chem. [Engl. Transl.] **40** [1966] 88/9).
[17] Morton, J. R.; Smith, D. R. (Can. J. Chem. **44** [1966] 1951/5).
[18] Koksal, F.; Cakir, O.; Gumrukcu, I.; Birey, M. (Z. Naturforsch. **40a** [1985] 903/5).
[19] Sorokin, Yu. A.; Kotov, A. G.; Pshezhetskii, S. Ya. (Dokl. Akad. Nauk SSSR **159** [1964] 1385/8; Dokl. Phys. Chem. [Engl. Transl.] **154/159** [1964] 1163/5).
[20] Suzuno, Y.; Shiotani, M.; Sohma, J. (Chem. Phys. Lett. **44** [1976] 177/9).
[21] Vansant, E. F.; Lunsford, J. H. (J. Phys. Chem. **76** [1972] 2716/8).
[22] Nagai, S. (Bull. Chem. Soc. Jpn. **46** [1973] 1144/8).

2.1.7.4.3 Photoelectron Spectrum

The He I photoelectron spectrum of the $NH_2(\tilde{X}\,^2B_1)$ radical, produced from the fast reaction $F + NH_3 \rightarrow NH_2 + HF$, shows three bands corresponding to the ionization of NH_2 to the $\tilde{X}\,^3B_1$, $\tilde{a}\,^1A_1$, and $\tilde{b}\,^1B_1$ states of NH_2^+ (see p. 240). The sharp band at 12.45 eV has two vibrational components, separated by 1350 ± 50 and 2900 ± 50 cm^{-1}, which were assigned to the excitation of both the ν_2 and ν_1 vibrations in the ionic state $\tilde{a}\,^1A_1$. The band at 12.00 eV shows regular vibrational spacings on the low ionization potential side which were attributed to the excitation of the ν_2 mode of $NH_2^+(\tilde{X}\,^3B_1)$. Studies of spectra of the deuterated radical supported the assignments.

Reference:

Dunlavey, S. J.; Dyke, J. M.; Jonathan, N.; Morris, A. (Mol. Phys. **39** [1980] 1121/35).

2.1.7.4.4 Microwave and Far-Infrared Spectra

NH_2

In the **microwave absorption** (MW) spectrum the fine and hyperfine components of the $3_{1,3} \leftarrow 2_{2,0}$ (12 lines) and $1_{1,0} \leftarrow 1_{0,1}$(9 lines) rotational transitions in the $\tilde{X}\,^2B_1(0,0,0)$ state were measured in the range 230 to 469 GHz (7.67 to 15.36 cm^{-1}). The lines which lie in the atmospheric windows near 231 GHz and between the water lines at 448 and 471 GHz, respectively, appear to be favorable for the detection of NH_2 in dense interstellar clouds [1]; see also [2]. A far-infrared tunable laser spectrum of ultracold (about 5 K) free NH_2 was recorded. The bands observed at 952.5 GHz were assigned to the $1_{1,1}(J=3/2) \leftarrow 0_{0,0}(J=1/2)$ transition in the $\tilde{X}\,^2B_1(0,0,0)$ state [3].

Magnetic-dipole-allowed transitions between the spin doublets $F_1(J)$ and $F_2(J)$ in the $\tilde{X}\,^2B_1$ ground and the $\tilde{A}\,^2A_1$ excited electronic states were observed in **microwave optical double resonance** (MODR) spectra in the frequency range <30 GHz (1 cm^{-1}). In addition, a few MODR bands were identified as electric-dipole-allowed transitions in the ground state $\tilde{X}\,^2B_1$ which have three to four orders of magnitude stronger transition rates than the magnetic-dipole-allowed MW transitions [4].

MODR signals were obtained by pumping the appropriate optical transitions, $\tilde{A}\,^2A_1(0,10,0) \leftarrow \tilde{X}\,^2B_1(0,0,0)$ at about 17500 cm^{-1} and $\tilde{A}\,^2A_1(0,9,0) \leftarrow \tilde{X}\,^2B_1(0,0,0)$ at about 16600 cm^{-1}, with a Rhodamine 6G dye laser. The majority of the MODR bands observed were assigned to the magnetic-dipole-allowed transitions between F_1 and F_2 spin components for a rotational state N_{K_a,K_c} in the vibronic ground state $\tilde{X}\,^2B_1(0,0,0)$ [4 to 7] and in the excited vibronic states $\tilde{A}\,^2A_1(0,9,0)$ with $K_a=0$ and $\tilde{A}\,^2A_1(0,10,0)$ with $K_a=1$ [8]. The electronic and vibrational states, the ranges of rotational states N and K_a, the number of rotational levels n_{rot} and assigned lines n_{line}, and the frequency range $\Delta\nu$ studied in MODR spectra are given in the following table:

state	(v_1,v_2,v_3)	N, K_a	n_{rot}	n_{line}	$\Delta\nu$ in GHz	Ref.
$\tilde{X}\,^2B_1$	(0,0,0)	N, $K_a \leq 8$	31	>200[a), 8[b)	1.8 to 18	[5][c)
		$N \leq 7$, $K_a \leq 2$	6	19[a)	7.3 to 11.5	[4]
		$3_{3,1}$, $4_{1,3}$	2	22[d)	5.2 to 11.7	[6]
		$2 \leq N \leq 9$, $K_a \leq 2$	11	35[a), 3[b)	3.2 to 9.7	[7]
$\tilde{A}\,^2A_1$	(0,10,0)	$N \leq 9$, $K_a = 1$	9	111[a)	2.0 to 22.3	[8]
	(0,9,0)	$3 \leq N \leq 8$, $K_a = 0$	4	30[a)	2.0 to 8.0	[8]

[a) Magnetic-dipole-allowed transitions. — [b) Electric-dipole-allowed transitions. — [c) MODR frequencies [4, 9, 10] were used for the fitting procedure. — [d) The transitions $3_{3,1}(J=7/2) \leftarrow 4_{1,3}(J=7/2)$ and $3_{3,1}(J=7/2) \leftarrow 4_{1,3}(J=9/2)$ are magnetic-dipole-allowed as a result of the spin-rotation interaction mixing of level $4_{1,3}(J=7/2)$ with level $3_{3,1}(J=7/2)$.

Strong MODR signals around 6.4 and 15.7 GHz were assigned to transitions between a rovibronic level $u2_{2,0}(J=3/2)$ and the spin-split levels of $\tilde{A}\,^2A_1(0,10,0)1_{1,0}$ [11, 12]. The level u was tentatively assigned to $\tilde{X}\,^2B_1(0,12,0)$ [11]. Following a theoretical treatment of the Renner effect [13] (see p. 171), the assignment of u was modified to $\tilde{X}\,^2B_1(0,13,0)$ [12]. MW transitions identified between rovibronic states were used to assign transitions in the optical spectrum. For example, one hyperfine component of the $\tilde{A}\,^2A_1(0,10,0)1_{1,0}(F_2) \leftrightarrow u(F_2)$ transition around 6.3 GHz was pumped, while the dye laser frequency was tuned in the range 17250 to 17600 cm^{-1} (580 to 568 nm) [14].

Saturation spectra of NH_2 were recorded using the intermodulated fluorescence technique. The saturation spectrum showed a hyperfine structure in the $\tilde{A}\,^2A_1(0,10,0) \leftarrow \tilde{X}\,^2B_1(0,0,0)$ band for the transitions $2_{1,1} \leftarrow 1_{0,1}$ $(J=3/2 \leftarrow 1/2$ and $5/2 \leftarrow 3/2)$, $4_{1,3} \leftarrow 3_{0,3}$ $(J=9/2 \leftarrow 7/2$ and $7/2 \leftarrow 5/2)$, $5_{1,4} \leftarrow 4_{2,2}$ $(J=9/2 \leftarrow 7/2)$, and $6_{1,5} \leftarrow 5_{2,3}$ $(J=13/2 \leftarrow 11/2)$ [15]. The hyperfine splitting of the $\tilde{A}\,^2A_1(0,10,0)1_{1,0}(J=3/2) \leftarrow \tilde{X}\,^2B_1(0,0,0)2_{2,0}(J=5/2)$ band was studied by using microwave-modulated saturation spectroscopy, a combination of the MODR and intermodulated fluorescence techniques [16].

Hyperfine splittings in the $\tilde{A}\,^2A_1(0,9,0)$ state were measured for the levels $2_{2,0}(J=5/2)$, $2_{0,2}(J=3/2,5/2)$, $4_{0,4}(J=9/2)$, and $6_{0,6}(11/2)$ using the **magnetic level-crossing** technique [17, 18]. The hyperfine splitting in the $\tilde{A}\,^2A_1(0,10,0)$ state [17] taken from the level-crossing spectrum is in good agreement with the results obtained from the saturation spectrum [15]. The studies of the hyperfine structure were supplemented by measurements of the optical-optical double resonance (OODR) spectra in weak magnetic fields, carried out for the $\tilde{A}\,^2A_1(0,9,0)0_{0,0}(J=1/2)$ level [19]. The hyperfine structures were also investigated in the **polarization spectra** obtained for the ${}^14_{1,3} \leftarrow {}^13_{0,3}$ and ${}^25_{1,4} \leftarrow {}^24_{2,2}$ subbands of the $\tilde{A}\,^2A_1(0,10,0) \leftarrow \tilde{X}\,^2B_1(0,0,0)$ transition and for the ${}^13_{0,3} \leftarrow {}^14_{1,3}$ and ${}^15_{0,5} \leftarrow {}^14_{1,3}$ subbands of the $\tilde{A}\,^2A_1(0,9,0) \leftarrow \tilde{X}\,^2B_1(0,0,0)$ transition [20].

In the **laser magnetic resonance** (LMR) spectrum the rotational transitions of NH_2 in the $\tilde{X}\,^2B_1(0,0,0)$ ground state were resonance-tuned with far-infrared lasers (H_2O, D_2O, and CO_2) in the frequency range 15 to 127 cm^{-1}, using magnetic fields up to 16 kG. LMR spectral data were assigned to Zeeman components of the rotational transitions $3_{3,1} \leftarrow 2_{2,0}$, $2_{2,0} \leftarrow 1_{1,1}$, and $7_{4,4} \leftarrow 7_{3,5}$, observed at frequencies of 127.48 cm^{-1} (H_2O laser), 84.32 cm^{-1}

(H_2O laser), and 92.82 cm^{-1} (D_2O laser), respectively [21]. Seventeen rotational transitions up to N=7 and K_a=5 were measured by using 15 far-infrared laser frequencies [22, 23].

ND_2

A frequency-modulated MW spectrum was recorded in the range 265 to 531 GHz. One-hundred-twenty spectral lines, indicating fine and hyperfine splitting, were assigned to seven rotational transitions in the $\tilde{X}\,^2B_1(0,0,0)$ ground state: $4_{2,2} \leftarrow 3_{3,1}$, $1_{1,0} \leftarrow 1_{0,1}$, $5_{2,4} \leftarrow 4_{3,1}$, $2_{1,1} \leftarrow 2_{0,2}$, $2_{0,2} \leftarrow 1_{1,1}$, $3_{1,2} \leftarrow 3_{0,3}$, and $1_{1,1} \leftarrow 0_{0,0}$ [24].

MODR signals were observed by pumping the appropriate optical transitions of ND_2 $\tilde{A}\,^2A_1(0,13,0) \leftarrow \tilde{X}\,^2B_1(0,0,0)$ and $\tilde{A}\,^2A_1(0,12,0) \leftarrow \tilde{X}\,^2B_1(0,0,0)$ with a Rhodamine 6G dye laser. More than 200 transitions involving twenty-one rotational levels $N_{1,N}$ and $N_{1,N-1}$ ($1 \leq N \leq 6$), $N_{2,N-1}$ ($3 \leq N \leq 8$), and $N_{2,N-2}$ (N=3, 4, and 7) for the $\tilde{X}\,^2B_1(0,0,0)$ ground state were studied in the range 1.5 to 7.0 GHz [25]. Twenty-one of these MODR frequencies were used to reevaluate the molecular parameters [26].

NHD

Both a-type rotational transitions between K doublet levels and magnetic-dipole-allowed transitions between spin doublets of NHD were observed using optical transitions from $\tilde{X}\,^2A''(0,0,0)$ to $\tilde{A}\,^2A'(0,11,0)$, (0,10,0), and (0,9,0) in MODR experiments. Ten frequencies ranging from 5.97 to 72.76 GHz were assigned to the three electric-dipole-allowed transitions $1_{1,0} \leftarrow 1_{1,1}$, $2_{2,0} \leftarrow 2_{2,1}$, and $4_{3,1} \leftarrow 4_{3,2}$ in the $\tilde{X}\,^2A''$ state. Magnetic-dipole-allowed transitions were studied for ten rotational states ($N \leq 6$ and $K_a \leq 2$) in the range 4.83 to 9.57 GHz [27]. The $1_{1,0} \leftarrow 1_{1,1}$ and $2_{2,0} \leftarrow 2_{2,1}$ electric-dipole-allowed transitions of NHD at about 73 and 10 GHz, respectively, are important in astrophysics, because the rotational levels involved are significantly populated at the low temperatures found in many interstellar clouds [28].

Far-IR LMR spectra of NHD in the ground state $\tilde{X}\,^2A''(0,0,0)$ were observed for eleven $3_{1,3} \leftarrow 2_{0,2}$ rotational transitions ($F_1 \leftarrow F_1$ and $F_2 \leftarrow F_2$) (laser line $\lambda = 47.3\ cm^{-1}$) and fourteen $4_{1,3} \leftarrow 3_{2,2}$ rotational transitions ($F_1 \leftarrow F_1$) ($\lambda = 26.7\ cm^{-1}$) by varying the magnetic field up to 5 kG [29].

References:

[1] Charo, A.; Sastry, K. V. L. N.; Herbst, E.; De Lucia, F. C. (Astrophys. J. **244** [1981] L111/L112).
[2] Charo, A. A. (Diss. Duke Univ. 1981, 188 pp. from Diss. Abstr. Int. B **42** [1982] 4466; C.A. **97** [1982] No. 46700).
[3] Cohen, R. C.; Busarow, K. L.; Schmuttenmaer, C. A.; Lee, Y. T.; Saykally, R. J. (Chem. Phys. Lett. **164** [1989] 321/4).
[4] Hills, G. W.; Cook, J. M.; Curl, R. F., Jr.; Tittel, F. K. (J. Chem. Phys. **65** [1976] 823/8).
[5] Cook, J. M.; Hills, G. W.; Curl, R. F., Jr. (J. Chem. Phys. **67** [1977] 1450/61).
[6] Hills, G. W.; Lowe, R. S.; Cook, J. M.; Curl, R. F., Jr. (J. Chem. Phys. **68** [1978] 4073/6).
[7] Hills, G. W.; Cook, J. M. (J. Mol. Spectrosc. **94** [1982] 456/60).
[8] Hills, G. W.; Brazier, C. R.; Brown, J. M.; Cook, J. M.; Curl, R. F., Jr. (J. Chem. Phys. **76** [1982] 240/52).
[9] Cook, J. M.; Hills, G. W.; Curl, R. F., Jr. (Astrophys. J. **207** [1976] L139/L140).
[10] Hills, G. W.; Cook, J. M. (Astrophys. J. **209** [1976] L157/L159).
[11] Hills, G. W.; Curl, R. F., Jr. (J. Chem. Phys. **66** [1977] 1507/13).
[12] Lowe, R. S.; Kasper, J. V. V.; Hills, G. W.; Dillenschneider, W.; Curl, R. F., Jr. (J. Chem. Phys. **70** [1979] 3356/61).

[13] Jungen, C.; Hallin, K. E. J.; Merer, A. J. (Mol. Phys. **40** [1980] 25/63).
[14] Hills, G. W. (J. Mol. Spectrosc. **93** [1982] 395/404).
[15] Hills, G. W.; Philen, D. L.; Curl, R. F., Jr.; Tittel, F. K. (Chem. Phys. **12** [1976] 107/11).
[16] Kasper, J. V. V.; Lowe, R. S.; Curl, R. F., Jr. (J. Chem. Phys. **70** [1979] 3350/5).
[17] Dixon, R. N.; Field, D. (Mol. Phys. **34** [1977] 1563/76).
[18] Dixon, R. N.; Field, D. (Lasers Chem. Proc. Conf., London 1977, pp. 145/9; C.A. **88** [1978] No. 81234).
[19] Dixon, R. N.; Field, D.; Noble, M. (Proc. 3rd Yamada Conf. Free Radicals, Sanda, Jpn., 1979, pp. 239/42; C.A. **93** [1980] No. 57382).
[20] Hemmerling, B.; Vervloet, M. (Chem. Phys. Lett. **150** [1988] 464/8).

[21] Davies, P. B.; Russell, D. K.; Thrush, B. A.; Wayne, F. D. (J. Chem. Phys. **62** [1975] 3739/42).
[22] Davies, P. B.; Russell, D. K.; Thrush, B. A.; Radford, H. E. (Chem. Phys. Lett. **42** [1976] 35/8).
[23] Davies, P. B.; Russell, D. K.; Thrush, B. A.; Radford, H. E. (Proc. R. Soc. London A **353** [1977] 299/318).
[24] Kanada, M.; Yamamoto, S.; Saito, S. (J. Chem. Phys. **94** [1991] 3423/8).
[25] Cook, J. M.; Hills, G. W. (J. Chem. Phys. **78** [1983] 2144/53).
[26] Muenchausen, R. E.; Hills, G. W.; Merienne-Lafore, M. F.; Ramsay, D. A.; Vervloet, M.; Birss, F. W. (J. Mol. Spectrosc. **112** [1985] 203/10).
[27] Steimle, T. C.; Brown, J. M.; Curl, R. F., Jr. (J. Chem. Phys. **73** [1980] 2552/8).
[28] Brown, J. M.; Steimle, T. C. (Astrophys. J. **236** [1980] L101/L103).
[29] Carrington, A.; Geiger, J. S.; Smith, D. R.; Bonnett, J. D.; Brown, C. (Chem. Phys. Lett. **90** [1982] 6/8).

2.1.7.4.5 Raman Spectrum

NH_2 radicals, generated by photolysis of NH_3 (0.5 mbar) with an ArF exciplex laser at 193 nm, were detected by time-resolved, coherent anti-Stokes Raman spectroscopy in the NH stretch regions at 3334 and 3220 cm^{-1} [1]. Two Raman bands of NH_2, $\nu_3 = 3276 \pm 2$ and $\nu_2 = 1471 \pm 2$ cm^{-1}, were observed in solid ammonia at 77 K after irradiation with an atomic aluminium beam [2].

References:

[1] Dreier, T.; Wolfrum, J. (Appl. Phys. B **33** [1984] 213/8).
[2] Klimov, V. D.; Mamchenko, A. V.; Nabiev, S. S.; Sukhanov, L. P. (Zh. Fiz. Khim. **65** [1991] 1819/25; Russ. J. Phys. Chem. [Engl. Transl.] **65** [1991] 966/70).

2.1.7.4.6 Infrared Spectrum

Infrared Absorption

The vibrational band (ν_i), the number of assigned lines (n), the corresponding rotational quantum numbers (N, K_a), and the frequency range ($\Delta\nu$) studied in high-resolution absorption spectra of NH_2 are given in the following table:

fundamental	n	N, K_a	$\Delta\nu$ in cm^{-1}	remark	Ref.
ν_1	139	$1 \leq N \leq 6$, $K_a \leq 4$	3126 to 3372	a)	[1]
	220	$N \leq 9$, $K_a \leq 5$	3018 to 3445	b)	[2]
	3	N, $K_a \leq 2$	near 3250	c)	[3]
ν_3	72	$1 \leq N \leq 5$, $K_a \leq 4$	3208 to 3399	a)	[1]
	39	$N \leq 7$, $K_a \leq 4$	3252 to 3436	b)	[2]
ν_2	336	$N \leq 10$, $K_a \leq 5$	1270 to 1726	b)	[4]
	19	$2 \leq N \leq 6$, $K_a \leq 4$	1342 to 1605	d)	[5]

a) Tunable difference-frequency laser spectrum. — b) Fourier transform spectrum. — c) Single-frequency, color-center laser spectrum of jet-cooled NH_2 (80 and 25 K) in He. — d) Tunable semiconductor laser spectrum.

The surprising result that the ν_1 band of NH_2 is considerably stronger than the ν_3 fundamental (the ratio of the vibrational transition moments is $|\mu_1/\mu_3| \approx 4$ in contrast to H_2O) could not be interpreted [1].

IR absorption spectra of NH_2, ND_2, and NHD, isolated in argon, nitrogen, and carbon monoxide matrices at 14 K, were recorded in the range of the fundamental vibrations. The relatively complex spectrum of NH_2 in an argon matrix indicates molecule rotation in the matrix. In an N_2 matrix the following positions of the ν_2 fundamental were observed [6]:

species	$^{14}NH_2$	$^{15}NH_2$	$^{14}ND_2$	^{14}NHD
ν_2 (in cm^{-1})	1499	1495.5	1110	1321

The absorption bands at 3215 and 3220 cm^{-1} were originally assigned to $\nu_3(^{15}NH_2)$ and $\nu_3(^{14}NH_2)$, respectively [6]. More recently, the 3220-cm^{-1} band was reassigned to ν_1 [7 to 9]. The absorptions at 3217 and 1496 cm^{-1} in an Ar matrix spectrum of NH_3, bombarded at 12 K with 50-eV electrons, were also assigned to NH_2 [10].

Laser Magnetic Resonance (LMR)

NH_2. Laser lines in resonance with tuned ν_2 transitions, frequency range ($\Delta\nu$), magnetic field (H_r), and corresponding rotational quantum numbers are given in the following table:

laser line	$\Delta\nu$ in cm^{-1}	H_r in kG	N and K_a	Ref.
^{13}CO: P(8,12,14)	1704 to 1762	up to 11	$3 \leq N \leq 5$, $3 \leq K_a \leq 5$ a)	[11]
$^{13}CO_2$: R(12,18,22,36,40)	1031 to 1108	2.5 to 7.7	$5 \leq N \leq 10$, $0 \leq K_a \leq 5$ b)	[12]
^{13}CO: P(10) to P(19)	1511 to 1777	0.5 to 18.5	$0 \leq N \leq 7$, $0 \leq K_a \leq 5$ c)	[13]

a) Zeeman components of the three transitions $5_{4,1} \leftarrow 4_{3,2}$, $5_{5,1} \leftarrow 4_{4,0}$, and $4_{4,1} \leftarrow 3_{3,0}$. — b) Only weak transitions for $\Delta N = -1$ and $\Delta K_a = -3$. — c) 160 Zeeman components from 30 vibration-rotation transitions; $\Delta K_a = 3$ is valid for $5_{3,3} \leftarrow 4_{0,4}$; for all other transitions, $\Delta K_a = \pm 1$.

ND_2. Over three hundred Zeeman resonances of the $\tilde{X}\,^2B_1(0,1,0) \leftarrow (0,0,0)$ transition ($\nu_2 = 1108.7493\ cm^{-1}$) were observed using thirty-five laser ($^{13}C^{16}O_2$, $^{12}C^{16}O_2$, and $^{12}C^{18}O_2$) lines in the range 998 to 1108 cm^{-1}. The magnetic field was varied between 0.2 and 11.0

kG. The spectral lines were assigned to transitions involving the rotational quantum numbers $1 \leq N \leq 7$ and $0 \leq K_a \leq 6$. Four LMR transitions were tentatively assigned to the $(0,2,0)2_{0,2} \leftarrow (0,1,0)1_{1,1}$ hot band (band origin: 1087.00 cm^{-1}) [14]. The assignments for 13 LMR transitions were revised, including the four hot band transitions [15]. A time-resolved LMR technique was used for measuring the vibrational relaxation of ND_2 in the $\tilde{X}\,^2B_1(0,1,0)$ state; see p. 192. The ν_2 vibrational transition moment was estimated to be 0.056 ± 0.02 D [16].

Infrared Optical Double Resonance (IODR)

NH_2. A Rhodamine 6G dye laser and a CO_2/NO_2 laser were combined with the LMR technique to study the vibration-rotation transitions of NH_2 in the $\tilde{A}\,^2A_1$ state which were observed with sub-Doppler resolution. When the $\tilde{A}\,^2A_1(0,9,0)2_{2,0} \leftarrow \tilde{X}\,^2B_1(0,0,0)1_{1,0}$ transition was pumped with a dye laser, thirty-nine lines could be assigned to the $2_{1,1} \leftarrow 2_{2,0}$ transition and five lines to the $1_{1,1} \leftarrow 2_{2,0}$ transition by IR excitation of the $\tilde{A}\,^2A_1(0,10,0) \leftarrow \tilde{A}\,^2A_1(0,9,0)$ transition and by sweeping the magnetic field to as high as 14.2 kG. Four additional lines were assigned to IR transitions from the $\tilde{A}\,^2A_1(0,9,0)$ state to a highly excited vibrational state, tentatively assigned to $\tilde{X}\,^2B_1(2,8,0)$ [17]. By extending the IODR technique, CO_2/N_2O laser magnetic resonance spectra of the $\tilde{A}\,^2A_1(0,12,0) \leftarrow (0,11,0)$ band with $N_{1,N-1} \leftarrow N_{0,N}$ (N = 1 to 7 and ν range 945 to 965 cm^{-1}) were observed. Furthermore, transitions between the $3_{0,3}$ and $4_{0,4}$ levels of $\tilde{A}\,^2A_1(0,11,0)$ and highly excited vibrational states of $\tilde{X}\,^2B_1$ were identified with laser lines between 952 and 955 cm^{-1}. The $\tilde{A}\,^2A_1(0,11,0)$ levels were excited with a dye laser via the transitions $\tilde{A}\,^2A_1(0,11,0)N_{0,N} \leftarrow \tilde{X}\,^2B_1(0,0,0)N_{1,N}$ [18].

ND_2. The IR transitions $5_{0,5}(F_1) \leftarrow 5_{1,4}(F_1)$ and $2_{1,2}(F_2) \leftarrow 2_{2,1}(F_2)$ of the $\tilde{X}\,^2B_1(0,1,0) \leftarrow (0,0,0)$ band were pumped with $^{12}C^{18}O_2$ P(10) and $^{12}C^{16}O_2$ R(26) laser lines at 1076.57 and 1082.30 cm^{-1}, respectively. IODR signals were observed by exciting various rovibronic transitions of $\tilde{A}\,^2A_1 \leftarrow \tilde{X}\,^2B_1$ (vibrational excitations: from (0,1,0) to (0,14,0), (0,15,0), and from (0,0,0) to (0,12,0), (0,13,0), (1,9,0)) with a Rhodamine 6G dye laser. The energy levels obtained from this study [19] and LMR experiments [14] show significant differences for $N \geq 6$.

References:

[1] Amano, T.; Bernath, P. F.; McKellar, A. R. W. (J. Mol. Spectrosc. **94** [1982] 100/13).
[2] McKellar, A. R. W.; Vervloet, M.; Burkholder, J. B.; Howard, C. J. (J. Mol. Spectrosc. **142** [1990] 319/35).
[3] Curl, R. F., Jr.; Murray, K. K.; Petri, M.; Richnow, M. L.; Tittel, F. K. (Chem. Phys. Lett. **161** [1989] 98/102).
[4] Burkholder, J. B.; Howard, C. J.; McKellar, A. R. W. (J. Mol. Spectrosc. **127** [1988] 415/24).
[5] Krivtsun, V. M.; Nadezhdin, B. B.; Britov, A. D.; Zasavitskii, I. I.; Shotov, A. P. (Opt. Spektrosk. **60** [1986] 1162/4; Opt. Spectrosc. [Engl. Transl.] **60** [1986] 720/1).
[6] Milligan, D. E.; Jacox, M. E. (J. Chem. Phys. **43** [1965] 4487/93).
[7] Jacox, M. E. (J. Phys. Chem. Ref. Data **13** [1984] 945/1068).
[8] Jacox, M. E. (J. Phys. Chem. Ref. Data **17** [1988] 269/511, 282/3).
[9] Jacox, M. E. (J. Phys. Chem. Ref. Data **19** [1990] 1387/546, 1396/7).
[10] Suzer, S.; Andrews, L. (J. Chem. Phys. **89** [1988] 5347/9).

[11] Brown, J. M.; Buttenshaw, J.; Carrington, A.; Parent, C. R. (Mol. Phys. **33** [1977] 589/92).
[12] Hills, G. W.; McKellar, A. R. W. (J. Mol. Spectrosc. **74** [1979] 224/7).
[13] Kawaguchi, K.; Yamada, C.; Hirota, E.; Brown, J. M.; Buttenshaw, J.; Parent, C. R.; Sears, T. J. (J. Mol. Spectrosc. **81** [1980] 60/72).

[14] Hills, G. W.; McKellar, A. R. W. (J. Chem. Phys. **71** [1979] 3330/7).
[15] Muenchausen, R. E.; Hills, G. W.; Merienne-Lafore, M. F.; Ramsay, D. A.; Vervloet, M.; Birss, F. W. (J. Mol. Spectrosc. **112** [1985] 203/10).
[16] Chichinin, A. I.; Krasnoperov, L. N. (Chem. Phys. Lett. **115** [1985] 343/8).
[17] Amano, T.; Kawaguchi, K.; Kakimoto, M.; Saito, S.; Hirota, E. (J. Chem. Phys. **77** [1982] 159/67).
[18] Kawaguchi, K.; Suzuki, T.; Saito, S.; Hirota, E. (J. Opt. Soc. Am. B Opt. Phys. **4** [1987] 1203/11).
[19] Muenchausen, R. E.; Hills, G. W. (Chem. Phys. Lett. **99** [1983] 335/41).

2.1.7.4.7 Near-Infrared, Visible, and Ultraviolet Spectra

General Remarks

Absorption, emission, and laser-induced fluorescence (LIF) techniques were applied to the $\tilde{A}\,^2A_1 \leftrightarrow \tilde{X}\,^2B_1$ electronic transition, giving abundant information about the ground and excited rotational and vibrational states of the NH_2 and ND_2 radicals in both electronic states. Many levels of the upper electronic state were characterized in absorption experiments, while emission studies gave precise values for various ground-state levels. One reason for the complexity of electronic spectra is the different structure of NH_2 in the ground and excited states; see p. 184. Absorptions assigned to transitions from $\tilde{X}\,^2B_1(0,0,0)$ to highly excited vibrational states $\tilde{X}\,^2B_1(0,v_2',0)$ were identified in the electronic spectrum on account of the Renner-Teller effect. Combinations of Fermi resonances and Renner-Teller interactions are illustrated in the following scheme using the bent-molecule notation (see p. 195) [1]:

$$
\begin{array}{ccccc}
\tilde{A}\,^2A_1(0,v_2,0) & \longleftarrow & \text{Fermi} & \longrightarrow & \tilde{A}\,^2A_1(1,v_2-2,0) \\
\uparrow & & & & \uparrow \\
\text{Renner-Teller} & & & & \text{Renner-Teller} \\
\downarrow & & & & \downarrow \\
\tilde{X}\,^2B_1(0,v_2+9,0) & \longleftarrow & \text{Fermi} & \longrightarrow & \tilde{X}\,^2B_1(1,v_2+7,0)
\end{array}
$$

Quantum-chemical calculations of term values on the basis of the Renner-Teller effect employed ab initio treatments [2 to 4] or empirical methods [5 to 7].

Absorption Spectrum

Gaseous NH_2. The absorption spectra of NH_2 were recorded in the range 830 to 390 nm [8 to 15]. The complex vibronic band structure was explained by a strong electronic-vibrational coupling which causes the electronic state $^2\Pi$ to split to the states $\tilde{X}\,^2B_1$ and $\tilde{A}\,^2A_1$ (see p. 171). The observed long progression of sixteen NH_2 bands was identified as $\tilde{A}\,^2A_1(0,v_2',0) \leftarrow \tilde{X}\,^2B_1(0,0,0)$ transitions with $v_2' = 3$ to 18 involving the rotational quantum numbers $0 \leq N' \leq 7$ in the $\tilde{A}\,^2A_1$ state. In addition, four bands of a subsidiary progression, $\tilde{A}\,^2A_1(1,v_2',0) \leftarrow \tilde{X}\,^2B_1(0,0,0)$, were found. These bands gain most of their intensity from a Fermi-type resonance between $(0,v_2'+2,0)$ and $(1,v_2',0)$ levels in the excited $\tilde{A}\,^2A_1$ state [15].

The absorption bands in a spectrum ranging from 950 to 635 nm were assigned to transitions from $\tilde{X}\,^2B_1(0,0,0)$ to the vibronic levels $\tilde{A}\,^2A_1(0,1,0)(K_a'=0)$, $(0,2,0)(K_a'=1)$, $(0,3,0)(K_a'=0,2)$, $(0,4,0)(K_a'=1,3)$, $(0,5,0)(K_a'=0,2,4)$, $(0,6,0)(K_a'=1,3,5)$, $(0,7,0)(K_a'=0,2,4,6)$, and

(0,8,0)(K_a' = 1,3,5,7). The absorption lines arising from $\tilde{X}\,^2B_1$(0,1,0) were also identified. Large perturbations (~200 cm^{-1}) were observed between some $\tilde{A}\,^2A_1(0,v_2',0)$ levels and the high vibrational levels of the $\tilde{X}\,^2B_1$ state [16]. In another absorption study at much higher resolution, nearly all rotational levels up to N' = 8 and K_a' = 8 were identified as well as a few additional levels up to N' = 10 and K_a' = 4 [17]. In the absorption studies [15 to 17] about 3500 band assignments were given [18]. Absorption bands ranging from 680 to 530 nm, which were assigned to transitions from $\tilde{X}\,^2B_1$(0,0,0) to $\tilde{X}\,^2B_1(0,v_2',0)$ with v_2' = 9 to 13, appeared in the spectrum on account of the Renner-Teller effect [1].

Frequencies of assigned absorption bands of NH_2 in the gas phase are given in Table 14.

The absorption **oscillator strength** f was measured and calculated for a few $\tilde{A}\,^2A_1 \leftarrow \tilde{X}\,^2B_1$ transitions. Studies of the $^RQ_{0,N=4}$ transition of the (0,12,0) ← (0,0,0) band at 516.6 nm [19, 20], $^PQ_{1,N=7}$ transition of the (0,9,0) ← (0,0,0) band at 597.4 nm [21 to 23], and $^PR_{1,N-1}$ transition of the (0,9,0) ← (0,0,0) band ($N \leq 6$) at 596 nm [24] were reported. The absorption [19, 20] and laser absorption [21, 22] were measured in NH_3-O_2-N_2 flames. Laser absorption spectroscopy was also used to study NH_2 in NH_3-O_2 flames [25]. Shock tube laser absorption experiments were carried out on NH_2, generated by photolysis and pyrolysis of NH_3 [23]. The oscillator strength f was calculated from the respective vibronic transition moment given in [6] and the wavelength of the $\tilde{A}\,^2A_1(0,v_2,0) \leftarrow \tilde{X}\,^2B_1$(0,0,0) transition for $3 \leq v_2 \leq 15$ [26].

The measured and calculated absolute oscillator strengths are given in the following table:

transition $\tilde{A}\,^2A_1 \leftarrow \tilde{X}\,^2B_1$	oscillator strength $f \times 10^5$ measured	calculated [a]
$(0,9,0)7_{0,7} \leftarrow (0,0,0)7_{1,7}$	6.35 ± 1.90 [23] 5.3 ± 1.6, 3.9 ± 1.2 [27] [b), c)]	19.6 ± 2.0 [27, 28]
$(0,12,0)4_{1,4} \leftarrow (0,0,0)4_{0,4}$	28.1 ± 4.7, 38.0 ± 8.5 [27] [b), d)]	28.4 ± 2.8 [27, 28]

a) Using a semiempirical method [29]. — b) Measurements for various flame compositions. — c) Revised from data given in [21, 22]. — d) Revised from data in [19, 20].

Matrix-Isolated NH_2. Absorption spectra of NH_2 radicals, which were produced in a microwave discharge of Ar mixed with a small amount of NH_3 or N_2H_4 and trapped on a 4.2 K surface [30, 31], indicated nearly free rotation of NH_2 in an Ar matrix [32]; for frequencies of assigned absorption bands, see Table 14.

UV absorption of NH_2 was also observed after photolysis of ammonia in an argon matrix at 4.2 K [33] or in nitrogen or carbon monoxide matrices at 14 K [34]. The $\tilde{A}\,^2A_1$(0,8,0) ← $\tilde{X}\,^2B_1$(0,0,0) band of NH_2 trapped in a rare gas matrix at 4.2 K showed line shifts of +28.7 cm^{-1} in Ar, −5.8 cm^{-1} in Kr, and −40.3 cm^{-1} in Xe relative to the gas phase spectrum [35]. More recently, absorption spectra of NH_2 and its isotopomers, isolated in Ne, Ar, Kr, and Xe matrices at 5 K, were measured in the visible range 370 to 880 nm [36].

Table 14

NH_2. Absorption Lines ν from Gas-Phase and Argon-Matrix Spectra.
All transitions to $\tilde{A}\,^2A_1(v_1',v_2',0)$ or $\tilde{X}\,^2B_1(v_1',v_2',0)$ originate in the $\tilde{X}\,^2B_1(0,0,0)$ state. For the notation, see p. 195.

$\tilde{A}(v_1',v_2',0)$ v_2'(linear)	v_2'(bent)	$\tilde{X}(v_1',v_2',0)$ v_2'(bent)	$^{F'}N'_{K_a'K_c'}-{}^{F''}N''_{K_a''K_c''}$	ν in cm^{-1} gas	Ar matrix[a]	I[b]
(0,22,0)	(0,10,0)	–	$(1_{10}-0_{00})$?	–	28982[c]	3
(0,20,0)	(0,9,0)	–	$1_{10}-0_{00}$	–	27022.3	7
(0,18,0)	(0,8,0)	–	$1_{10}-0_{00}$	25059.3[a]	25056.8	14
(0,17,0)	(0,8,0)	–	$1_{01}-0_{00}$	24106.23[d]	–	–
(0,16,0)	(0,7,0)	–	$1_{10}-0_{00}$	–	23145.8	26
(0,15,0)	(0,7,0)	–	$0_{00}-0_{00}$	22175.91[d]	–	–
(0,14,0)	(0,6,0)	–	$1_{10}-0_{00}$	21216.73[e]	21220.7	75
(0,13,0)	(0,6,0)	–	$0_{00}-1_{10}$	20275.04[e]	20292.2	4
(1,9,0)	(1,4,0)	–	$1_{01}-0_{00}$	20106.36[d]	–	–
(0,12,0)	(0,5,0)	–	$1_{10}-0_{00}$	19393.77[e]	19412.2	96
(1,8,0)	(1,3,0)	–	$1_{10}-0_{00}$	19227.59[e]	19244.3	39
(1,7,0)	(1,3,0)	–	$0_{00}-0_{00}$	18583.93[d]	–	–
(0,11,0)	(0,5,0)	–	$0_{00}-0_{00}$	18430.23[d]	–	–
(1,6,0)	(1,2,0)	–	$^{1}1_{10}-{}^{1}0_{00}$	17753.462[f]	17774.5	22
(0,10,0)	(0,4,0)	–	$^{1}1_{10}-{}^{1}0_{00}$	17558.728[f]	17588.0	97
–	–	(0,13,0)	$^{1}1_{10}-{}^{1}0_{00}$	16826.903[f]	16800.9[g]	3
(0,9,0)	(0,4,0)	–	$^{1}0_{00}-{}^{2}1_{10}$	16704.787[f]	16736.6	5
–	–	(1,10,0)	$1_{11}-0_{00}$	–	16355.6[g]	3
(0,8,0)	(0,3,0)	–	$^{1}1_{10}-{}^{1}0_{00}$	15898.778[f]	15931.0	100
–	–	(0,12,0)	$^{2}1_{11}-{}^{1}0_{00}$	15454.169[f]	15442.4[g]	5
(0,7,0)	(0,3,0)	–	$0_{00}-0_{00}$	15120.10[h]	–	–
(0,7,0)	(0,2,0)	–	$^{1}2_{21}-{}^{1}1_{11}$	15012.742[f]	15040.3	2
(0,6,0)	(0,2,0)	–	$^{1}1_{10}-{}^{1}0_{00}$	14362.12[f]	14388.4	38
–	–	(0,11,0)	$^{1}1_{10}-{}^{1}0_{00}$	–	13964.1[g]	7
(0,5,0)	(0,2,0)	–	$0_{00}-0_{00}$	13618.89[h]	–	–
–	–	(0,10,0)	$^{1}1_{11}-{}^{1}0_{00}$	13034.41[f]	13052.8[g]	12
–	–	(1,7,0)	$1_{11}-0_{00}$	–	12864.0[g]	13
(0,4,0)	(0,1,0)	–	$^{1}1_{10}-{}^{1}0_{00}$	12647.74[f]	12669.1[g]	24
(0,3,0)	(0,1,0)	–	$0_{00}-0_{00}$	12280.52[h]	–	–
(0,2,0)	(0,0,0)	–	$^{1}1_{10}-0_{00}$	11329.09[h]	–	–
(0,1,0)	(0,0,0)	–	$1_{01}-0_{00}$	11139.86[h]	–	–

[a] From [32]. – [b] Relative intensity from Ar matrix spectrum [32]. – [c] Average of the observed frequencies 28951 and 29012 cm^{-1}. – [d] Calculated term value [15]. – [e] From [15]. – [f] From [1]. – [g] Frequency from [32], assignment from [1]. – [h] Calculated term value [16].

Gaseous ND_2. The following ND_2 absorption lines ν [15] were assigned to the transitions $\tilde{A}\,^2A_1(v_1',v_2',0)N'_{K_a'K_c'} \leftarrow \tilde{X}\,^2B_1(0,0,0)0_{0,0}$:

$(v_1',v_2',0)N'_{K_a',0}$	$(0,17,0)0_{0,0}$	$(0,15,0)0_{0,0}$	$(0,14,0)1_{1,0}$	$(1,9,0)0_{0,0}$	$(0,13,0)0_{0,0}$
ν in cm^{-1}	19924.77	18588.41	17948.55	17361.00	17224.44

$(v_1',v_2',0)N'_{K_a',0}$	$(0,12,0)1_{1,0}$	$(0,11,0)0_{0,0}$	$(0,10,0)1_{1,0}$	$(0,9,0)0_{0,0}$
ν in cm^{-1}	16604.80	16014.42	15405.54	14844.28

Approximately 1200 bands in an ND_2 absorption spectrum were assigned in the range 485 to 630 nm [37].

NHD. Absorption spectra of NHD, prepared by flash photolysis of an NH_3+D_2O mixture, were recorded in the range 570 to 650 nm. Approximately 350 lines were assigned to $\tilde{A}\,^2A'(0,v_2',0) \leftarrow \tilde{X}\,^2A''(0,0,0)$ transitions with $v_2'=9$, 10, and 11 [38]. Three transitions which originate in the $\tilde{X}\,^2A''(0,0,0)0_{0,0}$ state are given below:

$(0,v_2',0)N'_{K_a',0}$	$(0,11,0)0_{0,0}$	$(0,10,0)1_{1,0}$	$(0,9,0)0_{0,0}$
ν in cm^{-1}	17313.254	16545.994	15868.071

Stark splittings were observed for the corresponding transitions $\tilde{A}\,^2A'(0,11,0)2_{2,1} \leftarrow \tilde{X}\,^2A''(0,0,0)3_{3,1}$ (F_1 line), $\tilde{A}\,^2A'(0,10,0)1_{1,0} \leftarrow \tilde{X}\,^2A''(0,0,0)2_{2,0}$ (F_1 and F_2 lines), and $\tilde{A}\,^2A'(0,9,0)2_{2,1} \leftarrow \tilde{X}\,^2A''(0,0,0)3_{3,1}$ (F_2 line) [39].

Emission Spectrum

The complex $\tilde{A}\,^2A_1 \rightarrow \tilde{X}\,^2B_1$ emission spectrum of NH_2, obtained by radiofrequency discharge through flowing ammonia, was recorded over a wide spectral range from the IR to the UV. Emission lines in the range 470 to 480 nm were assigned to $\tilde{A}\,^2A_1(0,14,0) \rightarrow \tilde{X}\,^2B_1(0,0,0)$ transitions for $1 \leq N' \leq 7$ and $K_a'=1$ or 3 [40]. Approximately 27000 lines were recorded between 400 and 890 nm. For the $\tilde{X}\,^2B_1(0,2,0)$ level, most of the rotational term values were determined up to $N''=8$ and $K_a''=5$. For the (0,1,0) and (0,0,0) levels of the $\tilde{X}\,^2B_1$ state, higher rotational term values (up to N'', $K_a''=16$ for the (0,0,0) level) were observed [41]. In the near-IR range 0.9 to 1.3 μm, 453 bands were assigned to transitions to the $\tilde{X}\,^2B_1(0,2,0)$ state for N and $K_a \leq 8$. Twenty-eight intense lines around 700 nm were assigned to transitions to the $\tilde{X}\,^2B_1(1,0,0)$ state [42]. For the $\tilde{A}\,^2A_1(0,v_2',0) \rightarrow \tilde{X}\,^2B_1(0,3,0)$ band system (400 to 800 nm), the strongest subbands originate from $\tilde{A}\,^2A_1(0,21,0)_{K_a=0,2}$ and $\tilde{A}\,^2A_1(0,20,0)_{K_a=1,3}$ levels. The transitions from (0,19,0), (0,18,0), (0,14,0), and (0,13,0) levels gave less intense lines [43].

In the emission spectrum of NH_2, formed by ArF excimer laser photolysis of NH_3 at 193 nm, the following lines λ could be identified as $\tilde{A}\,^2A_1(0,v_2',0) \rightarrow \tilde{X}\,^2B_1(0,0,0)$ transitions [44]:

$(0,v_2',0)_{K_a}$	$(0,8,0)_5$	$(0,6,0)_1$	$(0,8,0)_5$ and $(0,7,0)_4$
λ in nm	686	697	710
$(0,v_2',0)_{K_a}$	$(0,7,0)_4$	$(0,5,0)_0$	$(0,6,0)_3$ and $(0,7,0)_6$
λ in nm	730 to 733	741.5	751

For the study of high angular momentum states of NH_2 (K_a up to 22), electronic spectra were recorded threefold: 1) as a dispersed emission spectrum following laser excitation of photolysis products of NH_3, 2) as emission spectrum from a discharge through ammonia with an FT spectrometer, and 3) as LIF spectrum (see below). The dispersed emission spectrum was recorded over the range 400 to 740 nm and the FT emission spectrum from IR (3570 nm) to UV (435 nm) [45]. The rovibrational state distribution of nascent $NH_2(\tilde{A}\,^2A_1)$, generated by 193.3 nm photolysis of a room-temperature NH_3 sample, was determined by analyzing the $NH_2(\tilde{A}\,^2A_1 \rightarrow \tilde{X}\,^2B_1)$ near-IR emission (13000 to 6000 cm^{-1}) obtained by time-resolved FT spectroscopy. $NH_2(\tilde{A})$ was formed predominantly in the zero-point vibrational level with the most populated level at $N'=K_a'=5$ [46].

For an assignment of the NH_2 emission from 640 to 510 nm to the transitions $\tilde{A}\,^2A_1 \rightarrow \tilde{X}\,^2B_1$ and $\tilde{B}\,^2B_2 \rightarrow \tilde{A}\,^2A_1$, see [47 to 49]. NH_2 emission spectra were recorded from extraterrestrial sources. For example, twenty-one lines of the $\tilde{A}\,^2A_1(0,8,0) \rightarrow \tilde{X}\,^2B_1(0,0,0)$ band could be identified in a high-resolution spectrum at 630 nm of Comet P/Halley [50].

Laser-Induced Fluorescence (LIF)

Vibronic states $\tilde{A}\,^2A_1(v_1',v_2',v_3')$ and $\tilde{X}\,^2B_1(v_1'',v_2'',v_3'')$ which were characterized by LIF spectra are listed in the following table:

$\tilde{A}\,^2A_1(v_1',v_2',0)$	$\tilde{X}\,^2B_1(v_1'',v_2'',v_3'')$	Ref.
$(0,v_2',0)$, $v_2'=15$, 14, 13, 12	(0,1,0)	[51]
$(0,v_2',0)$, $v_2'=14$, 12, 10, 8	(0,0,0)	[52]
(0,14,0)	(0,1,0), (0,0,0)	[40]
(0,11,0)	(1,0,0), (0,1,0), (0,0,0)	[29]
$(1,v_2',0)$, $v_2'=8$, 6	(0,0,0)	[52]
(0,10,0)	(0,4,0), (1,1,0), (1,0,0), (0,2,0), (0,1,0), (0,0,0)	[53]
(0,10,0)	(1,0,0), (0,2,0), (0,01)	[54]
(0,10,0), (0,9,0)	$(0,v_2'',0)$, $v_2''=4$ to 10; $(1,v_2'',0)$, $v_2''=2$ to 5; $(2,v_2'',0)$, $v_2''=0$ to 3; (0,1,2), (0,2,2), (0,3,2), (3,0,0), (3,1,0)	[55]

The LIF technique was applied to radiative lifetime and collisional deactivation measurements [52, 56 to 60] or to (time-resolved) measurements of the rotational energy transfer within the $\tilde{A}\,^2A_1(0,9,0)$ state [61 to 63]; see p. 191.

References:

[1] Ross, S. C.; Birss, F. W.; Vervloet, M.; Ramsay, D. A. (J. Mol. Spectrosc. **129** [1988] 436/70).
[2] Perić, M.; Peyerimhoff, S. D.; Buenker, R. J. (Mol. Phys. **49** [1983] 379/400).
[3] Buenker, R. J.; Perić, M.; Peyerimhoff, S. D.; Marian, R. (Mol. Phys. **43** [1981] 987/1014).
[4] Perić, M.; Buenker, R. J.; Peyerimhoff, S. D. (Mol. Phys. **59** [1986] 1283/303).
[5] Pople, J. A.; Longuet-Higgins, H. C. (Mol. Phys. **1** [1958] 372/83).
[6] Jungen, C.; Hallin, K. E. J.; Merer, A. J. (Mol. Phys. **40** [1980] 25/63).
[7] Duxbury, G.; Dixon, R. N. (Mol. Phys. **43** [1981] 255/74).
[8] Herzberg, G.; Ramsay, D. A. (J. Chem. Phys. **20** [1952] 347).
[9] Herzberg, G.; Ramsay, D. A. (Discuss. Faraday Soc. No. 14 [1953] 11/6).
[10] Ramsay, D. A. (J. Phys. Chem. **57** [1953] 415/7).

[11] Ramsay, D. A. (J. Chem. Phys. **25** [1956] 188/9).
[12] Ramsay, D. A. (Ann. N. Y. Acad. Sci. **67** [1957] 485/98).
[13] Dyne, P. J. (Can. J. Phys. **31** [1953] 453/6).
[14] Ramsay, D. A. (Mem. Soc. R. Sci. Liege [4] **18** [1957] 471/9).
[15] Dressler, K.; Ramsay, D. A. (Philos. Trans. R. Soc. London A **251** [1959] 553/604).
[16] Johns, J. W. C.; Ramsay, D. A.; Ross, S. C. (Can. J. Phys. **54** [1976] 1804/14).
[17] Birss, F. W.; Ramsay, D. A.; Ross, S. C.; Zauli, C. (J. Mol. Spectrosc. **78** [1979] 344/6).
[18] Ramsay, D. A. (Proc. 3rd Yamada Conf. Free Radicals, Sanda, Jpn., 1979, pp.; C.A. **93** [1980] No. 57008).
[19] Nadler, M. P.; Wang, V. K.; Kaskan, W. E. (J. Phys. Chem. **74** [1970] 917/22).
[20] Fisher, C. J. (Combust. Flame **30** [1977] 143/9).

[21] Chou, M. S.; Dean, A. M.; Stern, D. (J. Chem. Phys. **76** [1982] 5334/40).

[22] Dean, A. M.; Chou, M. S.; Stern, D. (Int. J. Chem. Kinet. **16** [1984] 633/53).
[23] Kohse-Hoeinghaus, K.; Davidson, D. F.; Chang, A. Y.; Hanson, R. K. (J. Quant. Spectrosc. Radiat. Transfer **42** [1989] 1/17).
[24] Bykov, Y. V.; Gitlin, M. S.; Novikov, M. A.; Polushkin, I. N.; Khanin, Y. I.; Shcherbakov, A. I. (Zh. Tekh. Fiz. **54** [1984] 1310/4; Sov. Phys. Tech. Phys. [Engl. Transl.] **29** [1984] 755/7).
[25] Green, R. M.; Miller, J. A. (J. Quant. Spectrosc. Radiat. Transfer **26** [1981] 313/21).
[26] Tegler, S.; Wyckoff, S. (Astrophys. J. **343** [1989] 445/9).
[27] Anderson, W. R. (J. Phys. Chem. **93** [1989] 530/6).
[28] Anderson, W. R. (AD-A213837 [1990] 24 pp.; C.A. **114** [1991] No. 26773).
[29] Wong, K. N.; Anderson, W. R.; Vanderhoff, J. A.; Kotlar, A. J. (J. Chem. Phys. **86** [1987] 93/101).
[30] Robinson, G. W.; McCarty, M., Jr. (J. Chem. Phys. **28** [1958] 349/50).

[31] Robinson, G. W.; McCarty, M., Jr. (Can. J. Phys. **36** [1958] 1590/1).
[32] Robinson, G. W.; McCarty, M., Jr. (J. Chem. Phys. **30** [1959] 999/1005).
[33] Schnepp, O.; Dressler, K. (J. Chem. Phys. **32** [1960] 1682/6).
[34] Milligan, D. E.; Jacox, M. E. (J. Chem. Phys. **43** [1965] 4487/93).
[35] McCarty, M., Jr.; Robinson, G. W. (4th Int. Symp. Free Radical Stab. Trapped Radicals Low Temp., Washington, D. C., 1959, pp. F-III-1/F-III-20; C.A. **57** [1962] 5425).
[36] Blindauer, C.; Perić, M.; Schurath, U. (J. Mol. Spectrosc. **158** [1992] 177/200).
[37] Muenchausen, R. E.; Hills, G. W.; Merienne-Lafore, M. F.; Ramsay, D. A.; Vervloet, M.; Birss, F. W. (J. Mol. Spectrosc. **112** [1985] 203/10).
[38] Ramsay, D. A.; Wayne, F. D. (Can. J. Phys. **57** [1979] 761/6).
[39] Brown, J. M.; Chalkley, S. W.; Wayne, F. D. (Mol. Phys. **38** [1979] 1521/37).
[40] Vervloet, M.; Merienne-Lafore, M. F. (J. Chem. Phys. **69** [1978] 1257/62).

[41] Birss, F. W.; Merienne-Lafore, M. F.; Ramsay, D. A.; Vervloet, M. (J. Mol. Spectrosc. **85** [1981] 493/5).
[42] McKellar, A. R. W.; Vervloet, M.; Burkholder, J. B.; Howard, C. J. (J. Mol. Spectrosc. **142** [1990] 319/35).
[43] Merienne-Lafore, M. F.; Vervloet, M. (J. Mol. Spectrosc. **108** [1984] 160/2).
[44] Donnelly, V. M.; Baronavski, A. P.; McDonald, J. R. (Chem. Phys. **43** [1979] 283/93).
[45] Dixon, R. N.; Irving, S. J.; Nightingale, J. R.; Vervloet, M. (J. Chem. Soc. Faraday Trans. **87** [1991] 2121/33).
[46] Woodbridge, E. L.; Ashfold, M. N. R.; Leone, S. R. (J. Chem. Phys. **94** [1991] 4195/204).
[47] Proisy, P. (C. R. Hebd. Seances Acad. Sci. **244** [1957] 2784/5).
[48] Proisy, P. (C. R. Hebd. Seances Acad. Sci. **243** [1956] 1305/7).
[49] Proisy, P. (Mem. Soc. Roy. Sci. Liege [4] **18** [1957] 454/70).
[50] Combi, M. R.; McCrosky, R. E. (Icarus **91** [1991] 270/9).

[51] Vervloet, M.; Merienne-Lafore, M. F.; Ramsay, D. A. (Chem. Phys. Lett. **57** [1978] 5/7).
[52] Mayama, S.; Hiraoka, S.; Obi, K. (J. Chem. Phys. **80** [1984] 7/12).
[53] Kroll, M. (J. Chem. Phys. **63** [1975] 319/25).
[54] Vervloet, M.; Merienne-Lafore, M. F. (Can. J. Phys. **60** [1982] 49/55).
[55] Vervloet, M. (Mol. Phys. **63** [1988] 433/49).
[56] Lenzi, M.; McNesby, J. R.; Mele, A.; Nguyen Xuan, C. (J. Chem. Phys. **57** [1972] 319/23).
[57] Halpern, J. B.; Hancock, G.; Lenzi, M.; Welge, K. H. (J. Chem. Phys. **63** [1975] 4808/16).
[58] Kroll, M. (J. Chem. Phys. **63** [1975] 1803/9).
[59] Fuyuki, T.; Allain, B.; Perrin, J. (J. Appl. Phys. **68** [1990] 3322/37).
[60] Wysong, I. J.; Jeffries, J. B.; Crosley, D. R. (J. Chem. Phys. **93** [1990] 237/41).

[61] Dixon, R. N.; Field, D. (Proc. R. Soc. London A **366** [1979] 247/76).
[62] Dearden, S. J.; Dixon, R. N.; Field, D. (J. Chem. Soc. Faraday Trans. II **78** [1982] 1423/32).
[63] Alwahabi, Z. T.; Harkin, C. G.; McCaffery, A. J.; Whitaker, B. J. (J. Chem. Soc. Faraday Trans. II **85** [1989] 1003/15).

2.1.7.5 Reactions of NH_2

Walter Hack
Max-Planck-Institut für Strömungsforschung
Göttingen

The elementary reactions of imidogen and amidogen (i.e. NH_i, i = 1, 2) radicals are essential for understanding the nitrogen chemistry in the gas phase. Of predominant interest, however, is the unwanted formation of NO during technical combustion. NO forms from N_2 in the air and in addition from N-containing fuel. The NH_2 radical reactions play a critical role in the conversion of fuel nitrogen to NO and on the other hand in the depletion of NO. To understand and to model the combustion chemistry, the elementary reactions of NH_2 have to be known over wide temperature and pressure ranges with reference to the reaction rates and the reaction products. Kinetic studies were done in complex systems like flames or gas pyrolysis processes, but mainly in simple systems in which NH_2 radicals were produced by selected reactions and detected in specific ways. Publications on the reactivity and kinetic properties of NH_2 radicals have been covered up to 1984 in [1].

2.1.7.5.1 Elementary Reactions of NH_2 in the Gas Phase

The elementary reactions of NH_2 are divided into N-H-, N-H-O-, N-H-C-, and N-H-C-O-containing systems and systems in which other elements are involved.

NH_2 Reactions in the N-H System

In the N-H system the following reactions are described:

$$NH_2 + \{H, H_2\}, \{N, NH, NH_2\}, \text{ and } \{N_2H_4\}$$

The reaction with H atoms has two pathways

$$NH_2 + H \xrightarrow{M} NH_3 \qquad (1a)$$
$$\rightarrow NH + H_2 \qquad (1b)$$

The recombination of H atoms with NH_2 (1a) is expected to be the dominant reaction pathway at high pressures [2, 3], whereas the reaction channel (1b) is important at high temperatures. At low pressures reaction (1a) is of minor importance [4 to 8] as observed in the flash photolysis of ammonia. In the pressure range 330 to 2020 mbar the recombination channel (1a) was studied by pulse radiolysis of NH_3 [5]; k_{1a} was found to be in the fall-off region at these pressures. The rate constant at zero pressure was determined to be $k_{1a}(0) = 2.2 \times 10^{18}$ $cm^6 \cdot mol^{-2} \cdot s^{-1}$ ($M = NH_3$). The high-pressure limit $k_{1a}(\infty)$ seems to be too high with respect to the results obtained from the reverse reaction $NH_3 \rightarrow NH_2 + H$, the thermal dissociation of ammonia.

Using the ESR and LMR detection techniques for H, D, and NH_2, a room-temperature rate constant of $k = (3.2 \pm 0.8) \times 10^{13}$ $cm^3 \cdot mol^{-1} \cdot s^{-1}$ was obtained for the isotope exchange reaction $NH_2 + D \rightarrow NHD + H$ [9]. This value is expected to be equivalent to the high-pressure limit of reaction (1a).

For the reaction pathway (1b), a rate constant $k_{1b}(T) = 3 \times 10^{13}$ exp(−4600/T) $cm^3 \cdot mol^{-1} \cdot s^{-1}$ was determined in the temperature range 1800 to 3000 K [10, 15]. An extrapolation to low temperatures suggests that reaction (1b) is slow at low temperatures. A recent determination yielded $k_{1b}(T) = 4 \times 10^{13}$ exp(−15.3 $kJ \cdot mol^{-1}$/RT) $cm^3 \cdot mol^{-1} \cdot s^{-1}$ in the temperature range 2200 to 2800 K [12]. The high-temperature value is in good agreement with other high-temperature results obtained in shock tube experiments [10, 13]. In an N_2H_4 shock tube experiment ($2230 \leq T/K \leq 3460$), however, a significantly larger activation energy, $E_A = 43.6$ kJ/mol, was found [15]. The reaction has been investigated theoretically [16]. The different forces acting on the approaching H atom in the different stages of the reaction were calculated with the electrostatic force (ESF) theory. For large NH_2-H distances, the induced dipole interactions were found to be dominant, whereas an attractive exchange force was found to be important for the stabilization [16].

The reaction

$$NH_2 + H_2 \rightarrow NH_3 + H \qquad (3),$$

the reverse reaction of H with ammonia, has been discussed as an NH_2 source and studied at high temperatures. A rate constant of $k(T) = 1.26 \times 10^{12}$ exp[−(35.5 ± 1.7) $kJ \cdot mol^{-1}$/RT] $cm^3 \cdot mol^{-1} \cdot s^{-1}$ was determined in a flash photolysis system in the temperature range 295 to 500 K [17 to 19]. Also, ab initio calculations were done on this hydrogen abstraction reaction [20 to 22] giving a similar activation energy as that obtained experimentally [20]. Flash photolysis shock tube studies gave a value of $k(T) = 3.2 \times 10^{13}$ exp(−54 $kJ \cdot mol^{-1}$/RT) $cm^3 \cdot mol^{-1} \cdot s^{-1}$ in the temperature range 900 to 1620 K [23]. This value is in good agreement with a value of $k(1900\ K) = 8.9 \times 10^{11} cm^3 \cdot mol^{-1} \cdot s^{-1}$ from shock tube experiments, following the kinetics of the UV absorption of NH_3 and the NH_2 emission [24]. Discharge flow reactor experiments gave a value of $k(T) = 3.6 \times 10^{12}$ exp[−(38 ± 3) $kJ \cdot mol^{-1}$/RT] $cm^3 \cdot mol^{-1} \cdot s^{-1}$ in the temperature range 673 to 1003 K [25]. The reverse reaction $H + NH_3 \rightarrow NH_2 + H_2$ was measured in the same experimental arrangement. The rate constants (k_3, k_{-3}) measured in these two independent experiments, gave thermodynamic data, which were in very good agreement with the thermodynamic data obtained by other methods. The higher preexponential factor and activation energy obtained at higher temperatures in [23] indicate that the Arrhenius plot is, as expected from tunneling contributions at low temperatures, not linear.

The reaction $NH_2 + D_2 \rightarrow NHD + D$ was measured in an isothermal flow reactor [26] under pseudo-first-order conditions ($[NH_2]_0 \ll [D_2]$). In the temperature range 639 to 1140 K a rate constant of $k(T) = 2.9 \times 10^{12}$ exp[−(47.6 ± 2) $kJ \cdot mol^{-1}$/RT] $cm^3 \cdot mol^{-1} \cdot s^{-1}$ was obtained. Direct measurements of the kinetic isotope effect gave $k(H_2)/k(D_2) = 3.3$ at 740 K and 2.4 at 1140 K, indicating that tunneling significantly contributes to the reaction [26].

The reaction of N atoms with NH_2 has several exothermic pathways:

$$\begin{aligned} NH_2 + N &\rightarrow N_2 + H_2 && (2a) \\ &\rightarrow N_2 + 2\ H && (2b) \\ &\rightarrow N_2H + H && (2c) \end{aligned}$$

The existence of N_2H is still under discussion [27]. For reaction (2) a rate constant of k(298 K) = $(7.3 \pm 8) \times 10^{13}$ $cm^3 \cdot mol^{-1} \cdot s^{-1}$ was determined [28]. NH(X) was observed as a reaction product, although the pathway $NH_2 + N \rightarrow 2\ NH$ is endothermic by about 40 kJ/mol. In a later investigation a prompt production of H atoms has been observed with Lymann-α fluorescence during pulsed, 193-nm photolysis of NH_3; it increased when N atoms were introduced. This observation was interpreted to be an H atom production via the reaction $N + NH_2 \rightarrow N_2 + 2\ H$ [29]. In a discharge flow reactor with LMR detection of NH_2 (NH_2 source $F + NH_3$) and ESR detection of H and N atoms (the N atoms were produced in an N_2 microwave discharge) NH_2 profiles were measured under pseudo-first-order conditions ($[NH_2]_0 \ll [N]_0$).

A rate constant of $k=6.9\times10^{13}$ $cm^3\cdot mol^{-1}\cdot s^{-1}$ was determined. From the observed and calibrated H atom production, a reaction mechanism $N+NH_2\rightarrow N_2+2\ H$ was concluded [30].

Two of the other reactions to be discussed in this section are the radical-radical reactions

$$NH+NH_2 \rightarrow \text{products} \qquad (4)$$
$$NH_2+NH_2 \rightarrow \text{products} \qquad (5)$$

The rate of reaction (4) has been estimated from computer models of complex systems. Thus, a rate constant of $k(349\ K)=7\times10^{13}$ $cm^3\cdot mol^{-1}\cdot s^{-1}$ was obtained in NH_3 pulse radiolysis experiments [31]. A direct measurement in a flow reactor, in which NH and NH_2 were detected by LMR and LIF, yielded the rate constant $k(296\ K)=(8\pm3)\times10^{13}$ $cm^3\cdot mol^{-1}\cdot s^{-1}$ [32]. At high temperatures ($2200\leq T/K\leq2800$) NH_3 pyrolysis was observed in reflected shock waves. A rate constant of $k(2500\ K)=3\times10^{13}$ $cm^3\cdot mol^{-1}\cdot s^{-1}$ was obtained from the best fit of measured and calculated concentration profiles [12]. A temperature dependence of $T^{-1/2}$ was concluded from the fact that reaction (4) is a radical-radical reaction [12]. The measured room-temperature value is consistent with this small negative temperature dependence. Besides the recombination $NH+NH_2+M\rightarrow N_2H_3+M$, four exothermic reaction pathways are possible (N_2H+H_2, N_2+H+H_2, N_2H_2+H, and NH_3+N) as discussed in [32].

The other radical-radical reaction mentioned above already appeared in the literature in 1927 [33]. The formation of hydrazine after the decompositon of ammonia has been attributed to the reaction

$$NH_2+NH_2+M \rightarrow N_2H_4+M \qquad (5a)$$

[2, 33 to 35]. A detailed investigation of reaction (5) was done in a pulsed rf NH_3 discharge system [7]. A pressure-independent ($0.5\leq p/mbar\leq1$) rate constant of $k_5=(2.4\pm0.2)\times10^{12}$ $cm^3\cdot mol^{-1}\cdot s^{-1}$ was measured. This rate constant is attributed to one of the disproportionation pathways

$$NH_2+NH_2 \rightarrow NH_3+NH \qquad (5b)$$
$$\rightarrow N_2H_2+H_2 \qquad (5c)$$
$$\rightarrow N_2+2\ H_2 \qquad (5d)$$

[36]. In the pressure range 6.6 to 13 mbar a pressure dependence of k_5 was observed leading to a zero-pressure value of $k_5(0)=4.6\times10^{11}$ $cm^3\cdot mol^{-1}\cdot s^{-1}$ which is expected to describe reaction (5b) [37]. A high-pressure ($330\leq p/mbar\leq1660$) pulse radiolysis study of NH_3 yielded the NH_2 recombination rate constant $k_{5a}(\infty)=6.3\times10^{13}$ $cm^3\cdot mol^{-1}\cdot s^{-1}$ [5]. This value was later modified by taking a corrected initial concentration of NH_2 into account [38], and then it agreed well with the values $k_5=(4.7\pm2)\times10^{13}$ $cm^3\cdot mol^{-1}\cdot s^{-1}$ obtained by the rotating sector method at 400 mbar C_2F_6 [39] and $k_5=(3.6\pm1.5)\times10^{13}$ $cm^3\cdot mol^{-1}\cdot s^{-1}$ determined by flash photolysis [40, 41]. Using flash photolysis of NH_3, k_5 was measured over a wide pressure range ($0.4\leq p/mbar\leq1330$) with the buffer gases NH_3, N_2, and Ar [4]. The termolecular, low-pressure limits of $k_5=1.0\times10^{19}$ $cm^6\cdot mol^{-2}\cdot s^{-1}$ (NH_3), $k_5=2.5\times10^{18}$ $cm^6\cdot mol^{-2}\cdot s^{-1}$ (N_2), and $k_5=1.0\times10^{18}$ $cm^6\cdot mol^{-2}\cdot s^{-1}$ (Ar) were obtained for the three buffer gases. Bimolecular NH_2 depletion was observed as an intercept at zero pressure; it is interpreted as a reaction pathway other than (5a) [6 to 8, 37, 42, 43]. Experiments, however, in which products other than N_2H_4 are observed at high pressures, would give much stronger evidence for the existence of reaction pathways other than (5a).

The high-pressure rate constant $k_{5a}(\infty)=(1.5\pm50\%)\times10^{13}$ $cm^3\cdot mol^{-1}\cdot s^{-1}$ was obtained below 1.3 bar (N_2) at room temperature [4]. This value was confirmed in an O_3-NH_3 photolysis study [44]. The low-pressure limit $k_5 = (8.5 \pm 50\%) \times 10^{11}\ cm^3 \cdot mol^{-1} \cdot s^{-1}$ is interpreted to be consistent with the disproportionation reaction of amidogen radicals (5b) as the principal reaction at pressures below 1.3 mbar. The same interpretation was given to the results of mass-spectrometric flow reactor studies on the reaction $H + N_2H_3 \rightarrow N_2H_4^{\ddagger} \rightarrow 2\ NH_2(NH + NH_3)$ [43]. However, no NH(X) was observed during pulsed vacuum UV photolysis of NH_3; thus, reaction pathway (5b) was concluded to be negligible [45]. This is confirmed by the observation that no NH(X) ($\leq 1\%$) is detected, if $N_2H_4^{\ddagger}$ is formed with the electronic chemical activation method via $NH(a\ ^1\Delta) + NH_3 \rightarrow N_2H_4^{\ddagger} \rightarrow 2\ NH_2$ [46].

A small negative temperature effect was observed in the temperature range 300 to 500 K at 27 mbar N_2. Also, the low- and high-pressure limits are nearly temperature-independent [4].

The reaction $NH_2 + NH_2$ was studied at high temperatures in shock tubes [12, 36, 47]. The mechanism of NH_3 pyrolysis was established using the time-history of the species NH and NH_2 spectroscopically measured behind reflected shock waves in the temperature range 2200 to 2800 K. The best fit of experimental data by computer simulations gave the rate coefficient for $NH_2 + NH_2 \rightarrow NH + NH_3$. $k_{5b}(T) = 5 \times 10^{13} \exp(-41.8\ kJ \cdot mol^{-1}/RT)\ cm^3 \cdot mol^{-1} \cdot s^{-1}$ was obtained with an estimated uncertainty of $\pm 50\%$ [12]. At high temperatures N_2H_4 can decompose via $NH_2 + NH_2 \rightarrow N_2H_4^{\ddagger} \rightarrow N_2H_3 + H$. A simulation of the kinetics in rich ammonia flames yielded an estimate for the rate constant of $k(T) = 7.4 \times 10^{11} \exp(-10.5\ kJ \cdot mol^{-1}/RT)\ cm^3 \cdot mol^{-1} \cdot s^{-1}$ [48].

The reaction

$$NH_2 + N_2H_4 \rightarrow NH_3 + N_2H_3$$

was studied in a flow reactor with mass-spectrometric detection giving the rate constant $k(300\ K) = (3.1 \pm 0.4) \times 10^{11}\ cm^3 \cdot mol^{-1} \cdot s^{-1}$ [49].

NH_2 Reactions in the N-H-O System

In the N-H-O system the following groups of reactions are described:

$$NH_2 + \{O, O_2(X^3\Sigma_g^-), O_2(a^1\Delta_g), O_3\}, \{NO, NO_2\}, \text{ and } \{OH, H_2O, HO_2, H_2O_2\}$$

The reaction with O atoms including the pathways

$$\begin{aligned} NH_2 + O &\rightarrow HNO + H \\ &\rightarrow NH + OH \\ &\rightarrow NO + H_2 \\ &\rightarrow H_2O + N \end{aligned}$$

has been studied experimentally [32, 43, 50 to 55] and theoretically [54 to 57]. The overall rate constant at room temperature, $k(296\ K) = (5.3 \pm 1.5) \times 10^{13}\ cm^3 \cdot mol^{-1} \cdot s^{-1}$, reported in [32] was obtained in a flow reactor with ESR-LMR and LIF detection devices. The main reaction products are NH + OH and HNO + H [32]. In ESR-MS studies in flow reactors the production of $NO + H_2$ and $H_2O + N$ was found to be insignificant. No H_2O as reaction product was found in MS flow reactor studies [52]. The ratio of the two rate constants $k(HNO + H) = (4.6 \pm 1.2) \times 10^{13}\ cm^3 \cdot mol^{-1} \cdot s^{-1}$ and $k(NH + OH) = (7 \pm 3) \times 10^{12}\ cm^3 \cdot mol^{-1} \cdot s^{-1}$ [32] is in good agreement with results obtained from RRKM calculations [56]. The rotational state distribution of OD(v = 0,1) measured in crossed-beam experiments ($O + ND_2$) is significantly colder than predicted by statistical theories (PST and SACM) [55]. A very cold rotational state distribution was observed also for HNO in beam experiments and associated with a barrier for the formation of HNO + H [54]. Theoretical calculations predicted that the formation of OH + NH proceeds via an abstraction reaction which is dominant at high temperatures, whereas the products HNO + H resulting from the decomposition of an intermediate complex dominate at lower temperatures [57].

The reaction

$$NH_2 + O_2 \rightarrow \text{products}$$

is the rate-controlling reaction during the induction period of the ammonia-oxygen combustion. The reaction has been studied at room temperature [44, 51, 58 to 65] and at high temperatures [66 to 68]. The following individual channels are feasible:

$$\begin{aligned} NH_2 + O_2 &\rightarrow NO + H_2O \\ &\rightarrow NO_2 + H_2 \\ &\rightarrow HNO + OH \\ &\rightarrow HNO_2 + H \\ &\rightarrow H_2NO + O \\ + M &\rightarrow NH_2O_2 + M \end{aligned}$$

At room temperature the $NH_2 + O_2$ reaction seems to be very slow. When ammonia was photolyzed in a flash photolysis system in the presence of O_2, the NH_2 radicals (detected by dye laser absorption) were mainly consumed by the reaction $NH_2 + HO_2$. An upper limit for the NH_2 depletion by O_2 in these experiments (218 K, 500 K, 670 mbar) was $k < 1.8 \times 10^6$ $cm^3 \cdot mol^{-1} \cdot s^{-1}$ [51]. The same upper limit was determined in a very similar experiment [62, 63] and in pulse radiolysis experiments at 1 bar (Ar) and 350 K [31]. In order to avoid complications by H atoms via $H + O_2 + M \rightarrow HO_2 + M$, $HO_2 + NH_2 \rightarrow$ products, C_2H_4 was added to consume the H atoms in the reaction $H + C_2H_4 \rightarrow C_2H_5$. The upper limits determined in these NH_3/O_2 flash photolysis experiments at 33 mbar (Ar) were $k(245\ K) < 2.8 \times 10^7$ $cm^3 \cdot mol^{-1} \cdot s^{-1}$, $k(298\ K) < 4.6 \times 10^6$ $cm^3 \cdot mol^{-1} \cdot s^{-1}$, and $k(459\ K) < 3.4 \times 10^8$ $cm^3 \cdot mol^{-1} \cdot s^{-1}$ [64]. The H atom-free NH_2 source $F + NH_3 \rightarrow NH_2 + HF$ was applied in a flow reactor with LIF-NH_2 detection [61]. In these experiments a significantly higher rate was observed with positive pressure dependence and a negative temperature dependence, which was explained by the addition reaction $NH_2 + O_2 + M \rightarrow NH_2O_2 + M$ with a rate constant of $k(T) = (1.3 \pm 0.5) \times 10^{15}\ (T/295)^{-2}$ $cm^6 \cdot mol^{-2} \cdot s^{-1}$. This result is in conflict with all other determinations; apparently heterogeneous effects are involved in the NH_2 depletion. (Either NH_2 reacts with O_2 (or a compound formed with O_2) on the wall or NH_2O_2 is formed initially and reacts in an unusually fast reaction with the wall.) The other experiment, which was done to avoid the reaction $H + O_2$ and thus the $HO_2 + NH_2$ reaction, was the O_3 photolysis in the presence of NH_3 and O_2 ($O_3 + h\nu \rightarrow O_2 + O(^1D)$; $O(^1D) + NH_3 \rightarrow NH_2 + OH$; $OH + NH_3 \rightarrow NH_2 + H_2O$). At 66 mbar in the temperature range 272 to 348 K, an upper limit $k < 1.5 \times 10^6$ $cm^3 \cdot mol^{-1} \cdot s^{-1}$ was obtained [44].

The existence of the NH_2O_2 radical was observed in ammonia solutions at 210 K by UV absorption. The aminoperoxy radical was produced in the reaction $KO_3 + NH_3(l) + h\nu \rightarrow KOH + NH_2O_2$ [69]. NH_2O_2 was also detected (via IR absorption) when an $NH_3/O_2/Ar$ matrix (4.2 K) was photolyzed with a mercury lamp [70]. A theoretical study showed that the NH_2O_2 complex is barely stable (NH_2-O_2 bond strength 21 kJ/mol) and that the rate constant for the $NH_2 + O_2$ reaction is very small, the high-temperature products probably being $NH_2O + O$. The barrier for $NH_2O + O$ formation was calculated to be 126 kJ/mol; the barrier for the reaction pathway to form HNO + OH through the 1,3-H shift transition state was calculated to be only a little higher (148 kJ/mol) [57]. An estimate for the bond strength of NH_2-O_2 of 52.3 kJ/mol [56] also indicates that an aminoperoxy radical, formed via an addition reaction in the vapor phase, is highly improbable [71]. Above 600 K the primary product channel of the reaction $NH_2 + O_2$ was calculated to be $NH_2O + O$ [56]. Fourier transform IR (FTIR) spectroscopy showed the product channel $NH_2 + O_2 \rightarrow NO_x$ to be very slow at 296 K with $k < 3.6 \times 10^3$ $cm^3 \cdot mol^{-1} \cdot s^{-1}$ [65].

The reaction $NH_2 + O_2$ was studied at high temperatures in shock waves [66 to 68]. The induction period for the appearance of OH absorption and the consumption of NH_3 was measured in shock-heated NH_3-O_2 mixtures in the temperature range 1500 to 2800 K; an activation energy of $E_A = 178$ kJ/mol was obtained for the reaction $NH_2 + O_2 \rightarrow NH + HO_2$ [67]. The kinetics of ammonia oxidation in reflected shock waves was followed by observing the induction period. Computer simulation of the NH_3-O_2 system with the reaction scheme

$$NH_2 + O_2 \underset{k_r}{\overset{k_f}{\rightleftharpoons}} (NH_2O_2)^{\ddagger} \xrightarrow{k_d} HNO$$
$$\xrightarrow{k_s\,(M)} NH_2O_2$$

in the temperature range 1550 to 1800 K gave the rate constants $k_f = 3.2 \times 10^{12} \exp[-(63 \pm 20)\ kJ \cdot mol^{-1}/RT]\ cm^3 \cdot mol^{-1} \cdot s^{-1}$, $k_d = (1.3 \pm 0.3) \times 10^{10}\ s^{-1}$, and $k_r = (1.0 \pm 0.3) \times 10^{10}\ s^{-1}$ [68]. Computer modeling of the kinetics of NH_3 oxidation in the temperature range 1279 to 1323 K enabled one to estimate the rate constant for the reaction $NH_2 + O_2 \rightarrow HNO + OH$ to be $k = 5.1 \times 10^{13} \exp(-125\ kJ \cdot mol^{-1}/RT)\ cm^3 \cdot mol^{-1} \cdot s^{-1}$ [66]. The data in shock tubes were indirectly obtained, and it is necessary to obtain NH_2 radical profiles in order to determine the rate constant directly.

Compared to the analogous system $H + O_2$, in which HO_2 is formed at low temperatures and $OH + O$ at high temperatures with a significant activation energy, the $NH_2 + O_2$ system is much less defined probably due to the low binding energy in NH_2-O_2. The reaction, equivalent to $H + O_2 \rightarrow HO_2$, is not important in the $NH_2 + O_2$ system at and near room temperature. A value of $k < 1.8 \times 10^6\ cm^3 \cdot mol^{-1} \cdot s^{-1}$ is recommended by CODATA 1992 [72] at these temperatures. The determinations at high temperatures are too indirect to make any definite recommendation. Extrapolating these high-temperature values to room temperature leads to values that are much lower than the experimentally determined upper limits.

The reaction of NH_2 with electronically excited, metastable O_2 molecules ($O_2(a\ ^1\Delta_g)$) has been measured in a discharge flow reactor. In the temperature range 295 to 353 K a value of $k = (6 \pm 2) \times 10^9\ cm^3 \cdot mol^{-1} \cdot s^{-1}$ was determined [73] which was found to be independent of temperature and pressure in the range 2.6 to 5.2 mbar. OH radicals were observed as reaction products, and other products were not detected; thus, it was concluded that $NH_2 + O_2(a\ ^1\Delta_g) \rightarrow HNO + OH$ is the main reaction pathway [73].

The reaction with ozone

$$NH_2 + O_3 \rightarrow NH_2O + O_2$$

may be an important step for NH_2 depletion and thus for the oxidation of ammonia in urban atmospheres, since the reaction of NH_2 with O_2 is very slow at ambient temperatures. There have been four independent measurements of the rate constant as a function of temperature near room temperature as shown in **Fig. 8** [44, 74 to 77] and two at room temperature [78, 79]. The value obtained by ozone photolysis ($O_3 + h\nu \rightarrow O_2 + O(^1D)$, $O(^1D) + NH_3 \rightarrow OH + NH_2$, $OH + NH_3 \rightarrow NH_2 + H_2O$) is nearly a factor of four higher than the value obtained when NH_3 is photolyzed to produce NH_2 [74, 75]. It is possible that in the O_3 photolysis not only $O_2(X\ ^3\Sigma_g^-)$ is formed but also electronically excited O_2 which might react with the NH_2 radicals. The values obtained in the temperature range 200 to 350 K were also obtained with the reaction $O(^1D) + NH_3$ as the NH_2 source [76]. The absolute value obtained at room temperature by the same group [79], however, is significantly higher than the value reported later for the same temperature. The flow system measurements have the great advantage that a clean NH_2 source ($F + NH_3 \rightarrow NH_2 + HF$) is available [77].

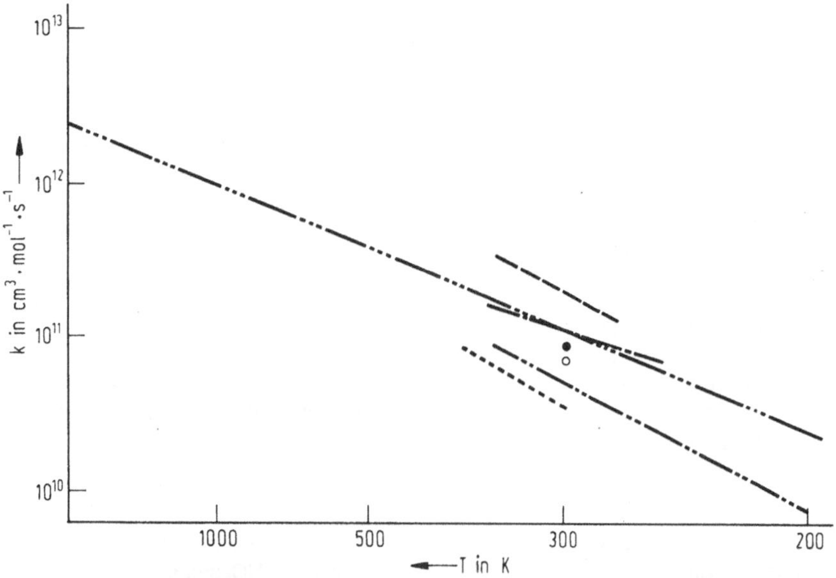

Fig. 8. Rate constant for the reaction NH_2+O_3 in an Arrhenius diagram.
– – – – Patrick and Golden [44]; ——-—— Hack et al. [77];
---- Kursawa and Lesclaux [74, 75]; ——--—— Bulatov et al. [76]; ● Cheskis et al. [78];
○ Bulatov et al. [79]; ——---—— recommended value $200 \leq T/K \leq 400$.

It is reasonable to assume, in particular in comparison with the isoelectronic reactions of OH and CH_3 with O_3, that the absolute value of the rate constant is accurate. If a preexponential factor of $A = 2.5 \times 10^{12}$ $cm^3 \cdot mol^{-1} \cdot s^{-1}$ is accepted, the activation energy becomes $E_A = 8$ kJ/mol. These rate data (as indicated in Fig. 8) are reasonable in the temperature range 200 to 400 K. It is not unlikely that the reaction dynamics changes at higher temperatures, leading to higher values; there is no direct and simple way to measure the rate under such conditions, since O_3 decomposes significantly at higher temperatures.

At room temperature and in the temperature range given above, the reaction pathway $NH_2+O_3 \rightarrow NH_2O+O_2$ is equivalent to the elementary reactions of the isoelectronic radicals OH and CH_3 with ozone and is assumed to be dominant. NH_2O has been detected as the main product in the NH_2+O_3 reaction by FIR LMR ($299 \leq \lambda/\mu m \leq 742$) in a discharge flow reactor [80, 81].

The reaction

$$NH_2 + NO \rightarrow \text{products}$$

has been studied extensively since the early photolysis experiments in 1939 [82]. The interest in this reaction derives from its importance in ammonia combustion [83 to 91] (see p. 231) and ammonia oxidation in urban atmospheres [87 to 89] (see p. 233) as well as in $DeNO_x$ processes [90 to 92] (see p. 233). The reaction was believed to be fast [93] before quantitative experiments were performed [94, 95]. The recent results on the kinetics of the NH_2+NO reaction are summarized in Table 15, pp. 216/8.

Table 15
Summary of Recent Experimental Results on the Reaction $NH_2 + NO \rightarrow$ Products.

experimental method	experimental conditions	results k in $cm^3 \cdot mol^{-1} \cdot s^{-1}$ E_A in kJ/mol	Ref.
pulse radiolysis [NH](t) measured by absorption	298 K 0.03 to 1 mbar (NO)	$k = 1.6 \times 10^{13}$	[5, 96]
discharge flow reactor mass spectrometer $NH_3 + OH(H + NO_2) \rightarrow$ $NH_2 + H_2O$ IR chemiluminescence	298 K 1.3 to 13 mbar	$k = 5 \times 10^{12}$ products: $N_2 + H_2O$, H_2O with high vibrational energy	[43]
flame	2010 K	$k = 1.2 \times 10^{10}$	[84]
incident shock waves induction periods observed, NO, OH, NH_3, NH monitored in absorption	1450 to 3250 K	quantitative modeling of the NH_3-NO system $k = 1 \times 10^{13}$	[97]
pulse photolysis of NH_3 NH_2, LIF detection	298 K 1.3 mbar (Ar)	$k = (1.26 \pm 0.12) \times 10^{13}$	[98]
flash photolysis of NH_3	298 to 500 K 266 to 931 mbar (N_2)	$k(298\ K) = (1.1 \pm 0.2) \times 10^{13}$ negative temperature dependence, $E_A = -4.4 \pm 0.8$ k not pressure-dependent	[99, 100]
flash photolysis of NH_3	298 K 1.3 mbar	$k = 1.7 \times 10^{13}$	[101]
NH_3 photolysis (213.9 nm)	298 K	relative rates	[59]
discharge flow reactor $F + NH_3 \rightarrow NH_2 + HF$ $[NH_2](t)$, LIF detection	210 to 503 K 0.8 to 5.3 mbar	$k = 7.2 \times 10^{12}\ (T/298K)^{-1.85}$ k(T) not pressure-dependent	[102, 103]
flash photolysis of NH_3 $[NH_2]$ intracavity absorption	296 K 0.13 to 1.3 mbar	$k = 1.0 \times 10^{13}$ k independent of pressure	[104]
flash photolysis $[NH_2](t)$, LIF detection	298 K 4 to 13.3 mbar	$k = 1.14 \times 10^{13}$	[105]
shock tube experiments detection of NH_2 emission (538 nm)	1680 to 2850 K	$k = 7 \times 10^{13}\ \exp(-11.6/RT)$ for $NH_2 + NO \rightarrow N_2O + H_2$, $k = 3 \times 10^{13}\ \exp(-23.7/RT)$ for $NH_2 + NO \rightarrow \{N_2O + H_2$, $N_2 + H + OH$, $N_2 + H_2O\}$	[14]

Table 15 (continued)

experimental method	experimental conditions	results k in $cm^3 \cdot mol^{-1} \cdot s^{-1}$ E_A in kJ/mol	Ref.
exciplex laser photolysis of NH_3 (193 nm) mass spectrometer	298 K 70 to 1000 mbar	formation of H and OH observed, stoichiometry determined	[106]
flow reactor $F+NH_3 \rightarrow NH_2+HF$ NH_2detection H atom detection Lyman-α absorption	294 to 1215 K $[NH_3]:[F]=3$ to 50 for the rate constant, $[NH_3]:[F]=3$ for the products	$k=(5.36 \pm 0.9) \times 10^{13}$ $(T/298)^{-(2.3 \pm 0.02)}$ $\exp[-(5.68 \pm 0.5)/RT]$ two major product channels $a=\{N_2+H_2O\}$, $b=\{N_2H+OH\}$ no H atom detected $\alpha=k_b/(k_a+k_b)=0.4$	[107]
reevaluation of the branching ratio, $NH_2+F \rightarrow NH+HF$ and $NH+NO \rightarrow OH+N_2$ taken into account		$\alpha=0.1$	[108]
NH_3 flash photolysis NH_2, LIF detection	216 to 480 K 3.3 to 13.3 mbar	$k=(1.3 \pm 0.2) \times 10^{13}$ $(T/298K)^{-1.67}$ k independent of p, $[NO]_0$, and $[NH_2]_0$	[109, 110]
flow reactor NH_3 laser photolysis (193 nm) $H_2O^{\ddagger}$ IR fluorescence	300 to 1150 K 1.3 mbar	$k(295\ K)=(1.3 \pm 0.3) \times 10^{13}$ from $[H_2O^{\ddagger}](t)$ no H atoms observed $\{N_2H+OH\} \geq 0.65$ $\{N_2+H_2O\} \geq 0.29$ $\{N_2O+H_2\} \leq 0.01$ $\{N_2+H+OH,\ N_2OH+H\} \leq 0.05$	[111]
NH_3 laser photolysis (193 nm) NH_2, LIF detection	room temperature 1.3 to 2.7 mbar	$k=(1.1 \pm 0.07) \times 10^{13}$	[28]
NH_2 generated by IR multiple photon dissociation, precursor N_2H_4, CH_3NH_2 $NH_2(v_2''=0,1)$ by LIF	298 K 2.7 mbar	$k(v_2''=0)=8.4 \times 10^{12}$ $k(v_2''=1)=1.9 \times 10^{13}$	[112]
flow reactor IR laser multiphoton dissociation precursor N_2H_4 + SF_6	298 K 1.7 mbar $2 \leq [NO]/10^{12}\ cm^{-3} \leq 20$	$k=5.4 \times 10^{12}$	[113]
NH_3 laser photolysis (193 nm)	295 K 10 mbar	$k=1.0 \times 10^{13}$ from $[NH_2](t)$ $k=(7.6 \pm 2.5) \times 10^{12}$	[114]

Table 15 (continued)

experimental method	experimental conditions	results k in $cm^3 \cdot mol^{-1} \cdot s^{-1}$ E_A in kJ/mol	Ref.
NH_2, N_2 by CARS detection $H_2O^{\neq}$ by IR fluorescence		from $[H_2O^{\neq}](t)$ $T_{vib}(H_2O) = 10^4$ K $N_2(v)$ lower temperature	
NH_3 laser photolysis (193 nm)	298 K 2.9 to 20 mbar (He, SF_6)	H_2O in very high vibrational states product channels: $[OH] = (13 \pm 2)\%$ $[H_2O] = (85 \pm 9)\%$ or $(66 \pm 3)\%$	[115, 116]
flow reactor mass spectrometer $F + NH_3 \rightarrow NH_2 + HF$	300 K 1.34 mbar $[NH_3]:[F] \cong 50$	product channels: $N_2 + H_2O$ (a) $N_2H + OH$ (b) $k_b/k_a < 0.15$	[117]
NH_3 photolysis (184.9 nm) FTIR spectroscopy	4.2 K (matrix) low pressure $^{15}NH_3$ and $^{14}NH_3$ used	N_2O as a direct product $^{15}N^{14}NO$ from $NO + {}^{15}NH_2$	[118]
NH_3 photolysis Xe lamp (>190 nm) $[NH_2]$ intracavity absorption	295 to 620 K 1.5 to 42 mbar	$k(T) = 1.2 \times 10^{13}$ (T/298 K)$^{-2.2}$ product ratios: $\{N_2H + OH, N_2 + H + OH\}$ (a) $k_a/k = 0.1 \pm 0.02$ (295 K) $k_a/k = 0.14 \pm 0.03$ (470 K) $k_a/k = 0.2 \pm 0.04$ (620 K)	[119, 120]
NH_3 laser photolysis (193 nm) [OH] by LIF	294 to 1027 K 13 mbar (N_2) photolysis laser intensity $= 10^{17}$ photons$\cdot cm^{-2} \cdot s^{-1}$ OH calibration with OH from H_2O_2 4 to 133 mbar (N_2) ≤ 1.6 mbar H_2O addition	$k(T) = 1 \times 10^{13}$ (T/298)$^{-1.17}$ $\alpha = [OH]_{\infty}/[NH_2]_0$ $\alpha(300\ K) = 0.1$ $\alpha(1000\ K) = 0.19$ no pressure dependence of α no dependence on H_2O addition	[121]
NH_3 photolysis NH_2 depletion	300 to 1200 K	product branching ratios	[122]
pulse radiolysis of $NH_3/SF_6/Ar$ $[NH_2](t)$ absorption	298 K	$k = (1.3 \pm 0.2) \times 10^{13}$ $H_2O + N_2$ is the most important product channel	[123]

There is a general agreement on the reaction rate. It is a fast reaction, the rate has a negative temperature dependence, and the reaction rate constant is independent of pressure in a wide pressure range ($0.5 \leq p/mbar \leq 500$). It is noticeable, however, that all values obtained in flow reactors, using $NH_3 + F \rightarrow NH_2 + HF$ as a radical source, are significantly lower (by about a factor of two) than the values determined by photolysis, where NH_2 is obtained via $NH_3+h\nu \rightarrow NH_2+H$ (independent of the photolysis wavelength). The difference between the two methods indicates that the influence of the H atoms in the NH_3-NO system is not fully understood; in all the photolysis experiments, no attempt was made to add a scavenger for the H atoms. The constant $k(298\ K)=9.6\times10^{12}\ cm^3\cdot mol^{-1}\cdot s^{-1}$, recommended in CODATA [124], is an average of all recent direct determinations. The factor $(T/298\ K)^{-1.75}$ can be recommended for the temperature dependence. Thus, the rate constant $k(T)=9.6\times10^{12}\ (T/298\ K)^{-1.75}\ cm^3\cdot mol^{-1}\cdot s^{-1}$ was obtained with a high degree of confidence in the temperature range 200 to 500 K.

After a long discussion [107 to 111, 116, 117, 121] $NH_2+NO \rightarrow N_2+H_2O^{\neq}$ (a) is accepted as the main reaction pathway, as already pointed out in an earlier mass-spectrometric investigation [43], contributing about 85 to 90%. The water obtained in this highly exothermic reaction was found to be vibrationally excited [43, 111]. The production of OH radicals (reaction pathway (b)) contributes about 10 to 15% of the total reaction. This is in agreement with earlier determinations of a quantum yield of $\phi(N_2)=1$ for N_2 molecules [58, 59, 94, 95, 125]. A more pronounced contribution of the reaction channel (b) would lead to a significantly higher quantum yield of $\phi(N_2)=2$ due to secondary NH_2 formation, since OH is an efficient chain carrier at any temperature via $OH+NH_3 \rightarrow NH_2+H_2O$ (NH_3 is available in excess in the photolysis systems). The problem that is not yet completely solved is whether channel (b) should be written

$$NH_2+NO \rightarrow N_2+H+OH$$

as first proposed in [126, 127] or

$$NH_2+NO \rightarrow N_2H+OH,$$

since no H atoms could be detected in the system at temperatures as high as 900 K [107]. The N_2H radical, on the other hand, also was never detected neither in this reaction nor in any other system. Quantum-mechanical calculations indicate that N_2H may be stable due to a barrier in the dissociation coordinate, since N_2H in its electronic ground state does not correlate with N_2 and H in their electronic ground states [128 to 132]. The amount of OH produced in the reaction NH_2+NO is independent of pressure ($0.5\leq p/mbar\leq 500$) and increases with increasing temperature [121]. At 1250 K a value of 0.25 up to 0.5 was estimated [91]. The formation of N_2O via the reaction pathway $NH_2+NO \rightarrow N_2O+H_2$, which is also exothermic and was observed at high temperatures [14], was suggested to be produced in secondary reactions [133].

The reaction NH_2+NO has also been studied theoretically [134 to 140]. The radical-radical reaction $NH_2(\tilde{X}\ ^2A)+NO(X\ ^2\Pi)$ proceeds without any activation energy via a loose transition state to form the complex H_2NNO as indicated in **Fig. 9**, p. 220. This primary adduct is too short-lived (4×10^{-12} s [136]) to be stabilized by collisions, since the time between the collisions is in the order of 10^{-10} s, even at a pressure of one atmosphere; thus, it is understandable that no pressure dependence was observed in the low-pressure experiments ($p\leq1$ bar). The negative temperature dependence results from the stronger temperature dependence of k_a compared to the temperature dependence of k_b. The main barrier for the reaction pathway to the products $H_2O^{\neq}+N_2$ is the isomerization from trans-hydroxydiazene to cis-hydroxydiazene [137]; see Fig. 9.

For the reaction

$$NH_2+NO_2 \rightarrow \text{products}$$

six independent, direct measurements for the rate constant at room temperature were published [28, 102, 105, 119, 123, 141, 142]. All determinations (done with different methods:

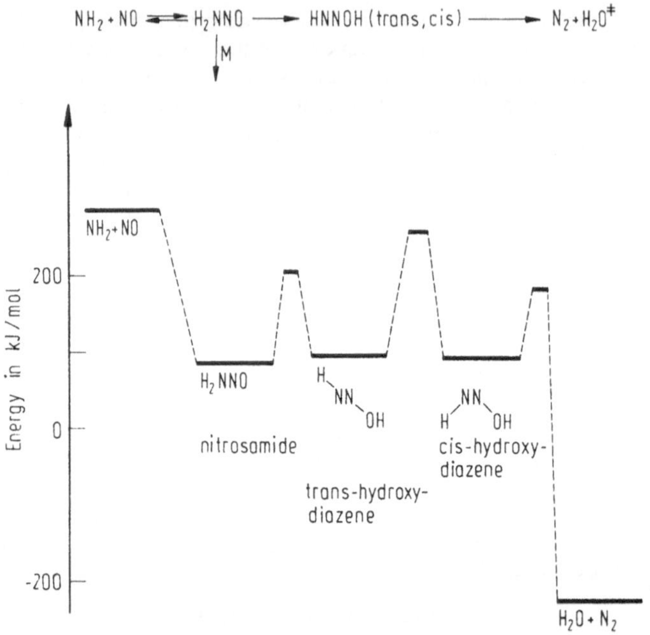

Fig. 9. Calculated reaction pathway for the reaction $NH_2(\tilde{X}) + NO(^2\Pi) \rightarrow N_2(X) + H_2O^{\ddagger}(\tilde{X})$.

flow reactor, flash photolysis, pulse radiolysis, IR MPD) are in good agreement; thus, an averaged value of $k(298\ K) = 1.2 \times 10^{13}\ cm^3 \cdot mol^{-1} \cdot s^{-1}$ can be given. The three groups, who have studied the temperature dependence [102, 105, 119, 142], report a negative temperature dependence. Thus, the rate as a function of temperature can be expressed by $k(T) = 1.2 \times 10^{13}\ (T/298\ K)^{-2}\ cm^3 \cdot mol^{-1} \cdot s^{-1}$. Also, a decrease of the reaction rate with vibrational bending mode excitation in NH_2 was observed [141]. The rate constant was found independent of pressure [142] consistent with the fact that at low pressures (around several mbar) and at one atmosphere [123] the same values for the rate constant were measured.

The products were followed in a flow reactor by mass-spectrometric detection [102]. The main reaction pathway was found to be $NH_2 + NO_2 \rightarrow N_2O + H_2O$; the pathway $N_2 + H_2O_2$ contributes less than 5%; also the formation of $HN + NHO_2$ was ruled out [59]. It can be assumed that the reaction mechanism is very similar to the one for $NH_2 + NO$. In the radical-radical reaction with $NH_2(\tilde{X}\ ^2A)$ a complex is formed [143, 144] which can redissociate or decompose to the products N_2O and water. The unstable molecule nitramide, NH_2NO_2, which is known to decompose into N_2O and H_2O, has been trapped in Ar and N_2 matrices at 12 K and detected by IR spectroscopy [145].

For the rate of the radical-radical reaction

$$NH_2 + OH \rightarrow \text{products}$$

a rate constant of $k = (1.1 \pm 0.4) \times 10^{13}\ cm^3 \cdot mol^{-1} \cdot s^{-1}$ was determined in a flow reactor at low pressures ($1.4 \leq p/mbar \leq 3.6$) following OH and NH_2 profiles with ESR and LMR detection devices. No O atoms were observed (ESR), which led to the conclusion that the reaction pathway $NH_2 + OH \rightarrow O + NH_3$ has a rate constant of $k < 6 \times 10^{10}\ cm^3 \cdot mol^{-1} \cdot s^{-1}$ in agreement with the rate constant determined from the reverse reaction $O + NH_3 \rightarrow OH + NH_2$ for this reaction pathway [9]. An estimate obtained by computer modeling of an NH_3-O_2 flash photolysis system [60] is in agreement with that value. A recent estimate [146], however, gives an upper limit which is lower by about a factor of two. The rate constant $k(T) = 2 \times 10^{10}\ T^{0.405} \exp(-250/T)\ cm^3 \cdot mol^{-1} \cdot s^{-1}$ obtained from the reverse reaction $O + NH_3$ is recommended in the temperature range 500 to 2500 K, assuming that no direct data are available [124].

The reaction

$$NH_2 + H_2O \rightleftharpoons NH_3 + OH$$

was measured directly in the forward and backward direction. The value from the NH_2 depletion (LIF, NH_2 detection) at 1000 K in an isothermal flow reactor is $k = 6 \times 10^9\ cm^3 \cdot mol^{-1} \cdot s^{-1}$ in good agreement with the value obtained from the reverse reaction at that temperature [26, 147], i.e., the rate constant confirms the thermodynamic data. The rate constant in the temperature range 300 to 1000 K can be described by the expression $k(T) = 5 \times 10^{12} \exp(-61.4\ kJ \cdot mol^{-1}/RT)\ cm^3 \cdot mol^{-1} \cdot s^{-1}$ [26]. The value of the activation energy is in good agreement with recent ab initio calculations [148].

For the radical-radical reaction

$$NH_2 + HO_2 \rightarrow \text{products}$$

four closely agreeing rate constants were published [31, 40, 60, 74, 75]. At room temperature a value of $k(300\ K) = 2 \times 10^{13}\ cm^3 \cdot mol^{-1} \cdot s^{-1}$ can be given. Flash photolysis experiments showed that the reaction proceeds to approximately equal extent in each of the two paths $NH_3 + O_2$ and $HNO + H_2O$ [1, 40]. Ab initio calculations for this reaction, however, predict that the reaction pathway $HNO + H_2O$ is of minor importance [149]; RRKM calculations indicate that $NH_2O + OH$ is the only important reaction channel [56].

The reaction of NH_2 with hydrogen peroxide via

$$\begin{aligned} NH_2 + H_2O_2 &\rightarrow NH_3 + HO_2 \\ &\rightarrow NH_2OH + OH \end{aligned}$$

was studied in an isothermal flow reactor ($1.65 \leq p/mbar \leq 5.61$) attached to an LMR spectrometer ($[NH_2]_0 \ll [H_2O_2]$). A rate constant of $k(298\ K) = (3.1 \pm 1) \times 10^{11}\ cm^3 \cdot mol^{-1} \cdot s^{-1}$ was determined. HO_2 was found to be the main reaction product, one order of magnitude larger than OH [9].

NH_2 Reactions in the N-H-C System

NH_2 Reactions with Saturated Hydrocarbons. NH_2 reacts with saturated hydrocarbons via H-atom abstraction

$$NH_2 + RH \rightarrow NH_3 + R$$

The reactions have been studied for C_nH_{n+2} with n in the range 1 to 5. The experiments were either performed in flash photolysis cells using laser resonance absorption to detect NH_2 [17 to 19] or in isothermal discharge flow reactors applying laser-induced fluorescence detection [150 to 153]. Moreover, high-temperature measurements were made behind incident shock waves [154]. The results are summarized in Table 16, p. 223, including the hydrogen abstraction reaction from *cyclo*-C_6H_{12}, a molecule which has only secondary H atoms and consists of a ring free of tension. The activation energy E_A in these reactions was found to be correlated with D(R-H), the dissociation energy of the R-H bond broken in the hydrocarbon. This is shown by an Evans-Polanyi plot [155] in **Fig. 10**, p. 222. Two other hydrocarbons with very weak CH bonds, CH_3CHO and $C_6H_5CH_3$, are included in Fig. 10 and described below. This plot can be used to estimate the activation energy of the reaction

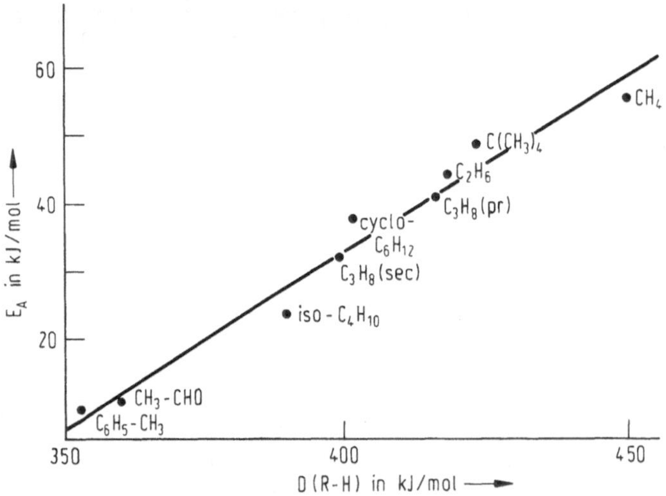

Fig. 10. Evans-Polanyi plot (activation energy E_A vs. R-H bond dissociation energy D(R-H)) for the hydrogen abstraction reactions $NH_2 + R\text{-}H \xrightarrow{k(T)} NH_3 + R$.

of NH_2 with other saturated hydrocarbons. Also bond-energy-bond-order estimations, as shown in [152], can be used to predict activation energies. It was suggested to estimate a rate constant composed of contributions of primary, secondary, and tertiary H atoms in the hydrocarbon [1]. The rate constants for abstracting a primary (k_p), secondary (k_s), and tertiary (k_t) H atom are related to each other by the ratio $k_p:k_s:k_t = 1:15.7:140$. By analyzing the reaction products (i.e. the amines resulting from the recombination $R + NH_2$) in the reaction NH_2 + isobutane it was shown that the tertiary hydrogen atom is significantly more reactive than the primary H atom [156]. The effect of this difference in the reactivity or activation energy on the rate constant was observed directly, when the rate constant k(T) was monitored over a large temperature range [150 to 153].

Among the saturated hydrocarbon reactions, the reaction $NH_2 + CH_4 \rightarrow NH_3 + CH_3$ is the most important one. Its activation energy was calculated from the reverse reaction [157]. With a semiempirical method proposed in [158], a value of $E_A = 41.8$ kJ/mol [19] was estimated. The direct measurement in an isothermal flow reactor yielded $E_A = 50$ kJ/mol (see Table 16). The hydrogen abstraction reaction $NH_2 + CH_4$ has been studied theoretically [162, 163]; a reaction barrier of 67 kJ/mol was calculated. Tunneling of H atoms through the reaction barrier can significantly contribute only at very low temperatures. The barrier heights were also calculated for other hydrocarbons [164]. A comparison with experimentally determined activation energies is not very satisfactory. All hydrocarbons exhibit pronounced curvature in the Arrhenius plots (ln k vs. 1/T). Thus, it is understandable why the activation energies and the preexponential factors measured at lower temperatures (see Table 16) are smaller than those obtained at higher temperatures.

The kinetics of hydrogen abstraction by NH_2 from saturated hydrocarbons can be compared to the kinetic behavior of other radicals such as OH and CH_3, giving $k(OH) \gg k(NH_2) > k(CH_3)$ [151, 157, 165]. The reactivity of H atoms lies between that of NH_2 and CH_3 [166, 167]. The activation energies increase in the order $E_A(OH) < E_A(NH_2) < E_A(H) < E_A(CH_3)$. The selectivity between different CH bonds expectedly changes in the same order as the activation energies.

Table 16
Summary of the Measured Rate Constants for NH_2 with Alkanes (estimated values [159] are not included in the table).

reactant	temperature range in K	rate constant in $cm^3 \cdot mol^{-1} \cdot s^{-1}$ activation energy in kJ/mol	Ref.
CH_4	1730 to 1950	$k=(4\pm2)\times10^{11}$	[154]
	743 to 1023	$k(T)=5.8\times10^{12}\exp(-55.16/RT)$	[150]
	740 to 2000	$k(T)=8.46\times10^{7}\,T^{1.54}\exp(-50/RT)$ (recommended)	[150]
C_2H_6	300 to 500	$k(T)=3.7\times10^{11}\exp[-(30\pm1)/RT]$	[17 to 19]
	598 to 973	$k(T)=9.7\times10^{12}\exp[-(44.3\pm3)/RT]$	[152]
C_3H_8	300 to 520	$k(T)=5.4\times10^{11}\exp[-(25.7\pm1)/RT]$	[17 to 19]
	550 to 1073	$k(T)=8.5\times10^{12}\exp[-(39.3\pm4)/RT]$	[152]
n-C_4H_{10}	300 to 520	$k(T)=7.1\times10^{11}\exp[-(25.5\pm1)/RT]$	[17 to 19]
iso-C_4H_{10}	300 to 520	$k(T)=7.1\times10^{11}\exp[-(25.5\pm1)/RT]$	[17 to 19] [156, 161]
	470 to 973	$k(T)=4.9\times10^{11}\exp[-(22\pm2)/RT]$ $+\ 8.8\times10^{12}\exp[-(43.5\pm3)/RT]$	[150]
	295 to 543	$k(T)=1.9\times10^{11}\exp(-19.8/RT)$	[32]
n-C_5H_{12}	254 to 354	$k(T)=3.7\times10^{10}\exp(-10.2/RT)$	[153]
$C(CH_3)_4$	673 to 1003	$k(T)=2.1\times10^{13}\exp[-(48.6\pm4)/RT]$	[151]
c-C_6H_{12}	544 to 973	$k(T)=2.7\times10^{13}\exp[-(37.5\pm5)/RT]$	[150]

$NH_2+C_nH_{2n+1}$ (alkyl radicals). The reaction of NH_2 with alkyl radicals, C_nH_{2n+1}, was studied for $2\leq n\leq4$ in ammonia flash photolysis systems. The alkyl radicals were produced via the $H+C_nH_{2n}$ association reaction [17, 18, 168] or via radiolysis of the saturated hydrocarbon. The room-temperature rate constants (in $10^{13}\ cm^3 \cdot mol^{-1} \cdot s^{-1}$), $k(NH_2+C_2H_5)=2.5\pm0.5$, $k(NH_2+\textit{iso}\text{-}C_3H_7)=2.0\pm0.4$, and $k(NH_2+\textit{t}\text{-}C_4H_9)=2.5\pm0.5$, were obtained at low pressure (4 mbar), probably already sufficiently high to be in the high-pressure regime. This high-pressure limit assumption is supported by the experimental finding that each alkyl radical leads to the same rate constant independent of the number of degrees of freedom in the adduct formed.

NH_2 Reactions with Unsaturated Hydrocarbons. The reactivity of NH_2 radicals towards all olefines is very similar, i.e., the activation energies, in contrast to those of saturated hydrocarbons, are not very sensitive to molecular parameters. In addition to the simplest olefine C_2H_4, which was extensively studied experimentally [18, 102, 161, 169 to 173] and theoretically [174 to 176], several other unsaturated hydrocarbons, C_3H_6 [18, 102, 103, 161, 169], C_3H_4 [102, 103], C_4H_8 [161], C_4H_6 [61, 177], and C_5H_{10} [1], have been investigated. The activation energies are all around 17 kJ/mol, indicating that a similar addition reaction occurs for all cases. A slight dependence of the addition rate constant on the alkyl substitution of the double bond suggests a very low electrophilicity of NH_2 in these reactions. A more pronounced change in the rate constant or the activation energy can be expected, if, for example, Cl or F atoms are substituting the H atoms next to the double bond. Cumulated double bonds, as in the case of allene, CH_2=C=CH_2, do not significantly enhance the reactivity. The room-temperature rate constant was found to be $k(298\ K) \leq 5 \times 10^8\ cm^3 \cdot mol^{-1} \cdot s^{-1}$ compared to that of propylene of $k(298\ K) = 2.2 \times 10^8\ cm^3 \cdot mol^{-1} \cdot s^{-1}$ [161]. Two conjugated double bonds as in 1,3-butadiene, however, lower the activation energy, $E_A(1,3\text{-}C_4H_6) =$ 9.5 kJ/mol [61] compared to E_A(1-butene) = 17.1 kJ/mol [161]. The activation barrier, obtained recently from ab initio SCF CI calculations for the NH_2-C_2H_4 approach, is with 4.7 kJ/mol [176] much lower than that obtained in earlier SCF CI calculations [175], but also lower than expected from the experimentally determined activation energy ($E_A(C_2H_4) = 16.5$ kJ/mol) [18, 161]. The preexponential factor for the various olefines listed above was found to be in the range (1 to 6) $\times 10^{11}\ cm^3 \cdot mol^{-1} \cdot s^{-1}$. These reaction rates are similar to the reaction rates of CH_3 radicals [178] and about one order of magnitude lower than those of OH radicals [179, 180]. The OH radicals react with an activation energy $E_A = 0$, and the CH_3 radicals with an activation energy in the range $E_A \cong 30$ to 35 kJ/mol [178, 179]. The rate constant was found to be independent of pressure at low temperatures ($T \leq 350$ K) above 10 mbar and at higher temperatures above 50 mbar, indicating that the high-pressure limits were obtained under these conditions.

The reaction mechanism of NH_2 + olefines was studied by photolyzing NH_3-C_2H_4 mixtures and analyzing the products [173]. The first step is $NH_2 + C_2H_4 \rightarrow NH_2\text{-}C_2H_4$ [173]. In the reaction $NH_2 + C_2H_4$ the hydrogen abstraction pathway is endothermic; even if the hydrogen abstraction is exothermic as in the case of butene, the addition to the double bond is still an important reaction pathway. This statement is based on the NH_3 depletion quantum yield determined in the NH_2-butene system [1]. The abstraction reaction contributes about 50%. It can be concluded that the contribution of hydrogen abstraction becomes more important, if H atoms with sufficiently low bond energies (see Fig. 10, p. 222) are available. The ratio of the contributions of the abstraction and double bond addition pathways can be estimated taking into account the number of H atoms available for abstraction and their activation energies which can be obtained from an Evans-Polanyi plot (Fig. 10). For OH reactions with olefines, addition as well as abstraction reaction pathways have also been observed [181, 182].

The rate constant $k(T) = 1.1 \times 10^{11} \exp[-(9.4 \pm 2)\ kJ \cdot mol^{-1}/RT]\ cm^3 \cdot mol^{-1} \cdot s^{-1}$ for the aromatic hydrocarbon $C_6H_5CH_3$ was measured in an isothermal flow reactor in the temperature range 297 to 542 K and interpreted to be due to hydrogen abstraction, since the activation energy fits an Evans-Polanyi plot for saturated hydrocarbons [151]. The effect of the aromatic ring on the reactivity is mainly due to the reduction of the CH bond energy in the methyl group.

The reactivity of a triple bond, studied experimentally on acetylene [102, 103, 170, 172, 177], seems to be less than that of a double bond, such as in e.g. C_2H_4. In the temperature range 340 to 510 K an activation energy of $E_A = 23.1$ kJ/mol was determined, which is significantly higher than the activation energies obtained for olefines. The preexponential factor, $4.9 \times 10^{11}\ cm^3 \cdot mol^{-1} \cdot s^{-1}$, was found to be independent of pressure in the range 40 to 133 mbar, indicating that the high-pressure limit was at least nearly reached in these experiments [177].

NH_2 Reactions in the N-H-C-O System

Several elementary reactions were investigated in the system containing C and O atoms. The reaction NH_2 + CO is slow with an upper limit for the room-temperature rate constant being $k(298\ K) < 10^9\ cm^3 \cdot mol^{-1} \cdot s^{-1}$ [32] or $k(298\ K) = 6 \times 10^6\ cm^3 \cdot mol^{-1} \cdot s^{-1}$ [183]. The reaction of NH_2 with acetaldehyde, which has a rate constant of $k(T) = 2.1 \times 10^{11}$

$\exp[-(10.4\pm2)\ kJ\cdot mol^{-1}/RT]\ cm^3\cdot mol^{-1}\cdot s^{-1}$ in the temperature range 297 to 543 K, proceeds probably via the H atom abstraction reaction $NH_2+CH_3CHO\rightarrow NH_3+CH_3CO$ [151]. The activation energy correlates with the CH bond energy as shown in the Evans-Polanyi plot (Fig. 10). For the reaction of NH_2 with formamide, NH_2CHO, a rate constant at only one temperature is known, $k(573\ K)=8.4\times10^9\ cm^3\cdot mol^{-1}\cdot s^{-1}$ [39].

An upper limit for the rate constant of the reaction $NH_2+HNCO\rightarrow NH_3+NCO$, which is of importance for the NH_2 kinetics in flames, has been determined by following the NH_2 concentration profiles in shock-heated HNCO-Ar mixtures in the temperature range 2340 to 2680 K, $k\leq5\times10^{11}\ cm^3\cdot mol^{-1}\cdot s^{-1}$ [184].

The reaction of NH_2 with CH_3OH is much faster than expected for a hydrogen abstraction reaction. A room-temperature value of $k(293\ K)=9.0\times10^8\ cm^3\cdot mol^{-1}\cdot s^{-1}$ was measured [185] which is more than two orders of magnitude higher than the k value of the reaction with e.g. C_2H_6. Probably a different mechanism is valid for methanol, and thus the rate constant can not be compared to hydrogen abstraction rate constants.

NH_2 Reactions in Systems Involving Other Elements

The reaction $NH_2+F\rightarrow$ products has to be considered, if hydrogen abstraction of ammonia ($F+NH_3$) is used as a means to generate NH_2 (see p. 161). It is a fast reaction [186, 187] with a room-temperature rate constant of $k=2.3\times10^{13}\ cm^3\cdot mol^{-1}\cdot s^{-1}$ [187]. The products are exclusively NH and HF [187]. This opens up the possibility to calibrate NH and NH_2 relative to each other by measuring the NH_2 depletion and the NH formation. With the IR-chemiluminescence technique HF was observed to be vibrationally excited ($v\leq4$) with an inverted energy distribution [188]; this energy distribution indicates that the reaction proceeds via direct abstraction on the triplet surface.

The hydrogen abstraction from PH_3 and SiH_4 was also investigated experimentally. For NH_2+PH_3 a rate constant of $k(T)=9.2\times10^{11}\exp(-7.7\ kJ\cdot mol^{-1}/RT)\ cm^3\cdot mol^{-1}\cdot s^{-1}$ was determined in the temperature range 218 to 456 K [189]. The hydrogen abstraction from SiH_4 via $NH_2+SiH_4\rightarrow NH_3+SiH_3$ is with $k(300\ K)=(9\pm6)\times10^{10}\ cm^3\cdot mol^{-1}\cdot s^{-1}$ [190] much faster than the NH_2 reaction with the isovalent CH_4.

The reaction of NH_2 with sulfur dioxide can be described as a combination process, $NH_2+SO_2+M\rightarrow NH_2SO_2+M$. The high-pressure value of the rate constant, $k_\infty(T)=9.0\times10^{10}\ (T/289\ K)^{-1.3}\ cm^3\cdot mol^{-1}\cdot s^{-1}$, was determined in the temperature range 298 to 363 K, and the low-pressure values for NH_3 and N_2 as collision partners are $k_0(NH_3)=(7.6\pm1.5)\times10^{16}$ and $k_0(N_2)=(1.1\pm0.2)\times10^{16}\ cm^6\cdot mol^{-2}\cdot s^{-1}$, respectively [191, 192].

The reaction of NH_2 with NOCl via $NH_2+NOCl\rightarrow NO+NH_2Cl$ was found to be slow with $k(298\ K)\leq1.5\times10^{11}\ cm^3\cdot mol^{-1}\cdot s^{-1}$ [160].

References:

[1] Lesclaux, R. (Rev. Chem. Intermed. **5** [1984] 347/92).
[2] McDonald, C. C.; Gunning, H. E. (J. Chem. Phys. **23** [1955] 532/41).
[3] Anderson, W. H.; Zwolinski, B. J.; Parin, R. B. (Ind. Eng. Chem. **51** [1959] 527/30).
[4] Khe, P. V.; Soulignac, J.; Lesclaux, R. (J. Phys. Chem. **81** [1977] 210/4).
[5] Gordon, S.; Mulac, W.; Nangia, P. (J. Phys. Chem. **75** [1971] 2087/93).
[6] Hanes, M. H. (Diss. Indian Univ. 1962).
[7] Hanes, M. H.; Bair, E. J. (J. Chem. Phys. **38** [1963] 672/6).
[8] Hanes, M. H.; Bair, E. J. (PB-155623 [1961] 3pp.; C.A. **58** [1963] 9443).

[9] Dransfeld, P. (Dipl.-Arbeit Göttingen 1980).
[10] Roose, T. R.; Hanson, R. K.; Kruger, C. H. (Shock Tubes Waves Proc. 12th Int. Symp., Jerusalem 1979 [1980], pp. 476/85; C.A. **93** [1980] No. 102053).
[11] Yumara, M.; Asaba, T.; Matsumoto, Y.; Matsui, H. (Int. J. Chem. Kinet. **12** [1980] 439/50).
[12] Davidson, D. F.; Kohse-Höinghaus, K.; Chang, A. Y.; Hanson, R. K. (Int. J. Chem. Kinet. **22** [1990] 513/35).
[13] Dove, J. E.; Nip, W. S. (Can. J. Chem. **57** [1979] 689/701).
[14] Roose, T. R.; Hanson, R. K.; Kruger, C. H. (Symp. Int. Combust. Proc. **18** [1981] 853/62).
[15] Yumura, M.; Asaba, T. (Symp. Int. Combust. Proc. **18** [1981] 863/72; 14th Int. Symp. Shock Tubes Waves, Sydney 1983, p. 678).
[16] Nakatsuji, H. T.; Koga, T.; Kondo, K.; Yonezawa, T. (J. Am. Chem. Soc. **100** [1978] 1029/36).
[17] Lesclaux, R.; Demissy, M. (J. Photochem. **9** [1978] 110/2).
[18] Lesclaux, R.; Khe, P. V. (NBS Spec. Publ. [U.S.] No. 526 [1978] 331/3).
[19] Demissy, M.; Lesclaux, R. (J. Am. Chem. Soc. **102** [1980] 2897/902).
[20] Cardy, H.; Liotard, D.; Dargelos, A.; Poquet, E. (Nouv. J. Chim. **4** [1980] 751/6).

[21] Gordon, M. S.; Gano, D. R.; Boatz, J. A. (J. Am. Chem. Soc. **105** [1983] 5771/5).
[22] Leroy, G.; Sana, M.; Tinant, A. (Can. J. Chem. **63** [1985] 1447/56).
[23] Sutherland, J. W.; Michael, J. V. (J. Chem. Phys. **88** [1988] 830/4).
[24] Holzrichter, K. (Diss. Göttingen [1980]).
[25] Hack, W.; Rouveirolles, P.; Wagner, H. Gg. (J. Phys. Chem. **90** [1986] 2505/11).
[26] Bätz, P.; Ehbrecht, J.; Hack, W.; Rouveirolles, P.; Wagner, H. Gg. (Symp. Int. Combust. Proc. **22** [1988] 1107/15).
[27] Walch, S. P. (J. Chem. Phys. **95** [1991] 4277/83).
[28] Whyte, A. R.; Phillips, L. F. (Chem. Phys. Lett. **102** [1983] 451/4).
[29] Whyte, A. R.; Phillips, L. F. (J. Phys. Chem. **88** [1984] 5670/3).
[30] Dransfeld, P.; Wagner, H. Gg. (Z. Phys. Chem. [Munich] **153** [1987] 89/97).

[31] Pagsberg, P. B.; Eriksen, J.; Christensen, H. C. (J. Phys. Chem. **83** [1979] 582/90).
[32] Dransfeld, P.; Hack, W.; Kurzke, H.; Wagner, H. Gg. (Symp. Int. Combust. Proc. **20** [1984] 655/63).
[33] Bates, J. R.; Taylor, H. S. (J. Am. Chem. Soc. **49** [1927] 2438/56).
[34] Gedye, G. R.; Rideal, E. K. (J. Chem. Soc. **135** [1932] 1160/9).
[35] McDonald, C. C.; Kahn, A.; Gunning, H. E. (J. Chem. Phys. **22** [1954] 908/16).
[36] Diesen, R. W. (J. Chem. Phys. **39** [1963] 2121/8).
[37] Salzman, J. D.; Bair, E. J. (J. Chem. Phys. **41** [1964] 3654/5).
[38] Boyd, A. W.; Willis, C.; Miller, O. A. (Can. J. Chem. **49** [1971] 2283/9).
[39] Back, R. A.; Yokota, T. (Int. J. Chem. Kinet. **5** [1973] 1039/46).
[40] Lozovskii, V. A.; Nadtochenko, V. A.; Sarkisov, O. M.; Cheskis, S. G. (Kinet. Katal. **20** [1979] 1118/23; Kinet. Catal. [Engl. Transl.] **20** [1979] 918/22).

[41] Mulenko, S. A.; Smirnov, V. N. (Kratk. Soobshch. Fiz. **1980** No. 12, pp. 24/31).
[42] Duncanson, J. A.; Geridu, L. M.; Guillory, W. A. (Int. Conf. Photochem. **11** 1983).
[43] Gehring, M.; Hoyermann, K.; Schacke, H.; Wolfrum, J. (Symp. Int. Combust. Proc. **14** [1973] 99/105).
[44] Patrick, P.; Golden, D. M. (J. Phys. Chem. **88** [1984] 491/5).
[45] Zetzsch, C.; Stuhl, F. (Ber. Bunsen-Ges. Phys. Chem. **85** [1981] 564/8).
[46] Hack, W.; Rathmann, K. (Z. Phys. Chem. [Munich] **176** [1992] 151/60).
[47] Genich, A. P.; Zhirnov, A. A.; Manelis, G. B. (Kinet. Katal. **16** [1975] 841/5; Kinet. Catal. [Engl. Transl.] **16** [1975] 729/32).

[48] Dean, A. M.; Chou, M. S.; Stern, D. (Int. J. Chem. Kinet. **16** [1984] 633/53).
[49] Gehring, M.; Hoyermann, K.; Wagner, H. Gg.; Wolfrum, J. (Ber. Bunsen-Ges. Phys. Chem. **75** [1971] 1287/94).
[50] Albers, E. A; Hoyermann, K; Wagner, H. Gg.; Wolfrum, J. (Symp. Int. Combust. Proc. **12** [1969] 313/21).

[51] Lesclaux, R.; Demissy, M. (Nouv. J. Chim. **1** [1977] 443/4).
[52] Walther, C. D. (Diss. Göttingen 1981).
[53] Temps, F. (Ber. Max-Planck-Inst. Strömungsforsch. **1983** No. 4, 140 pp.).
[54] Patel-Misra, D.; Dagdigian, P. J. (Chem. Phys. Lett. **185** [1991] 387/92).
[55] Patel-Misra, D.; Saunder, D. G.; Dagdigian, P. J. (J. Chem. Phys. **95** [1991] 955/62).
[56] Bozzelli, J. W.; Dean, A. M. (J. Phys. Chem. **93** [1989] 1058/65).
[57] Melius, C. F.; Binkley, J. S. (ACS Symp. Ser. **249** [1984] 103/15).
[58] Jayanty, R. K. M.; Simonaitis, R.; Heicklen, J. (J. Phys. Chem. **80** [1976] 433/7).
[59] Jayanty, R. K. M.; Simonaitis, R.; Heicklen, J. (NASA-CR 148598 [1976] 21 pp.; C.A. **86** [1977] No. 83055).
[60] Cheskis, S. G.; Sarkisov, O. M. (Chem. Phys. Lett. **62** [1979] 72/6).

[61] Hack, W.; Schröter, M. R.; Wagner, H. Gg. (Ber. Bunsen-Ges. Phys. Chem. **86** [1982] 326/30).
[62] Lozovsky, V. A.; Ioffe, M. A.; Sarkisov, O. M. (Chem. Phys. Lett. **110** [1984] 651/4).
[63] Lozovsky, V. A.; Ioffe, M. A.; Sarkisov, O. M. (Khim. Fiz. **4** [1985] 931/5).
[64] Michael, J. V.; Sutherland, J. W.; Klemm, R. B. (Int. J. Chem. Kinet. **17** [1985] 315/26).
[65] Tyndall, G. S.; Orlando, J. J.; Nickerson, K. E.; Cantrell, C. A.; Calvert, J. G. (J. Geophys. Res. **96** [1991] 20761/8).
[66] Dean, A. M.; Hardy, J. E.; Lyon, R. K. (Symp. Int. Combust. Proc. **19** [1983] 97/105).
[67] Takeyama, T.; Miyama, H. (Symp. Int. Combust. Proc. **11** [1967] 845/52).
[68] Fujii, N.; Miyama, H.; Koshi, M.; Asaba, T. (Symp. Int. Combust. Proc. **18** [1981] 873/83).
[69] Giguere, P. A.; Herman, K. (Chem. Phys. Lett. **44** [1976] 273/6).
[70] Crowley, J. N.; Sodeau, J. R. (J. Phys. Chem. **93** [1989] 4785/90).

[71] Pouchan, C.; Chaillet, M. (Chem. Phys. Lett. **90** [1982] 310/6).
[72] Baulch, D. L.; Cobos, C. J.; Cox, R. A.; Esser, C.; Frank, P.; Just, Th.; Kerr, J. A.; Pilling, H. J.; Troe, J.; Walker R. W.; Warnatz, J. (J. Phys. Chem. Ref. Data **21** [1992] 411/734, 621).
[73] Hack, W.; Kurzke, H. (Ber. Bunsen-Ges. Phys. Chem. **89** [1985] 86/93).
[74] Kurasawa, H.; Lesclaux, R. (14th Informal Conf. Photochem., Newport Beach, Calif., 1980).
[75] Kurasawa, H.; Lesclaux, R. (Chem. Phys. Lett. **72** [1980] 437/42).
[76] Bulatov, V. P.; Buloyan, A. A.; Iogansen, A. A.; Sarkisov, O. M.; Cheskis, S. (Khim. Fiz. **1982** No. 4, pp. 513/5).
[77] Hack, W.; Horie, O.; Wagner, H. Gg. (Ber. Bunsen-Ges. Phys. Chem. **85** [1981] 72/8).
[78] Cheskis, S. G.; Iogansen, A. A.; Sarkisov, O. M.; Titov, A. A. (Chem. Phys. Lett. **120** [1985] 45/9).
[79] Bulatov, V. P.; Buloyan, A. A.; Cheskis, S. G.; Kozliner, M. Z.; Sarkisov, O. M.; Trostin, A. I. (Chem. Phys. Lett. **74** [1980] 288/92).
[80] Davies, P. B.; Dransfeld, P.; Temps, F.; Wagner, H. Gg. (Ber. Max-Planck-Inst. Strömungsforsch. **1984** No. 6, pp. 1/25; C.A. **101** [1984] No. 182810).

[81] Davies, P. B.; Dransfeld, P.; Temps, F.; Wagner, H. Gg. (J. Chem. Phys. **81** [1984] 3763/5).
[82] Bamford, C. H. (Trans. Faraday Soc. **35** [1939] 568/76).
[83] Fenimore, C. P.; Jones, G. W. (J. Phys. Chem. **65** [1961] 298/303).

[84] Kaskan, W. E.; Hugues, D. E. (Combust. Flame **20** [1973] 381/8).
[85] Haynes, B. S. (Combust. Flame **28** [1977] 81/91).
[86] Haynes, B. S. (Combust. Flame **28** [1977] 113/21).
[87] Stuhl, F. (J. Chem. Phys. **59** [1973] 635/7).
[88] McConnell, J. C. (J. Geophys. Res. **78** [1973] 7812/22).
[89] McConnell, J. C.; McElroy, M. B. (J. Atmos. Sci. **30** [1973] 1465/80).
[90] Lyon, R. K. (U.S. 3900 [1975] 544).

[91] Lyon, R. K. (Int. J. Chem. Kinet. **8** [1976] 315/8).
[92] Lyon, R. K.; Benn, D. (Symp. Int. Combust. Proc. **17** [1979] 601/10).
[93] Wise, H.; Frech, M. F. (J. Chem. Phys. **22** [1954] 1463/4).
[94] Serewicz, A.; Noyes, W. A., Jr. (J. Phys. Chem. **63** [1959] 843/5).
[95] Srinivasan, R. (J. Phys. Chem. **64** [1960] 679/80).
[96] Gordon, S.; Mulac, W. A. (Int. J. Chem. Kinet. Symp. No. 1 [1975] 289/99).
[97] Duxbury, J.; Pratt, N. H. (Symp. Int. Combust. Proc. **15** [1974] 843/55).
[98] Hancock, G.; Lange, W.; Lenzi, M.; Welge, K. H. (Chem. Phys. Lett. **33** [1975] 168/72).
[99] Lesclaux, R.; Khe, P. V.; Soulignac, J. C.; Joussot-Dubien, J. (Int. Conf. Photochem., Edmonton, Can., 1975, Vol. 7).
[100] Lesclaux, R.; Khe, P. V.; Dezauzier, P.; Soulignac, J. C. (Chem. Phys. Lett. **35** [1975] 493/7).

[101] Welge, K. H. (Int. J. Chem. Kinet. Symp. No. 1 [1975] 508).
[102] Hack, W.; Schacke, H.; Schröter, M.; Wagner, H. Gg. (Symp. Int. Combust. Proc. **17** [1978] 505/13).
[103] Hack, W.; Schröter, M. R.; Wagner, H. Gg. (Ber. Max-Planck-Inst. Strömungsforsch. **1979** No. 13).
[104] Sarkisov, O. M.; Cheskis, S. G.; Sviridenkov, E. A. (Izv. Akad. Nauk SSSR Ser. Khim. **1978** 2612; Bull. Acad. Sci. USSR Div. Chem. Sci. [Engl. Transl.] **27** [1978] 2336/8).
[105] Kurasawa, H.; Lesclaux, R. (Chem. Phys. Lett. **66** [1979] 602/7).
[106] Jacobs, A. (Ber. Max-Planck-Inst. Strömungsforsch. **1981** No. 12).
[107] Silver, J. A.; Kolb, C. E. (J. Phys. Chem. **86** [1982] 3240/6).
[108] Silver, J. A.; Kolb, C. E. (J. Phys. Chem. **91** [1987] 3713/4).
[109] Stief, L. J.; Brobst, W. D.; Nava, D. F.; Borkowski, R. P.; Michael, J. V. (J. Chem. Soc. Faraday Trans. II **78** [1982] 1391/401).
[110] Stief, L. J.; Brobst, W. D.; Nava, D. F.; Borkowski, R. P.; Michael, J. V. (NASA-TM-83928 [1982] 31pp.; C.A. **97** [1982] No. 169820).

[111] Andresen, P.; Jacobs, A.; Kleinermanns, C.; Wolfrum, J. (Symp. Int. Combust. Proc. **19** [1982] 11/22).
[112] Gericke, K. H.; Torres, L. M.; Guillory, W. A. (J. Chem. Phys. **80** [1984] 6134/40).
[113] Jeffries, J. B.; McCaulley, J. A.; Kaufman, F. (Chem. Phys. Lett. **106** [1984] 111/6).
[114] Dreier, Th.; Wolfrum, J. (Symp. Int. Combust. Proc. **20** [1984] 695/702).
[115] Hall, J. L.; Zeitz, D.; Stephens, J. W.; Curl, R. F.; Kasper, J. V. V.; Tittel, F. K. (Springer Ser. Opt. Sci. **49** [1985] 377/8).
[116] Hall, J. L.; Zeitz, D.; Stephens, J. W.; Kasper, J. V. V.; Glass, G. P.; Curl, R. F.; Tittel, F. K. (J. Phys. Chem. **90** [1986] 2501/5).
[117] Dolson, D. A. (J. Phys. Chem. **90** [1986] 6714/8).
[118] Crowley, J. N.; Sodeau, J. R. (J. Phys. Chem. **91** [1987] 2024/6).
[119] Bulatov, V. P.; Ioffe, A. A.; Lozovsky, V. A.; Sarkisov, O. M. (Chem. Phys. Lett. **161** [1989] 141/6).
[120] Bulatov, V. P.; Ioffe, A. A.; Lozovsky, V. A.; Sarkisov, O. M. (Khim. Fiz. **9** [1990] 1368/74).

[121] Atakan, B.; Jacobs, A.; Wahl, M.; Weller, R.; Wolfrum, J. (Chem. Phys. Lett. **155** [1989] 609/13).
[122] Curl, R. F.; Glass, G. P. (DOE-ER-12439 [1990] 6).
[123] Pagsberg, P. B.; Sztuba, B.; Ratajczak, E.; Sillesen, A. (Acta Chem. Scand. **45** [1991] 329/34).
[124] Baulch, D. L.; Cobos, C. J.; Cox, R. A.; Esser, C.; Frank, P.; Just, Th.; Kerr, J. A.; Pilling, H. J.; Troe, J.; Walker, R. W.; Warnatz, J. (J. Phys. Chem. Ref. Data **21** [1992] 411/734, 622).
[125] Kagiya, T.; Ogita, T.; Hatta, H. (Nippon Kagaku Kaishi **1982** 520/7).
[126] Adams, G. K.; Parker, W. G.; Wolfhard, H. G. (Discuss. Faraday Soc. No. 14 [1953] 97/103).
[127] Poole, D. R.; Graven, W. M. (J. Am. Chem. Soc. **83** [1961] 283/6).
[128] Baird, N. C. (J. Chem. Phys. **62** [1975] 300/1).
[129] Casewit, C. J.; Goddard, W. A., III (J. Am. Chem. Soc. **104** [1982] 3280/7).
[130] Willis, C.; Back, R. A.; Parsons, J. M. (J. Photochem. **6** [1976/77] 253/64).

[131] Vasudevan, K.; Peyerimhoff, S. D.; Buenker, R. J. (J. Mol. Struct. **29** [1975] 285/97).
[132] Curtiss, L. A.; Drapcho, D. L.; Pople, J. A. (Chem. Phys. Lett. **103** [1984] 437/42).
[133] Miller, J. A.; Mitchell, R. E. (Symp. Int. Combust. Proc. **18** [1981] 860/1).
[134] Miller, J. A.; Branch, M. C.; Kee, R. J. (SAND-80-8635 [1980]).
[135] Boettner, J. C.; James, H. (6th World Congr. Air Quality 1983).
[136] Abou-Rachid, H.; Pouchan, C.; Chaillet, M. (Chem. Phys. **90** [1984] 243/55).
[137] Melius, C. F.; Binkley, J. S. (Symp. Int. Combust. Proc. **20** [1985] 575/83).
[138] Gilbert, R.; Whyte, A. R.; Phillips, L. F. (Int. J. Chem. Kinet. **18** [1986] 721/37).
[139] Harrison, J. A.; Maclagan, R. G. A. R.; Whyte, A. R. (J. Phys. Chem. **91** [1987] 6683/6).
[140] Phillips, L. F. (Chem. Phys. Lett. **135** [1987] 269/74).

[141] Xiang, T. X.; Torres, L. M.; Guillory, W. A. (J. Chem. Phys. **83** [1985] 1623/9).
[142] Bulatov, V. P.; Ioffe, A. A.; Lozovsky, V. A.; Sarkisov, O. M.; Buloyan, A. A. (Khim. Fiz. **9** [1990] 370/4).
[143] Saxon, R. P.; Joshimine, M. (J. Phys. Chem. **93** [1989] 3130/5).
[144] Roszak, S.; Kaufman, J. J. (Chem. Phys. **160** [1992] 1/9).
[145] Nonella, M.; Müller, R. P.; Huber, J. R. (J. Mol. Spectrosc. **112** [1985] 142/52).
[146] Diau, E. W. G.; Tso, T. L.; Lee, Y. P. (J. Phys. Chem. **94** [1990] 5261/5).
[147] Dransfeld, P.; Hack, W.; Jost, W.; Rouveirolles, P.; Wagner, H. Gg. (Khim. Fiz. **6** [1987] 1668/76).
[148] Giménez, X.; Moreno, M.; Lluch, J. M. (Chem. Phys. **165** [1992] 41/6).
[149] Pouchan, C.; Lam, B.; Bishop, D. M. (J. Phys. Chem. **91** [1987] 4809/13).
[150] Hack, W.; Kurzke, H.; Rouveirolles, P.; Wagner, H. Gg. (Symp. Int. Combust. Proc. **21** [1986] 905/11).

[151] Hack, W.; Kurzke, H.; Rouveirolles, P.; Wagner, H. Gg. (Ber. Bunsen-Ges. Phys. Chem. **90** [1986] 1210/9).
[152] Ehbrecht, J.; Hack, W.; Rouveirolles, P.; Wagner, H. Gg. (Ber. Bunsen-Ges. Phys. Chem. **91** [1987] 700/8).
[153] Hack, W.; Horie, O.; Wagner, H. Gg. (Ber. Max-Planck-Inst. Strömungsforsch. **1980** No. 8).
[154] Möller, W.; Wagner, H. Gg. (Z. Naturforsch. **39a** [1984] 846/52).
[155] Evans, M. G.; Polanyi, M. (Trans. Faraday Soc. **34** [1938] 11/29).
[156] Lesclaux, R.; Khe, P. V. (J. Chim. Phys. **70** [1973] 119/25).
[157] Gray, P.; Herod, A. A.; Jones, A. (Chem. Rev. **71** [1971] 247/94).
[158] Zavitsas, A. A.; Melikian, A. A. (J. Am. Chem. Soc. **97** [1975] 2757/63).

[159] Rozenberg, A. S.; Voronkov, V. G. (Zh. Fiz. Khim. **46** [1972] 744/6; Russ. J. Phys. Chem. [Engl. Transl.] **46** [1972] 425/6).
[160] Schröter, M. (Ber. Max-Planck-Inst. Strömungsforsch. **1981** No. 2, p. 106).

[161] Khe, P. V.; Lesclaux, R. (J. Phys. Chem. **83** [1979] 1119/22).
[162] Sana, M.; Leroy, G.; Killaveces, J. G. (Theor. Chim. Acta **65** [1984] 109/25).
[163] Lee, I.; Song, C. H.; Park, B. S. (Tachan Hwahakhoe Chi. **30** [1986] 166/71).
[164] Leroy, G.; Sana, M. (J. Mol. Struct. **136** [1986] 283/301 [THEOCHEM **29**]).
[165] Hampson, R. F. (FAA/E-80-17 [1980] 488 pp.; C.A. **94** [1981] No. 212100).
[166] Jones, W. E.; Hacknight, S. D.; Teng, L. (Chem. Rev. **73** [1973] 407/40).
[167] Kerr, J. A. (Compr. Chem. Kinet. **18** [1976] 39/109).
[168] Demissy, M.; Lesclaux, R. (Int. J. Chem. Kinet. **14** [1982] 1/12).
[169] Lesclaux, R.; Soulignac, J. C.; Khe, P. V. (Chem. Phys. Lett. **43** [1976] 520/3).
[170] Lesclaux, R.; Khe, P. V. (Informal Conf. Photochem. **12** [1976] P 3, 1/3).

[171] Bosco, S. R. (NASA Tech. Memo. No. 83968 [1982] 1/184; C.A. **98** [1983] No. 63228).
[172] Bosco, S. R.; Nova, D. F.; Brobst, W. D.; Stief, L. J. (J. Chem. Phys. **81** [1984] 3505/11).
[173] Schurath, U.; Tiedemann, P.; Schindler, R. N (J. Phys. Chem. **73** [1969] 456/9).
[174] Nagase, S.; Fueno, T. (Theor. Chim. Acta. **41** [1976] 59/70).
[175] Shih, S.; Buenker, R. J.; Peyerimhoff, S. D.; Michejda, C. J. (J. Am. Chem. Soc. **74** [1972] 7620/7).
[176] Gonbeau, D.; Guimon, M. G.; Ollivier, J.; Pfister-Guillouzo, G. (Chem. Phys. **120** [1988] 399/409).
[177] Lesclaux, R.; Veyret, B.; Roussel, P. (Ber. Bunsen-Ges. Phys. Chem. **89** [1985] 330/5).
[178] Cvetanovic, R. J.; Irwin, R. S. (J. Chem. Phys. **46** [1967] 1694/702).
[179] Atkinson, R.; Pitts, J. N., Jr. (J. Chem. Phys. **63** [1975] 3591/5).
[180] Michael, J. V.; Nava, D. F.; Borkowksi, R. P.; Payne, W. A.; Stief, L. J. (J. Chem. Phys. **73** [1980] 6108/16).

[181] Henri, J. P. L.; Carr, R. W., Jr. (J. Photochem. **5** [1975] 69/73).
[182] Slagle, I. R.; Gilbert, J. R.; Graham, R. E.; Gutman, D. (Int. J. Chem. Kinet. **7** [1974] 317/28).
[183] Bulatov, V. P.; Ioffe, A. A.; Lozovsky, V. A.; Sarkisov, O. M.; Poroikova, A. I. (Fotokhim. Protsessy Zemnoi Atmos. Dokl. Vses. Simp., Moscow 1987 [1990], p. 11).
[184] Mertens, J. D.; Kohse-Höinghaus, K.; Hanson, R. K.; Bowman, C. T. (Int. J. Chem. Kinet. **23** [1991] 655/68).
[185] Glebov, E. M.; Plyusnin, V. F.; Bazhin N. M.; Grivin, V. P. (Khim. Fiz. **10** [1991] 1495/502).
[186] Mayer, S. W.; Schieler, L. (J. Phys. Chem. **72** [1968] 236/40).
[187] Walther, C. D.; Wagner, H. Gg. (Ber. Bunsen-Ges. Phys. Chem. **87** [1983] 403/9).
[188] Donaldson, D. J.; Sloan, J. J.; Goddard, J. D. (J. Chem. Phys. **82** [1985] 4524/36).
[189] Bosco, S. R.; Brobst, W. D.; Nava, D. F.; Stief, L. J. (J. Geophys. Res. C **88** [1983] 8543/9).
[190] Burgers, D. (Mater. Res. Soc. Symp. Proc. **168** [1990] 137/42).

[191] Ioffe, A. A.; Bulatov, V. P.; Lozovsky, V. A.; Goldenberg, M. Y.; Sarkisov, O. M.; Umansky, S. Y. (Chem. Phys. Lett. **156** [1989] 425/32).
[192] Bulatov, V. P.; Vereshchuk, S. I.; Ioffe, A. A.; Poroikova, A. I.; Lozovsky, V. A. (Arm. Khim. Zh. **41** [1988] 26/31).

2.1.7.5.2 NH_2 Reactions in Complex Systems

NH_2 Radicals in Flames

Flames are complex kinetic systems with many interesting aspects. This paragraph is restricted to describe the role of NH_2 radicals in flames as a small part of the chemistry of nitrogen in combustion. Based on the origin of the nitrogen in NH_2, this overview is divided into three parts: (i) pure fuel nitrogen-oxygen flames (NH_3-O_2, N_2H_4-O_2) and nitrogen-containing fuel-oxygen flames (RNH_2, pyridine...-O_2), (ii) fuel-oxygen flames with N-containing fuel added (H_2, CH_4, C_2H_4, C_2H_2...-O_2+{NH_3, N_2H_4, RNH_2, pyridine..., or NO}, (iii) pure fuel-air flames (H_2, CH_4...-O_2-N_2). In most technical combustion systems the nitrogen originates from N-containing impurities in the fuel as well as from the N_2 in the air. The NH_2 chemistry in flames is influenced by burner operating conditions, such as the equivalence ratio (rich, stoichiometric, or lean flames), temperature (low-temperature, diluted flames, e.g. with Ar, raising the flame temperature by preheating the inlet air), fuel types (H_2, CH_4, C_2H_4, C_2H_2, etc.), pressure in the combustion chamber, amount of fuel nitrogen, premixed or diffusion flames, turbulent or laminar flames. Despite the many parameters, it will be shown that the chemistry of NH_2 in flames can be broadly understood, from which the NH_2 behavior, *cum grano salis*, can be predicted.

Nitrogen in combustion is of significant technological interest due to NO formation and destruction. The impact of nitrogen compounds emitted from combustion sources on the environment has motivated an extensive research activity in this field. Mechanisms and models for nitrogen combustion are given in [1].

Ammonia-Oxygen Flames

Ammonia-oxygen flames have been extensively studied during the last twenty-five years [1 to 13]. The NH_2 radical is an essential component in these flames. The following scheme outlines the important role of NH_2 [14]:

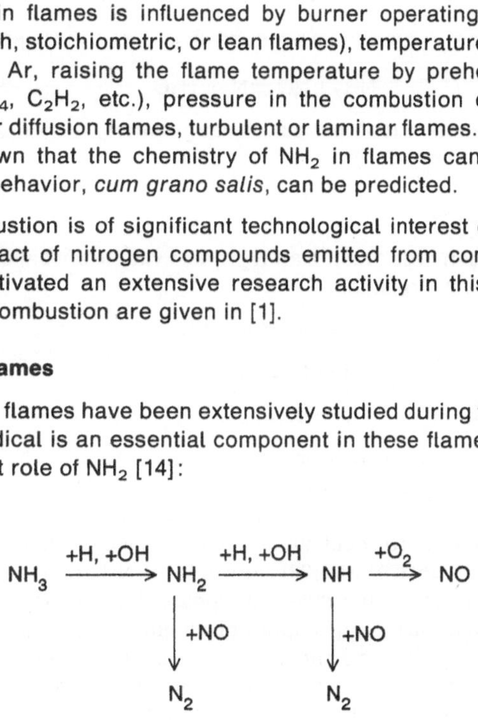

The NH_2 radical is formed via H abstraction from NH_3 by H atoms and/or OH radicals. The reaction $O+NH_3 \rightarrow NH_2+OH$ is too slow at the flame temperature encountered under most conditions. The competition between the reactions of NH_2 with NO and H or OH determines whether N_2 or NO is the final product. During the combustion of NH_3 and N_2H_4 with oxygen, the main products are N_2, H_2O, H_2, and NO as well as N_2O [15]. The main N-containing final product is N_2. The amount of NO decreases when the equivalence ratio increases [2].

Fuel-Oxygen Flames Doped with Nitrogen-Containing Fuel

The nitrogen chemistry in hydrocarbon flames with N-containing fuel added has been extensively studied, mainly to explain the NO formation and destruction in these systems [1, 9, 16 to 27]. The general scheme for fuel-oxygen flames is

$$\text{fuel-N} \longrightarrow HCN \longrightarrow NH_i \xrightarrow{+NO} N_2$$
$$NH_i \downarrow NO$$

NH_i includes the radicals NH_2 and NH and N atoms (i = 0, 1, 2). NH_2 is formed at a later stage from HCN regardless of the chemical nature of the fuel-nitrogen (NH_3, RNH_2, C_5H_5N, or HCN). The nitrogen-containing compounds are all very rapidly converted into HCN or CN. The NH_2 radical is then formed in rich flames via the reaction sequence [18]

$$OH + HCN \rightleftharpoons HNCO + H$$

$$H + HNCO \rightleftharpoons NH_2 + CO$$

The NH_2 radical is of great importance because it is the critical species for destruction of NO via

$$NH_2 + NO \rightleftharpoons N_2 + H_2O$$

in the so-called $DeNO_x$ reaction (see p. 233). In lean flames, N atoms are more important, particularly for the formation of NO.

Nitrogen-Free Fuel-Air Flames

In air flames with nitrogen-free fuel the air N_2 reacts via the so-called extended Zeldovich mechanism which involves the reactions

$$O + N_2 \rightleftharpoons NO + N$$

$$N + O_2 \rightleftharpoons NO + O$$

$$N + OH \rightleftharpoons NO + H$$

The NO thus formed reacts with hydrocarbon radicals to form HCN or CN radicals which then convert to NH_i, similarly as described above. Due to the high activation energy of the reaction $O + N_2$ of $E_A = 335$ kJ/mol, the NO (the NH_2 precursor molecule) appears relatively slowly in the high-temperature, post-flame "thermal NO". There also is a direct HCN production channel, $CH + N_2 \rightleftharpoons HCN + N$ [28]. The fate of NH_2 depends on the equivalence ratio. The reaction mechanisms in the different types of flames are more complicated than can be indicated here, since it is very important whether the radicals are present in the burning zone or in the burnt gases. The importance of a specific elementary reaction depends on the concentration of both reaction partners as much as on other kinetic parameters like temperature. The amount of NO produced in these flames is generally smaller than in combustion systems in which fuel nitrogen is present.

In technical combustion processes, N-containing fuel impurities are always present besides the air N_2. In fact, the principal source of nitrogen oxide emission in the combustion of fossil fuels is the nitrogen chemically bound in the fuel (0.5 to 2% N by weight). Adding N-containing molecules can destroy NO via the reaction $NH_2 + NO \rightarrow N_2 + H_2O$ ($DeNO_x$; see above).

Reduction of NO in Burnt Gas

The main concern of nitrogen in combustion is the unwanted production of nitric oxide. In an urban atmosphere the initially formed NO is converted into NO_2 and other nitrogen oxides which are usually summarized as NO_x. The NO in combustion results from nitrogen fuel impurities in technical fuel and from nitrogen molecules in the air applied as oxidizer (see p. 231).

The NO emission in technical combustion can be reduced by modifying the combustion processes (staged combustion, reburning) or by posttreatment of the flue gases. The catalytic NO reduction on surfaces like V_2O_5 [29] is not part of this paragraph in which the role of NH_2 radicals is placed in the foreground, although the same NH_2-NO complex responsible for NO reduction in the homogeneous gas phase reaction may form on the catalytic surface.

The thermal $DeNO_x$ [30] and $RAPRENO_x$ (rapid reduction of NO_x) [31] processes, which are based on a posttreatment of combuster exhaust products, are described in this paragraph, since NH_2 is an active species in these processes. In the thermal $DeNO_x$ process, NH_3 is injected into the exhaust gas of a stationary combustor. The NO reduction in this process has been intensively investigated [21, 30 to 39]. The process works best in the temperature range 1100 to 1300 K. At substantially lower temperatures, NO reduction is obtained, if NH_3-H_2 mixtures are injected rather than NH_3 alone [40]. The addition of H_2O_2 to NH_3-NO mixtures resulted in NO reduction even at temperatures as low as 773 K [41]. The homogeneous gas phase process responsible for the NO attack is $NH_2 + NO \rightarrow N_2 + H_2O$.

A detailed chemical kinetic model for the selective reduction of NO by NH_3 is given in [33]. The narrow temperature window (see above) for the process can be explained with a simplified reaction path diagram for the thermal $DeNO_x$ process as follows

$$NH_3 \xrightarrow{OH,\,O} NH_2 \xrightarrow{NO} N_2$$

At temperatures below 1100 K the process fails, because the formation of the active NH_2 radicals becomes too slow. At temperatures above 1400 K, NH_3 is consumed without destroying a comparable amount of NO. The addition of H_2 or H_2O_2 shifts the effective temperature window to lower temperatures due to a more efficient NH_2 production at low temperatures, but at high temperatures the NH_3 destruction is accelerated, thus the width of the temperature window is not changed significantly [33]. Also, other NH_2 precursor molecules like urea, $(NH_2)_2CO$, or hydrazine, N_2H_4, can be injected [42].

If cyanuric acid, $(HOCN)_3$, is injected into the exhaust stream, the NO_x abatement scheme is called $RAPRENO_x$ [31, 33]. At the exhaust temperature, the cyanuric acid decomposes to HNCO. The NH_2 formation step in this posttreatment system is $H + HNCO \rightleftharpoons NH_2 + CO$ [31]. Again, the reaction $NH_2 + NO \rightarrow H_2O + N_2$ is one of the processes responsible for the NO reduction.

NH_2 Radicals in Urban (Oxidizing) and Reducing Atmospheres

NH_2 radicals in an urban atmosphere [43] result from NH_3 photolysis by sun light in the stratosphere via

$$NH_3 + h\nu \rightarrow NH_2 + H \quad (\lambda < 230\ nm)$$

The lifetime of NH_3 under solar radiation is very short $(2 \leq \tau/h \leq 3 \times 10^4)$ [44]. The other NH_2 source involves the attack of ammonia by OH radicals in the troposphere via

$$NH_3 + OH \rightarrow NH_2 + H_2O$$

Ammonia, on the other hand, is one of the important reduced gases of biological origin like CH_4 and H_2S. It is released to the atmosphere in large amounts from soils [45] and oceans [46]. The NH_2 radical is the first species in the photooxidation of ammonia. Subsequent reactions of NH_2 radicals present a source or sink for nitrogen oxides. The oxidation can proceed via

$$NH_2 + O_2 \rightarrow NO_x + \ldots$$

$$NH_2 + O_3 \rightarrow NO_x + \ldots$$

leading to NO_x formation or resulting via $NH_2+\{NO, NO_2\}$ in NO_x consumption. These four molecules (O_2, O_3, NO, NO_2) can all be present in the atmosphere in sufficient concentrations to react with NH_2. Their concentrations, however, depend on many parameters like altitude and photon flux from the sun [47].

In the reducing atmospheres of Jupiter and Saturn, NH_3 is one of the constituents; NH_2 radicals are formed by photolysis of NH_3 and depleted by reaction with itself and other radicals, also produced by photolysis, and stable unsaturated hydrocarbons like C_2H_4 [48, 49]. The reaction NH_2+PH_3 is proposed to be a possible channel regenerating ammonia in the upper atmosphere of Jupiter [50].

References:

[1] Miller, J. A.; Bowman, C. T. (Prog. Energy Combust. Sci. **15** [1989] 287/338).
[2] McLean, D. I.; Wagner, H. Gg. (Symp. Int. Combust. Proc. **11** [1967] 871/8).
[3] Nadler, M. P.; Wang, V. K.; Kaskan, W. E. (J. Phys. Chem. **74** [1970] 917/22).
[4] Kaskan, W. E.; Hugues, D. E. (Combust. Flame **20** [1973] 381/8).
[5] Fischer, C. J. (Combust. Flame **30** [1977] 143/9).
[6] Carlap, F.; Tinel, C.; Loirat, H. (Combust. Flame **33** [1978] 299/303).
[7] Carlap, F.; Loirat, H. (J. Chem. Res. Synop. **1981** No. 8, pp. 240/1).
[8] Dasch, C. J.; Blint, R. J. (Chem. Phys. Processes Combust. Paper No. 71 [1982] 4).
[9] Miller, J. A.; Smoke, M. D.; Green, R. M.; Kee, R. J. (SAND 82-8609 [1983] 1/41).
[10] Miller, J. A.; Smoke, M. D.; Green, R. M.; Kee, R. J. (Combust. Sci. Technol. **34** [1983] 149/76).

[11] Morley, C. (Symp. Int. Combust. Proc. **18** [1981] 23/32).
[12] Dean, A. M.; Chou, M. S.; Stern, D. (Int. J. Chem. Kinet. **16** [1984] 633/53).
[13] Dean, A. M.; Chou, M. S.; Stern, D. (ACS Symp. Ser. **249** [1984] 71/86).
[14] Bian, J.; Vandooren, J.; Van Tiggelen, P. J. (Symp. Int. Combust. Proc. **21** [1986] 953/63).
[15] Wagner, H. Gg. (Symp. Int. Combust. Proc. **14** [1973] 27/36).
[16] Fenimore, C. P. (Combust. Flame **26** [1976] 249/56).
[17] Haynes, B. S. (Combust. Flame **28** [1977] 81/91).
[18] Haynes, B. S. (Combust. Flame **28** [1977] 113/21).
[19] Fenimore, C. P. (Symp. Int. Combust. Proc. **17** [1978] 661/70).
[20] Takagi, T.; Tatsumi, T.; Ogasawara, M. (Combust. Flame **35** [1979] 17/25).

[21] Fenimore, C. P. (Combust. Flame **37** [1980] 245/50).
[22] Debrou, G. B.; Goodings, J. M.; Bohme, D. K. (Combust. Flame **39** [1980] 1/19).
[23] Foster, D. E.; Keck, J. C. (Combust. Flame **38** [1980] 199/209).
[24] Puechberty, D.; Cotteraui, M. J. (Combust. Flame **51** [1983] 299/311).
[25] Blint, R. J.; Dasch, C. J. (ACS Symp. Ser. **249** [1984] 87/101).
[26] Martin, R. J.; Brown, N. J. (Combust. Flame **78** [1989] 365/76).
[27] Miller, J. A.; Lee, R. J.; Westbrook, C. K. (Annu. Rev. Phys. Chem. **41** [1990] 345/87).
[28] Fenimore, C. P. (Symp. Int. Combust. Proc. **13** [1971] 373/80).
[29] Faber, M.; Harries, S. P. (J. Phys. Chem. **88** [1984] 680/2).
[30] Lyon, R. K. (Environ. Sci. Technol. **21** [1987] 231/6).

[31] Perry, R. A.; Siebers, D. L. (Nature **324** [1986] 657/8).
[32] Lyon, R. K. (U.S. 3900 [1975] 544).
[33] Miller, J. A.; Branch, M. C.; Kee, R. J. (Combust. Flame **43** [1981] 81/98).
[34] Muzio, L. J.; Arand, J. K.; Teixeira, D. P. (Symp. Int. Combust. Proc. **16** [1977] 199/208).
[35] Salimian, S.; Hanson, R. K. (Combust. Sci. Technol. **23** [1980] 225/30).
[36] Silver, J. A. (Combust. Flame **53** [1983] 17/21).

[37] Kimball-Linne, M. A.; Hanson, R. K. (Combust. Flame **64** [1986] 337/51).
[38] Dean, A. M.; Hardy, J. E.; Lyon, R. K. (Symp. Int. Combust. Proc. **19** [1983] 97/105).
[39] Lucas, D.; Brown, N. J. (Combust. Flame **47** [1982] 219/34).
[40] Lyon, R. K.; Hardy, J. E. (Ind. Eng. Chem. Fundam. **25** [1986] 19/24).
[41] Azuhata, S.; Kaji, R.; Akimoto, H.; Hishinuma, Y. (Symp. Int. Combust. Proc. **18** [1981] 845/52).
[42] Ota, M.; Kamiguchi, T.; Arita, S.; Ito, M.; Miura, K. (Jpn. Kokai Tokkyo Koho No. 79-99076 [1979]; C.A. **92** [1980] No. 46748).
[43] Wayne, R. P. (J. Photochem. Photobiol. A **62** [1992] 379/96).
[44] Kuhn, W. R.; Atreya, S. K. (Icarus **37** [1979] 207/13).
[45] Dawson, G. A. (J. Geophys. Res. **82** [1977] 3125/33).
[46] Graedel, T. E. (J. Geophys. Res. C **84** [1979] 273/86).
[47] Nicolet, M. (Can. J. Chem. **52** [1974] 1381/96).
[48] Strobel, D. F. (Int. Rev. Phys. Chem. **3** [1984] 145/76).
[49] Kaye, J. A.; Strobel, D. F. (Icarus **54** [1983] 417/33).
[50] Strobel, D. F. (Astrophys. J. **214** [1977] L97/L99).

2.1.7.5.3 NH_2 Radicals in Condensed Media

Most of the NH_2 studies were done in the gas phase; however, NH_2 radicals can also be produced and studied in condensed media, in aqueous solutions [1 to 11], in liquid ammonia [12 to 16], in low-temperature matrices [17 to 27], or in salts at room temperature [28].

In aqueous solutions NH_2 radicals are produced by the redox reaction between Ti^{3+} ions [1] or V^{3+} ions [5] and hydroxylamine in acid media

$$Ti^{3+} + NH_2OH + H^+ \rightarrow NH_2 + H_2O + Ti^{4+}$$

Relative reactivities of NH_2 radicals towards a substrate (SH) and towards Ti^{3+} ions were obtained by comparing the rates of

$$NH_2 + Ti^{3+} + H^+ \rightarrow Ti^{4+} + NH_3$$

and

$$NH_2 + SH \rightarrow NH_3 + S$$

As substrates, the alcohols $CH_3(CH_2)_nOH$, where n is in the range 0 to 4 (i.e. methanol to pentanol), were measured. The reactivity of the alcohols increases with increasing n [4]. The reactivity of NH_2 radicals towards amines increases in the same way with increasing n in the substances $CH_3(CH_2)_nNH_3^+$, with n ranging from 2 to 7 [4].

The reactions in the liquid phase are in many respect very similar to the gas phase reactions at their high-pressure limit. Therefore, addition on a double bond is expected, and the amine radical thus formed is stabilized by collisions. This was observed directly for the reaction of NH_2 with maleic acid via

$$NH_2 + HO_2C\text{-}CH{=}CH\text{-}CO_2H \rightarrow HO_2C\text{-}HC(NH_2)\text{-}CH\text{-}CO_2H$$

[6]. Also, the unsaturated substrates

$$CH_3CH{=}CH\text{-}CO_2H,\ CH_3CH{=}CH\text{-}CONH_2,\ CH_3CH{=}CH\text{-}CHO,\ CH_2{=}C(CH_3)\text{-}CO_2H,$$
$$CH_3CH{=}CHCN,\ CH_2{=}CH\text{-}CO_2H,$$

and N-vinylsuccinimide were investigated with the Ti^{3+}-NH_2OH method [3]. The V^{3+}- and Ti^{3+}-NH_2OH systems were applied to study the NH_2 reaction in aqueous solution ($0 \leq pH \leq 7$) with substrates like benzenesulfonic acid or phenol [5].

Another method to produce NH_2 radicals in aqueous solution is the pulse radiolysis of NH_3-H_2O solutions. H_2O dissociates and the OH radicals react with ammonia via $OH + NH_3 \rightarrow NH_2 + H_2O$. A rate constant of $k = 1 \times 10^{11}$ $cm^3 \cdot mol^{-1} \cdot s^{-1}$ was determined for this reaction from the NH_2 absorption under pseudo-first-order conditions [29]; for the reaction $NH_2 + H_2O$ a rate constant of $k = 9 \times 10^{10}$ $cm^3 \cdot mol^{-1} \cdot s^{-1}$ was determined by the same method [29]. It is interesting to note that NH_2 reacts with O_2 in aqueous solution with a rate of $k = 3 \times 10^{11}$ $cm^3 \cdot mol^{-1} \cdot s^{-1}$ [29]. This indicates that in solution the complex NH_2-O_2 is formed and may also form in the gas phase, in particular at high pressures.

In liquid ammonia the NH_2 radicals are produced by γ radiolysis [12] or pulse radiolysis with high-energy electrons [14]. In the photolytic system $NH_3(l) + h\nu$ ($\lambda_{max} = 254$ nm) and in the radiolytic system, solvated electrons are produced [15] which react with NH_2 radicals. The NH_2^- ion, however, has not yet been observed directly [16].

NH_2 radicals in matrixes are either formed in the gas phase and then trapped on a cooled surface [18] (see p. 169) or produced in a matrix by radiolysis [20], electron impact [26], or chemical reaction (Al atoms + NH_3 reaction) [27]. The radicals were detected by ESR (see p. 168), by UV absorption [17, 18], or by IR spectroscopy [26]. In general, the mobility in a low-temperature matrix is insufficient for chemical reactions. A diffusion coefficient of $D \cong 10^{-14}$ cm^2/s and an activation energy of diffusion $E_A = 20.9$ kJ/mol were determined for NH_2 in a matrix in the temperature range 77 to 98 K [24]. Only the reaction $NH_2 + OH$ has been studied in a matrix at 4.2 K. The products HONO + H were detected [25].

NH_2 radicals were detected even at room temperature in a solid state environment. When a single crystal of hydroxylammonium cloride was irradiated with γ rays, ESR spectra of NH_2 were observed, even a long time after irradiation [28].

References:

[1] Farnia, G.; Anselmi, D.; Vianello, E. (Ric. Sci. **38** [1968] 1211/8).
[2] Lati, J.; Meyerstein, D. (Inorg. Chem. **11** [1972] 2393/7).
[3] Anselmi, D.; Farnia, G.; Vianello, E. (Chim. Ind. [Milan] **54** [1972] 1081/6).
[4] Farnia, G.; Tomat, R.; Vianello, E. (J. Chem. Soc. Perkin Trans. II **1975** 763/8).
[5] Tomat, R.; Rigo, A. (Electroanal. Chem. Interfacial Electrochem. **63** [1975] 329/37).
[6] Farnia, G.; Sandona, G.; Vianello, E. (J. Electronanal. Chem. Interfacial Electrochem. **88** [1978] 147/9).
[7] Kumar, A.; Neta, P. (J. Phys. Chem. **83** [1979] 3091/5).
[8] Gadzhiev, T. A.; Mkrtycheva, E. M. (Vopr. Razvit. Neftekhim. Khim. Prisadok Azerb. **1980** 74/7).
[9] Menkin, V. B.; Makarov, I. E.; Pikaev, A. K. (Khim. Vys. Energ. **22** [1988] 298/401).
[10] Neta, P.; Huie, R. E.; Ross, A. B. (J. Phys. Chem. Ref. Data **17** [1988] 1027/284).

[11] Ogura, K.; Migita, C. T.; Yamada, T. (J. Photochem. Photobiol. A **49** [1989] 53/61).
[12] Zaitsev, A. A.; Maiboroda, V. D.; Petryaev, E. P.; Salamatov, I. I.; Shadyro, O. I. (Vestn. Beloruss. Gos. Univ. Ser. 2 **1973** No. 1, pp. 6/9; C.A. **80** [1974] No. 42865).
[13] Marx, R.; Mauclaire, G. (Int. J. Mass Spectrom. Ion. Phys. **10** [1973] 213/26).
[14] Khaikin, G. I.; Zhigunov, V. A. (Khim. Vys. Energ. **9** [1975] 211/3).
[15] Belloni, J.; Billiau, F.; Cordier, P.; Delaire, J. A.; Delcourt, M. O. (J. Phys. Chem. **82** [1978] 532/6).

[16] Belloni, J.; Billiau, F.; Cordier, P.; Delaire, J. A.; Delcourt, M. O. (J. Phys. Chem. **82** [1978] 537/9).
[17] Robinson, G. W.; Mc Carty, M., Jr. (J. Chem. Phys. **28** [1958] 349/50).
[18] Robinson, G. W.; Mc Carty, M., Jr. (J. Chem. Phys. **30** [1959] 999/1005).
[19] Tupikov, V. I.; Pshezhetskii, S. Ya. (Zh. Fiz. Khim. **37** [1963] 1900/3; Russ. J. Phys. Chem. [Engl. Transl.] **37** [1963] 1031/4).
[20] Tupikov, V. I.; Tsivenko, V. I.; Pshezhetskii, S. Ya.; Kotov, A. G.; Milinchuk, V. K. (Zh. Fiz. Khim. **37** [1963] 138/42; Russ. J. Phys. Chem. [Engl. Transl.] **37** [1963] 65/7).

[21] Jacox, M. E.; Milligan, D. E. (Appl. Opt. **3** [1964] 873/6).
[22] Al-Naimy, B. S.; Moorthey, P. N.; Weiss, J. J. (J. Chem. Phys. **70** [1966] 3654/60).
[23] Fischer, P. H. H.; Charles, S. W.; McDorwell, E. A. (J. Chem. Phys. **46** [1967] 2162/6).
[24] Klochikhin, V. L.; Pshezhetskii, S. Ya.; Slavinskaya, N. A.; Trakhtenberg, L. I. (Zh. Fiz. Khim. **59** [1985] 2841/3; Russ. J. Phys. Chem. [Engl. Transl.] **59** [1985] 1698/9).
[25] Crowley, J. N.; Sodeau, J. R. (Risoe-M-2630 [1988]).
[26] Suzer, S.; Andrews, L. (J. Chem. Phys. **89** [1988] 5347/9).
[27] Klimov, V. D.; Mamchenko, A. V.; Nabiev, S. S.; Sukhanov, L. P. (Zh. Fiz. Khim. **65** [1991] 1819/25; Russ. J. Phys. Chem. [Engl. Transl.] **65** [1991] 966/70).
[28] Koksal, F.; Cakir, O.; Gumrukcu, I.; Birey, M. (Z. Naturforsch. **40a** [1985] 903/5).
[29] Pagsberg, P. B. (Dan. A. E. C., Risoe Rep. 256 [1972] 209/21; C.A. **77** [1972] No. 171183; C.A. **80** [1974] No. 89477).

2.1.8 The Amidogen Cation, NH_2^+

CAS Registry Numbers: NH_2^+ *[15194-15-7]*, NHD^+ *[54842-54-5]*, ND_2^+ *[54842-55-6]*

2.1.8.1 Formation

From N_2 and H_2. NH_2^+ cations were generated in a liquid-nitrogen-cooled ac discharge of He (7 Torr), N_2 (60 mTorr), and H_2 (100 mTorr) [1].

From N^+ and H_2; N and D_2^+. The formation of NH_2^+ [2] and ND_2^+ ions [3] by the reactions $N^+ + H_2$ [2] and $N + D_2^+$ [3] was shown in molecular beam experiments for low initial relative energies (<1 eV). Ab initio calculations of the $N^+ + H_2$ reaction indicated an intersystem crossing of the $\tilde{X}\,^3B_1$ and $1\,^3A_2$ potential surfaces of NH_2^+ (see p. 241), revealing a barrier-free pathway from $N^+(^3P) + H_2(^1\Sigma_g^+)$ to $NH_2^+(\tilde{X}\,^3B_1)$ [4 to 8].

From NH_2. NH_2^+ cations were formed by photoionization of NH_2 radicals [9, 10]; see p. 175.

From NH_3. NH_2^+ and ND_2^+ cations are the predominant fragment ions which were observed in photoionization and electron ionization mass spectra of NH_3 and ND_3, respectively [11].

NH_2^+ cations were produced in an air-cooled ac positive glow discharge of He (6 Torr) and NH_3 (50 Torr) [1]. NH_2^+ cations were formed by the dissociative photoionization of ammonia, $NH_3 + h\nu \rightarrow NH_2^+ + H + e^-$, in the range $\lambda = 795$ to 780 Å [12] and from $\lambda = 788$ Å (onset of ionization) to 600 Å [13]. The observed threshold values, 15.71 ± 0.04 eV [14] and 15.73 ± 0.02 eV [13], are in good agreement. Photoionization of NH_3 and isotopomers at 160 and 298 K gave the following appearance potentials (AP), extrapolated to 0 K [12, 15]:

ammonia/cation	NH_3/NH_2^+	NH_2D/NHD^+	NHD_2/NHD^+	NHD_2/ND_2^+	ND_3/ND_2^+
AP in eV	15.768(4)	15.79(1)	15.90(1)	15.79(1)	15.89(1)

NH_2^+ (ND_2^+) ions were formed by the dissociative electron ionization of ammonia NH_3 (ND_3) [16 to 19]. The threshold for the appearance of NH_2^+ according to $NH_3 + e^- \rightarrow NH_2^+(\tilde{X}\,^3B_1) + H(^2S) + 2\,e^-$ was determined as 15.97 ± 0.1 eV [16]. Appearance potentials of NH_2^+ from this process were calculated by ab initio methods [20]. Studies of the electron ionization from the threshold up to 180 eV gave a maximum cross section of 8.63×10^{-15} cm^2 at 90.5 eV [17].

From NH_3^+. The formation of NH_2^+ and ND_2^+ by the dissociation of NH_3^+ to $NH_2^+ + H$ [21] and ND_3^+ to $ND_2^+ + D$ [22] was shown using a photoelectron-photoion coincidence technique. An ArF excimer laser radiation with the energy of 6.42 eV (193 nm) was used to form ND_2^+ by photodissociation of ND_3^+. The reaction $NH_3^+ \rightarrow NH_2^+ + H$ is endothermic by 5.69 eV [23]. Mass spectrometric studies indicated the formation of NH_2^+ cations by the collision of NH_3^+ with Kr atoms [24].

From NH_3^{2+}. NH_2^+ cations were formed by the fragmentation of NH_3^{2+} cations, which were produced by photoionization [25] or electron ionization [26] of NH_3 molecules. The dissociation of $NH_3^{2+}(^1A_1)$ involves a singlet-triplet curve crossing to $NH_2^+(\tilde{X}\,^3B_1) + H^+$ or a tunneling mechanism to $NH_2^+(\tilde{a}\,^1A_1) + H^+$ [26].

From N_2H_4. NH_2^+ cations were formed by the photodissociation of N_2H_4. The appearance potential of NH_2^+ from $N_2H_4 + h\nu \rightarrow NH_2^+ + NH_2 + e^-$ at 0 K was determined thermochemically to be 13.98 eV [9]. NH_2^+ cations were formed by collision reactions of N_2H_4 with O^+, Ar^+, or Kr^+ in a low pressure chamber (0.1 Pa). Studies using a time-of-flight technique gave the reaction energies (ΔE) for $N_2H_4 + O^+ \rightarrow NH_2^+ + NH + OH$ (ΔE = 0.3 eV), $N_2H_4 + O^+ \rightarrow NH_2^+ + NH_2 + O$ (ΔE = 0.4 eV), $N_2H_4 + Ar^+ \rightarrow NH_2^+ + NH_2 + Ar$ (ΔE = −1.7 eV), and $N_2H_4 + Kr^+ \rightarrow NH_2^+ + NH_2 + Kr$ (ΔE = 0.1 eV) [27].

From CH_3NH_2. The formation of NH_2^+ was shown by a mass spectrometric analysis of photoproducts after irradiation of methylamine with an ArF laser at 193 nm [28].

References:

[1] Okumura, M.; Rehfuss, B. D.; Dinelli, B. M.; Bawendi, M. G.; Oka, T. (J. Chem. Phys. **90** [1989] 5918/23).
[2] Fair, J. A.; Mahan, B. H. (J. Chem. Phys. **62** [1975] 515/9).
[3] McClure, D. J.; Douglass, C. H.; Gentry, R. W. (J. Chem. Phys. **66** [1977] 2079/93).
[4] Bender, C. F.; Meadows, J. H.; Schaefer, H. F., III (Faraday Discuss. Chem. Soc. No. 62 [1977] 59/66).
[5] Bender, C. F.; Meadows, J. H.; Schaefer, H. F., III (LBL-5114 [1976], 23 pp.; C.A. **86** [1977] No. 146291).
[6] Hirst, D. M. (Mol. Phys. **35** [1978] 1559/68).
[7] Gittins, M. A.; Hirst, D. M.; Guest, M. F. (Faraday Discuss. Chem. Soc. No. 62 [1977] 67/76).
[8] Gittins, M. A.; Hirst, D. M. (Chem. Phys. Lett. **35** [1975] 534/6).
[9] Gibson, S. T.; Greene, J. P.; Berkowitz, J. (J. Chem. Phys. **83** [1985] 4319/28).
[10] Dunlavey, S. J.; Dyke, J. M.; Jonathan, N.; Morris, A. (Mol. Phys. **39** [1980] 1121/35).

[11] Locht, R.; Momigny, J. (Chem. Phys. Lett. **138** [1987] 391/6).
[12] McCulloh, K. E. (Int. J. Mass Spectrom. Ion Phys. **21** [1976] 333/42).
[13] Dibeler, V. H.; Walker, J. A.; Rosenstock, H. M. (J. Res. Natl. Bur. Stand. A **70** [1966] 459/63).
[14] Samson, J. A. R.; Haddad, G. N.; Kilcoyne, L. D. (J. Chem. Phys. **87** [1987] 6416/22).
[15] McCulloh, K. E. (Vac. Ultraviolet Radiat. Phys. Proc. 4th Int. Conf., Hamburg 1974, pp. 195/7).

[16] Locht, R.; Servais, C.; Ligot, M.; Derwa, F.; Momigny, J. (Chem. Phys. **123** [1988] 443/54).
[17] Maerk, T. D.; Egger, F.; Cheret, M. (J. Chem. Phys. **67** [1977] 3795/802).
[18] Maerk, T. D. (Proc. 7th Int. Vac. Congr., Vienna 1977, Vol. 2, pp. 1341/5).
[19] Morrison, J. D.; Traeger, J. C. (Int. J. Mass Spectrom. Ion Phys. **11** [1973] 277/88).
[20] Power, D.; Brint, P.; Spalding, T. R. (J. Mol. Struct. **110** [1984] 155/66 [THEOCHEM **19**]).

[21] Powis, I. (J. Chem. Soc. Faraday Trans. II **77** [1981] 1433/47).
[22] Powis, I. (Chem. Phys. **68** [1982] 251/4).
[23] Kutina, R. E.; Edwards, A. K.; Pandolfi, R. S.; Berkowitz, J. (J. Chem. Phys. **80** [1984] 4112/9).
[24] Manvelyan, R. V.; Perov, A. A.; Kupriyanov, S. E.; Potapov, V. K. (Teor. Eksp. Khim. **14** [1978] 831/4; Theor. Exp. Chem. [Engl. Transl.] **14** [1978] 645/7).
[25] Winkoun, D.; Dujardin, G. (Z. Phys. D At. Mol. Clusters **4** [1986] 57/64).
[26] Boyd, R. K.; Singh, S.; Beynon, J. H. (Chem. Phys. **100** [1985] 297/314).
[27] Gardner, J. A.; Dressler, R. A.; Salter, R. H.; Murad, E. (J. Phys. Chem. **96** [1992] 4210/7).
[28] Nishi, N.; Shinohara, H.; Hanazaki, I. (Chem. Phys. Lett. **73** [1980] 473/7).

2.1.8.2 Geometric Structure. Electron Configuration. Electronic States. Potential Energy Functions. Ionization Potential

Geometric Structure. Electron Configuration. Electronic States

A few experimental investigations, but predominantly ab initio calculations, supplied the data on the geometry and electron configurations of NH_2^+ in the electronic ground state $\tilde{X}\,^3B_1$ and seven electronically excited states presented in Table 17. Additional electronic states $1\,^3A_1$, $2\,^3B_1$, $3\,^1A_1$, $2\,^1B_1$, $4\,^1A_1$ [1], and $1\,^5A_2$ [1, 2] were treated by ab initio methods. Semiempirical calculations were reported for NH_2^+ in low-lying electronic states [3 to 13].

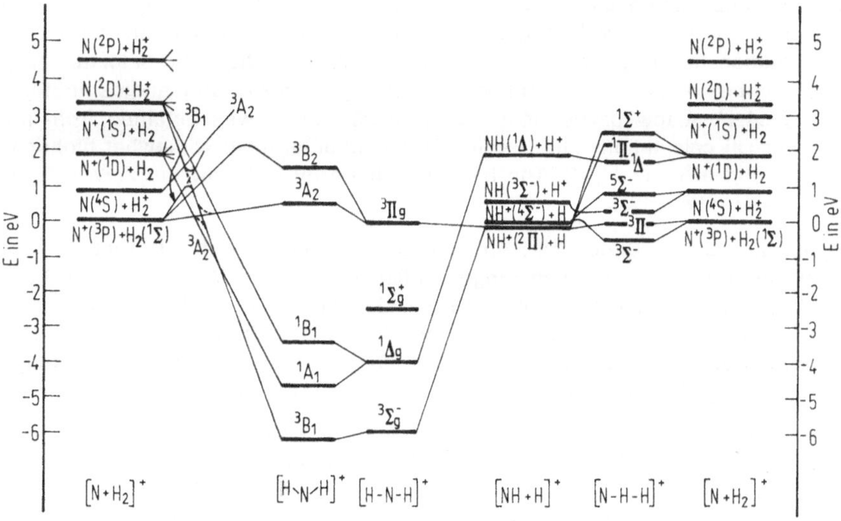

Fig. 11. Electronic state correlation diagram for the low-lying states of NH_2^+. At the right, a collinear approach of N^+ to H_2 is assumed. At the left, N^+ approaches H_2 along the perpendicular bisector of the hydrogen molecule [14].

Fig. 11 (p. 239) gives the electronic correlation diagram for the $[N+H_2]^+$ reactive system [14]. Additional remarks are given below for the $\tilde{X}\,^3B_1$, $\tilde{a}\,^1A_1$ and $\tilde{b}\,^1B_1$ electronic states.

Table 17
Electronic Ground State and Excited States of NH_2^+ and ND_2^+.
T_0 are the experimental and T_e the calculated term values; r_e and α_e are ab initio-calculated bond lengths and angles. Renner-Teller states are combined by large brackets.

state	bent configuration [1]	T_0 in cm^{-1}	T_e in cm^{-1}	r_e in Å	α_e	linear configuration [1]
$1\,^1B_2$	$...(1b_2)\,(3a_1)\,(1b_1)^2$	–	74525 a)	–	98° a)	$^1\Pi_g...(2\sigma_u)\,(\pi_u)^3$
$1\,^1A_2$	$...(1b_2)\,(3a_1)^2(1b_1)$	–	45489 a)	–	63° a)	
$1\,^3B_2$	$...(1b_2)\,(3a_1)\,(1b_1)^2$	–	59846 a)	–	89° a)	$^3\Pi_g...(2\sigma_u)\,(\pi_u)^3$
$1\,^3A_2$	$...(1b_2)\,(3a_1)^2(1b_1)$	–	33230 a)	– b)	60° a) b)	
$2\,^1A_1$	$...(1b_2)^2(1b_1)^2$	–	27826 a)	–	180° a)	$^1\Sigma_g^+...(2\sigma_u)^2\,(\pi_u)^2$
$\tilde{b}\,^1B_1$	$...(1b_2)^2\,(3a_1)\,(1b_1)$	≤20490(160) c)	15397 d)	1.034 d)	164° d)	$^1\Delta_g...(2\sigma_u)^2(\pi_u)^2$
$\tilde{a}\,^1A_1$	$...(1b_2)^2\,(3a_1)^2$	10530(80) e)	10475 d)	1.051 d)	108.4° d)	
$\tilde{X}\,^3B_1$	$...(1b_2)^2\,(3a_1)\,(1b_1)$	0.0	0.0	1.034 d)	153.2° d)	$^3\Sigma_g^-...(2\sigma_u)^2(\pi_u)^2$

a) From ab initio SCF calculations with constant bond length $r_e=1.027$ Å [1]. – b) An additional ab initio calculation gave $r_e=1.224$ Å and $\alpha_e=44.17°$ [15]. – c) Cited in [16], based on a photoelectron spectrum [15] revised in [17]. – d) From ab initio second-order CI calculations [18]. – e) Cited in [16], based on a photoionization mass spectrum [17]. The low value $T_0=7985(160)$ cm^{-1} from a photoelectron spectrum [15] was criticized; see the ionization potential of NH_2, p. 175.

Electronic Ground State $\tilde{X}\,^3B_1$. The geometric structure could not be determined from spectroscopic constants because of the quasi-linear structure of NH_2^+; the large amplitude of the bending vibration prevented a determination of the rotational constant A [19]; see p. 242. In addition to the geometric data in Table 17, other ab initio calculations [15, 20 to 29] gave bond lengths in the range 1.019 Å$\le r_e \le$1.042 Å and angles in the range $143.1° \le \alpha_e \le 154.4°$. A low barrier to linearity of 209 cm^{-1} was obtained from ab initio (second-order CI) calculations [18]. Earlier CI calculations gave somewhat higher values, 330 cm^{-1} [1] and 323 cm^{-1} [30]. The barrier to linearity is below the zero point energy of the bending vibration [19].

Electronically Excited State $\tilde{a}\,^1A_1$. Ab initio calculations gave the geometry (r_e, α_e) and the term value (T_e) in the following ranges: 1.029 Å$\le r_e \le$1.051 Å, $107.1° \le \alpha_e \le 110.5°$ [15, 18, 20 to 22, 25 to 28], and 9533 $cm^{-1} \le T_e \le$ 11332 cm^{-1} [1, 4, 20, 22, 26, 27, 31]. The barrier to linearity was calculated as 5071 cm^{-1} [18] or 6200 cm^{-1} [32] by ab initio CI methods. Earlier ab initio-calculated values of the barrier to linearity are significantly lower, 1371 cm^{-1} [29] and 2798 cm^{-1} [33], or higher, 8184 cm^{-1} [33].

Electronically Excited State $\tilde{b}\,^1B_1$. The barrier to linearity, 220 cm^{-1} [32], and similar values as in Table 17, $r_e=1.0261$ Å [15], $\alpha_e=163.4°$ [15], and $T_e=16373$ cm^{-1} [1, 30], were calculated with ab initio methods. The energy differences between the lowest-lying vibronic K=0 levels (not influenced by Renner-Teller effect) of the $\tilde{a}\,^1A_1$ and $\tilde{b}\,^1B_1$ states, $\Delta E_0=6313$ (NH_2^+), 6224 (NHD^+), and 6137 cm^{-1} (ND_2^+), were calculated on the basis of ab initio potential curves for the two states [32]. The vertical energy of the $\tilde{b}\,^1B_1$ state above the equilibrium point of the $\tilde{a}\,^1A_1$ state was ab initio-calculated as 11650 cm^{-1} [18].

Potential Energy Functions

Ab initio-calculated bending potential curves are presented for the $\tilde{X}\,^3B_1$ [1, 15, 24, 30, 33], $\tilde{a}\,^1A_1$ [1, 15, 24, 30, 33], $\tilde{b}\,^1B_1$ [1, 15, 30, 33], and higher excited electronic states [1, 15, 30, 33]. The bending potential curves of the Renner-Teller states $\tilde{a}\,^1A_1$ and $\tilde{b}\,^1B_1$ were plotted together with the K = 0 vibronic levels and intensity distributions for transitions in NH_2^+, NHD^+, and ND_2^+ [32]. Bending potential coefficients were calculated semiempirically [10]. Symmetric stretch potential curves are shown for vibrations in low-lying electronic states [1, 15]. Coefficients of potential energy functions (in terms of stretching and bending vibrations) were calculated for NH_2^+ in the three lowest electronic states by ab initio second-order CI methods [18].

Potential energy surfaces of NH_2^+ with C_{2v} symmetry in the $\tilde{X}\,^3B_1$ and $1\,^3A_2$ states are illustrated for the variables r(H···H) and R(N···H_2) (R is the distance between the center of mass of H_2 and N) based on theoretical studies [26, 27, 34]. The $\tilde{X}\,^3B_1$ surface has a deep well which is not adiabatically accessible at low relative energies [35, 36]. The potential barrier was calculated as 14280, 16050, 16860 cm^{-1} (extended diatomics-in-molecules model) [34] or 23390 cm^{-1} (CI method) [35]. The potential energy surface of the $1\,^3A_2$ state has a deep potential minimum relative to separated $N^+(^3P)+H_2(^1\Sigma_g^+)$ [26, 27, 37]. A barrier-free (adiabatic) pathway is discussed from the $1\,^3A_2$ to the $\tilde{X}\,^3B_1$ surfaces [35, 36]. Cuts through the potential surfaces illustrate the case as N^+ approaches H_2 with a fixed H-H distance. Such plots were shown for two $^3\Sigma^-$ states of linear NH_2^+, for three triplet states with C_{2v} symmetry ($1\,^3A_2$, $\tilde{X}\,^3B_1$, and $2\,^3A_2$), and for three $^3A''$ states with C_s symmetry [35, 36, 38 to 40] based on theoretical studies of the reaction $N^+ + H_2 \rightarrow NH^+ + H$. Cuts through surfaces of quintet states $^5\Sigma^-$ (linear NH_2^+), $^5A''$ (C_s symmetry), and $1\,^5A_2$ (C_{2v} symmetry) show a minimum for the $^5\Sigma^-$ state and repulsive potentials for C_{2v} and near-C_{2v} geometries [2].

Ionization Potential

Ionization of $NH_2^+(\tilde{X}\,^3B_1)$ leads to the dication NH_2^{2+} in the electronic ground state $\tilde{X}\,^2\Pi_u$; see p. 248. The adiabatic ionization potential $E_i(ad) = 24.5 \pm 0.5$ eV was determined following the change of the translational energy of the NH_2^{2+} ions in the single-electron capture process $NH_2^{2+} + He \rightarrow NH_2^+ + He^+$ [41]. $E_i(ad) = 23.3$ eV was measured in an investigation of the charge-stripping reaction $NH_2^+ + N \rightarrow NH_2^{2+} + N + e^-$ [42]. Ab initio calculations gave $E_i(ad) = 23.8$ to 24.6 eV [22, 43, 44] and a vertical ionization potential of $E_i(vert) = 25.2$ eV [44].

The ionization potentials of NH_2^+ in the $\tilde{a}\,^1A_1$ state, $E_i(vert) = 24.7$ and $E_i(ad) = 23.2$ eV, were obtained using the 4th-order Møller-Plesset perturbation theory [44]. Semiempirical calculations are given in [8]. Additional $E_i(vert)$ values of $NH_2^+(\tilde{a}\,^1A_1)$ were determined applying Koopman's theorem [28].

References:

[1] Peyerimhoff, S. D.; Buenker, R. J. (Chem. Phys. **42** [1979] 167/76).
[2] Hirst, D. M. (Chem. Phys. Lett. **53** [1978] 125/7).
[3] Falvey, D. E.; Cramer, C. J. (Tetrahedron Lett. **33** [1992] 1705/8).
[4] Glover, S. A.; Scott, A. P. (Tetrahedron **45** [1989] 1763/76).
[5] Siddarth, P.; Gopinathan, M. S. (J. Am. Chem. Soc. **110** [1988] 96/104).
[6] Siddarth, P.; Gopinathan, M. S. (Proc. Indian Acad. Sci. Chem. Sci. **99** [1987] 91/5).
[7] Ford, G. P.; Scribner, J. D. (J. Am. Chem. Soc. **103** [1981] 4281/91).
[8] Bews, J. R.; Glidewell, C. (Inorg. Chim. Acta **39** [1980] 217/25).

[9] Bews, J. R.; Glidewell, C. (J. Mol. Struct. **67** [1980] 141/50).
[10] Glidewell, C. (J. Mol. Struct. **65** [1980] 231/8).

[11] Takahata, Y. (Chem. Phys. Lett. **59** [1978] 472/7).
[12] Dewar, M. J. S.; Haddon, R. C.; Li, W. K.; Thiel, W.; Weiner, P. K. (J. Am. Chem. Soc. **97** [1975] 4540/5).
[13] Zeller Stenkamp, L.; Davidson, E. R. (Theor. Chim. Acta **30** [1973] 283/314).
[14] Fair, J. A.; Mahan, B. H. (J. Chem. Phys. **62** [1975] 515/9).
[15] Dunlavey, S. J.; Dyke, J. M.; Jonathan, N.; Morris, A. (Mol. Phys. **39** [1980] 1121/35).
[16] Jacox, M. E. (J. Phys. Chem. Ref. Data **17** [1988] 269/511, 281).
[17] Gibson, S. T.; Greene, J. P.; Berkowitz, J. (J. Chem. Phys. **83** [1985] 4319/28).
[18] Jensen, P.; Bunker, P. R.; McLean, A. D. (Chem. Phys. Lett. **141** [1987] 53/7).
[19] Okumura, M.; Rehfuss, B. D.; Dinelli, B. M.; Bawendi, M. G.; Oka, T. (J. Chem. Phys. **90** [1989] 5918/23).
[20] Ford, G. P.; Herman, P. S. (J. Am. Chem. Soc. **111** [1989] 3987/96).

[21] Pople, J. A.; Curtiss, L. A. (J. Phys. Chem. **91** [1987] 155/62).
[22] Pope, S. A.; Hillier, I. H.; Guest, M. F. (Faraday Symp. Chem. Soc. No. 19 [1984] 109/23).
[23] Zhang, B. (Kexue Tongbao **28** [1983] 511/2; C.A. **99** [1983] No. 28252).
[24] Wasilewski, J. (J. Mol. Struct. **52** [1979] 281/91).
[25] Bell, S. (J. Chem. Phys. **69** [1978] 3879/80).
[26] Bender, C. F.; Meadows, J. H.; Schaefer, H. F., III (Faraday Discuss. Chem. Soc. No. 62 [1977] 59/66).
[27] Bender, C. F.; Meadows, J. H.; Schaefer, H. F., III (LBL-5114 [1976], 23 pp.; C.A. **86** [1977] No. 146291).
[28] Heaton, M. M.; Cowdry, R. (J. Chem. Phys. **62** [1975] 3002/9).
[29] Harrison, J. F.; Eakers, C. W. (J. Am. Chem. Soc. **95** [1973] 3467/72).
[30] Chu, S. Y.; Siu, A. K. Q.; Hayes, E. F. (J. Am. Chem. Soc. **94** [1972] 2969/72).

[31] Pople, J. A.; Schleyer, P. R. (Chem. Phys. Lett. **129** [1986] 279/81).
[32] Perić, M.; Buenker, R. J.; Peyerimhoff, S. D. (Astrophys. Lett. **24** [1984] 69/73).
[33] Lee, S. T.; Morokuma, K. (J. Am. Chem. Soc. **93** [1971] 6863/6).
[34] Polák, R. (Chem. Phys. **153** [1991] 91/7).
[35] Hirst, D. M. (Mol. Phys. **35** [1978] 1559/68).
[36] Gittins, M. A.; Hirst, D. M.; Guest, M. F. (Faraday Discuss. Chem. Soc. No. 62 [1977] 67/76).
[37] González, M.; Aguilar, A.; Fernández, Y. (Chem. Phys. **104** [1986] 57/66).
[38] Hirst, D. M. (NATO Adv. Study Inst. Ser. B **40** [1979] 24/9).
[39] Hirst, D. M. (DL/SCI/R [1978] 9/14; C.A. **91** [1979] No. 44805).
[40] Gittins, M. A.; Hirst, D. M. (Chem. Phys. Lett. **35** [1975] 534/6).

[41] Hamdan, M.; Brenton, A. G. (Int. J. Mass Spectrom. Ion Processes **84** [1988] 211/4).
[42] Proctor, C. J.; Porter, C. J.; Ast, T.; Bolton, P. D.; Beynon, J. H. (Org. Mass Spectrom. **16** [1981] 454/8).
[43] Pope, S. A.; Hillier, I. H.; Guest, M. F.; Kendric, J. (Chem. Phys. Lett. **95** [1983] 247/9).
[44] Koch, W.; Schwarz, H. (Int. J. Mass Spectrom. Ion Processes **68** [1986] 49/56).

2.1.8.3 Rotational and Vibrational Constants

Rotational and Spin-Rotation Coupling Constants. Centrifugal Distortion Constants

The IR absorption spectrum of the quasi-linear NH_2^+ ion (showing large amplitudes of the bending vibration) was analyzed applying an asymmetric top formalism. Due to the selection rule $\Delta K_a=0$, only differences of the rotational constant A, A(0,0,1)−A(0,0,0) = 15.0137(25) cm^{-1} (a axis ⊥ C_2, in the molecular plane), were obtained [1]. Estimates of A in the vibrational ground state gave the large values, 203 or 280 cm^{-1} [1], consistent with theoretical predictions [2]. The analysis of the IR spectrum yielded the following rotational constants B (b axis ∥ C_2) and C, the spin-rotation constants A_s, B_s, and C_s, and the centrifugal distortion constants Δ_{NK}, Δ_N, δ_N (all in cm^{-1}) for the vibrational ground and an excited state [1].

constant	(0,0,0)	(0,0,1)	constant	(0,0,0)	(0,0,1)
B	8.27299(34)	8.09746(33)	$\delta_N \times 10^5$	4.27(13)	4.361(93)
C	7.64443(34)	7.47436(33)	$A_s \times 10^2$	0.89(27)	1.41(34)
$\Delta_{NK} \times 10^2$	−9.736(20)	−10.603(19)	$B_s \times 10^2$	−0.879(80)	−0.860(89)
$\Delta_N \times 10^4$	2.390(44)	2.440(34)	$C_s \times 10^2$	−1.658(80)	−1.661(89)

The rotational constants of NH_2^+, A = 146.118, B = 8.6348, and C = 8.1530 cm^{-1} [3], were obtained from the ab initio-calculated bond length r_e = 1.021 Å and angle α_e = 150.86° [4]. The rotational constants of ND_2^+ A = 57.6 and B = 4.3 cm^{-1} were also calculated from the structural data [5].

Vibrational Frequencies

The antisymmetric stretching vibration in the electronic ground state $\tilde{X}\,^2B_1$, ν_3 = 3359.932(2) cm^{-1}, was obtained from an IR absorption spectrum [1]. Additional vibrational frequencies of NH_2^+ in the three lowest electronic states were determined from photoelectron spectra (PES) of $NH_2(\tilde{X}\,^2B_1)$ [6]. Vibrational frequencies of NH_2^+ and isotopomers (in cm^{-1}) from experiments (see data collections [7 to 9]) and ab initio calculations are listed in the following table:

state	cation	ν_2 PES [6]	ν_2 ab initio	ν_1 PES [6]	ν_1 ab initio	ν_3 IR [1]	ν_3 ab initio
$\tilde{X}\,^3B_1$	NH_2^+	840 ± 50	918 a) b)	–	3118 a) b)	3359.93	3363 a) b)
	NHD^+	–	772 a)	–	3246 a)	–	2363 a)
	ND_2^+	660 ± 50	608 a)	–	2229 a)	–	2536 a)
$\tilde{a}\,^1A_1$	NH_2^+	1350 ± 50	1289 a) c) d)	2900 ± 50	3027 a) c)	–	3111 a) c)
	NHD^+	–	1140 a) d)	–	2270 a) e)	–	3069 a) e)
	ND_2^+	940 ± 50	965 a) d)	2210 ± 50	2214 a)	–	2324 a)
$\tilde{b}\,^1B_1$	NH_2^+	920 ± 150	1453 f)	–	3509 g)	–	–
	NHD^+	–	1230 f)	–	–	–	–
	ND_2^+	–	983 f)	–	2487 g)	–	–

a) From [2]. — b) Additional calculations: ν_2 = 820, ν_1 = 3342, ν_3 = 3701 [10], and ν_2 = 814, ν_1 = 3463, ν_3 = 3717 [4]. — c) Additional calculations: ν_2 = 1512, ν_1 = 3462, ν_3 = 3566 [10], and ν_2 = 1562, ν_1 = 3490, ν_3 = 3594 [4]. — d) Additional calculations: ν_2 = 1450 (NH_2^+), 1274 (NHD^+), 1069 (ND_2^+) [11]. — e) Values of ν_1 and ν_3 are exchanged in the original. — f) From [11]. — g) From [6].

The harmonic vibrational frequencies ω ($\Delta G_{1/2}$) of NH_2^+ in the electronic ground state $\tilde{X}\,^3B_1$ were ab initio-calculated as ω_2 = 792 (725), ω_1 = 3330 (3082), and ω_3 = 3593 (3308) cm^{-1} using the MP2 method [12].

Overtone and combination frequencies of NH_2^+, ND_2^+, and NHD^+ were ab initio-calculated for (0,2,0), (0,3,0), (0,4,0), (1,1,0), and (0,1,1) [2].

The force constant due to stretch, $k_e = 1520$ N/m, was determined from an ab initio-calculated energy curve at $r_e = 1.07$ Å [13].

References:

[1] Okumura, M.; Rehfuss, B. D.; Dinelli, B. M.; Bawendi, M. G.; Oka, T. (J. Chem. Phys. **90** [1989] 5918/23).
[2] Jensen, P.; Bunker, P. R.; McLean, A. D. (Chem. Phys. Lett. **141** [1987] 53/7).
[3] Gibson, S. T.; Greene, J. P.; Berkowitz, J. (J. Chem. Phys. **83** [1985] 4319/28).
[4] Pople, J. A.; Curtiss, L. A. (J. Phys. Chem. **91** [1987] 155/62).
[5] Powis, I. (Chem. Phys. **68** [1982] 251/4).
[6] Dunlavey, S. J.; Dyke, J. M.; Jonathan, N.; Morris, A. (Mol. Phys. **39** [1980] 1121/35).
[7] Jacox, M. E. (J. Phys. Chem. Ref. Data **19** [1990] 1387/546, 1396).
[8] Jacox, M. E. (J. Phys. Chem. Ref. Data **17** [1988] 269/511, 281).
[9] Jacox, M. E. (J. Phys. Chem. Ref. Data **13** [1984] 945/1068, 952).
[10] Glover, S. A.; Scott, A. P. (Tetrahedron **45** [1989] 1763/76).

[11] Perić, M.; Buenker, R. J.; Peyerimhoff, S. D. (Astrophys. Lett. **24** [1984] 69/73).
[12] DeFrees, D. J.; McLean, A. D. (J. Chem. Phys. **82** [1985] 333/41).
[13] Peyerimhoff, S. D.; Buenker, R. J.; Whitten, J. L. (J. Chem. Phys. **46** [1967] 1707/16).

2.1.8.4 Dissociation Energy. Atomization Energy

Dissociation energies D according to $NH_2^+ \rightarrow N^+(^3P) + H_2(^1\Sigma_g^+)$ were calculated for the ground state $\tilde{X}\,^3B_1$ and excited states of NH_2^+ with ab initio procedures.

state	$\tilde{X}\,^3B_1$	$\tilde{a}\,^1A_1$	$\tilde{b}\,^1B_1$	$2\,^1A_1$	$1\,^3A_2$	$1\,^1A_2$
D_e in kJ/mol [1]	622	498	426	291	269	131
[2]	577	–	–	–	254	–

Additional ab initio calculations gave $D_e = 590$ [3] and 546 kJ/mol [4] for the $\tilde{X}\,^3B_1$ state, and $D_e = 261$ [3], 251 [5, 6], and 170 kJ/mol [4] for the $1\,^3A_2$ state. NH_3^+ photodissociation experiments yielded the limits $D \leq 561$ kJ/mol for $NH_2^+ \rightarrow NH^+ + H$ and $613.4 \leq D \leq 638.5$ kJ/mol for $NH_2^+ \rightarrow NH + H^+$ [7].

The **atomization energies** 1200.9 and 1086.2 kJ/mol of NH_2^+ ($\rightarrow N^+ + 2\,H$) in the $\tilde{X}\,^3B_1$ and $\tilde{a}\,^1A_1$ states, respectively, were calculated by ab initio SCF methods [8].

References:

[1] Peyerimhoff, S. D.; Buenker, R. J. (Chem. Phys. **42** [1979] 167/76).
[2] Hirst, D. M. (Mol. Phys. **35** [1978] 1559/68).
[3] Polák, R. (Chem. Phys. **153** [1991] 91/7).
[4] Hirst, D. M. (Faraday Discuss. Chem. Soc. No. 62 [1977] 138).
[5] Bender, C. F.; Meadows, J. H.; Schaefer, H. F., III (Faraday Discuss. Chem. Soc. No. 62 [1977] 59/66).
[6] Bender, C. F.; Meadows, J. H.; Schaefer, H. F., III (LBL-5114 [1976], 23 pp.; C.A. **86** [1977] No. 146291).
[7] Kutina, R. E.; Edwards, A. K.; Pandolfi, R. S.; Berkowitz, J. (J. Chem. Phys. **80** [1984] 4112/9).
[8] Glover, S. A.; Scott, A. P. (Tetrahedron **45** [1989] 1763/76).

2.1.8.5 Enthalpy of Formation

$\Delta_fH_0^\circ = 1266.5 \pm 0.4$ [1], 1266.1 ± 0.8 [2], and 1257.3 kJ/mol [3] was determined from the threshold energy of the dissociative photoionization of ammonia (1521.4 ± 0.4 kJ/mol [4, 5]) in combination with the enthalpies of formation for NH_3 and H. Earlier measurements gave $\Delta_fH^\circ_0 = 1261.9$ kJ/mol using the same procedure [6].

Combining literature data of $\Delta_fH^\circ(NH_2)$ with the adiabatic ionization potential of NH_2 gave $\Delta_fH^\circ_0(NH_2^+) = 1276.9$ [7], 1275.3 [8, 9], and 1273.6 to 1298.3 kJ/mol [2]. Semiempirically calculated values of Δ_fH were reported in [10 to 14].

References:

[1] Gibson, S. T.; Greene, J. P.; Berkowitz, J. (J. Chem. Phys. **83** [1985] 4319/28).
[2] Kutina, R. E.; Edwards, A. K.; Pandolfi, R. S.; Berkowitz, J. (J. Chem. Phys. **80** [1984] 4112/9).
[3] Baumgaertel, H.; Jochims, H. W.; Ruehl, E.; Bock, H.; Dammel, R.; Minkwitz, J.; Nass, R. (Inorg. Chem. **28** [1989] 943/9).
[4] McCulloh, K. E. (Int. J. Mass Spectrom. Ion Phys. **21** [1976] 333/42).
[5] McCulloh, K. E. (Vac. Ultraviolet Radiat. Phys. Proc. 4th Int. Conf., Hamburg 1974, pp. 195/7).
[6] Dibeler, V. H.; Walker, J. A.; Rosenstock, H. M. (J. Res. Natl. Bur. Stand. A **70** [1966] 459/63).
[7] Power, D.; Brint, P.; Spalding, T. R. (J. Mol. Struct. **110** [1984] 155/66 [THEOCHEM **19**]).
[8] Pople, J. A.; Curtiss, L. A. (J. Phys. Chem. **91** [1987] 155/62).
[9] Pople, J. A.; Curtiss, L. A. (J. Phys. Chem. **91** [1987] 3637/9).
[10] Ford, G. P.; Scribner, J. D. (J. Am. Chem. Soc. **103** [1981] 4281/91).
[11] Bews, J. R.; Glidewell, C. (Inorg. Chim. Acta **39** [1980] 217/25).
[12] Bews, J. R.; Glidewell, C. (J. Mol. Struct. **67** [1980] 141/50).
[13] Dewar, M. J. S.; Haddon, R. C.; Li, W. K.; Thiel, W.; Weiner, P. K. (J. Am. Chem. Soc. **97** [1975] 4540/5).
[14] Glidewell, C. (J. Mol. Struct. **65** [1980] 231/8).

2.1.8.6 Spectrum

Only one investigation of the absorption spectrum of NH_2^+ in the electronic ground state $\tilde{X}\,^2B_1$, using a tunable difference-frequency laser spectrometer in the 3 μm region, was reported. Fifty-three rovibronic transitions in the ν_3 band were measured in the limits from 3510.677 cm^{-1} (assigned to the transition $11_{0,11}(J'=10) \leftarrow 10_{0,10}(J''=9)$) and 3186.295 cm^{-1} ($9_{0,9}(J'=8) \leftarrow 10_{0,10}(J''=9)$) [1].

Oscillator strengths and vertical energy differences were calculated for transitions between electronic states of the same multiplicity using ab initio (SCF and multireference double CI) methods [2].

References:

[1] Okumura, M.; Rehfuss, B. D.; Dinelli, B. M.; Bawendi, M. G.; Oka, T. (J. Chem. Phys. **90** [1989] 5918/23).
[2] Peyerimhoff, S. D.; Buenker, R. J. (Chem. Phys. **42** [1979] 167/76).

2.1.8.7 Reactions

Reactions of NH_2^+ and ND_2^+ with a number of molecules were investigated using various methods. The results (partly reviewed in [1]) are listed in Table 18.

The NH_2^+ cations were generated, for example, in a low pressure electron-impact ion source (electron energy about 70 eV) containing ammonia at a pressure of about 1 mTorr [2, 3]. The measurements were carried out using the flowing-afterglow (FA) technique [4, 5], the ion cyclotron resonance (ICR) method [6 to 13], a modified mass spectrometer (MMS) [14 to 18], a selected ion flow tube (SIFT) apparatus [2, 3], a drift-tube mass spectrometer (TDSM) [19 to 23], and the technique of trapping ions within a quadrupole ion store (QIS) attached to a quadrupole mass filter [24]. E_{kin} is the mean relative kinetic energy between the ion and the reactant species.

Table 18
Reactions of NH_2^+ and $ND_2^{+\,(*)}$.
Total rate constant 10^{10} k in $cm^3 \cdot molecule^{-1} \cdot s^{-1}$ at room temperature if not otherwise stated. Δ_rH (product ion) is the enthalpy of reaction for NH_2^+ + reactant → product ion + neutral.

reactant	product ion (percentage)	10^{10} k	method	Ref.
CH_2O	CH_3O^+ (80), NH_3^+ (20)	28	SIFT	[2]
CHOOH	NH_4^+, ($CHOOH_2^+$?)	27 ± 8	FA [a)]	[5]
CH_4	NH_3^+ (100)	9.2	SIFT [b)]	[2]
CH_3OH	$CH_3OH_2^+$ (85), NH_3^+ (15)	31	SIFT	[2]
CH_3OH	$CH_3OH_2^+$ (87), NH_3^+ (13)	30	TDSM [c)]	[21]
CH_3NH_2	$CH_3NH_2^+$ (50), $CH_3NH_3^+$ (20), H_4CN^+ (20), NH_4^+ (10)	18	SIFT	[2]
CH_3NH_2	$CH_3NH_2^+$ (55), $CH_3NH_3^+$ (20 ± 5), H_4CN^+ (20), NH_4^+ (5)	20 ± 6	TDSM [d)]	[20]
CO	$NH_2^+ \cdot CO$ (100)	0.24	SIFT [e)]	[2]
COS	H_2NS^+ (80), H_2NCO^+ (15), $HCOS^+$ (5)	15	SIFT	[2]
CO_2	products not given	< 0.01	SIFT	[2]
C_2H_4	$C_2H_4^+$ (30), H_4CN^+ (30), $C_2H_5^+$ (20), $H_5C_2N^+$ (20)	15 ± 3	SIFT	[3]
C_2D_4	$C_2D_4^+$ (30), $H_2D_2CN^+$ (30), $C_2HD_4^+$ (20), $H_2D_3C_2N^+$ (20)	16 ± 3	SIFT	[3]
$C_2H_4^{(*)}$	$C_2H_4^+$ (30), $H_2D_2CN^+$ (30), $C_2H_4D^+$ (20), $H_3D_2C_2N^+$ (20)	14 ± 3	SIFT	[3]
$C_2D_4^{(*)}$	$C_2D_4^+$ (30), D_4CN^+ (30), $C_2D_5^+$ (20), $D_5C_2N^+$ (20)	14 ± 3	SIFT	[3]
C_2H_4O	$C_2H_4O^+$ (42), HCO^+ (33.5), CH_3CO^+ (24.5)	35.8	TDSM [f)]	[19]
H_2	NH_3^+ (100)	2.7	SIFT [g)]	[2]
H_2O	H_3O^+ (95), NH_4^+ (5)	29	SIFT	[2]
H_2O	H_3O^+ (92), NH_4^+ (2), NH_3^+ (6)	17 ± 0.5	ICR	[6]
H_2O	H_3O^+ (95 ± 5), NH_3^+ (5 ± 5)	50	TDSM [h)]	[19]
H_2S	H_2S^+ (59), H_3S^+ (41)	5.8	ICR [i)]	[9]
H_2S	H_2S^+ (40), NH_3^+ (25), H_3S^+ (15), NH_4^+ (10), SH^+ (10)	18	SIFT	[2]
NH_3	NH_4^+ (50), NH_3^+ (50)	22 ± 4	ICR	[13]
NH_3	NH_4^+ (30), NH_3^+ (70)	30 ± 5.5	ICR [k)]	[11]
NH_3	NH_4^+ (21), NH_3^+ (79)	35 ± 2	QIS [l)]	[24]
NH_3	NH_4^+, NH_3^+	18	MMS	[14]
NH_3	NH_4^+ (53), NH_3^+ (47)	29 ± 1	MMS	[16]
NH_3	NH_4^+	6.5	MMS [m)]	[17]
NH_3	NH_4^+	7.99	MMS	[18]
NH_3	NH_4^+ (70), NH_3^+ (30)	23	SIFT	[2]
NO	NO^+ (100)	7	SIFT	[2]
N_2	products are not given	< 0.005	SIFT	[2]
O_2	H_2NO^+ (85), HNO^+ (15)	14	SIFT	[2]
SiH_4	SiH_3^+ (70), SiH_2^+ (30)	44.4	MMS	[15]

a) $\Delta_rH(NH_4^+) = -425$ and $\Delta_rH(CHOOH_2^+) = -161$ kJ/mol. — b) No reaction was observed between NH_2^+ and CH_4 in an ICR study [10]. — c) For $E_{kin} = 0.046$ eV; k decreases with increasing E_{kin}; $\Delta_rH(NH_3^+) = -159$ and $\Delta_rH(CH_3OH_2^+) = -113$ kJ/mol. — d) For $E_{kin} = 0.049$ eV; k increases with increasing E_{kin}; $\Delta_rH(CH_3NH_3^+) = -272$, $\Delta_rH(CH_3NH_2^+) = -226$, $\Delta_rH(CH_2NH_2^+) = -556$ kJ/mol. — e) The reaction $NH_2^+ + CO \rightarrow HCO^+ + NH$ did not occur, because it is nearly thermoneutral [2]. — f) For $E_{kin} = 0.05$ eV. — g) $k \approx 10 \times 10^{-10}$ $cm^3 \cdot molecule^{-1} \cdot s^{-1}$ (FA) [4], $k = 1.2 \times 10^{-10}$ $cm^3 \cdot molecule^{-1} \cdot s^{-1}$ (ICR), $\Delta_rH(NH_3^+) = 121$ kJ/mol [8]. — h) For $E_{kin} = 0.05$ eV; $k = 29 \times 10^{-10}$ $cm^3 \cdot molecule^{-1} \cdot s^{-1}$ for $E_{kin} = 0.28$ eV. — i) $\Delta_rH(H_2S^+) = -96$ and $\Delta_rH(H_3S^+) = -121$ kJ/mol. — k) Additional rate constants are quoted: $k = (18.8 \pm 3) \times 10^{-10}$ [25] and $\sim 11.4 \times 10^{-10}$ $cm^3 \cdot molecule^{-1} \cdot s^{-1}$ [26]. $\Delta_rH(NH_4^+) = -267$ and $\Delta_rH(NH_3^+) = -121$ kJ/mol [25]. — l) E_{kin} in the range 1 to 3 eV. — m) For T = 373 K and thermal ion exit energy (E'); $k = 9.7 \times 10^{-10}$ $cm^3 \cdot molecule^{-1} \cdot s^{-1}$ for E' = 3.4 eV.

Significant discrepancies are apparent [2] between ICR [7 to 11] and SIFT [2] data especially concerning the product ion distribution.

The reaction of NH_2^+ with H_2 and D_2 was studied by measuring the angular and energy distribution of the product ions NH_3^+, NH_2D^+, and NHD_2^+ at collision energies from 0.4 to 10 eV [27]. The energy distribution of (long-lived) collision complexes with high cross section was measured for low-energy (<1.5 eV) collisions of NH_2^+ with H_2 [28]. The exothermic reaction $NH_2^+ + SiH_4 \rightarrow SiH_5^+ + NH$ was indicated by tandem mass spectrometric studies [29].

A population analysis of singlet and triplet states, performed by an ab initio (SCF CI) treatment, suggested that the singlet $NH_2^+(^1A_1)$ will react like a carbonium ion and the triplet $NH_2^+(^3B_1)$ will react like a triplet methylene [30].

References:

[1] Ikezoe, Y.; Matsuoka, S.; Takebe, M.; Viggiano, A. (Gas Phase Ion-Molecule Rate Constants Through 1986, Ion Reaction Research Group of the Mass Spectroscopy Society of Japan, Toyko 1987, pp. 1/224, 97/8).

[2] Adams, N. G.; Smith, D.; Paulson, J. F. (J. Chem. Phys. **72** [1980] 288/97).

[3] Smith, D.; Adams, N. G. (Chem. Phys. Lett. **76** [1980] 418/23).

[4] Fehsenfeld, F. C.; Schmeltekopf, A. L.; Ferguson, E. E. (J. Chem. Phys. **46** [1967] 2802/8).

[5] Freeman, C. G.; Harland, P. W.; McEwan, M. J. (Austral. J. Chem. **31** [1978] 2593/9).

[6] Anicich, V. G.; Kim, J. K.; Huntress, W. T., Jr. (Int. J. Mass Spectrom. Ion Phys. **25** [1977] 433/8).

[7] Huntress, W. T., Jr. (Astrophys. J. Suppl. Ser. **33** [1977] 495/514).

[8] Kim, J. K.; Theard, L. P.; Huntress, W. T., Jr. (J. Chem. Phys. **62** [1975] 45/52).

[9] Laudenslager, J. B.; Huntress, W. T., Jr. (Int. J. Mass Spectrom. Ion Phys. **14** [1974] 435/48).

[10] Huntress, W. T., Jr.; Pinizzotto, R. F., Jr.; Laudenslager, J. B. (J. Am. Chem. Soc. **95** [1973] 4107/15).

[11] Marx, R.; Mauclaire, G. (Int. J. Mass Spectrom. Ion Phys. **10** [1973] 213/26).

[12] Huntress, W. T., Jr.; Pinizzotto, R. F., Jr. (J. Chem. Phys. **59** [1973] 4742/56).

[13] Huntress, W. T., Jr.; Mosesman, M. M.; Elleman, D. D. (J. Chem. Phys. **54** [1971] 843/9).

[14] Derwish, G. A. W.; Galli, A.; Giardini-Guidoni, A.; Volpi, G. G. (J. Chem. Phys. **39** [1963] 1599/605).

[15] Haller, I. (J. Phys. Chem. **94** [1990] 4135/7).

[16] Fluegge, R. A.; Landman, D. A. (J. Chem. Phys. **54** [1971] 1576/91).

[17] Gupta, S. K.; Jones, E. G.; Harrison, A. G.; Myher, J. J. (Can. J. Chem. **45** [1967] 3107/17).

[18] Matsumoto, A.; Okada, S.; Taniguchi, S.; Hayakawa, T. (Bull. Chem. Soc. Jpn. **48** [1975] 3387/8).
[19] Barassin, J.; Reynaud, C.; Barassin, A. (Chem. Phys. Lett. **123** [1986] 191/6).
[20] Barassin, J.; Barassin, A.; Thomas, R. (Int. J. Mass Spectrom. Ion Phys. **49** [1983] 51/60).
[21] Thomas, R.; Barassin, J.; Barassin, A. (Int. J. Mass Spectrom. Ion Phys. **41** [1981] 95/107).
[22] Barassin, J.; Thomas, R.; Barassin, A. (Symp. Proc. 5th Int. Symp. Plasma Chem., Edinburgh 1981, Vol. 1, pp. 47/51).
[23] Thomas, R.; Barassin, J.; Barassin, A. (Contrib. Symp. At. Surf. Phys. **1982** 285/90).
[24] Lawson, G.; Bonner, R. F.; Mather, R. E.; Todd, J. F. J. (J. Chem. Soc. Faraday Trans. I **72** [1976] 545/57).
[25] Ryan, K. R. (J. Chem. Phys. **53** [1970] 3844/8).
[26] Harrison, A. G.; Thynne, J. C. J. (Trans. Faraday Soc. **62** [1966] 2804/14).
[27] Eisele, G.; Henglein, A.; Botschwina, P.; Meyer, W. (Ber. Bunsen-Ges. Phys. Chem. **78** [1974] 1090/7).
[28] Mayer, T. M. (Ber. Bunsen-Ges. Phys. Chem. **79** [1975] 352/6).
[29] Cheng, T. M. H.; Lampe, F. W. (Chem. Phys. Lett. **19** [1973] 532/4).
[30] Lee, S. T.; Morokuma, K. (J. Am. Chem. Soc. **93** [1971] 6863/6).

2.1.9 Adducts of NH_2^+ with NH_3

A minor quantity of $[NH_2(NH_3)_n]^+$ ions (n not specified) forms besides the main product $[NH_4(NH_3)_m]^+$ upon ionization of the clusters in jet-expanded NH_3 via $(NH_3)_{n+1}^+$ and loss of hydrogen.

Reference:

Shinohara, H.; Nishi, N.; Washida, N. (J. Chem. Phys. **83** [1985] 1939/47; Chem. Phys. Lett. **106** [1984] 302/6).

2.1.10 The Amidogen Dication, NH_2^{2+}

CAS Registry Number: *[85420-11-7]*

NH_2^{2+} dications were formed by ionizing NH_2^+ by collision with N atoms according to $NH_2^+ + N \rightarrow NH_2^{2+} + N + e^-$ [1]. NH_2^{2+} ions were also formed in a 100-eV electron-impact ion source containing NH_3 maintained at a pressure of 10^{-6} Torr and a temperature of 450 K [2]. NH_2^{2+} ions were formed by dissociative electron ionization of ammonia at electron energies of 20 to 183 eV [3]. NH_2^{2+} ions were mass spectrometrically detected in the unimolecular fragmentation of NH_3^{2+}, formed by electron ionization of ammonia, using techniques of ion translational energy spectroscopy [4].

The five-valence-electron dication NH_2^{2+} is linear ($D_{\infty h}$ symmetry) in the electronic ground state $\tilde{X}\,^2\Pi_u$. The electron configuration is given by $(1\sigma_g)^2$ $(2\sigma_g)^2$ $(2\sigma_u)^2$ (π_u) [5]. The internuclear distance $r_e = 1.155$ [6], 1.145, and 1.168 Å [5],and the harmonic vibrational frequencies $\omega_2(\Pi_u) = 436$, $\omega_3(\Sigma_u) = 2010$, and $\omega_1(\Sigma_g) = 2073$ cm^{-1} [6] were calculated by ab initio methods.

The excitation energy $T_0 = 0.9 \pm 0.2$ eV from the ground state $\tilde{X}\,^3B_1$ to the first excited metastable state $^2\Sigma_g^+$ was obtained from the translational energy spectrum of 6-keV NH_2^{2+}

ions scattered off He atoms. The translational energy spectrum of NH^+ fragment ions resulting from the unimolecular dissociation of NH_2^{2+} to NH^+ and H^+ indicated two dissociative energy levels of NH_2^{2+} separated by 2.4 eV [2].

The high value of the enthalpy of formation $\Delta_fH°(NH_2^{2+})=3636$ kJ/mol estimated from thermodynamic data indicates an extreme thermodynamic instability [6]. Ab initio calculations of the deprotonation $NH_2^{2+}(^2\Pi_u) \rightarrow NH^+(^2\Pi) + H^+$ gave a highly exothermic energy of deprotonation, −452.3 [6], −423.0 to −454.4 kJ/mol [5, 7], and the respective values of the activation barrier of 41.0 [6], 60.5 to 47.7 kJ/mol [5, 7]. The relatively large activation barrier for deprotonation indicates a kinetically stable NH_2^{2+} [6].

References:

[1] Proctor, C. J.; Porter, C. J.; Ast, T.; Bolton, P. D.; Beynon, J. H. (Org. Mass Spectrom. **16** [1981] 454/8).
[2] Hamdan, M.; Brenton, A. G. (Int. J. Mass Spectrom. Ion Processes **84** [1988] 211/4).
[3] Maerk, T. D.; Egger, F.; Cheret, M. (J. Chem. Phys. **67** [1977] 3795/802).
[4] Boyd, R. K.; Singh, S.; Beynon, J. H. (Chem. Phys. **100** [1985] 297/314).
[5] Pope, S. A.; Hillier, I. H.; Guest, M. F. (Faraday Symp. Chem. Soc. No. 19 [1984] 109/23).
[6] Koch, W.; Schwarz, H. (Int. J. Mass Spectrom. Ion Processes **68** [1986] 49/56).
[7] Pope, S. A.; Hillier, I. H.; Guest, M. F.; Kendric, J. (Chem. Phys. Lett. **95** [1983] 247/9).

2.1.11 The Amide Ion, NH_2^-

CAS Registry Numbers: NH_2^- *[17655-31-1]*, $^{15}NH_2^-$ *[102363-01-9]*, NHD^- *[15117-75-6]*, ND_2^- *[22856-00-4]*

2.1.11.1 Formation

2.1.11.1.1 In the Gas Phase

The gaseous NH_2^- ion forms from NH_3 and electrons with an energy of ~15 eV or less via resonance capture and decomposition of the intermediately formed NH_3^- [1]. Amide can be obtained by exposing NH_3 to the electron beam of a mass spectrometer; details are given in the next paragraph. Electrons of appropriate energy are also present at high current density in glow discharges through NH_3. An NH_2^- concentration of 4×10^{11} ions/cm^3 was estimated in a water-cooled, alternating-current discharge (yielding a discharge current of 400 mA) through 3 Torr of NH_3, after the walls of the discharge cell had been sputtered with iron, platinum, or copper [2]. A beam of NH_2^- ions was produced by exposing NH_3 to an He plasma [3]. A duoplasmatron-type, direct extraction ion source with a hollow-cathode arc running on Ar yielded an intense, bright stream of NH_2^- ions from NH_3 [4]. The application of a potassium surface-ionization source did not lead to satisfactory results with NH_3 [5].

The minimum electron energy required for the production of NH_2^- is in the range of 4 to 5.5 eV [6 to 8]; the value for ND_2^- is ~5.5 eV [8]. The highest NH_2^- current was reached at 5.60 [9] to 5.95±0.15 eV [10]; a value of 6.3±0.2 eV was reported in [11]. Maximum ND_2^- formation (5.80 eV given in [9]) was reached at a slightly higher energy than that for NH_2^-. The energy difference between the two isotopomers is attributed to their different zero-point energies. An NHD^- signal at intermediate energy is observed in NH_3-ND_3 mixtures and stems from H-D exchange between the reactants [9]. An investigation of plots of the $NH(D)_2^-$ yield as a function of energy shows that at higher resolution the resulting curves exhibit a number of narrow peaks, thought to arise from a vibrational progression of the $NH(D)_3^-$ intermediate [12].

An appearance potential of 4.6 eV was measured for NH_2^- when generated from NH_2OH by dissociative electron detachment and of 4.7 eV when generated from N_2H_4; the maxima of the narrow resonances were found at about 6 eV [3]. The main signal at 6.42±0.10 eV was found for the formation of NH_2^- from CH_3NH_2 by a resonance process. A single resonance peak of ND_2^- was found in the negative ion mass spectrum of CH_3ND_2 at 6.42±0.30 eV [10].

Kinetic energies of close to 0 eV for NH_2^- and more than 1 eV for the simultaneously formed H^- were measured for low-energy electron impact on NH_3 [13]. Experiments using the retarding-field technique showed that the amide ion is formed in the electronic ground state at ~5.7 eV from ammonia. A weak, second maximum at a higher energy can be assigned to the electronically excited anion [8]. This signal for NH_2^- is broad, peaks at 9.68±0.10 eV [10], and has a relative intensity of 0.03 with respect to the main signal. The corresponding signal of ND_2^- from ND_3 has a maximum at ~10 eV and a relative intensity of 0.04 [9]. A weak, third resonance peak at 12.05±0.10 eV was also assigned to the excited anion [10]. The weakness of the signals of the excited amide ion at ~10 eV indicates that the reaction channel $NH_3^- \rightarrow NH_2^{-*} + H$ is unimportant; the products in this reaction form mostly via $NH_3^- \rightarrow NH_2^* + H^-$. The cross sections σ for amide ion formation from NH_3 at incident electron energies E_e with respect to the total production σ_t of positive ions at 85 eV are as follows [8]:

	$E_e = 5.65$ eV NH_3	ND_3	$E_e = 10.5$ eV NH_3	ND_3
σ in cm^2	1.5×10^{-18}	1.2×10^{-18}	4.6×10^{-20}	3.2×10^{-20}
σ/σ_t	$(6.6 \pm 0.3) \times 10^{-3}$	$(5.0 \pm 0.2) \times 10^{-3}$	$(2.1 \pm 0.1) \times 10^{-4}$	$(1.4 \pm 0.1) \times 10^{-4}$

The formation of electronically excited NH_2^- from CH_3NH_2 by resonance capture led to additional, weak signals at 9.96±0.30 and 11.86±0.30 eV. Ion pair processes produced NH_2^- signals at electron energies of >16 eV in NH_3 and CH_3NH_2. The formation of ND_2^- from CH_3ND_2 is highly efficient, but ion pair processes could not be detected [10].

Amide ion formation by the reaction $NH_3 + H^- \rightleftharpoons NH_2^- + H_2$ after electron impact on NH_3 is less important than the formation via resonance capture of electrons by NH_3 as described above. The reaction with H^- has a rate constant of $(9.0 \pm 0.6) \times 10^{-13}$ $cm^3 \cdot molecule^{-1} \cdot s^{-1}$ and an equilibrium constant of 27±9 at 297 K [14]. An ab initio calculation predicted for the reaction an exothermicity of 57.3 kJ/mol, when zero-point energies are considered [15].

A very small amount of NH_2^- was identified among the products of the autocatalytical decomposition of NH_3 at a Pt filament at 1500 K. The NH_2^- concentration seems to increase exponentially with temperature above 1500 K [16]. The possible presence of NH_2^- besides O^- in the products of C_2H_2-O_2-N_2 flames was deduced from a signal in the mass spectrum and from the experimentally observed formation of NH_3 [17].

References:

[1] Mann, M. M.; Hustrulid, A.; Tate, J. T. (Phys. Rev. [2] **58** [1940] 340/7).
[2] Tack, L. M.; Rosenbaum, N. H.; Owrutsky, J. C.; Saykally, R. J. (J. Chem. Phys. **85** [1986] 4222/7).
[3] Yalcin, T.; Suzer, S. (J. Mol. Struct. **266** [1992] 353/6).

[4] Richards, H. T.; Klody, G. M. (Proc. 2nd Int. Conf. Ion Sources, Vienna 1972 [1973], pp. 804/11; C.A. **83** [1975] No. 106907).
[5] Middleton, R.; Adams, C. T. (Nucl. Instrum. Methods **118** [1974] 329/36).
[6] Melton, C. E. (J. Chem. Phys. **45** [1966] 4414/24).
[7] Kraus, K. (Z. Naturforsch. **16a** [1961] 1378/85).
[8] Sharp, T. E.; Dowell, J. T. (J. Chem. Phys. **50** [1969] 3024/35).
[9] Compton, R. N.; Stockdale, J. A.; Reinhard, P. W. (Phys. Rev. [2] **180** [1969] 111/20).
[10] Collin, J. E.; Hubin-Franskin, M. J.; D'Or, L. (Adv. Mass Spectrom. **4** [1968] 713/26).

[11] Dillard, J. G.; Franklin, J. L. (J. Chem. Phys. **48** [1968] 2353/8).
[12] Stricklett, K. L.; Burrow, P. D. (J. Phys. B **19** [1986] 4241/53).
[13] Tronc, M.; Azria, R.; Arfa, M. B. (J. Phys. B **21** [1988] 2497/506).
[14] Bohme, D. K.; Hemsworth, R. S.; Rundle, H. W. (J. Chem. Phys. **59** [1973] 77/81).
[15] Ritchie, C. D.; King, H. F. (J. Am. Chem. Soc. **90** [1968] 838/43).
[16] Melton, C. E.; Emmet, P. H. (J. Phys. Chem. **68** [1964] 3318/24).
[17] Hayhurst, A. N.; Kittelson, D. B. (Combust. Flame **31** [1978] 37/51).

2.1.11.1.2 In Condensed Phases

Solutions of amide ions in liquid NH_3 are obtained by dissolving the ionic amides of alkali and alkaline earth metals. However, all solutions are only weak electrolytes. The amides $M(NH_2)_2$ with M = Ca, Sr, Ba dissociate slightly less than the alkali amides. Solutions of KNH_2 are usually preferred because of the good solubility of the salt which is better than that of $NaNH_2$, whereas $LiNH_2$ is exceptional among the alkali amides by being rather insoluble in liquid NH_3 [1]. The amides of the alkali and alkaline earth metals can be prepared by either or both of the well-known reactions of the metals with gaseous NH_3 at elevated temperatures or with liquid NH_3 in homogeneous solution via reduction of the solvent by solvated electrons. The metal solutions in liquid NH_3 are quite stable in some cases. Their rate of decomposition into metal amides and H_2 increases with temperature and upon photolysis. The most effective catalysts for the formation of the metal amides are elemental Pt and Fe salts [2]. The catalytic formation of alkali amides by glass and its deactivation by treatment with NH_4F or potassium vapor are described in [3]. The amides are sensitive to air and moisture. Details are given in "Lithium" Erg.-Bd., 1960, p. 279, "Natrium" 1928, pp. 253/4, "Natrium" Erg.-Bd. 3, 1966, pp. 918/9, "Kalium" 2, 1937, p. 251, "Rubidium" 1937, pp. 114/5, "Caesium" 2, 1938, p. 116, "Magnesium" B, 1939, pp. 73/4, "Calcium" B 2, 1957, p. 332, "Strontium" 1931, p. 90, "Barium" 1932, pp. 138/9, and "Barium" Erg.-Bd., 1960, p. 171. Ionic amides of the other metals are rare. A few examples are $Eu(NH_2)_2$, $Yb(NH_2)_2$, and $Yb(NH_2)_3$ which result from the reaction of the metals with NH_3, whereas $Y(NH_2)_3$ has to be prepared either by reacting YI_3 with KNH_2 or $NaY(NH_2)_4$ with NH_4I; see "Seltenerdelemente" C 2, 1974, pp. 182/5 for details. The ionic $Zn(NH_2)_2$ is obtained by the reaction of $Zn(C_2H_5)_2$ with gaseous NH_3; see "Zink" Erg.-Bd., 1956, pp. 184/5.

The electrochemical formation of NH_2^- by reduction of NH_3 at Pt electrodes at 213 K was observed between −2.2 and −2.5 V in the presence of NaI as electrolyte [4]. Thermodynamic data for the formation of NH_2^- from NH_3 by reduction with solvated electrons in NH_3 solution at 233 K and estimated rate constants are as follows [5]:

reaction	ΔH° in kJ/mol	ΔS° in $J\cdot K^{-1}\cdot mol^{-1}$	k in $L\cdot mol^{-1}\cdot s^{-1}$
$2\,NH_3+2\,e^- \rightarrow 2\,NH_2^- + H_2$	−168	−466	$<10^{-7}$
$NH_3+e^- \rightarrow NH_2^- + H$	125	−272	$<10^{-17}$
$NH_3+e^- + H \rightarrow NH_2^- + H_2$	−293	−194	$<10^{2}$

An absolute reduction potential of $E^\circ = 0.22$ eV was calculated from the ionization potential of NH_2^- for the reaction $NH_2(aq) + 1/2\ H_2 \rightarrow NH_2^-(aq) + H^+(aq)$ in a 1 M solution at 298 K [6].

The formation of NH_2^- during the autoprotolysis of liquid NH_3 via $2\ NH_3 \rightleftharpoons NH_2^- + NH_4^+$ is negligible as shown by the extrapolated equilibrium constant $K = 8.8 \times 10^{-31}$ at 298 K. Thermodynamic data for the forward reaction are $\Delta_{diss}H^\circ = 94 \pm 6$ kJ/mol and $\Delta_{diss}S^\circ = -260$ $J \cdot K^{-1} \cdot mol^{-1}$. These data are based on electrochemical measurements on NH_2^-, NH_4^+, and Na^+ in liquid NH_3 which led to K values in the range of 10^{-37} ($\Delta G^\circ = 148$ kJ/mol) at 208 K and of 10^{-34} ($\Delta G^\circ = 156$ kJ/mol) at 238 K [7]. Thermodynamic data for NH_2^- yielded the equilibrium constants $K = 3.2 \times 10^{-33}$ at 240 K and 2.2×10^{-28} at 298.15 K [8].

Samples of NH_2^- were matrix-isolated by bombarding a diluted mixture of NH_3 in Ar continuously with electrons during condensation at 12 K. The quantity of NH_2^- and NH_2 increased upon annealing [9]. Alkali halide crystals doped with NH_2^-, NHD^-, and ND_2^- can be prepared either by treating the alkali halide crystal first with potassium vapor (generation of F centers) and then with ammonia, by reduction of ammonium incorporated in the crystals during their growth, or by crystallizing the alkali halide from the melt while regularly adding small amounts of an alkali amide [10].

The formation of NH_2^- during chemisorption of NH_3 on dehydratyed γ-Al_2O_3 occurs only to a small extent and is attributed to the surface reaction $O^{2-} + NH_3 \rightarrow NH_2^- + OH^-$; most of NH_3 acts as a donor to Al centers; see [11] and the literature cited there. NH_2^- ions at low concentrations were identified mass-spectrometrically after exposing pure Cu and Fe samples to high-frequency sparks. The ions are formed together with other cations and anions from impurities in the metals [12].

Deuterated amides can be prepared by reacting the metals with ND_3 instead of NH_3. A sample of NaNHD was obtained by melting together approximately equal amounts of $NaNH_2$ and $NaND_2$ [13]. The reaction of Ca with a mixture of NH_3 and ND_3 yielded a sample of $Ca(NHD)_2$ [14].

References:

[1] Jander, J. (Chemie in nichtwässrigen ionisierenden Lösungsmitteln, Vol. 1, Pt. 1, Anorganische und allgemeine Chemie in flüssigem Ammoniak, Vieweg, Braunschweig 1966, 1/561, 298).
[2] Jander, J. (from [1], pp. 403/10).
[3] Jackman, D. C.; Keenan, C. W. (J. Inorg. Nucl. Chem. **30** [1968] 2047/57).
[4] Herlem, M. (Bull. Soc. Chim. Fr. **1965** 3329/34).
[5] Schindewolf, U. (Ber. Bunsen-Ges. Phys. Chem. **86** [1982] 887/94).
[6] Pearson, R. G. (J. Am. Chem. Soc. **108** [1986] 6109/14).
[7] Werner, M.; Schindewolf, U. (Ber. Bunsen-Ges. Phys. Chem. **84** [1980] 547/50).
[8] Coulter, L. V.; Sinclair, J. R.; Cole, A. G.; Roper, G. C. (J. Am. Chem. Soc. **81** [1959] 2986/9).
[9] Suzer, S.; Andrews, L. (J. Chem. Phys. **89** [1988] 5347/9).
[10] Windheim, R.; Fischer, F. (Z. Phys. **197** [1966] 309/27; see also Proc. Int. Conf. Lumin., Budapest 1966 [1968], Vol. 1, pp. 877/82; C.A. **70** [1969] No. 52327).
[11] Dunken, H.; Fink, P. (Acta Chim. Acad. Sci. Hung. **53** [1967] 179/92).
[12] Schuy, K. D.; Franzen, J.; Hintenberger, H. (Z. Naturforsch. **19a** [1964] 153/5).
[13] Nibler, J. W.; Pimentel, G. C. (Spectrochim. Acta **21** [1965] 877/82).
[14] Bouclier, P.; Novak, A.; Portier, J.; Hagenmuller, P. (C. R. Seances Acad. Sci. C **263** [1966] 875/8).

2.1.11.2 Molecular Properties and Spectra

Structure of the Ion

The Walsh rules predict C_{2v} symmetry for the amide ion just as for the isoelectronic water molecule [1]. Experimental results confirm the theoretical predictions:

r(NH(D)) in Å	∠HNH(DND)	comment, source of the structural parameters
1.0367(154)	102.0(3.3)°	r_0 structure, from rotational constants of gaseous NH_2^-
1.028	101.9°	estimated r_e structure, derived from the r_0 structure [2]
0.91(3)	113(11)°	average results from an X-ray investigation of $Mg(NH_2)_2$ [3]
1.015(50)	109(18)°	quasi-elastic neutron scattering by solid KNH_2 [4]
0.99(4)	109(6)°	neutron diffraction by $LiND_2$ [5]
0.941(4)	95°	neutron diffraction by $NaND_2$ [6]
0.99(2)	104°	neutron diffraction by KND_2 at 31 K, angle assumed [7]
1.10(2)	107.3(3.7)°	average results from neutron diffraction by $Sr(ND_2)_2$ between 31 and 570 K [8]

Similar structural data for the amide ion were derived from the IR bands of $LiNH_2$ [9] and from the ^1H NMR spectrum of solid KNH_2 taken at 90 K [10]. Calculations based on crystal parameters were done for $NaNH_2$ (see "Natrium" Erg.-Bd. 3, 1966, p. 921), $Ca(NH_2)_2$, and $Sr(NH_2)_2$ [11].

Electronic Structure. Detachment Energy

The ion in the electronic ground state with the electron configuration $(1a_1)^2\ (2a_1)^2\ (1b_2)^2\ (3a_1)^2\ (1b_1)^2$ has $^1A_1$ symmetry. The highest occupied molecular orbital is a nonbonding, lone-pair orbital of nitrogen [1, 2, 12]; see also p. 170.

The energy required to remove an electron from NH_2^- has been determined in several gas-phase photodetachment studies. The experimentally measured, vertical photodetachment energy of NH_2^- is identical with the adiabatic electron affinity of the NH_2 radical, because both species have similar structures and are in their electronic ground state [13, 14]. Numerical values are given in the section on the electron affinity of NH_2 on p. 176.

Dipole and Quadrupole Moment

The dipole moment of amide ions in alkali halide crystals was calculated from the dependence of the UV absorption on the direction of an applied electric field, giving 1.104 ± 0.048 D for NH_2^- and 1.224 ± 0.072 D for ND_2^- in KCl and KBr [15]. Values for KI and Rb salts are two to three times higher, supposedly due to the increased size of the occupied holes [15, 16].

A dipole moment of 0.5491 D was calculated for NH_2^- at the ab initio SCF CI level [17]; similar results are given in [12, 18]. Quadrupole moments, derivatives of dipole moments, and the polarizability were also calculated [17, 18].

Proton and Lithium Cation Affinities

A proton affinity of 1701 ± 11 kJ/mol was derived for gaseous NH_2^- with the photodetachment energy [14]. The experimentally determined standard enthalpy at 298 K of the reaction of NH_2^- with H_2 yielded a proton affinity of 1689 ± 4 kJ/mol for NH_2^- [19]. An earlier calculation with a Born-Haber cycle gave 1590 kJ/mol [20].

The affinity of NH_2^- for Li^+ was calculated at the ab initio SCF level to be −797 kJ/mol [21].

Spin-Spin Coupling

The nuclear spin-spin coupling tensors in NH_2^- were calculated by the equations-of-motion (EOM) approach, yielding the constants $^1J(NH) = -29.31$ and $^2J(HH) = -13.82$ Hz [22]. A coupled Hartree-Fock (CHF) perturbed scheme yielded average values of $^1J(NH) = 37.513$ and $^2J(HH) = -13.709$ Hz (a comparison of the experimental and calculated coupling constants for other species showed that the calculated values are usually 10 to 30% higher than the experimental ones) [23].

Constants of Molecular Rotation and Vibration

Rotational and Centrifugal Distortion Constants. The rotational constants of gaseous NH_2^- were obtained by fitting the rovibrational bands of the high-resolution IR spectrum to a Watson-type S-reduced Hamiltonian. The results in cm^{-1} for the ground state of the ion are $A_0 = 23.0508 \pm 0.0019$, $B_0 = 13.0684 \pm 0.0015$, $C_0 = 8.11463 \pm 0.00048$, $D_J = 0.001082 \pm 0.000022$, $D_{JK} = -0.00381 \pm 0.00012$, $D_K = 0.02065 \pm 0.00013$, $D_1 = -0.000492 \pm 0.000014$, and $D_2 = -0.0000461 \pm 0.0000054$. The corresponding values for the $v_1 = 1$ and the $v_3 = 1$ states and the estimated equilibrium rotational constants of the ground state are also listed. The analysis was restricted to quartic distortion terms, because the inclusion of sextic terms did not result in a better fit of the bands. The Hamiltonian used does not include the effects of the rotational interactions which are noticeable in some bands [2]. Rotational constants of $^{14}NH_2^-$, $^{15}NH_2^-$, $^{14}ND_2^-$, and $^{14}NHD^-$ from CEPA calculations are given in [24].

Fundamentals. The band origins of the symmetric stretch ν_1 at 3121.9306 ± 0.0061 cm^{-1} and of the antisymmetric stretch ν_3 at 3190.291 ± 0.014 cm^{-1} were identified for gaseous NH_2^- in a high-resolution IR spectrum. Differences between the fundamentals of NH_2^- in different salts are attributable to matrix effects in the crystal lattice [2]. The bending vibration $\nu_2 = 1462 \pm 20$ cm^{-1} was predicted for gaseous $^{14}NH_2^-$ from the CEPA equilibrium structure which is close to the experimental one; the fundamentals for $^{15}NH_2^-$, $^{14}ND_2^-$, and $^{14}NHD^-$ were also calculated [24, 25]. An ab initio calculation at the SCF level yielded the harmonic fundamentals of NH_2^- and the deuterated ions, and confirmed the generally low relative intensity of ν_2 and the low intensity of ν_3 of NHD^- [17]; see also [12]. The fundamentals in an Ar matrix at 12 K are ν_2 at 1523 cm^{-1} for $^{14}NH_2^-$ and at 1519 cm^{-1} for $^{15}NH_2^-$ and ν_3 at 3152 and 3145 cm^{-1}, respectively [26].

The fundamentals of the amide ion in some salts are given in Table 19. They were determined mostly at ambient temperature; the band positions of the alkali amides change only slightly upon cooling to 90 K [27]. The bands were assigned assuming a bent NH_2^- ion with C_{2v} geometry [28]. The assignment of the fundamentals, especially in the case of ν_1 and ν_3, is based on the magnitude and direction of the resulting interaction force constant [29] and on a comparison of the bands of protonated amides with those of the deuterated ones [27, 30].

Differences in the band positions of the fundamentals are attributed to the influence of the various cations and also to the environment in the crystal lattices [29]. The occasional split of the fundamentals is probably due to site effects in the crystals [27, 30]. The vibrational bands of amide ions in alkali halide matrices agree qualitatively with those of the pure salts [36].

Table 19
Fundamental Vibrations of Solid and Molten Amides in cm^{-1}.

salt	conditions	ν_1 (A_1, ν_s)	ν_2 (A_1, δ)	ν_3 (B_2, ν_{as})	Ref.
$LiNH_2$	IR in Nujol, Fluorolube	3258 s	1564, 1538 m	3313 m	[27]
	IR in KBr	3261	–	3315	[9]
$NaNH_2$	Raman of the solid	3218	1531	3267	[28]
	Raman, melt at 493 K	3218 p	1550 p	3267 dp	[28]
	IR in Nujol, Fluorolube	3206 s	1529 m	3256 s	[27]
	IR in Nujol	3212.5	1539.5	3263.0	[29]
KNH_2	IR in Nujol, Fluorolube	3210 s	1546 m	3258 s	[27]
$Mg(NH_2)_2$	IR in KBr	3277 s	1577 s	3332, 3326 vs	[31]
$Ca(NH_2)_2$ [a)]	IR in Nujol, Fluorolube	3228 s	1509 s	3290 s	[32]
$Sr(NH_2)_2$	IR in Nujol, Fluorolube	3234 w, 3206 s	1567 w, 1505 w	3292 w, 3267 s	[30]
	IR in KBr	3207 s	1511 m	3262 m	[33]
$Ba(NH_2)_2$	IR in Nujol, Fluorolube	3192 s, 3180 m	1545 s, 1520 sh	3245 s, 3235 m	[30]
$Eu(NH_2)_2$	IR in KBr	3200 vs	1503 s	3263 m	[34]
$Yb(NH_2)_2$	IR in KBr	3210 m	1504 s	3273 s	[34]
	IR in KBr	3370 m	1510 s	3440 m	[35]
	IR in Nujol	3280 m	1510 s	3340 m	[35]
$Yb(NH_2)_3$	IR in KBr	3277 m	1529 s, 1519 sh	3342 m	[34]
NaNHD	IR in Nujol	2389.0	1350.0	3234.5	[29]
$Ca(NHD)_2$ [a)]	IR in Nujol, Fluorolube	2390 s [b)]	1333 s	3261 s	[32]
$LiND_2$	IR in Nujol, Fluorolube	2392 w	1153, 1137 m	2477 w	[27]
$NaND_2$	IR in Nujol	2358.0	1130.5	2428.0	[29]
$Ca(ND_2)_2$ [a)]	IR in Nujol, Fluorolube	2370 w	1114 s	2449 m, 2428 m [b)]	[32]

[a)] Bands of the α phase; the bands of the β phase are also listed in the original paper; they are split and broader. — [b)] Bands of the β phase, not observed for the α phase.

Force Constants. A calculation of the valence bond force constants of the amide ion was based on the fundamentals of the sodium salts of NH_2^-, NHD^-, and ND_2^- and on the estimated structural parameters r(NH) = 1.03 Å and ∠HNH = 104.6°. The results, $f_r = 5.724 \times 10^5$ dyn/cm for the NH valence force constant, $f_\alpha = 0.755 \times 10^{-11}$ erg/rad² for the bending force constant, $f_{rr} = -0.112 \times 10^5$ dyn/cm, and $f_{r\alpha} = -0.313 \times 10^{-3}$ dyn/rad are similar to the values for the isoelectronic water molecule [29]. The value of f_r for the amide ion is intermediate between those for NH_3 and the imide ion [38]. Other calculations on the force constants of NH_2^- in alkali and alkaline earth amides are reported in [32, 37, 46]. Harmonic force constants were calculated with the ab initio SCF method [17].

Quantum-Chemical Calculations

The amide ion has been the subject of several quantum-chemical studies. The more important ones are referenced in conjunction with the individual properties of NH_2^-. For a more complete survey of quantum-chemical calculations and especially those covering computational aspects, consult the bibliography of ab initio calculations given on p. 31.

Spectra

Infrared and Raman Spectra. A high-resolution IR spectrum of gaseous NH_2^- was recorded between 2950 and 3350 cm^{-1} by velocity modulation laser spectroscopy which suppressed the absorption of concomitant neutral species. The rovibrational assignments of 117 bands are listed in [2]. The IR and Raman investigations of solid and molten amides are given in Table 19; the sharp bands indicate that there is no hydrogen bonding among the amide ions [27, 28, 30].

Scattering in the range of lattice vibrations from 280 to 790 cm^{-1} was found to persist in the Raman spectrum after melting $NaNH_2$. Deconvolution of the depolarized, broad band gave three peaks which were attributed to hindered rotational modes of the amide ion about its three principal axes [28].

Bands in the spectra of $Sr(NH_2)_2$ at 4760 cm^{-1} and of $Eu(NH_2)_2$ at 4725 cm^{-1} are attributed to the combination $\nu_2+\nu_3$ [34]. The $\nu_2+\nu_3$ band of $NaND_2$ was observed at 3551.0 cm^{-1} [29]. The near-infrared spectrum of KNH_2 in NH_3 contains a band at 6250 cm^{-1} which was assigned to $2\nu_1$ of NH_2^- [39]. A number of additional overtone and combination bands of NH_2^- were obtained by CEPA calculations [40]. A moderately strong Fermi resonance between $2\nu_2$, calculated to appear at 2894 cm^{-1}, and ν_1 was predicted from CEPA results [25].

Ultraviolet Spectra. Only one absorption above 300 nm is clearly visible in the UV spectrum of amides in NH_3 solution. A second band at 240 nm or shorter wavelengths is obscured by the onset of the absorption of the solvent [41, 42]. At a given temperature the absorption shifts to higher wavelengths as the size of the cation increases from Li to K and then remains constant. A similar solvation of the larger cations (K, Rb, Cs) indicates electrostatically associated ion pairs without appreciable perturbation of the solvent sphere around the anion. The smaller cations, however, have a stronger influence on the solvent sphere of the amide ion [43]. The positions of the maxima of the absorption and extinction coefficients in NH_3 are as follows:

salt	maximum absorption (temperature)	extinction ε in $L\cdot mol^{-1}\cdot cm^{-1}$	Ref.
$LiNH_2$	298 nm (209.2 K), 304 nm (231.8 K)	–	[43]
$NaNH_2$	317 nm (214.5 K), 322 nm (237.9 K)	–	[43]
KNH_2	328 nm (208 K), 357 nm (293 K)	3180 at 293 K	[41]
	~334 nm (213 K), ~339 nm (240 K)	3480 at 224 K	[44]
	330 nm (223 K), 348 nm (293 K)	~3100 at 223 K, ~2600 at 293 K	[42, 45]
$RbNH_2$	330 nm (212.7 K), 337 nm (239.2 K)	–	[43]
$CsNH_2$	329 nm (213.7 K), 337 nm (238.8 K)	–	[43]
KND_2	326 nm (223 K), 344 nm (293 K)	~3600 at 223 K, ~3100 at 293 K	[42]

The maximum of absorption of dissolved KNH_2 shifts by ~0.33 to 0.42±0.02 nm/K to longer wavelengths with increasing temperature. A similar shift supposedly occurs for the obscured second band [41, 42]. The extinction coefficient of KNH_2 remains unchanged upon varying the amide concentration at constant temperature, indicating that the free and the paired anions have the same spectral characteristics [44]. The extinction coefficient of KNH_2 decreases by ~0.4% per degree with increasing temperature [41]. The value of $\varepsilon=2600$ $L\cdot mol^{-1}\cdot cm^{-1}$ at 293 K seems to be the most accurate one, because the concentration of amide was determined by measuring the amount of H_2 evolved [42]. The spectral data indicate that the UV band arises from a process involving a charge transfer to the solvent (charge-transfer-to-solvent transition), i.e., the removal of an electron from the amide ion and transferring it to the first solvation layer [44].

The UV spectra of the amide ion in crystalline alkali halides at 21 K consist of a main band with a shoulder at shorter wavelengths. The maximum of the main band shifts to shorter wavelengths with decreasing lattice constant of the crystal matrix. The positions for NH_2^- (values for ND_2^- in parentheses) in Na and K salts [36] and Rb salts [16] are as follows:

matrix material	NaBr	KCl	KBr	KI	RbCl	RbBr	RbI
main band in nm	246	254 (252)	272 (269)	295 (294)	268	289	308
shoulder in nm	233	239 (241)	255 (256)	273 (277)	254	268	284

The asymmetric shape of the main band results from a third, intermediate band. However, its exact position could not be determined. The positions of the bands and their intensities indicate that the main band results from an electronic excitation of the amide ion, whereas the shoulder originates from a simultaneous excitation of a stretching vibration. The deduced intermediate band probably results from the simultaneous excitation of the deformational vibration of amide [36]. The influence of mechanical strain and exposure to an external electric field on the intensities of the UV bands of amide ions in crystalline alkali halides at helium temperature were also investigated. The amide ion is optically anisotropic; light is absorbed more strongly, when its vector is perpendicular to the plane of the ion than when the vector is in the plane of the ion [15, 16]. The reestablishment of the amide equilibrium orientation after switching off the electric field was studied by electro-birefringence [16].

The reflectance spectra of $Sr(NH_2)_2$ and $Eu(NH_2)_2$ in MgO at ambient temperature exhibit maxima at ~268 and ~278 nm which are accompanied by shoulders at ~247 and ~250 nm, respectively. The bands were assigned to electronic excitation of the amide ion and a combination of electronic and vibrational excitation [34] in agreement with the assignment in [36].

Excitation in the UV bands generates fluorescence in the visible region. There are two principal bands whose relative intensity changes with temperature. They were assigned to purely electronic transitions of the amide ion in different positions within the anion vacancy [36].

References:

[1] Hansen, K. H. (Theor. Chim. Acta **6** [1966] 437/44).
[2] Tack, L. M.; Rosenbaum, N. H.; Owrutsky, J. C.; Saykally, R. J. (J. Chem. Phys. **85** [1986] 4222/7; J. Chem. Phys. **84** [1986] 7056/7).
[3] Jacobs, H. (Z. Anorg. Allg. Chem. **382** [1971] 97/109).
[4] Tielemans, L.; Wegener, W.; Dianoux, A.; Vorderwisch, P.; Van Gerven, L. (AIP Conf. Proc. **89** [1982] 270/2).
[5] Nagib, M.; Jacobs, H. (Atomkernenergie **21** [1973] 275/8).
[6] Nagib, M.; Kistrup, H.; Jacobs, H. (Atomkernenergie **26** [1975] 87/90).
[7] Nagib, M.; Jacobs, H.; von Osten, E. (Atomkernenergie **29** [1977] 303/4).
[8] Nagib, M.; Jacobs, H.; Kistrup, H. (Atomkernenergie **33** [1979] 38/42).
[9] Mason, S. F. (J. Phys. Chem. **61** [1957] 384).
[10] Freeman, R.; Richards, R. E. (Trans. Faraday Soc. **52** [1956] 802/6).
[11] Dufourcq, J.; Chézeau, J. M.; Lemanceau, B. (J. Mol. Struct. **4** [1969] 15/21).
[12] Lee, T. J.; Schaefer, H. F., III (J. Chem. Phys. **83** [1985] 1784/94).
[13] Celotta, R. J.; Bennett, R. A.; Hall, J. L. (J. Chem. Phys. **60** [1974] 1740/5).
[14] Smyth, K. C.; Brauman, J. I. (J. Chem. Phys. **56** [1972] 4620/5).
[15] Windheim, R. (Z. Phys. **215** [1968] 152/76).
[16] Günther, K. (Z. Phys. B **25** [1976] 237/46).
[17] Swanton, D. J.; Bacskay, G. B.; Hush, N. S. (Chem. Phys. **107** [1986] 25/31).

[18] Bishop, D. M.; Pouchan, C. (Chem. Phys. Lett. **139** [1987] 531/4).
[19] Mackay, G. I.; Hemsworth, R. S.; Bohme, D. K. (Can. J. Chem. **54** [1976] 1624/42).
[20] Juza, R. (Z. Anorg. Allg. Chem. **231** [1937] 121/35).

[21] Würthwein, E. U.; Sen, K. D.; Pople, J. A.; Von Ragué Schleyer, P. (Inorg. Chem. **22** [1983] 496/503).
[22] Fronzoni, G.; Galasso, V. (J. Mol. Struct. **122** [1985] 327/31 [THEOCHEM **23**]).
[23] Lazzeretti, P.; Rossi, E.; Taddei, F.; Zanasi, R. (J. Chem. Phys. **77** [1982] 2023/7).
[24] Botschwina, P. (J. Mol. Spectrosc. **117** [1986] 173/4).
[25] Botschwina, P. (NATO ASI Ser. C **193** [1987] 261/70).
[26] Suzer, S.; Andrews, L. (J. Chem. Phys. **89** [1988] 5347/9).
[27] Novak, A.; Portier, J.; Bouclier, P. (C. R. Hebd. Seances Acad. Sci. **261** [1965] 455/7).
[28] Cunningham, P. T.; Maroni, V. A. (J. Chem. Phys. **57** [1972] 1415/8).
[29] Nibler, J. W.; Pimentel, G. C. (Spectrochim. Acta **21** [1965] 877/82).
[30] Bouclier, P.; Novak, A.; Portier, J.; Hagenmuller, P. (C. R. Seances Acad. Sci. C **263** [1966] 875/8).

[31] Linde, G.; Juza, R. (Z. Anorg. Allg. Chem. **409** [1974] 199/214).
[32] Bouclier, P.; Portier, J.; Turrell, G. (J. Mol. Struct. **4** [1969] 1/13).
[33] Linde, G. (Dipl.-Arbeit Univ. Kiel 1966, ref. 15 in [34]).
[34] Hadenfeldt, C.; Jacobs, H.; Juza, R. (Z. Anorg. Allg. Chem. **379** [1970] 144/56).
[35] Warf, J. C.; Gutmann, V. (J. Inorg. Nucl. Chem. **33** [1971] 1583/7).
[36] Windheim, R.; Fischer, F. (Z. Phys. **197** [1966] 309/27; see also Proc. Int. Conf. Lumin., Budapest 1966 [1968], Vol. 1, pp. 877/82; C.A. **70** [1969] No. 52327).
[37] Volka, K.; Ksandr, Z. (Sb. Vys. Sk. Chem.-Technol. Praze Anal. Chem. H **9** [1973] 93/8).
[38] Bouclier, P.; Portier, J.; Hagenmuller, P. (C. R. Seances Acad. Sci. C **268** [1969] 720/3).
[39] Saito, E. (Electrons Fluids Nat. Met.-Ammonia Solutions 3rd Colloq. Weyl, Kibbutz Hanita, Israel, 1972 [1973], pp. 139/44; C.A. **81** [1974] No. 179518).
[40] Botschwina, P. (J. Chem. Soc. Faraday Trans. II **84** [1988] 1263/76).

[41] Corset, J.; Lepoutre, G. (J. Chim. Phys. Phys.-Chim. Biol. **63** [1966] 659/62).
[42] Belloni, J.; Billiau, F.; Saito, E. (Nouv. J. Chim. **3** [1979] 157/61).
[43] Caruso, J. A.; Takemoto, J. H.; Lagowski, J. J. (Spectrosc. Lett. **1** [1968] 311/6).
[44] Cuthrell, R. E.; Lagowski, J. J. (J. Phys. Chem. **71** [1967] 1298/301).
[45] Billiau, F.; Belloni, J.; Saito, E. (Nature [London] **263** [1976] 47/8).
[46] Müller, A.; Kebabcioglu, R.; Krebs, B.; Bouclier, P.; Portier, J.; Hagenmuller, P. (Z. Anorg. Allg. Chem. **368** [1969] 31/5).

2.1.11.3 Thermodynamic Data of Formation. Entropy

The standard enthalpy of formation of gaseous NH_2^-, $\Delta_fH_0^\circ = 117.2 \pm 1.7$ kJ/mol, is based on the most recent photodetachment energy and a value of $\Delta_fH_0^\circ(NH_2) = 191.6 \pm 1.3$ kJ/mol [1]. A value of 106.3 kJ/mol at 298 K was calculated from thermodynamic data of NH_3 and H^+ [2]. A standard entropy of 159 $J \cdot mol^{-1} \cdot K^{-1}$ for gaseous NH_2^- at 240 K was calculated from estimated structural parameters [3]. A value of $S_{298}^\circ = 188.7 \pm 2.1$ $J \cdot mol^{-1} \cdot K^{-1}$ was given in [4]. For data of the NH_3 solution, see p. 268.

References:

[1] Wickham-Jones, C. T.; Ervin, K. M.; Ellison, G. B.; Lineberger, W. C. (J. Chem. Phys. **91** [1989] 2762/3).

[2] Bartmess, J. E.; McIver, R. T., Jr. (in: Bowers, M. T.; Gas Phase Ion Chemistry, Vol. 2, Academic, New York 1979, pp. 87/121, 109).
[3] Coulter, L. V.; Sinclair, J. R.; Cole, A. G.; Roper, G. C. (J. Am. Chem. Soc. **81** [1959] 2986/9).
[4] Benson, S. W. (private communication, ref. 11 in Bohme, D. K.; Hemsworth, R. S.; Rundle, H. W.; J. Chem. Phys. **59** [1973] 77/81).

2.1.11.4 Chemical Behavior

Reactions of the amide ion can be investigated in the gas phase. In the liquid phase, amides in most cases are just applied as standard reactants; experimental proof usually is missing as to whether the amide ion or the undissociated, dissolved salt or maybe even the suspended salt is the active species. The influence of the counterion of amide is, for example, demonstrated by the reaction of solid alkali and alkaline earth amides with gaseous ^{15}NNO, where the ratio of the $^{15}NNN^-$ and $N^{15}NN^-$ formed changes drastically with the size of the cation [1]. A detailed description of the chemistry of amide in the following chapters is restricted to that of the gaseous amide ion for the aforementioned reasons. Reactions in the liquid phase were surveyed because of the possible participation of the amide ion, whereas reactions of solid amides were considered to be attributable to the individual salts and were usually omitted. The older literature on the chemical behavior of the individual amide salts and their reactions with elements, inorganic, and organic compounds is reviewed in the corresponding volumes of the Gmelin Handbook (see p. 251) and in [2 to 4].

References:

[1] Clusius, K.; Schumacher, H. (Helv. Chim. Acta **41** [1958] 972/82).
[2] Bergstrom, F. W.; Fernelius, W. C. (Chem. Rev. **12** [1933] 43/179 and **20** [1937] 413/81).
[3] Levine, R.; Fernelius, W. C. (Chem. Rev. **54** [1954] 449/573).
[4] Jander, J. (Chemie in nichtwässrigen ionisierenden Lösungsmitteln, Vol. 1, Pt. 1, Anorganische und allgemeine Chemie in flüssigem Ammoniak, Vieweg, Braunschweig 1966).

2.1.11.4.1 Thermal Decomposition. Photolysis. Radiolysis

The thermal stability of solid ionic amides varies widely for different cations. Examples are given for each of the principal modes of behavior. KNH_2 distills without decomposition; see "Kalium" 2, 1937, p. 252. Heating $LiNH_2$ below the melting point yields the imide, Li_2NH [1]. The nitride, Ca_3N_2, is obtained by heating $Ca(NH_2)_2$; see "Calcium" B 2, 1957, p. 332. Thermal decomposition of $NaNH_2$ leads to NaH, N_2, and H_2 at normal pressure and to a mixture of the elements at reduced pressure; see "Natrium" Erg.-Bd. 3, 1966, p. 922.

The photolysis of gaseous NH_2^- yields NH_2. The reaction was used to determine the detachment energy; see p. 253. Irradiation of NH_2^- in NH_3 solution at 240 to 400 nm produces solvated electrons and NH_2 which disappears in two stages. The recombination to NH_2^- is fast, and the residual reaction to N_2H_4 or other products is slow. An additional transient of unknown identity was observed during photolysis with unfiltered UV light [2]. The irradiation of amide in alkali halide crystals at the wavelenghts of the UV absorption band results in slow decomposition at 78 K which becomes faster with increasing temperature. The process seems to involve a reaction with H_2 which is also present in the crystal matrix (from amide formation via K-doped alkali halide with NH_3) and probably leads to NH_3, H^-, and the formation of F and U centers. The decomposition of NH_2^- to NH^- and H apparently does not occur [3].

The γ radiolysis of liquid ND_3 containing $NaND_2$ or KND_2 first generates solvated electrons. They react probably with formation of the elemental alkali metals [4]. The yield of H_2 in the presence of amides is smaller than that from pure NH_3, and the experimental ratio of H_2 and N_2 is unsatisfactory [5]. The presence of NH_2^- during γ irradiation of liquid NH_3 leads to a higher stationary concentration of N_2H_4, other products being H_2 and N_2; their yields increase with radiation dosage. The yield of N_2H_4 increases with the concentration of NH_2^- and the pressure of H_2 and N_2 over the solution [6, 7].

References:

[1] Juza, R.; Opp, K. (Z. Anorg. Allg. Chem. **266** [1951] 325/30).
[2] Ottolenghi, M.; Linschitz, H. (Adv. Chem. Ser. **50** [1965] 149/62).
[3] Windheim, R.; Fischer, F. (Z. Phys. **197** [1966] 309/27; see also Proc. Int. Conf. Lumin., Budapest 1966 [1968], Vol. 1, pp. 877/82; C.A. **70** [1969] No. 52327).
[4] Seldon, W. A.; Fletcher, J. W.; Jevcak, J.; Sopchyshin, F. C. (Can. J. Chem. **51** [1973] 3653/61).
[5] Dainton, F. S.; Skwarski, T.; Smithies, D.; Wezranowski, E. (Trans. Faraday Soc. **60** [1964] 1068/86).
[6] Belloni, J.; Fradin de la Renaudière, J. (Int. J. Radiat. Phys. Chem. **5** [1973] 23/30).
[7] Fradin de la Renaudière, J.; Belloni, J. (Int. J. Radiat. Phys. Chem. **5** [1973] 31/9).

2.1.11.4.2 Hydrogen Exchange Reactions

The hydrogen exchange between amide and ammonia in the gas phase at 300±2 K was investigated by the selected-ion flow tube (SIFT) technique. The following reactions and rate constants of amide consumption were found [1]:

reaction	k in $cm^3 \cdot molecule^{-1} \cdot s^{-1}$	ΔH in kJ/mol
$NH_2^- + ND_3 \rightarrow NDH^- + ND_2H$ (56%)	$(5.2 \pm 0.4) \times 10^{-10}$	+3.3
$\quad ND_2^- + NDH_2$ (44%)		+1.7
$ND_2^- + NH_3 \rightarrow NDH^- + NDH_2$ (27%)	$(9.0 \pm 0.8) \times 10^{-10}$	−1.7
$\quad NH_2^- + ND_2H$ (73%)		−3.3

The hydrogen exchange between NH_2^- and liquid NH_3 has a rate constant of $k = 1.5 \times 10^7$ $L \cdot mol^{-1} \cdot s^{-1}$ at 298 K which was determined from the change of the relaxation time in the ^1H NMR spectra with temperature and amide concentration. The activation parameters are $\Delta H = 16.7 \pm 2.1$ kJ/mol and $\Delta S = -51.5 \pm 8.4$ $J \cdot K^{-1} \cdot mol^{-1}$. The large entropy indicates a considerable rearrangement in the solvation sphere when the reactants form the activated complex [2].

References:

[1] Grabowski, J. J.; DePuy, C. H.; Bierbaum, V. M. (J. Am. Chem. Soc. **107** [1985] 7384/9).
[2] Swift, T. J.; Marks, S. B.; Sayre, W. G. (J. Chem. Phys. **44** [1966] 2797/801).

2.1.11.4.3 Electrolysis

Electrolytic oxidation of NH_2^- in NH_3 at 213 K yields N_2H_4 via an unknown intermediate, when a polished Pt electrode is used. The onset of the oxidation wave shifts from −0.50 V for a 0.005 molar solution to −0.90 V for a 1 molar solution. Using a platinized Pt electrode or adding Fe to the solution induces faster decomposition of the unknown intermediate, thus preventing the formation of N_2H_4 [1]. The electrolysis of the molten $NaNH_2$-KNH_2 eutectic mixture with graphite electrodes yields the alkali metals at the cathode. Products at the anode are NH_3, N_2, and a trace amount of N_2H_4 [2].

References:

[1] Herlem, M.; Thiébault, A.; Minet, J.-J. (J. Electroanal. Chem. Interfacial Electrochem. **26** [1970] 343/51).

[2] Macdonald, M. H.; Hill, R. D. (J. Inorg. Nucl. Chem. **15** [1960] 105/9).

2.1.11.4.4 Reactions with Hydrogen

The equilibrium $NH_2^- + H_2 \rightleftharpoons NH_3 + H^-$ in the gas phase was investigated in a flowing afterglow system giving a rate constant of $k = (2.3 \pm 0.1) \times 10^{-11}$ $cm^3 \cdot molecule^{-1} \cdot s^{-1}$ at 297 ± 2 K for the forward reaction. An equilibrium constant of $K = 27 \pm 9$ was calculated using the rate constant of the reverse reaction. This corresponds to $\Delta G^{\circ}_{297} = -7.9 \pm 0.8$ kJ/mol for the forward reaction; $\Delta S^{\circ}_{298} = -18.0 \pm 2.1$ $J \cdot mol^{-1} \cdot K^{-1}$ was calculated from tabulated and estimated entropies. The enthalpy of the reaction is then $\Delta H^{\circ}_{298} = -13.4 \pm 1.3$ kJ/mol [1]. A refined value for the equilibrium constant, $K = 26 \pm 6$ at 296 K, and the same constant for the forward reaction were given later in [2].

Equilibrium constants for $NH_2^- + 1/2\ H_2 \rightleftharpoons e^- + NH_3$ in liquid NH_3 ranging from 1.1×10^{-9} $mol \cdot L^{-1} \cdot bar^{-1}$ at 208 K to 6.6×10^{-7} $mol \cdot L^{-1} \cdot bar^{-1}$ at 238 K were determined electrochemically; a value of $K = 5 \times 10^{-3}$ was extrapolated to 298 K. The standard free enthalpies ΔG° of the forward reaction range from 36 ± 2.5 kJ/mol at 208 K to 28 ± 2.5 kJ/mol at 238 K. Values of $\Delta H^{\circ} = 88 \pm 4$ kJ/mol and $\Delta S^{\circ} = 250 \pm 5$ $J \cdot mol^{-1} \cdot K^{-1}$ were determined from a ΔG°-T plot [3]. Spectrophotometric and ESR investigations [4] and recently measured activity coefficients of KNH_2 solutions in NH_3 were used to calculate an equilibrium constant of $(2 \pm 0.8) \times 10^{-6}$ at 298 K after extrapolation to zero concentration. For the forward reaction, $\Delta G^{\circ}_{298} = 34.3 \pm 2.1$ kJ/mol follows [5]. Earlier determined reaction enthalpies for the forward reaction are 65.7 [4] and 50 ± 4 kJ/mol [6]. The differences in the enthalpies may be due to the temperature dependence of ΔH [3]; experiments in [4] were done at or below ambient temperature and those in [6] at higher temperatures. The solvated electrons, observed by UV spectroscopy [7] in the NH_2^--H_2 equilibrated solution, reduce potassium ions to potassium which can be isolated from the solution as an alloy with Hg or Zn [5].

The hydrogen exchange between liquid NH_3 and dissolved H_2 is catalyzed by amide ions [8]. The dissociated ion was identified to be the active particle on account of the fact that the reaction rate depends on the concentration of the free anion which varies with the counterion and other alkali salts added [9]. Intermediates of the exchange reaction were deduced by the interacting-bond method [10]. The amide ion in liquid NH_3 also catalyzes the conversion of *para*-H_2 to *ortho*-H_2 [8].

The enthalpy and entropy of neutralization of NH_2^- with H^+ in NH_3 at 233 K, -94 kJ/mol and 260 $J \cdot mol^{-1} \cdot K^{-1}$, were calculated from the results of electrochemical and calorimetric measurements [11].

References:

[1] Bohme, D. K.; Hemsworth, R. S.; Rundle, H. W. (J. Chem. Phys. **59** [1973] 77/81).

[2] Mackay, G. I.; Hemsworth, R. S.; Bohme, D. K. (Can. J. Chem. **54** [1976] 1624/42).

[3] Werner, M.; Schindewolf, U. (Ber. Bunsen-Ges. Phys. Chem. **84** [1980] 547/50).

[4] Kirschke, E. J.; Jolly, W. L. (Inorg. Chem. **6** [1967] 855/62).

[5] Warf, J. C. (Inorg. Chem. **21** [1982] 4125/7).

[6] Schindewolf, U.; Vogelsgesang, R.; Böddecker, K. W. (Angew. Chem. **79** [1967] 1064/5; Angew. Chem. Int. Ed. Engl. **6** [1967] 1076/7).
[7] Saito, E. (Electrons Fluids Nat. Met.-Ammonia Solutions 3rd Colloq. Weyl, Kibbutz Hanita, Israel, 1972 [1973], pp. 139/44; C.A. **81** [1974] No. 179518).
[8] Wilmarth, W. K.; Dayton, J. C. (J. Am. Chem. Soc. **75** [1953] 4553/6).
[9] Delmas, R.; Courvoisier, P.; Ravoire, J. (Adv. Chem. Ser. **89** [1969] 25/39).
[10] Medvinskii, A. A.; Bulgakov, N. N.; Baikov, Y. M. (React. Kinet. Catal. Lett. **20** [1982] 267/71).

[11] Schindewolf, U. (Ber. Bunsen-Ges. Phys. Chem. **86** [1982] 887/94).

2.1.11.4.5 Reactions with Inorganic and Organoelement Compounds

2.1.11.4.5.1 Reactions in the Gas Phase

The ion-molecule reactions of collisionally relaxed NH_2^- at 297 to 300 K in the gas phase with inorganic compounds are summarized in Table 20 and those with organoelement compounds in Table 21. The reactions were analyzed with a mass spectrometer [1, 2] in earlier experiments and later by Fourier transform (ion cyclotron resonance) mass spectrometry [3 to 7]. In other, more recent investigations, the flowing afterglow technique was applied and occasionally the newer selected-ion flow tube (SIFT) technique [8, 9], where the NH_2^- formed is separated from the other ions before the reaction. The methods allow the mass spectrometric identification of the anions only; the other products have to be deduced from the mass balance.

Table 20
Ion-Molecule Reactions of NH_2^- with Inorganic Compounds.
Rate constants k in 10^{-10} $cm^3 \cdot molecule^{-1} \cdot s^{-1}$ and estimated heats of reaction Δ_rH in kJ/mol at room temperature.

reactant	main product	by-products, k, Δ_rH	Ref.
CO_2	NCO^- (100%)	k=9.3, $\Delta_rH=-176$	[8, 10]
		k=8.6±1.0	[3]
C_3O_2	NCO^- (60%)	and HC_2O^- (30%), NC_3O^- (10%)	[11]
OCS	H_2NS^- (52%)	and HS^- (42%), NCO^- (6%), k=19	[12]
		$\Delta_rH=\leq 0, -172, -210$, respectively; early results in [1]	[8]
CS_2	NCS^- (54%)	and HS^- (46%), k=18, $\Delta_rH=-293, -176$	[8]
		CS_2^- additionally, k=13.4; see also [2]	[1]
HCN	CN^- (100%)	k=48±5, $\Delta_rH=-226\pm8$	[13]
NH_3	$NH_3 \cdot NH_2^-$	by-product of $NH_3+H^- \rightleftharpoons NH_2^-+H_2$, cf. p. 261	[14]
N_2O	N_3^- (72%)	and OH^- (28%), k=2.9, $\Delta_rH=-234, -38$	[8]
		k≈1	[10]
PH_3	PH_2^-	k=20	[15]
PF_3	$HNPF_2^-$, NPF^-	k=7.9 and 6.6, respectively	[4]
OPF_3	$HN(O)PF_2^-$	k=16.5	[4]
H_2O	OH^-	k=26±2, $\Delta_rH=-50$	[16]
D_2O	OD^- (83%)	and OH^-(17%), k=26±2, $\Delta_rH=-55.6, -57.7$	[9]
SO_2	SO_2^- (66%)	and NSO^- (26%), OH^- (8%), k=29, $\Delta_rH=-29, \leq -192, \leq 0$; early results in [2]	[8]
SF_6	SF_6^-	k=2.0±0.6	[17]
$Fe(CO)_5$	$(CO)_3FeNH_2^-$	k=23	[18]

The products form by charge exchange, proton abstraction, or elimination. Reaction mechanisms were frequently discussed; see for example [4, 8, 9]. Multiple H-D exchange, as in the formation of OH^- from D_2O, was occasionally observed, but its minor importance indicates a short lifetime of the reactive intermediate complex [19]. The average dipole orientation (ADO) theory usually predicts the rate constants satisfactorily; an exception is the value calculated for the reaction with N_2O which is much too small [8]. Calculated structural parameters of the adduct of NH_2^- with NH_3 are given on p. 269.

The primary products from the reactions of thermally equilibrated amide ions with organoelement compounds form via nucleophilic substitution or proton abstraction by amide, or by competition between both reaction channels. Elimination reactions and other secondary processes are observable in some cases; see for example [20 to 22].

Table 21
Ion-Molecule Reactions of NH_2^- with Organoelement Compounds.
Rate constants k in 10^{-9} $cm^3 \cdot molecule^{-1} \cdot s^{-1}$ and estimated heats of reaction Δ_rH in kJ/mol at room temperature.

reactant M	main product	by-products, k, Δ_rH	Ref.
$Si(CH_3)_4$	$Si(CH_3)_2CH_2^-$	$Si(CH_3)_3CH_2^-$	[23]
$Si(CH_3)_3CHCH_2$	$[M-C_2H_4]^-$	$[M-H]^-$, $[M-CH_4]^-$	[23]
$Si(CH_3)_3CH_2CHCH_2$	$[M-H]^-$	exclusive product	[23]
$Si(CH_3)_3OC(CH_3)CH_2$	$CH_3C(CH_2)O^-$	80%, $[M-H]^-$ (20%)	[24]
$P(CH_3)_3$	$P(CH_3)_2CH_2^-$	forms exclusively and irreversibly	[6]
		rapid and quantitative reaction	[25]
$OP(OCH_3)_3$	$O_2P(OCH_3)_2^-$	75%, $OP(OCH_3)_2^-$ (20%), CH_3O^- (5%)	[7]
$S(CH_3)_2$	$CH_3SCH_2^-$	$k=2.7\pm0.4$, $\Delta_rH=-44$	[5]
CD_3SCH_3	$[M-H(D)]^-$	$k_H/k_D=2.43\pm0.07$	[5]
$C_6H_5SCD_3$	$[M-H(D)]^-$	about equal abstraction of H^+ and D^+	[5]
CH_3SSCH_3	$CH_3SSCH_2^-$	67%, CH_3S^- (33%)	[21]
$CH_3OS(O)OCH_3$	$CH_3OSO_2^-$	43.8%, CH_3OSO^- (43.7%), NOS^- (5.4%), SO_2^- (5.2%), $CH_4NO_2S^-$ (1.9%), $k=4.52\pm0.92$	[22]
$CH_3OS(O)OC_2H_5$	$CH_3OSO_2^-$	33.6%, and $C_2H_5OSO_n^-$ with n=1 (22.1%) and n=2 (20.9%), CH_3OSO^- (9.8%), SO_2^- (7.2%), NSO^- (6.4%)	[20]

References:

[1] Dillard, J. G.; Franklin, J. L. (J. Chem. Phys. **48** [1968] 2353/8).
[2] Kraus, K. A.; Müller-Duysing, W.; Neyert, H. (Z. Naturforsch. **16a** [1961] 1385/7).
[3] Wight, C. A.; Beauchamp, J. L. (J. Phys. Chem. **84** [1980] 2503/6).
[4] Sullivan, S. A.; Beauchamp, J. L. (Inorg. Chem. **17** [1978] 1589/95).
[5] Ingemann, S.; Nibbering, N. M. M. (Can. J. Chem. **62** [1984] 2273/81).
[6] Ingemann, S.; Nibbering, N. M. M. (J. Chem. Soc. Perkin Trans. II **1985** 837/40).
[7] Hodges, R. V.; Sullivan, S. A.; Beauchamp, J. L. (J. Am. Chem. Soc. **102** [1980] 935/8).
[8] Bierbaum, V. M.; Grabowski, J. J.; DePuy, C. H. (J. Phys. Chem. **88** [1984] 1389/93).
[9] Grabowski, J. J.; DePuy, C. H.; Bierbaum, V. M. (J. Am. Chem. Soc. **107** [1985] 7384/9).
[10] Bierbaum, V. M.; DePuy, C. H.; Shapiro, R. H. (J. Am. Chem. Soc. **99** [1977] 5800/2).

[11] Roberts, C. R.; DePuy, C. H. (Zh. Obshch. Khim. **59** [1989] 2153/61; J. Gen. Chem. USSR [Engl. Transl.] **59** [1989] 1931/8).
[12] DePuy, C. H.; Bierbaum, V. M. (Tetrahedron Lett. **22** [1981] 5129/30).
[13] Mackay, G. I.; Betowski, L. D.; Payzant, J. D.; Schiff, H. I.; Bohme, D. K. (J. Phys. Chem. **80** [1976] 2919/22).
[14] Bohme, D. K.; Hemsworth, R. S.; Rundle, H. W. (J. Chem. Phys. **59** [1973] 77/81).
[15] Anderson, D. R.; Bierbaum, V. M.; DePuy, C. H. (J. Am. Chem. Soc. **105** [1983] 4244/8).
[16] Betowski, D.; Payzant, J. D.; Mackay, G. I.; Bohme, D. K. (Chem. Phys. Lett. **31** [1975] 321/4).
[17] Streit, G. E. (J. Chem. Phys. **77** [1982] 826/33).
[18] Lane, K. R.; Squires, R. R. (J. Am. Chem. Soc. **108** [1986] 7187/94).
[19] DePuy, C. H. (NATO ASI Ser. C **118** [1984] 227/41).
[20] Lum, R. C.; Grabowski, J. J. (J. Am. Chem. Soc. **110** [1988] 8568/70).

[21] Grabowski, J. J.; Zhang, L. (J. Am. Chem. Soc. **111** [1989] 1193/203).
[22] Grabowski, J. J.; Lum, R. C. (J. Am. Chem. Soc. **112** [1990] 607/20).
[23] DePuy, C. H.; Bierbaum, V. M.; Flippin, L. A.; Grabowski, J. J.; King, G. K.; Schmitt, R. J.; Sullivan, S. A. (J. Am. Chem. Soc. **102** [1980] 5012/5).
[24] Squires, R. R.; DePuy, C. H. (Org. Mass Spectrom. **17** [1982] 187/91).
[25] Grabowski, J. J.; Roy, P. D.; Leone, R. (J. Chem. Soc. Perkin Trans. II **1988** 1627/32).

2.1.11.4.5.2 Survey of the Reactions in Solution

The chemistry of dissolved ionic amides is only briefly reviewed and illustrated by some typical examples because of the uncertainty in identifying the reacting species as explained on p. 259.

The solution of ionic amides in NH_3 is strongly basic and even capable of removing only slightly acidic protons. A solution of $CsNH_2$ deprotonates $[(CH_3)_3SiNH]_2Si(CH_3)_2$ quantitatively with formation of $[(CH_3)_3SiNCs]_2Si(CH_3)_2$ [1]. Initial deprotonation of S_7NH and $S_4N_4H_4$ by KNH_2 in NH_3 solution is the first step in their degradation to $S_2N_2H^-$ [2]. A number of platinum-group complexes with chelating ligands is deprotonated by KNH_2 in NH_3; for example, $[Pt(idn)_2]Cl_2$ (idn = iminodiacetonitrile, $HN(CH_2CN)_2$) is converted to $Pt(idn-H)_2$ [3]. The neutralization of NH_2^- by NH_4^+ salts in NH_3 at 240 K has an average enthalpy of −110.9 kJ/mol and was determined with KNH_2 [4].

The nucleophilicity of ionic amides is demonstrated by substitution and addition reactions. The addition of KNH_2 to CO in NH_3 solution yields $HCONH_2$, HCONHK, KOCN, or $(KNH)_2CO$ depending on the reaction conditions [5]. The reaction of excess MNH_2 (M = K, Rb, Cs) in NH_3 with α-quartz yields crystals of $M_2[SiO_2(NH_2)_2]$ with isolated, monomeric anions [6]. Alkoxides like $Al(OC_2H_5)_3$ or $Ti(OCH_3)_4$ add KNH_2 and yield $KAl(OC_2H_5)_3NH_2$ and $K_2Ti(OCH_3)_4(NH_2)_2$ in NH_3 solution [7]. Substitution by KNH_2 converts Bi(O)I to $Bi(O)NH_2$ in NH_3 solution [8]. Equimolar quantities of KNH_2 and alkoxyphosphorus compounds form the following substitution products [9]:

$$KNH_2 + P(OC_2H_5)_3 \rightarrow KOC_2H_5 + [(C_2H_5O)_2PNH_2] \rightarrow \text{polymer} + NH_3,\ C_2H_5OH$$

$$KNH_2 + OP(OC_2H_5)_3 \rightarrow KO_2P(OC_2H_5)_2 + C_2H_5NH_2$$

Addition besides substitution during the titration of $SiCl_4$ with KNH_2 in NH_3 leads to $K_2Si(NH)_3$ at a reactant ratio of 1:6 and to $K_4Si(NH)_4$ at a reactant ratio of 1:8 [10]. The formation of $Th(NH_2)_2I_2 \cdot NH_3$ from equimolar quantities of KNH_2 and ThI_4 in NH_3 is attributed to initial ammonolysis with formation of $Th(NH_2)I_3$ as an intermediate [11].

The reducing properties of amides in NH_3 are demonstrated by the formation of MnO_2 from $KMnO_4$ with KNH_2 [12] and by the reduction of S_4N_4 with $NaNH_2$ yielding $S_3N_3^-$ and $S_4N_5^-$ [13]. The reduction of NI_3 to N_2 and I^- is attributed to the amide ion, because the sparingly soluble $LiNH_2$ and $Zn(NH_2)_2$ yield only black precipitates, whereas with the more highly soluble $NaNH_2$ and KNH_2 the precipitate forms only intermediately and is quickly converted to a red solution which yields the reduced, final products [14].

The addition of amides to NH_3 solutions induces ammonolysis of phenyl derivatives of elements of the fourth, fifth, and sixth main groups; a survey is given in [15]. Twice the molar quantity of KNH_2 is required to completely convert $Sn(C_6H_5)_4$ to $K_2Sn(NH_2)_6$ [16]. An equimolar quantity of KNH_2 suffices for the ammonolysis of $Pb(C_6H_5)_3c\text{-}C_6H_{11}$; $KPb(NH_2)_3$ can be isolated and results from an additional redox reaction with liberation of N_2 [17]. The addition of $NaNH_2$ to a solution of sulfur in NH_3 induces ammonolysis, deprotonation, and disproportionation equilibria which finally yield $S_3N_3^-$ and S^{2-} [18].

References:

[1] Brauer, D. J.; Bürger, H.; Geschwandtner, W.; Liewald, G. R.; Krüger, C. (J. Organomet. Chem. **248** [1983] 1/15).
[2] Chivers, T.; Schmidt, K. J. (J. Chem. Soc. Chem. Commun. **1990** 1342/4).
[3] Watt, G. W.; Javora, P. H. (J. Inorg. Nucl. Chem. **36** [1974] 1745/50).
[4] Mulder, H. D.; Schmidt, F. C. (J. Am. Chem. Soc. **73** [1951] 5575/7).
[5] Behrens, H.; Ruyter, E. (Z. Anorg. Allg. Chem. **349** [1967] 258/68).
[6] Jacobs, H.; Mengis, H. (Z. Anorg. Allg. Chem. **619** [1993] 303/10).
[7] Schmitz-Du Mont, O. (Angew. Chem. **62** [1950] 560/7).
[8] Watt, G. W.; Fernelius, W. C. (J. Am. Chem. Soc. **61** [1939] 1692/4).
[9] Schmitz-Du Mont, O.; Reckhard, H. (Z. Anorg. Allg. Chem. **294** [1958] 107/12).
[10] Schenk, P. W.; Tripathi, J. B. P. (Angew. Chem. **74** [1962] 116).

[11] Watt, G. W.; Malhotra, S. C. (J. Inorg. Nucl. Chem. **11** [1959] 255/6).
[12] Inoue, T.; Takamoto, S.; Kurokawa, S. (Nippon Kagaku Zasshi **78** [1957] 274/6).
[13] Bojes, J.; Chivers, T.; Drummond, I.; MacLean, J. (Inorg. Chem. **17** [1978] 3668/72).
[14] Jander, J.; Schmid, E. (Z. Anorg. Allg. Chem. **292** [1957] 178/91).
[15] Schmitz-Du Mont, O. (Rec. Chem. Prog. **29** [1968] 13/23).
[16] Schmitz-Du Mont, O.; Müller, G.; Schach, W. (Z. Anorg. Allg. Chem. **332** [1964] 263/8).
[17] Jansen, W.; Nickels, K.-O.; Kessel, H.; Schmitz-Du Mont, O. (Z. Anorg. Allg. Chem. **425** [1976] 272/6).
[18] Dubois, P.; Lelieur, J. P.; Lepoutre, G. (Inorg. Chem. **27** [1988] 3032/8).

2.1.11.4.6 Survey of Reactions with Organic Compounds

2.1.11.4.6.1 Reactions in the Gas Phase

The ion-molecule reactions of collisionally relaxed NH_2^- with typical representatives of organic compounds in the gas phase at ambient temperature are compiled in Table 22. The anions were analyzed by mass spectrometry in early experiments and later by Fourier transform (ion cyclotron resonance) mass spectrometry. More recent investigations usually apply the flowing afterglow technique or its offspring, the selected-ion flow tube (SIFT) technique. These methods allow the identification of anions only; the other products have to be deduced from the mass balance. Rate constants were determined by the flowing afterglow and the SIFT techniques. The products frequently form by proton abstraction which may be followed by elimination or by nucleophilic substitution. Reaction enthalpies and rate constants calculated by the average dipole orientation (ADO) theory are frequently given in the papers.

Table 22
Reactions of NH_2^- with Organic Compounds in the Gas Phase and Rate Constants k of NH_2^- Consumption.

reactant M	main product	k in 10^{-9} $cm^3 \cdot molecule^{-1} \cdot s^{-1}$; by-products; remarks
CH_3F	F^- (100%)	[1]; slow reaction with k = 0.0176 [2]
CH_3Cl	CH_2Cl^- (68%)	Cl^- (32%) [1]; $k = 1.5 \pm 0.2$ [2]
CH_3Br	CH_2Br^- (55%)	Br^- (45%) [1]; $k = 1.1 \pm 0.1$ [2]
CH_3I	I^- (51%)	CH_2I^- [1]; k = 3.1 [3]
CH_3CF_3	F^- (52%)	$CH_2CF_3^-$ (36%), NH_3F^- (12%); the product composition depends on the fluoroethane constitutional isomer [4]
CH_3CN	CH_2CN^- (>90%)	$k = 5.1 \pm 0.2$ [5]
RONO	NO_2^- (100%)	for R = CH_3, n-C_4H_9 [6]
	HN_2O^-	for R = $(CH_3)_3CCH_2$ [6]
CH_3NO_2	$CH_2NO_2^-$ (>95%)	$k = 4.85 \pm 0.46$ [7]
CH_3OCH_3	–	very slow reaction, $k \leq 0.0003$ [8]
$C_2H_5OC_2H_5$	$C_2H_5O^-$ (100%)	k = 0.44, and 0.08 for $C_2D_5OC_2D_5$ [9]
ROR′	RO^-	by attack at a β-hydrogen of the larger alkyl group R′ and its elimination [8]
$C_6H_5OCH_3$	$[M-H]^-$	$C_6H_5^-$ [10]; for fluoroarene ethers, see [11, 12]
$C_6H_5OC_2H_5$	$C_6H_5O^-$ (100%)	[10]
1,4-dioxane	$C_2H_3O^-$ (86%)	by fragmentation, 14% $[M-H]^-$; these products are typical for cyclic ethers [8, 13] and thioethers [14]
ROH, RSH	RO^-, RS^-	with R = CH_3, C_2H_5 [15]
CH_3NH_2	CH_3NH^-	$k \geq 0.1$; equilibrium constant 2.4 ± 0.4 [16]
$(CH_3)_2NH$	$(CH_3)_2N^-$	$k \approx 3$; very slow reaction for $(CH_3)_3N$ [16]
C_2H_4	NHD^-	with ND_2^- at zero kinetic energy; k = 0.027; complete formation of $C_2H_3^-$ with NH_2^- of kinetic energy [17]
C_2H_2	C_2H^-	$k = 1.84 \pm 0.08$ [18]
6,6-dimethyl-fulvene	$[M-H]^-$ (100%)	k = 3.0 [19]; proton abstraction was used to form anions of alkenes and alkines [8]
CH_2O	H^- (>90%)	and possibly HCO^-; $k = 1.9 \pm 0.5$ [20]
C_6H_5CHO	–	$[M-H]^-$, $C_6H_5^-$, H_2NCO^- [21]
$HCOOCH_3$	$[M-H]^-$ (~50%)	CH_3O^- (~30%), $HCONH^-$ (~15%), $HCOO^-$ (3%); k = 2.5; proton abstraction from HCO part is favored [22, 23]
$DCON(CH_3)_2$	$[M-H]^-$ (93%)	7% $[M-D]^-$; k = 4 [22]

References:

[1] Ingemann, S.; Nibbering, N. M. M. (J. Chem. Soc. Perkin Trans. II **1985** 837/40).
[2] Tanaka, K.; Mackay, G. I.; Payzant, J. D.; Bohme, D. K. (Can. J. Chem. **54** [1976] 1643/59).
[3] Bierbaum, V. M.; Grabowski, J. J.; DePuy, C. H. (J. Phys. Chem. **88** [1984] 1389/93).
[4] Sullivan, S. A.; Beauchamp, J. L. (J. Am. Chem. Soc. **98** [1976] 1160/5).
[5] Mackay, G. I.; Betowski, L. D.; Payzant, J. D.; Schiff, H. I.; Bohme, D. K. (J. Phys. Chem. **80** [1976] 2919/22).

[6] King, G. K.; Maricq, M. M.; Bierbaum, V. M.; DePuy, C. H. (J. Am. Chem. Soc. **103** [1981] 7133/40).
[7] Mackay, G. I.; Bohme, D. K. (Int. J. Mass Spectrom. Ion Phys. **26** [1978] 327/43).
[8] DePuy, C. H.; Bierbaum, V. M. (J. Am. Chem. Soc. **103** [1981] 5034/8).
[9] Bierbaum, V. M.; Filley, J.; DePuy, C. H.; Jarrold, M. F.; Bowers, M. T. (J. Am. Chem. Soc. **107** [1985] 2818/20).
[10] Kleingeld, J. C.; Nibbering, N. M. M. (Tetrahedron **39** [1983] 4193/9).

[11] Ingemann, S.; Nibbering, N. M. M.; Sullivan, S. A.; DePuy, C. H. (J. Am. Chem. Soc. **104** [1982] 6520/7).
[12] Ingemann, S.; Nibbering, N. M. M. (J. Org. Chem. **48** [1983] 183/91).
[13] DePuy, C. H.; Beedle, E. C.; Bierbaum, V. M. (J. Am. Chem. Soc. **104** [1982] 6483/8).
[14] Bartmess, J. E.; Hays, R. L.; Khatri, H. N.; Misra, R. N.; Wilson, S. R. (J. Am. Chem. Soc. **103** [1981] 4746/51).
[15] Vogt, D.; Neuert, H. (Z. Phys. **199** [1967] 82/7).
[16] Mackay, G. I.; Hemsworth, R. S.; Bohme, D. K. (Can. J. Chem. **54** [1976] 1624/42).
[17] DePuy, C. H. (NATO ASI Ser. C **118** [1984] 227/41).
[18] Mackay, G. I.; Tanaka, K.; Bohme, D. K. (Int. J. Mass Spectrom. Ion Phys. **24** [1977] 125/36).
[19] Brickhouse, M. D.; Squires, R. R. (J. Am. Chem. Soc. **110** [1988] 2706/14).
[20] Bohme, D. K.; Mackay, G. I.; Tanner, S. D. (J. Am. Chem. Soc. **102** [1980] 407/9).

[21] Kleingeld, J. C.; Nibbering, N. M. M. (Tetrahedron **40** [1984] 2789/94).
[22] DePuy, C. H.; Grabowski, J. J.; Bierbaum, V. M.; Ingemann, S.; Nibbering, N. M. M. (J. Am. Chem. Soc. **107** [1985] 1093/8).
[23] Johlman, C. L.; Wilkins, C. L. (J. Am. Chem. Soc. **107** [1985] 327/32).

2.1.11.4.6.2 Reactions in Solution

Organic reactions in the liquid phase mainly are carried out with $LiNH_2$, $NaNH_2$, or KNH_2. The amides are used in NH_3 which dissolves appreciable quantities of the latter salts or as suspensions in inert hydrocarbons and ethers. A very reactive, pasty mass of $NaNH_2$ is obtained by grinding the amide together with an inert hydrocabon like benzene in a ball mill [1, 2]. The base properties of suspended $NaNH_2$ increase upon adding a half molar amount of alcoholate or enolate which interact coordinatively with the Na^+ ion; aprotic, polar solvents like ethers also enhance its activity as a base [3].

Strongly basic amides are used for many years in organic reactions which involve the removal of a proton from one of the reactants. Anions are obtained in reactions of amides with alcohols, primary and secondary amines, and amides of carboxylic acids. However, reactions involving carbanion intermediates are of greater preparative interest. Formation of enolates from carbonyl compounds and their condensation reactions with elimination of water were most intensively investigated. The enolates of aldehydes undergo self-condensation and yield β-hydroxyaldehydes. Ketone enolates react with esters (Claisen condensation) to form β-diketones, and enolates of esters yield β-ketoesters by self-condensation. Amides also generate carbanions from CH groups which are activated by phenyl substituents or a carbon-carbon triple bond. Both enolate anions and carbanions can be alkylated, arylated, acylated, carbonated, and carbethoxylated [1, 2].

The reaction of amides with alkyl or alkenyl halides leads to the elimination of hydrogen halide and formation of (more highly) unsaturated compounds in mixtures with primary and secondary amides; the composition depends on the structure of the organic moiety in the starting material and on the nature of the halide [1, 2]. Products originating from intramolecular cyclization, like cyclopropanes, can be obtained under appropriate conditions [4]. Aryl halides are converted to aminoarenes. Molecules other than hydrogen halides can also be eliminated; for example, water splits off with formation of the precursor indoxyl in the synthesis of indigo [1, 2].

Nucleophilic substitution by amides converts acid chlorides to acid amides and pyridines to 2-aminopyridines (Tschitchibabin reaction). The amination of quinolines belongs to the few reactions where $Ba(NH_2)_2$ gives better results than the alkali amides [1, 2].

Amides reduce aromatic nitro compounds to azo compounds in the presence of aromatic alcohols. Nitroso compounds are reduced to amines. Sulfones and sulfonates are reduced to sulfur compounds of various oxidation numbers [1, 2].

Nitriles react with amides by initial deprotonation at the α position or by addition of NH_2^- and formation of amidine salts. The extent of each reaction channel depends on the reaction conditions, the acidity of the nitrile, and the existence of fast, secondary reactions [1, 2].

Amides induce anionic polymerization of olefins like styrene, methyl methacrylate, acrylonitrile [5], cyclohexenone [6], and ethylene oxide [7] in solution. The amination of ethene and propene by NH_3 in the gas phase or in NH_3 solution is catalyzed by amides [8].

References:

[1] Bergstrom, F. W.; Fernelius, W. C. (Chem. Rev. **12** [1933] 43/179 and **20** [1937] 413/81).
[2] Levine, R.; Fernelius, W. C. (Chem. Rev. **54** [1954] 449/573).
[3] Caubère, P. (Top. Curr. Chem. **73** [1978] 49/103).
[4] Köster, R.; Arora, S.; Binger, P. (Angew. Chem. **82** [1970] 839/40; Angew. Chem. Int. Ed. Engl. **9** [1970] 810/1).
[5] Evans, M. G.; Higginson, W. C. E.; Wooding, N. S. (Recl. Trav. Chim. Pays-Bas **68** [1949] 1069/78).
[6] Longi, P.; Greco, F.; Mapelli, F. (Chim. Ind. [Milan] **47** [1965] 951/4).
[7] Bailey, F. R., Jr.; France, H. G. (Macromol. Syn. **3** [1968/69] 77/82).
[8] Pez, G. P.; Galle, J. E. (Pure Appl. Chem. **57** [1985] 1917/26).

2.1.11.5 The Amide Ion in Solution

Aqueous Solution. The standard free energy of hydration of NH_2^- is $\Delta G° = -389$ kJ/mol at 298 K and 1 atm for a 1 M solution, calculated from the proton affinity of NH_2^- and thermodynamic data of NH_3. The ionization potential of NH_2^-, 4.72 ± 0.15 eV, was calculated from the value in the gas phase and the free energies of the anion and NH_2 at 298 K [1].

Ammonia Solution. For the ammonolysis of gaseous NH_2^- at 240 K, a value of $\Delta S° = -239.6\ J \cdot K^{-1} \cdot mol^{-1}$ was determined from experiments with $NaNH_2$ and from estimated data [2]. A value of $\Delta H° = -289$ kJ/mol at 233 K for the solvolysis is given in [3]. Similar thermodynamic data for the amide formation in NH_3 solution at 298 K were calculated from data measured on ions in aqueous solution [4].

Born-Haber cycles and experimental data for $NaNH_2$ yield $\Delta_f G°$ values for NH_2^- in NH_3 solution of 127.6 kJ/mol at 240 K and 146.4 kJ/mol at 298 K [2]; a value of $\Delta_f G^o_{298} = 141.0$ kJ/mol was derived from the activity coefficients of KNH_2 in NH_3 [5]. A value of

$\Delta_fH°=41.0\pm1.2$ kJ/mol at 240 K was calculated from the measured heat of amide neutralization and was confirmed by the heat of dissolution of KNH_2 [2].

Partial molar thermodynamic data for the standard state (1 M solution, 1 bar), $\Delta H°=-242$ kJ/mol and $\Delta S°=-39$ $J\cdot K^{-1}\cdot mol^{-1}$ at 233 K, were calculated from electrochemical and calorimetric measurements [3]. Values for $\Delta S°$ of -80.8 $J\cdot K^{-1}\cdot mol^{-1}$ at 240 K and -61.5 $J\cdot K^{-1}\cdot mol^{-1}$ at 298.15 K in an NH_3 solution were derived from the entropy of the gaseous ion [2].

The mobility of NH_2^- in NH_3 at 240 K, 133.0 $cm^2\cdot\Omega^{-1}\cdot Val^{-1}$, is in the range of the mobilities of other anions and was calculated from mobility data and the equivalent ionic conductivity at infinite dilution [6].

References:

[1] Pearson, R. G. (J. Am. Chem. Soc. **108** [1986] 6109/14).

[2] Coulter, L. V.; Sinclair, J. R.; Cole, A. G.; Roper, G. C. (J. Am. Chem. Soc. **81** [1959] 2986/9).

[3] Schindewolf, U. (Ber. Bunsen-Ges. Phys. Chem. **86** [1982] 887/94).

[4] Latimer, W. M.; Jolly, W. M. (J. Am. Chem. Soc. **75** [1953] 4147/8).

[5] Warf, J. C. (Inorg. Chem. **21** [1982] 4125/7).

[6] Kraus, C. A. (The Properties of Electrically Conducting Systems, The Chemical Catalog Co., New York 1922, pp. 63/6 from Jander, J.; Chemie in nichtwässrigen ionisierenden Lösungsmitteln, Vol. 1, Pt. 1, Anorganische und allgemeine Chemie in flüssigem Ammoniak, Vieweg, Braunschweig 1966, p. 170).

2.1.12 Adducts of NH_2^- with NH_3

CAS Registry Number: *[56121-49-4]*

The interaction of low-energy electrons (around 20 V/cm) with NH_3 at 1 Torr results in the formation of an anion mixture containing $[NH_2(NH_3)_n]^-$ with n ranging from 1 to 4. The ion intensities are similar when n=1, 2 and rapidly decrease as n gets higher with an increasing size of the cluster ions [1]. The cluster ions $[NH_2(NH_3)_n]^-$ with n=1 to 3 and $[H(NH_3)_m]^-$ with m=1 to 4 were generated by electron impact on an expanding supersonic jet of NH_3. The signals of the first series were more intense for ions having the same number of nitrogen atoms; the signal intensities decreased rapidly with an increasing number of nitrogen atoms in the ions [2]. The formation of $[NH_2NH_3]^-$ upon adding NH_3 to an He-NH_3 plasma was attributed to the reaction with the intermediate NH_2^- which in turn results from NH_3 and the initial major negative ion H^- [3]. The occurrence of an adduct $[NH_2NH_3]^-$ in crystals of $Cs_4La(NH_2)_7\cdot NH_3$ was discussed because NH_3 and one of the NH_2^- ions occupy a single crystallographic site [4].

An ab initio SCF calculation of the structure of $[H_2N\cdots HNH_2]^-$ predicted a hydrogen-bonded, essentially linear ($\angle$NHN=176°) ion with C_1 symmetry. Internuclear distances of r(N-N)=2.956 and r(N-$H_{bridging}$)=1.047 Å were calculated. The binding energy between NH_2^- and NH_3 was calculated to be 56.5 kJ/mol using the second-order Møller-Plesset perturbation theory (MP2) [5]. Earlier calculations at the ab initio STO-3G and semiempirical MINDO/3 levels also predicted a linear, hydrogen-bridged anion [6]. An anion with the crystallographically observed N-N distance of 3.32 Å, but lacking a hydrogen bridge, was found by the MNDO method [4].

The intramolecular transfer of the bridging proton in $[H_2N\cdots HNH_2]^-$ via tunneling was investigated by an ab initio SCF calculation of the potential energy surface. The calculation yielded a relatively small barrier height of 20.9 kJ/mol; thus, a large tunneling effect can be expected. The accompanying tunneling splitting of the lowest vibrational energy level amounts to 35.5 cm^{-1} [7].

References:

[1] Wilson, J. F.; Davis, F. J.; Nelson, D. R.; Compton, R. N.; Crawford, O. H. (J. Chem. Phys. **62** [1975] 4204/12).
[2] Coe, J. V.; Snodgrass, J. T.; Freidhoff, C. B.; McHugh, K. M.; Bowen, K. H. (J. Chem. Phys. **83** [1985] 3169/70).
[3] Bohme, D. K.; Hemsworth, R. S.; Rundle, H. W. (J. Chem. Phys. **59** [1973] 77/81).
[4] Jacobs, H.; Schmidt, D.; Schmitz, D.; Fleischhauer, J.; Schleker, W. (J. Less-Common Met. **81** [1981] 121/33).
[5] Del Bene, J. E. (J. Comput. Chem. **10** [1989] 603/15).
[6] Minyaev, R. M.; Pavlov, V. I. (J. Mol. Struct. **92** [1983] 205/16 [THEOCHEM **9**]).
[7] Shida, N.; Almlöf, J.; Barbara, P. F. (J. Phys. Chem. **95** [1991] 10457/64).

2.1.13 The NH_2^{2-} Ion

CAS Registry Numbers: NH_2^{2-} *[11089-17-1]*, ND_2^{2-} *[11088-67-8]*

The ions NH_2^{2-} or ND_2^{2-} were obtained together with H_2^- or D_2^- in NH_3 or ND_3 matrices upon irradiation with electrons either during or after condensation at 20 K. Warming to 90 K (NH_3) and 115 K (ND_3) increases the resolution of the spectra. The low concentration of the ESR-active centers is demonstrated by the observation that they do not interact upon varying their concentration by a factor of 100 in the colorless samples. The g-values, splitting constants A in G, and relative electron densities with respect to $|\Psi_H|^2=1$ for H_2^- are as follows:

ion	g	A_H	A_N	$\|\Psi_H\|^2$	$\|\Psi_N\|^2$
NH_2^{2-}	2.0033(5)	22.6(5)	15.7(5)	0.94	9.1
ND_2^{2-}	~2.003	–	~15	–	–

Reference:

Beuermann, G. (Z. Phys. **247** [1971] 25/31).

2.1.14 The Ammonium Radical, NH_4

Other name: Nitrogen tetrahydride

CAS Registry Numbers: NH_4 *[92075-50-8]*, ND_4 *[83682-14-8]*, $^{15}ND_4$ *[92567-98-1]*, NT_4 *[97483-91-5]*, NH_3D *[83937-57-9]*, NH_2D_2 *[83937-56-8]*, NHD_3 *[83937-55-7]*

Literature on the ammonium radical up to February 1936 is covered in an early Gmelin volume "Ammonium" 1936, pp. 6/7. Some later work dealing with the possible existence of a neutral NH_4 radical has been reviewed [1]. NH_4 has been proposed as being formed as an intermediate in the reaction of NH_4^+ with the hydrated [2] or ammoniated [3] electron. The formation of a metallic phase of NH_4 was occasionally discussed; for a more recent paper, see [4]. An ion-beam target-gas study in 1980 gave the first conclusive evidence for the existence of the ammonium radical. In 1981, the radical was identified as the carrier of a band system in an emission spectrum that was obtained following discharge or electron impact excitation of ammonia. See two reviews on the chemical history and experimental and theoretical investigations of the ammonium radical [5] and on its electronic spectrum [6].

The radical has been the subject of about 20 quantum-chemical studies. Only selected ab initio studies that provide lacking or additional information are considered in the following text. For other studies, consult the Quantum Chemistry Bibliography listed on p. 31.

Formation

The first conclusive evidence for the formation of the ammonium radical was obtained in an ion-beam target-gas study. A mass-resolved beam of 5-keV NH_4^+ or ND_4^+ ions was neutralized in collisions with Na or K atoms (at 1 mTorr in the collision chamber), according to electron-transfer neutralization, $NH_4^+(ND_4^+) + M \rightarrow NH_4(ND_4) + M^+$, where M = Na and K. Neutral dissociation products of both NH_4 and ND_4 as well as undissociated ND_4 were detected. The cross sections for the electron-transfer neutralization of NH_4^+ by K, $\sigma = (99 \pm 8) \times 10^{-16}$ cm^2, and by Na, $\sigma = (83 \pm 12) \times 10^{-16}$ cm^2, were determined [7]. The complete series of radicals NH_nD_{4-n} ($n \leq 4$) were similarly generated in the neutralization of 5- to 16-keV ammonium ions $NH_nD_{4-n}^+$ by Na. The kinetic energy release in the fragmentation products was determined. The radicals were found to be bound with respect to a potential minimum and to be separated from the dissociation limit ($NH_3 + H$) by a barrier (0.33 to 0.40 eV). The ND_4 species was found to have a stable ground state at or below the dissociation limit (with a dissociative lifetime $\tau > 20$ μs) besides its unstable state ($\tau < 0.15$ μs) [8]; see also [9]. The ND_4 species generated in the electron-transfer neutralization of an ND_4^+ ion beam by alkali metal vapor (K or Cs) was later detected in emission [10].

A neutralization-reionization study reported the formation of NH_4 in the electron-transfer neutralization of a beam of NH_4^+ ions (6 to 20 keV) by Xe or Kr target-gas atoms. Reionization of the radical in a collision cell ($NH_4 + X \rightarrow NH_4^+ + X + e^-$, where X = Xe or Kr) was used for its detection. A lifetime $\tau(NH_4) > 1$ μs was reported [11] in contradiction to [8].

In additional spectroscopic studies the ammonium radical and its isotopomers were generated using the following methods: electron beam excitation of ammonia (NH_3 or ND_3) gas at pressures up to 1 atm [12, 13]; a flash discharge through a gaseous mixture of ND_3 and D_2 (steady flow of D_2 and ND_3, counter flow of argon, total pressure 0.8 Torr) [14]; a Tesla discharge through gaseous ammonia at pressures up to 0.5 atm [13, 15]; an ozonizer discharge through gaseous ammonia at 100 Torr [13]; a corona-like d.c. discharge through supersonically expanding argon gas containing traces of ND_3 [16]; and the Hg-photosensitized reaction of ND_3 (possibly via a $HgND_3^*$ exciplex) [17]. Spectroscopic observations indicated the dissociative lifetime $\tau(ND_4) < 80$ μs [17] and 30 μs [14]. Theoretical predictions of the lifetimes that were based on a one-dimensional tunneling model were even longer [18].

The formation of ammoniated radical clusters $ND_4(NH_3)_n$ indicates that the radical is stabilized by solvation [19, 20]; see p. 275.

Circumstantial evidence suggesting that NH_4 is generated when NH_3 gas is passed over a heated platinum catalyst [21, 22] has been questioned [5, 7].

The possible formation of NH_4 in the one- and two-photon ionization of ammonia clusters was considered in an ab initio (RHF) study [23].

Molecular Properties

The NH_4 radical is assumed to have a tetrahedral structure (symmetry group T_d, spherical top). The electronic structure is Rydberg in character, that is, resembling that of NH_4^+ with an extra electron in a Rydberg orbital.

The **ground state** will have the electron configuration $(1sa_1)^2$ $(2sa_1)^2$ $(2pf_2)^6$ $(3sa_1)^1$ yielding the electronic state $^2A_1$ that, because of the Rydberg character, is designated 3s $^2A_1$ [15].

The ionization potential $E_i(NH_4) = 4.73 \pm 0.06$ eV was obtained in an ion-beam target-gas experiment [8]. Ab initio values $E_i(NH_4) = 4.56$ to 4.75 eV, $E_i(ND_4) = 4.53$ to 4.57 eV, and $E_i(NT_4) = 4.50$ to 4.55 eV were obtained at the single-double CI level. The range of values results from the different values of the zero-point energies that were used [18]. A neutralization-reionization study showed that the efficiency of the collisional ionization of a fast beam of ND_4 by various target gases decreases in the order $NO_2 > O_2 > N_2 > He$ [24].

The following molecular constants (in cm^{-1}) for ND_4 were based on ab initio calculations [25]: the rotational constant, $B_0 = 3.0406$, and the centrifugal corrections, $D_S^0 = 3.072 \times 10^{-5}$ and $D_T^0 = 1.023 \times 10^{-6}$ [14] (see [26] for the defining equation). A determination of the constants from the spectral data of the observed transition 3p $^2F_2$ − 3s $^2A_1$ is prevented by the $\Delta R = 0$ selection rule [26].

The equilibrium NH bond distance $r_e(NH) = 1.038$ to 1.043 Å [18, 27 to 29] was calculated in ab initio studies at the MP2 (second-order Møller-Plesset perturbation theory) [27, 29], single-double CI (configuration interaction) [18], and CASSCF (complete active space SCF) [28] levels.

The following fundamental frequencies (in cm^{-1}) were obtained by analyses of the 3p $^2F_2$ − 3s $^2A_1$ band system: ND_4, $\nu_1(A_1) = 1964$ [26], $\nu_2(E) = 1080.255(74)$ [6]; NH_4, $\nu_1(A_1) = 2552$, $\nu_2(E) = 1581$ [13]. Theoretical values of the harmonic and anharmonic frequencies were obtained from MP2 [27, 29], single-double CI [18], and CASSCF [28] results.

The **lowest-lying excited state** is the 3p $^2F_2$ Rydberg state with the outer electron in a $3pf_2$ orbital. It is triply degenerate and thus Jahn-Teller active [26, 30].

Molecular constants for the 3p $^2F_2$ state of ND_4 were obtained, according to the theory of rotational structure given in [26], in a rotational analysis [14] of the 3p $^2F_2$ − 3s $^2A_1$ band observed in a rotationally-cooled emission spectrum [16] and in a flash-discharge absorption spectrum [14]. Theoretical values [14] of the ground state constants were used (see above). The term energy $T_0 = 14828.2853(36)$ cm^{-1} and the splitting between the spin-orbit components $T_0(G_{3/2}) - T_0(E_{5/2}) = 3/2\ \xi_0 = 9.15495(39)$ cm^{-1} were determined. Preliminary estimates of the Jahn-Teller constants indicate appreciable distortion. For NH_4, $T_0 \approx 15078$ cm^{-1} was estimated by scaling the molecular constants of ND_4 [14].

The lifetime of the 3p $^2F_2$ state of ND_4, $\tau = 4.2$ ns, was determined in an ion-beam target-gas study. Predicted values of the radiative lifetime, 22 to 28 ns, obtained [10] using ab initio results [25, 31, 32], are much longer, indicating that ND_4 decays primarily by predissociation [10].

The energy ordering of the **low-lying Rydberg states** of NH_4 [27, 33 to 35] were calculated in ab initio studies using methods that include correlation [27, 33] and an SCF scattered-wave procedure [34, 35]. In a manifold of states calculated at a single-double CI level, the 3p $^2F_2$, 4s $^2A_1$, 3d $^2F_2$, and 3d ^2E states were found to lie 1.72, 2.59, 2.74, and 2.90 eV, respectively, above the electronic ground state [27].

For the **dissociation** behavior, see experimental work cited in the section on Formation. For theoretical work, see [18, 36 to 38].

Spectra

Emission Spectrum. The first observations of the spectra of NH_4 and ND_4 were made by Schüler and coworkers in the emission obtained following electron impact excitation of NH_3 and of ND_3 gas at pressures up to 1 atm [12]. The observed band systems were assigned conclusively to the NH_4 and ND_4 radicals on the basis of high-resolution emission spectra that were obtained following discharge or electron beam excitation of ammonia and of ammonia isotope mixtures at pressures up to 0.66 atm [13, 15].

The band system is now assigned to the 3p $^2F_2 \rightarrow$ 3s $^2A_1$ electronic transition of the ammonium radical. The band system of ND_4 (540 to 790 nm) shows a sharp principal band that is located at ~674.8 nm. The principal band of NH_4 is diffuse and located at ~663.6 nm. The principal bands of NH_4 and ND_4, assigned to the 0-0 transition, show characteristic prominent double peaks, separated by 6.92 (NH_4) and by 6.24 cm^{-1} (ND_4) [15], that probably result from the spin splitting of the upper state. Above and below the principal band are other bands that show double or multiple peaks; see, for example [13]. In the case of ND_4, two of these bands, situated 1100 cm^{-1} below the principal band, were assigned to the 0-1 bands in the bending vibrational modes $\nu_2(E)$ and $\nu_4(F_2)$. The assignment of the band system is complicated by the Jahn-Teller distortion of the upper state [6].

A rotationally-cooled emission spectrum of ND_4 was obtained following a corona-like d.c. discharge through supersonically expanding argon gas containing a trace of ND_3 [16]. A relative high rotational temperature of 250 K was estimated for this spectrum [5].

The assignment of the ammonium radical to a second band system (principal band at 564 nm (NH_4), 580 nm (ND_4)) [15, 39] in an emission spectrum obtained following discharge through gaseous ammonia (first reported by Schuster in 1872) [12] has been discarded. In the meantime, ammonia was proposed as the carrier of this band system [40]. For further discussion, see also [41].

Absorption Spectrum. ND_4 absorption spectra showing the principal band were obtained [14, 17] using a combination of laser frequency modulation (FM) spectroscopy with photochemical modulation in the mercury-photosensitized reactions of ND_3 [17] or using a flash discharge technique in ND_3-D_2 mixtures at 0.8 Torr [14]. A spectrum with 250 lines [14], including the 21 lines of the laser FM spectrum [17], was obtained in the flash discharge study. Observed line frequencies and assignments are given [14]. A detailed rotational analysis of the spectrum [14], according to the theory of rotational structure of a p $^2F_2 \leftarrow$ s $^2A_1$ transition developed in [26, 30], corroborated the assignment to the electronic transition 3p $^2F_2 \leftarrow$ 3s $^2A_1$ [14].

Extensive ab initio calculations on the vibrational transitions of the 3p $^2F_2$-3s $^2A_1$ band were performed at the MP2 [27] and single-double CI [18] levels of calculation. The anharmonicity of the vibrations [18, 27] and the Jahn-Teller distortion in the upper state [27] were taken into account. The principal band of NT_4 was predicted to appear at ~14700 cm^{-1} [18].

Auger Spectrum. The Auger spectrum of NH_4 was calculated using a scattered-wave (discrete-variational Xα) method. In the low-energy region it was found to be similar to an experimental N KVV Auger spectrum of NH_4Cl [42, 43].

References:

[1] Wan, J. K. S. (J. Chem. Educ. **45** [1968] 40/3).
[2] Anbar, M. (Adv. Chem. Ser. No. **50** [1965] 55/81, 66).
[3] Brooks, J. M.; Dewald, R. R. (J. Phys. Chem. **75** [1971] 986/7).
[4] Johnson, D. A. (Chem. Soc. Dalton Trans. **1988** 445/50).
[5] Kaspar, J.; Smith, V. H., Jr.; McMaster, B. N. (Appl. Quantum Chem. Proc. Nobel Laureate Symp., Honolulu 1984, [1986], pp. 403/20).
[6] Watson, J. K. G. (NBS Spec. Publ. [U. S.] No. 716 [1986] 650/70).
[7] Williams, B. W.; Porter, R. F. (J. Chem. Phys. **73** [1980] 5598/604).
[8] Gellene, G. I.; Cleary, D. A.; Porter, R. F. (J. Chem. Phys. **77** [1982] 3471/7).
[9] Jeon, J. S.; Raksit, A. B.; Gellene, G. I.; Porter, R. F.; Garver, W. P.; Burkhardt, C. E.; Leventhal, J. J. (NBS Spec. Publ. [U. S.] No. 716 [1986] 364/85).
[10] Ketterle, W.; Grasshoff, P.; Figger, H.; Walther, H. (Z. Phys. D At. Mol. Clusters **9** [1988] 325/9).

[11] Griffiths, W. J.; Harris, F. M.; Beynon, J. H. (Int. J. Mass Spectrom. Ion Processes **77** [1987] 233/9).
[12] Schüler, H.; Michel, A.; Grün, A. E. (Z. Naturforsch. **10a** [1955] 1/2).
[13] Herzberg, G. (J. Astrophys. Astron. **5** [1984] 131/8).
[14] Alberti, K.; Huber, K. P.; Watson, J. K. G. (J. Mol. Spectrosc. **107** [1984] 133/43).
[15] Herzberg, G. (Faraday Discuss. Chem. Soc. No. 71 [1981] 165/73).
[16] Huber, K. P.; Sears, T. J. (Chem. Phys. Lett. **113** [1985] 129/34).
[17] Whittaker, E. A.; Sullivan, B. J.; Bjorklund, G. C.; Wendt, H. R.; Hunziker, H. E. (J. Chem. Phys. **80** [1984] 961/2).
[18] Kaspar, J.; Smith, V. H., Jr.; McMaster, B. N. (Chem. Phys. **96** [1985] 81/95).
[19] Gellene, G. I.; Porter, R. F. (J. Phys. Chem. **88** [1984] 6680/4).
[20] Jeon, S. J.; Raksit, A. B.; Gellene, G. I.; Porter, R. F. (J. Am. Chem. Soc. **107** [1985] 4129/33).

[21] Melton, C. E.; Joy, H. W. (J. Chem. Phys. **46** [1967] 4275/83).
[22] Melton, C. E.; Joy, H. W. (J. Chem. Phys. **48** [1968] 5286/7).
[23] Cao, H. Z.; Evleth, E. M.; Kassab, E. (J. Chem. Phys. **81** [1984] 1512/3).
[24] Gellene, G. I.; Porter, R. F. (Int. J. Mass Spectrom. Ion Processes **64** [1985] 55/66).
[25] Havriliak, S.; King, H. F. (J. Am. Chem. Soc. **105** [1983] 4/12).
[26] Watson, J. K. G. (J. Mol. Spectrosc. **107** [1984] 124/32).
[27] Cardy, H.; Liotard, D.; Dargelos, A.; Marinelli, F.; Roche, M. (Chem. Phys. **123** [1988] 73/83).
[28] Gutowski, M.; Simons, J. (J. Chem. Phys. **93** [1990] 3874/80).
[29] Boldyrev, A. I.; Simons, J. (J. Phys. Chem. **96** [1992] 8840/3; Erratum: J. Phys. Chem. **97** [1993] 1470).
[30] Watson, J. K. G. (J. Mol. Spectrosc. **103** [1984] 125/46).

[31] Raynor, S.; Herschbach, D. R. (J. Phys. Chem. **86** [1982] 3592/8).
[32] Hirao, K. (J. Am. Chem. Soc. **106** [1984] 6283/5).
[33] Havriliak, S.; Furlani, T. R.; King, H. F. (Can. J. Phys. **62** [1984] 1336/46).
[34] Kaspar, J.; Smith, V. H., Jr. (Chem. Phys. **90** [1984] 47/53).
[35] Broclawik, E.; Mrozek, J.; Smith, V. H., Jr. (Chem. Phys. **66** [1982] 417/23).
[36] Kassab, E.; Evleth, E. M. (J. Am. Chem. Soc. **109** [1987] 1653/61).
[37] Cardy, H.; Liotard, D.; Dargelos, A.; Poquet, E. (Chem. Phys. **77** [1983] 287/99).
[38] McMaster, B. N.; Mrozek, J.; Smith, V. H., Jr. (Chem. Phys. **73** [1982] 131/43).
[39] Herzberg, G.; Hougen, J. T. (J. Mol. Spectrosc. **97** [1983] 430/40).
[40] Watson, J. K. G.; Majewski, W. A.; Glownia, J. H. (J. Mol. Spectrosc. **115** [1986] 82/7).

[41] Ashfold, M. N. R.; Bennnett, C. L.; Dixon, R. N.; Fielden, P.; Rieley, H.; Stickland, R. J. (J. Mol. Spectrosc. **117** [1986] 216/27).
[42] Mikhailov, G. M.; Gutsev, G. L.; Borod'ko, Y. G. (Khim. Fiz. **3** [1984] 672/9).
[43] Gutsev, G. L. (Teor. Eksp. Khim. **21** [1985] 10/8; Theor. Expt. Chem. [Engl. Transl.] **21** [1985] 9/16).

2.1.15 Adducts of NH_4 with NH_3

Cluster radicals of the type $NH_4 \cdot (NH_3)_n$, n=1 to 3, and their deuterated analogs were generated using an ion-beam target-gas technique. A 6-keV, mass-resolved beam of $NH_4 \cdot (NH_3)_n^+$ ions, produced by chemical ionization of NH_3, was neutralized in collision with a target metal gas at a few mTorr, according to $NH_4 \cdot (NH_3)_n^+ + M(g) \rightarrow NH_4 \cdot (NH_3)_n + M^+$, where M = Na or K. Neutral beam scattering profiles and collisional reionization mass spectra were observed. Lifetimes of $NH_4 \cdot NH_3$ and $ND_4 \cdot NH_3$ were found to be greater than 1 µs, showing that NH_4 is stabilized by complexation (solvation). The estimated ionization potentials of the cluster radicals were found to decrease with increasing n [1, 2]: $NH_4 \cdot NH_3$, $E_i = 4.0 \pm 0.2$ eV; $NH_4 \cdot (NH_3)_2$, $E_i = 3.5 \pm 0.2$ eV [2]; and $NH_4 \cdot (NH_3)_3$, $E_i = 2.4$ to 2.6 eV [1].

The efficiency of the collisional ionization of a fast beam of $ND_4 \cdot NH_3$ by various target gases was found to decrease in the order $NO_2 > O_2 > N_2 > He$ [3].

Solvation energies for the reaction $NH_4 \cdot (NH_3)_{n-1} + NH_3 \rightarrow NH_4 \cdot (NH_3)_n$, n=1 to 6, were calculated in a quantum mechanical study at the Hartree-Fock level. The NH_4 radical was found to be more strongly solvated (0.3 to 0.4 eV) up to the first solvation shell ($n \leq 4$) with NH_3 than NH_3 is with itself (0.25 eV) [4, 5].

The protonation of $NH_4 \cdot NH_3$ to give $N_2H_8^+$ ($= NH_4^+NH_4$) was considered in a quantum-chemical study at the Hartree-Fock level [6].

References:

[1] Gellene, G. I.; Porter, R. F. (J. Phys. Chem. **88** [1984] 6680/4).
[2] Jeon, S. J.; Raksit, A. B.; Gellene, G. I.; Porter, R. F. (J. Am. Chem. Soc. **107** [1985] 4129/33).
[3] Gellene, G. I.; Porter, R. F. (Int. J. Mass Spectrom. Ion Processes **64** [1985] 55/66).
[4] Kassab, E.; Evleth, E. M. (J. Am. Chem. Soc. **109** [1987] 1653/61).
[5] Evleth, E. M.; Kassab, E. (Pure Appl. Chem. **60** [1988] 209/14).
[6] Kassab, E.; Fouquet, J.; Evleth, E. M. (Chem. Phys. Lett. **153** [1988] 522/6).

2.1.16 $(NH_4)_2$ and $(NH_4)_2^+$

CAS Registry Numbers: $(NH_4)_2$ *[144181-08-8]*, $(NH_4)_2^+$ *[144191-77-5]*

An extensive quantum-mechanical study that included correlation by second-order Møller-Plesset perturbation and quadratic configuration-interaction methods showed that the cluster species $(NH_4)_2$ is bound with respect to 2 NH_4 by 7.5 to 9.7 kcal/mol, but unstable with respect to 2 $NH_3 + H_2$ by 86 to 89 kcal/mol. The bonding results from the interaction of the $3sa_1$ Rydberg orbitals of NH_4. A minimum-energy structure of C_{2h} symmetry was predicted.

The same study found that the cation $(NH_4)_2^+$ is bound with respect to $NH_4^+ + NH_4$ by 20 kcal/mol. A minimum-energy structure of D_{3d} symmetry was predicted.

Reference:

Boldyrev, A. I.; Simons, J. (J. Phys. Chem. **96** [1992] 8840/3; Erratum: J. Phys. Chem. **97** [1993] 1470).

2.1.17 The Nitrogen Tetrahydride Anion, NH_4^-

CAS Registry Number: *[12325-21-2]*

The NH_4^- ion most probably exists both as a stable $H^- \cdot NH_3$ ion-molecule complex and as a tetrahedral species with a geometry very similar to that of NH_4^+, i.e., as a so-called double-Rydberg anion $\{NH_4^+\}^{2-}$.

The $H^- \cdot NH_3$ complex has long been thought to occur in liquid-ammonia solutions as an intermediate in the proton-transfer reaction $NH_2^- + H_2 \rightleftharpoons H^- + NH_3$ [1 to 4].

Gaseous NH_4^- ions were generated in a Fourier transform ion cyclotron resonance (ICR) spectrometer by the reaction of formaldehyde with NH_2^- [5] via the steps

$$NH_2^- + CH_2O \rightarrow NH_3 + CHO^-$$

$$CHO^- + NH_3 \rightarrow CO + NH_4^-$$

Deuterium-labeling experiments on the formation of NH_4^-, its reaction with formaldehyde ($NH_4^- + CH_2O \rightarrow NH_3 + CH_3O^-$) [5], and on the collision-induced dissociation and charge reversal of NH_4^- [6] indicated that the four H atoms were not equivalent and that the anion is best described as a H^- ion solvated by an ammonia molecule, i.e., $H^- \cdot NH_3$.

The gaseous NH_4^- ion was also investigated by negative-ion photoelectron spectroscopy (PES) [7, 8]. An expanding supersonic jet of ammonia was exposed to electron impact and a beam of NH_4^- ions was mass-selected and crossed with a laser beam. The PE spectrum was recorded at photon energies of 2.409, 2.497, 2.540, and 2.707 eV. It consisted of a large peak (A) with a smaller peak (B) on the side of lower electron kinetic energies and a third, very small peak (C) shifted to higher electron kinetic energies. The following electron binding energies E_i in (eV) were derived:

peak	C	A	B
E_i for NH_4^-	0.472	1.109	1.541
E_i for ND_4^-	0.474	1.121	1.427

The main peak A was ascribed to the removal of an electron from the hydride part of the $H^- \cdot NH_3$ ($D^- \cdot ND_3$) complex; it resembled the photoelectron spectrum of the free hydride ion shifted to higher electron binding energies due to the stabilizing effect of solvation. The energy separations between peaks A and B, 3480 ± 40 cm^{-1} for NH_4^- and 2470 ± 40 cm^{-1} for ND_4^-, are close to the stretching frequencies of NH_3 and ND_3. Peak B, therefore, was ascribed to the excitation of a stretching mode in the NH_3 (ND_3) part during photodetachment from H^- (D^-). The unusually narrow peak C, the position of which was not significantly affected upon deuteration, was ascribed to the detachment of an electron from a tetrahedral isomer of NH_4^- (ND_4^-) [8].

The structures and stabilities of the two NH_4^- isomers have been studied theoretically with various quantum-chemical methods. Geometry optimizations for the $H^- \cdot NH_3$ species by the CEPA [9], MP2 and SDQ-CI [10], MP2 [11, 12], SD-CI [13], and MBPT(2) [14] methods all agreed on a structure of C_s symmetry in which the H^- ion is bound nearly linearly to one of the N-H bonds of the slightly distorted NH_3 molecule with a small tilt (maximal 15°) toward the other hydrogen atoms. The relatively large H-H^- distance (around 2 Å) indicates a bond arising from dipole-dipole attraction. Values between 25 and 63 kJ/mol were obtained for the binding energy in the H-H$^-$ bond (or stability with respect to the isolated parts $H^- + NH_3$). The studies of [9, 13] gave only a stable $H^- \cdot NH_3$ species of C_s symmetry, whereas the remaining ones [10 to 12, 14] found also minima on the potential energy hypersurface for an NH_4^- ion of T_d symmetry. The latter isomer was indeed calculated to be less stable than the dissociation products $H^- + NH_3$ by 5.4 kJ/mol [11] or 26 kJ/mol [10] and less stable than the solvated-hydride structure by 45.3 kJ/mol [12] or 40.5 kJ/mol [14]. However, a high potential barrier of ~78 kJ/mol along the C_{3v} dissociation pathway from $NH_4^-(T_d)$ to $H^- + NH_3$ was found by [10], which implies a sufficiently long lifetime for the T_d species to be experimentally observable. Electron propagator theory was applied to calculate the vertical ionization energies of NH_4^- in the two stable geometries; the results, $E_i(C_s) = 1.20$ eV and $E_i(T_d) = 0.42$ eV, are within 0.1 eV of the PES peaks and thus confirm the assignments given above [14, 15].

Ab initio CI calculations on a few examples of double-Rydberg ions included the tetrahedral NH_4^- species. The ion was shown to be locally stable at a bond length of 1.039 Å with fundamental vibrations (in cm^{-1}) $\nu_1(A_1) = 3140$, $\nu_2(E) = 1717$, $\nu_3(T_2) = 3276$, $\nu_4(T_2) = 1434$, and to be electronically stable by 0.45 eV ($= E_i$) with respect to the neutral NH_4 and by 4.48 eV with respect to the cation NH_4^+ [16, 17]. An MBPT(2) calculation gave $r_e = 1.045$ Å and $\nu_1 = 2971$, $\nu_2 = 1579$, $\nu_3 = 3112$, and $\nu_4 = 1323$ cm^{-1} [15]. An earlier MINDO study of electron-rich hydrides included the tetrahedral NH_4^- ion [18].

The N KVV Auger transitions (V = valence orbitals $3a_1$, $1t_2$, and $2a_1$) in tetrahedral NH_4^- (and in NH_4 and NH_4^+) were calculated by the Xα method and used for interpreting the experimental Auger spectra of NH_4Cl [19].

References:

[1] Wilmarth, W. K.; Dayton, J. C. (J. Am. Chem. Soc. **75** [1953] 4553/6).

[2] Bar-Eli, K.; Klein, F. S. (J. Chem. Soc. **1962** 1378/85).

[3] Dirian, G.; Botter, F.; Ravoire, J.; Grandcollot, P. (J. Chim. Phys. Phys. Chim. Biol. **60** [1963] 139/47).

[4] Delmas, R; Courvoisier, P.; Ravoire, J. (J. Chim. Phys. Phys. Chim. Biol. **62** [1965] 1423/5; Adv. Chem. Ser. **89** [1969] 25/39).

[5] Kleingeld, J. C.; Ingemann, S.; Jalonen, J. E.; Nibbering, M. N. N. (J. Am. Chem. Soc. **105** [1983] 2474/5).

[6] De Lange, W.; Nibbering, M. N. N. (Int. J. Mass Spectrom. Ion Processes **80** [1987] 201/9).

[7] Coe, J. V.; Snodgrass, J. T.; Freidhoff, C. B.; McHugh, K. M.; Bowen K. H. (J. Chem. Phys. **83** [1985] 3169/70).

[8] Snodgrass, J. T.; Coe, J. V.; Freidhoff, C. B.; McHugh, K. M.; Bowen, K. H. (Faraday Discuss. Chem. Soc. No. 86 [1986] 241/56).

[9] Kalcher, J.; Rosmus, P.; Quack, M. (Can. J. Phys. **62** [1984] 1323/7).

[10] Cardy, H.; Larrieu, C.; Dargelos, A. (Chem. Phys. Lett. **131** [1986] 507/12).

[11] Cremer, D.; Kraka, E. (J. Phys. Chem. **90** [1986] 33/40).

[12] Kos, A. J.; Schleyer, P. v. R.; Pople, J. A.; Squires, R. R. (unpublished work, from [8, 11]).

[13] Hirao, K.; Kawai, E. (J. Mol. Struct. **149** [1987] 391/4 [THEOCHEM **34**]).

[14] Ortiz, J. V. (J. Chem. Phys. **87** [1987] 3557/62).

[15] Ortiz, J. V. (J. Phys. Chem. **94** [1990] 4762/3).

[16] Gutowski, M.; Simons, J.; Hernandez, R.; Taylor, H. L. (J. Phys. Chem. **92** [1988] 6179/82).

[17] Gutowski, M.; Simons, J. (J. Chem. Phys. **93** [1990] 3874/80).

[18] Glidewell, C. (J. Mol. Struct. **67** [1980] 121/32).

[19] Mikhailov, G. M.; Gutsev, G. L.; Borod'ko, Yu. G. (Khim. Fiz. **3** [1984] 672/9; C.A. **101** [1984] No. 140363).

2.1.18 Nitrogen Pentahydride, NH_5

CAS Registry Number: *[73655-04-6]*

The existence of pentacoordinated NH_5 as an intermediate was proposed in a study on the reaction of ammonium trifluoroacetate with lithium hydride in the melt. Observed isotopic scramblings of the reaction products from reactions carried out using deuterated reagents are consistent with the formation of pentacoordinated nitrogen hydride, according to $NH_4O_2CCF_3 + LiH \rightarrow LiO_2CCF_3 + [NH_5]$, followed by $[NH_5] \rightarrow NH_3 + H_2$ [1].

Semiempirical [1] and ab initio (SCF [2, 3], and MP2 (second-order Møller-Plesset perturbation theory) [4]) calculations have focused on the thermodynamic stability of NH_5 structures of different symmetries. All structures considered were predicted to be thermodynamically unstable with respect to $NH_3 + H_2$ [2, 3].

References:

[1] Olah, G. A.; Donovan, D. J.; Shen, J.; Klopman, G. (J. Am. Chem. Soc. **97** [1975] 3559/61).
[2] Gründler, W.; Schädler, H. D. (Z. Chem. **20** [1980] 111)
[3] Morosi, G.; Simonetta, M. (Chem. Phys. Lett. **47** [1977] 396/8).
[4] Ewig, C. S.; Van Wazer, J. R. (J. Am. Chem. Soc. **111** [1989] 4172/8).

2.1.19 The Nitrogen Pentahydride Cation, NH_5^+

CAS Registry Number: *[62377-38-2]*

Ab initio studies at the SCF level of calculation are contrary in their predictions on the thermodynamic stability of NH_5^+ [1, 2].

A stabilization of NH_5^+ by solvation was suggested in a study of reactions following electron impact ionization of neat cluster beams of ammonia and of heteroclusters of ammonia-benzene. Peaks in the mass spectrum were assigned to $[NH_5(NH_3)_n]^+$ adducts, where $n \geq 4$. Analogous results were obtained with deuterated ammonia [3]. An ab initio SCF MO study on ammoniated NH_5^+ adducts [2] indicated that the species observed in the mass spectrum [3] is possibly ammoniated $N_2H_8^+$ ($= NH_4^+NH_4$).

References:

[1] Bugaets, O. P.; Zhogolev, D. A. (Chem. Phys. Lett. **45** [1977] 462/5).
[2] Kassab, E.; Fouquet, J.; Evleth, E. M. (Chem. Phys. Lett. **153** [1988] 522/6).
[3] Garvey, J. F.; Bernstein, R. B. (Chem. Phys. Lett. **143** [1988] 13/8).

2.1.20 The Nitrogen Pentahydride Dianion, NH_5^{2-}

CAS Registry Number: *[76014-66-9]*

A C_{4v} structure, more stable than a D_{3h} structure, was predicted in a semiempirical MO study (MINDO/3) that assumed the non-bonding electron pair is localized on the ntirogen atom.

Reference:

Glidewell, C. (J. Mol. Strcut. **67** [1980] 121/32).

Physical Constants and Conversion Factors

Avogadro constant N_A (or L) $= 6.02214 \times 10^{23}$ mol^{-1}

Faraday constant $F = 9.64853 \times 10^{4}$ C/mol

molar gas constant $R = 8.31451$ J·mol^{-1}·K^{-1}

molar volume (ideal gas) $V_m = 2.24141 \times 10^{1}$ L/mol
(273.15 K, 101325 Pa)

Planck constant $h = 6.62608 \times 10^{-34}$ J·s

elementary charge $e = 1.60218 \times 10^{-19}$ C

electron mass $m_e = 9.10939 \times 10^{-31}$ kg

proton mass $m_p = 1.67262 \times 10^{-27}$ kg

1 kg = 2.205 pounds

1 m = 3.937×10^{1} inches = 3.281 feet

1 m^3 = 2.642×10^{2} gallons (U.S.)

1 m^3 = 2.200×10^{2} gallons (Imperial)

Force	N	dyn	kp
1 N	1	10^{5}	1.019716×10^{-1}
1 dyn	10^{-5}	1	1.019716×10^{-6}
1 kp	9.80665	9.80665×10^{5}	1

Pressure	Pa	bar	kp/m^2	at	atm	Torr	lb/in^2
1 Pa = 1 N/m^2	1	10^{-5}	1.019716×10^{-1}	1.019716×10^{-5}	9.86923×10^{-6}	7.50062×10^{-3}	1.450378×10^{-4}
1 bar = 10^{6} dyn/cm^2	10^{5}	1	1.019716×10^{4}	1.019716	9.86923×10^{-1}	7.50062×10^{2}	1.450378×10^{1}
1 kp/m^2 = 1 mm H_2O	9.80665	9.80665×10^{-5}	1	10^{-4}	9.67841×10^{-5}	7.35559×10^{-2}	1.422335×10^{-3}
1 at (technical)	9.80665×10^{4}	9.80665×10^{-1}	10^{4}	1	9.67841×10^{-1}	7.35559×10^{2}	1.422335×10^{1}
1 atm = 760 Torr	1.01325×10^{5}	1.01325	1.033227×10^{4}	1.033227	1	7.60×10^{2}	1.469595×10^{1}
1 Torr = 1 mm Hg	1.333224×10^{2}	1.333224×10^{-3}	1.359510×10^{1}	1.359510×10^{-3}	1.315789×10^{-3}	1	1.933678×10^{-2}
1 lb/in^2 = 1 psi	6.89476×10^{3}	6.89476×10^{-2}	7.03069×10^{2}	7.03069×10^{-2}	6.80460×10^{-2}	5.17149×10^{1}	1

Work, Energy, Heat	J	kW·h	kcal	Btu	eV
1 J = 1 W·s = 1 N·m = 10^7 erg	1	2.778×10^{-7}	2.39006×10^{-4}	9.4781×10^{-4}	6.242×10^{18}
1 kW·h	3.6×10^6	1	8.604×10^2	3.41214×10^3	2.247×10^{25}
1 kcal	4.1840×10^3	1.1622×10^{-3}	1	3.96566	2.6117×10^{22}
1 Btu (British thermal unit)	1.05506×10^3	2.93071×10^{-4}	2.5164×10^{-1}	1	6.5858×10^{21}
1 eV	1.602×10^{-19}	4.450×10^{-26}	3.8289×10^{-23}	1.51840×10^{-22}	1

$1\ cm^{-1} \mathrel{\hat{=}} 1.239842 \times 10^{-4}$ eV $1\ Hz \mathrel{\hat{=}} 4.135669 \times 10^{-15}$ eV

2 Rydberg (Ry) = 1 hartree = 27.2114 eV $1\ eV \mathrel{\hat{=}} 23.0578$ kcal/mol

Power	kW	hp	$kp \cdot m \cdot s^{-1}$	kcal/s
1 kW = 10^3 J/s	1	1.35962	1.01972×10^2	2.39006×10^{-1}
1 hp (horsepower, metric)	7.3550×10^{-1}	1	7.5×10^1	1.7579×10^{-1}
1 $kp \cdot m \cdot s^{-1}$	9.80665×10^{-3}	1.333×10^{-2}	1	2.34384×10^{-3}
1 kcal/s	4.1840	5.6886	4.26650×10^2	1

References:

Mills, I. (Ed.), International Union of Pure and Applied Chemistry, Quantities, Units and Symbols in Physical Chemistry, Blackwell Scientific Publications, Oxford 1988.

The International System of Units (SI), National Bureau of Standards Spec. Publ. 330 [1972].

Landolt-Börnstein, 6th Ed., Vol. II, Pt. 1, 1971, pp. 1/14.

ISO Standards Handbook 2, Units of Measurement, 2nd Ed., Geneva 1982.

Cohen, E. R., Taylor, B. N., Codata Bulletin No. 63, Pergamon, Oxford 1986.